Graduate Texts in Physics

Graduate Texts in Physics publishes core learning/teaching material for graduate and advanced-level undergraduate courses on topics of current and emerging fields within physics, both pure and applied. These textbooks serve students at the MS- or PhD-level and their instructors as comprehensive sources of principles, definitions, derivations, experiments and applications (as relevant) for their mastery and teaching, respectively. International in scope and relevance, the textbooks correspond to course syllabi sufficiently to serve as required reading. Their didactic style, comprehensiveness and coverage of fundamental material also make them suitable as introductions or references for scientists entering, or requiring timely knowledge of, a research field.

More information about this series at http://www.springer.com/series/8431

Thanu Padmanabhan

Quantum Field Theory

The Why, What and How

 Springer

Thanu Padmanabhan
Inter-University Centre for Astronomy and Astrophysics
Pune, India

ISSN 1868-4513   ISSN 1868-4521   (electronic)
Graduate Texts in Physics
ISBN 978-3-319-28171-1   ISBN 978-3-319-28173-5   (eBook)
DOI 10.1007/978-3-319-28173-5

Library of Congress Control Number: 2016930164

Springer Cham Heidelberg New York Dordrecht London

Printed on acid-free paper

Springer International Publishing AG Switzerland is part of Springer Science+Business Media (www.springer.com)

Preface

Paraphrasing the author of a quantum mechanics textbook, one might say that there is as much need for yet another book on QFT as there is for a New Revised Table of Integers! So let me explain why I am writing this book and how it will enrich the existing literature and fill a niche.

Standard textbooks in QFT go along a well-trodden path: Classical fields, their quantization, interactions and perturbation theory, tree-level computation of cross-sections using Feynman diagrams, divergences and renormalization as a way to handle them, etc. etc. Depending on the level of the textbook, the above topics will be discussed at different levels of sophistication and could be extended further towards e.g., gauge theories or the standard model. The calculational techniques you learn will be good enough to compute all the usual stuff in high energy physics, which is broadly the aim of most of these textbooks.

My interaction with bright graduate students, post-docs and theoretical physicists over the last three decades have made me realize some of the shortcomings of learning QFT exclusively along these lines. Let me list a few of them.

(1) Many smart students have asked me the following question: "If I know the classical action for a non-relativistic particle, I can construct the corresponding quantum theory using a path integral. I *do* know the classical action for a relativistic particle. Why can't I use it and construct the quantum theory of the relativistic particle and be done with it?"

This question goes to the core of QFT but — as far as I know — it is not answered directly in any of the existing textbooks. In fact, many high energy physicists (and at least one author of a QFT book) did not know that the exact, *non*quadratic, path integral for the relativistic particle *can* be calculated in closed form! Sure, every book pays lip service to why the single particle description is inadequate when you bring quantum theory and special relativity together, why the existence of antiparticles is a key new feature and all that. But this is done in a bit of hurry after which the author proceeds quickly to the classical field theory and its quantization! In other words, what these books do is to *start from fields and obtain particles* as their quanta rather than demonstrate that *if you start from a relativistic particle you will be led to the concept of fields*.

This is the first question I will address in Chapter 1. I will show, using the *exact evaluation of the relativistic path integral*, that if you start from the path integral quantization of the relativistic particle you will arrive at the notion of a field in a satisfactory manner. This route, from particles to fields (instead of from fields to particles) should be a welcome addition to the existing literature.

(2) The way renormalization is introduced in most of the textbooks is conceptually somewhat unsatisfactory. Many graduate students come away with the impression that renormalization is a trick to get meaningful answers out of divergent expressions, and do not understand clearly the distinction between regularization and renormalization. (The notable exceptions are students and textbooks with a condensed matter perspective, who do a better job in this regard.) Many authors feel that the Wilsonian perspective of quantum field theory is a bit too "advanced" for an introductory level course. In fact *it is not*, and students grasp it with a lot more ease (and with a lot less misunderstanding) than the more conventional approach. I will take the Wilsonian point of view as a backdrop right from the beginning, so that the student needs to re-learn very little as she progresses and will not have any fear for the so called "advanced" concepts.

(3) A closely related issue is the discussion of non-perturbative phenomena, which is *conspicuous by its absence* in almost all textbooks. This, in turn, makes students identify some concepts like renormalization too strongly with perturbation theory. Further, it creates difficulties for many students who are learning QFT as a prelude to specializing in areas where non-perturbative techniques are important. For example, students who want to work in gravitational physics, like QFT in external gravitational fields or some aspects of quantum gravity, find that the conventional perturbation theory approach — which is all that is emphasized in most QFT textbooks — leaves them rather inadequately prepared.

To remedy this, I will introduce some of the non-perturbative aspects of QFT (like, for example, particle production by external sources) *before* the students lose their innocence through learning perturbation theory and Feynman diagrams! Concepts like electromagnetic charge renormalization and running coupling constants, for example, will be introduced through the study of pair production in external electromagnetic fields, thereby divorcing them conceptually from the perturbation theory. Again, some of these topics (like e.g., the effective Lagrangian in QED) which are usually considered "advanced", are actually quite easy to grasp if introduced in an appropriate manner and early on. This will make the book useful for a wider class of readers whose interest may not be limited to just computing Feynman diagrams in the standard model.

(4) Most textbooks fail to do justice to several interesting and curious phenomena in field theory because of the rather rigid framework in which they operate. For example, the Davies-Unruh effect, which teaches us that concepts like "vacuum state" and "particle" can change in a non-inertial frame, is a beautiful result with far-reaching implications. Similarly, the Casimir effect is an excellent illustration of non-trivial consequences of free-field theory in appropriate circumstances. Most standard textbooks do not spend adequate amount of time discussing such fascinating topics which are yet to enter the mainstream of high energy physics. I will dip into such applications of QFT whenever possible, which should help students to develop a broader perspective of the subject.

Of course, a textbook like this is useless if, at the end of the day, the student cannot calculate anything! Rest assured that this book develops at adequate depth *all the standard techniques* of QFT as well. A student who reaches the last chapter would have computed the anomalous magnetic moment of the electron in QED, and would have worked out the two-loop renormalization of the $\lambda\phi^4$ theory. (You will find a description of the individual chapters of the book in the Chapter Highlights in pages v-vii, just after this Preface.)

In addition to nearly 80 Exercises sprinkled throughout the text in the marginal notes, I have also included 18 Problems (with solutions) at the end of the book. These vary significantly in their difficulty levels; some will require the student to fill in the details of the discussion in the text, some will illustrate additional concepts extending the ideas in the text, and some others will be applications of the results in the text in different contexts. This should make the book useful in self-study. So, by mastering the material in this book, the student will learn both the conceptual foundations as well as the computational techniques. The latter aspect is the theme of many of the excellent textbooks which are available and the student can supplement her education from any of them.

The readership for this book is very wide. Senior undergraduates, graduate students and researchers interested in high energy physics and quantum field theory will find this book very useful. (I expect the student to have some background in advanced quantum mechanics and special relativity in four-vector notation. A previous exposure to the nonrelativistic path integral will help, but is not essential.) The approach I have taken will also attract readership from people working in the interface of gravity and quantum theory, as well as in condensed matter theory. Further, the book can be easily adapted for courses in QFT of different durations.

The approach presented here has been tested out in the class room. I have been teaching selected topics in QFT for graduate students for about three decades in order to train them, by and large, to work in the area of interface between QFT and General Relativity. My lectures have covered many of the issues I have described above. The approach was well-appreciated and the students found it useful and enlightening. The feedback I got very often was that my course — taught at graduate student level — came a year too late!

Prompted by all these, in 2012, I gave a 50 hour course on QFT to *Masters-level* students in physics of the University of Pune. (In the Indian educational system, these are the students who will proceed to Ph.D graduate school in the next year.) The lectures were viedographed and made available from my institute website to a wider audience. (These are now available on YouTube; you can access a higher resolution video by sending a request to library@iucaa.in.) From the classroom feedback as well as from students who used the videos, I learnt that the course was a success. This book is an expanded version of the course.

Many people have contributed to this venture and I express my gratitude to them. To begin with, I thank the Physics Department of Pune University for giving me the opportunity to teach this course to their students in 2012, and many of these students for their valuable feedback. I thank Suprit Singh who did an excellent job of videographing the lectures. Sumanta Chakraborty and Hamsa Padmanabhan read through the entire first draft of the book and gave detailed comments. In addition, Swastik Bhattacharya, Sunu Engineer, Dawood Kothawala, Kinjalk Lochan, Aseem Paranjape, Sudipta Sarkar, Sandipan Sengupta, S. Shankaranarayanan and Tejinder Pal Singh read through several of the chapters and offered comments. I thank all of them. Angela Lahee of Springer initiated this project and helped me through its completion, displaying considerable initiative and accommodating my special formatting requirements involving marginal notes.

This book would not have been possible without the dedicated support from Vasanthi Padmanabhan, who not only did the entire LaTeXing and formatting but also produced most of the figures. I thank her for her help. It is a pleasure to acknowledge the library facilities available at IUCAA, which were useful in this task.

T. Padmanabhan
Pune, September 2015.

Chapter Highlights

1. **From Particles to Fields**

   The purpose of this chapter is to compute the path integral amplitude for the propagation of a free relativistic particle from the event x_1^i to the event x_2^i and demonstrate how the concept of a field emerges from this description. After introducing (i) the path integral amplitude and (ii) the standard Hamiltonian evolution in the case of a non-relativistic particle, we proceed to evaluate the propagator for a relativistic particle. An investigation of the structure of this propagator will *lead to* the concept of a field in a rather natural fashion. You will see how the standard unitary evolution, propagating forward in time, requires an infinite number of degrees of freedom for the proper description of (what you thought is) a *single* relativistic particle. In the process, you will also learn a host of useful techniques related to propagators, path integrals, analytic extension to imaginary time, etc. I will also clarify how the approach leads to the notion of the antiparticle, and why causality requires us to deal with the particle and antiparticle together.

2. **Disturbing the Vacuum**

   The purpose of this — relatively short — chapter is to introduce you to the key aspect of QFT, viz., that particles can be created and destroyed. Using an external, classical scalar source $J(x)$, we obtain the propagator for a relativistic particle from general arguments related to the nature of creation and destruction events. The discussion then introduces functional techniques and shows how the notion of the field again arises, quite naturally, from the notion of particles which can be created or destroyed by external sources.

   By the end of the first two chapters, you would have firmly grasped how and why combining the principles of relativity and quantum theory *demands* a concept like the field (with an infinite number of degrees of freedom), and would have also mastered several mathematical techniques needed in QFT. These include path integrals, functional calculus, evaluation of operator determinants, analytic properties of propagators and the use of complex time methods.

3. **From Fields to Particles**

   Having shown in the first two chapters how the quantum theory of a relativistic particle naturally leads to the concept of fields, we next address the complementary issue of how fields lead to particles. After rapidly reviewing the action principle in classical mechanics, we make a seamless transition from mechanics to field theory. This is followed by a description of the (i) real and (ii) complex scalar fields and (iii) the electromagnetic field. Two key concepts in modern physics — spontaneous symmetry breaking and the notion of gauge fields — are introduced early on and in fact, the electromagnetic field will come in as a classical $U(1)$ gauge field.

   I then describe the quantization of real and complex scalar fields — which is fairly straightforward — and connect up with the ideas introduced in chapters 1 and 2. The discussion will compare the transition from particles to fields vis-a-vis from fields to particles, thereby strengthening conceptual understanding of both perspectives. The idea of particles arising as excitations of the fields naturally brings in the notion of Bogoliubov transformations. Using this, it is easy to understand the Unruh-Davies effect, viz., that the vacuum state in an inertial frame appears as a thermal state in a uniformly accelerated frame.

We next take up the detailed description of the quantization of the electromagnetic field. I do this first in the radiation gauge in order to get the physical results quickly and to explain the interaction of matter and radiation. This is followed by the covariant quantization of the gauge field which provides an opportunity to introduce the Fadeev-Popov technique in the simplest possible context, and to familiarize you with the issues that arise while quantizing a gauge field. Finally, I provide a detailed description of the Casimir effect which is used to introduce — among other things — the notion of dimensional regularization.

4. **Real Life I: Interactions**

Having described the *free* quantum fields, we now turn to the description of interacting fields. The standard procedure in textbooks is to introduce perturbation theory, obtain the Feynman rules, calculate physical processes, and then introduce renormalization as a procedure to tackle the divergences in the perturbation theory, etc. For the reasons I described in the Preface, I think it is better to start from a non-perturbative approach, through the concept of effective action.

I will do this both for $\lambda\phi^4$ theory and for electromagnetic field coupled to a complex scalar field. In both the cases, one is led to the concept of renormalization group and that of running coupling constants. These, in turn, allow us to introduce the Wilsonian approach to QFT, which is probably the best language available to us today to understand QFT. The notion of effective action also leads to the Schwinger effect, viz., the production of charged particles by a strong electric field. This effect is non-analytic in the electromagnetic coupling constant, and hence cannot be obtained by perturbation theory.

After having discussed the non-perturbative effects, I turn to the standard perturbation theory for the $\lambda\phi^4$ case and obtain the usual Feynman diagrams (using functional integral techniques) and describe how various processes are calculated. This allows us to connect up themes like the effective Lagrangian and the running coupling constant from both perturbative and non-perturbative perspectives.

5. **Real Life II: Fermions and QED**

Upto this point, I have avoided fermions in order to describe the issues of QFT in a simplified setting. This last chapter is devoted to the description of fermions and, in particular, QED. The Dirac equation is introduced in a slightly novel way through the relativistic square root $\sqrt{p^2} = \gamma^a p_a$, after discussing the corresponding nonrelativistic square root $\sqrt{\mathbf{p}^2} = \boldsymbol{\sigma} \cdot \mathbf{p}$ and the Pauli equation. Having obtained the Dirac equation, I describe the standard lore related to Dirac matrices and obtain the magnetic moment of the electron. I then proceed to discuss the quantization of the Dirac field, paying careful attention to the role of causality in fermionic field quantization. The path integral approach to fermionic fields is introduced through Grassmannians (which is developed to the extent required) and once again, we will begin with non-perturbative features like the Schwinger effect for electrons, before discussing perturbation theory and the Feynman rules in QED.

Finally, I provide a detailed discussion of the one loop QED and renormalization. This will allow, as an example, the computation of the anomalous magnetic moment of the electron, which many consider to be the greatest triumph of QED. The discussion of one loop diagrams in QED also allows the study of renormalization in the perturbative context and connect up the "running" of the electromagnetic coupling constant computed by the perturbative and non-perturbative techniques.

Contents

A Potpourri of Problems

(with Solutions)

Chapter 1

From Particles to Fields

1.1 Motivating the Quantum Field Theory

Quantum field theory is a set of rules using which high energy physicists can "shut up and calculate" the behaviour of particles at high energies. The resulting predictions match incredibly well with experiments, showing that these rules must have (at least) a grain of truth in them. While a practitioner of the art might be satisfied in inventing and using these rules (and picking up the resultant Nobel prizes), some students might be curious to know the conceptual motivation behind these computational rules and their relationship to the non-relativistic quantum mechanics (NRQM) they are familiar with. One natural question such a student asks is the following:

Given the *classical* description of a non-relativistic particle in an external potential, one can obtain the corresponding *quantum* description in a fairly straightforward way[1]. Since we do have a well-defined *classical* theory of *relativistic* particles as well, why can't we just quantise it and get a relativistically invariant quantum theory? Why do we need a quantum *field* theory?

Most textbooks of quantum field theory raise this question in their early part and tell you that you need the quantum field theory because any theory, which incorporates quantum mechanics and relativity, has to be a theory in which number of particles (and even the identity of particles) is not conserved. If you collide an electron with a positron at high enough energies, you may end up getting a plethora of other particle-antiparticle pairs. Even at low energies, you can annihilate an electron with a positron and get a couple of photons. Clearly, if you try to write a (relativistically invariant version of) the Schrodinger equation for an interacting electron-positron pair, you wouldn't know what to do once they have disappeared and some other particles have appeared on the scene.

This fact, while quite important, is only one part of the story. Condensed matter physics is full of examples in which phonons, magnons and other nons keep appearing and disappearing and — in fact — field theory is the most efficient language to describe these situations. But, in principle, one could have also written down the non-relativistic Schrodinger equation for, say, all the electrons and lattice ions in a solid and described the same physics. This situation is quite different from relativistic quantum theory, wherein the basic principles *demand the existence of antiparticles*, to which there is no simple non-relativistic analogue. Obviously, this has extra implications that go beyond the fact that we need to deal with systems having variable number of particles — which, by itself, can be handled comparatively easily.

The second complication is the following: Combining relativity with quantum theory also forbids the localization of a particle in an arbitrarily small region of space. In non-relativistic mechanics, one can work either with momentum eigenstates $|\mathbf{p}\rangle$ or position eigenstates $|\mathbf{x}\rangle$ for a particle with equal ease. You can localize a particle in space with arbitrary accuracy, if you are willing to sacrifice the knowledge about its momentum. In relativistic quantum mechanics, all sorts of bad things happen if you try to localize a particle of mass m to a region smaller than $\hbar/mc$. This will correspond to an uncertainty in the momentum of the order of mc leading to an uncertainty in the energy of the order of mc^2. This — in turn — can lead to the production of particle-antiparticle pairs; so the single particle description breaks down if you try to localize the particles in a small enough region. Mathematically, the momentum eigenstates $|\mathbf{p}\rangle$ continue to be useful in relativistic quantum mechanics, but not the position eigenstates $|\mathbf{x}\rangle$. The concept of a precise position for a particle, within a single particle description, ceases to exist in relativistic quantum mechanics. (We will say more about this in Sect. 1.4.6.)

The third issue is the existence of spin. In the case of an electron, one might try to imagine its spin as arising out of some kind of intrinsic rotation, though it leads to a hopelessly incorrect picture. In the case of a *massless* spin-1 particle like the photon, there is no rest frame and it is not clear how to even define a concept like 'intrinsic rotation'; but we do know that the photon also has two spin states (in spite of being a spin-1 particle). Classically, the electromagnetic wave has two polarization states and one would like to think of them translating to the two spin states of photons. But since the photon is massless and is never at rest, it is difficult to understand how an individual photon can have an intrinsic rotation attributed to it. Spin, as we shall see, is a purely relativistic concept.

A feature closely related to spin is the Pauli exclusion principle. Particles with half-integral spins (called *fermions*) obey this principle which states that no two identical fermions can be put in the same state. If we have a Schrodinger wave function $\psi(\boldsymbol{x_1}, \boldsymbol{x_2}, \cdots \boldsymbol{x_n})$ in NRQM describing the state of a system of N fermions — say, electrons — then this wave function must be antisymmetric under the exchange of any pair of coordinates $\boldsymbol{x}_i \leftrightarrow \boldsymbol{x}_j$. Why particles with half-integral spin should behave in this manner cannot be explained within the context of NRQM. It is possible to understand this result in quantum field theory — albeit after introducing a fairly formidable amount of theoretical machinery — as arising due to the structure of the Lorentz group, causality, locality and a few other natural assumptions. Once again, adding relativity to quantum theory leads to phenomena which are quite counter-intuitive and affects the structure of NRQM itself.

There are at least two more good reasons to think of fields — rather than particles — as fundamental physical entities. The first reason is the existence of the electromagnetic field which has classical solutions that describe propagating waves.[2] Experiments, however, show that electromagnetic waves *also* behave like a bunch of particles (photons) each having a momentum ($\mathbf{p}$) and energy ($\omega_{\mathbf{p}} = |\mathbf{p}|$). We need to come up with 'rules' that will get us the photons starting from a card-holding field, if there ever

[2]Historically, we started with the wave nature of radiation and the particle nature of the electron. The complementary description of particles as waves arises even in the non-relativistic theory while the description of waves as particles requires relativity. This is why, even though quantum theory had its historical origins in the realization that light waves behave as though they are made of particles ('photons'), the initial efforts were more successful in describing the *wave nature of particles* like electrons — because you could deal with a non-relativistic system — rather than in describing the *particle nature of waves*.

was one, which requires one to somehow "quantise" the field.

Second, there are certain phenomena like the spontaneous symmetry breaking which are most easily understood in terms of fields and the ground states of fields which are described as condensates. (We will say more about this in Sect. 3.2.) This is very difficult to interpret completely in terms of the particle picture though we know that there is a formal equivalence between the two. In general, interactions between the particles are best described using the language of fields and more so when there is some funny business like spontaneous symmetry breaking going on. So one needs to be able to proceed from particles to fields and vice-versa to describe the real world, which is what quantum field theory attempts to do.

Having said something along these lines, we can, at this stage, jump to the description of classical fields and their quantization — which is what most text books do. *I will do it differently.*

I will actually show you how a straightforward attempt to describe relativistic free particles in a quantum mechanical language *leads to* the concept of fields. We will do this from different perspectives demonstrating how these perspectives merge together in a consistent manner. In Chapter 1, we will start with a single, relativistic spinless particle and — by evaluating the probability amplitude for this particle to propagate from an event $\mathcal{P}$ to an event $\mathcal{Q}$ in spacetime — we will be *led to* the concept of quantum fields as well as the notion of antiparticles. These ideas will be reinforced in Chapter 2 where we will take a closer look at the creation and subsequent detection of spinless particles by external agencies. Once again, we will be led to the notion of a quantum field and the action functional for the same, starting from the amplitude for propagation.

Having obtained the fields from the particles and established the *raison d'être* for the quantum fields, we will reverse the logic and obtain the (relativistic) particles as excitations of the quantum fields in Chapter 3. This is more in tune with the conventional textbook discussion and has the power to deal with particles which have a nonzero spin — like, for example, the photons. In Chapter 4 we will introduce the interaction between the fields in the simplest possible context, still dealing with scalar and electromagnetic fields. Finally, Chapter 5 will introduce you to the strange features of fermions and QED.

The conceptual consistency of this formalism — viewed from different perspectives — as well as the computational success, leading to results which are in excellent agreement with the experiments, are what make one believe that the overall description cannot be far from the truth, in spite of several counterintuitive features we will come across (and tackle) as we go along.

1.2 Quantum Propagation Amplitude for the Non-relativistic Particle

Given the classical dynamics of a free, spin-zero non-relativistic particle of mass m, we can proceed to its quantum description using either the action functional A or the Hamiltonian H. In the case of a free particle, nothing much happens and the only interesting quantity to compute is the amplitude $\mathcal{A}$ for detecting a particle at an event $x_2 \equiv x_2^a \equiv (t_2, \mathbf{x}_2)$ if it was created at an event $x_1 \equiv x_1^a \equiv (t_1, \boldsymbol{x}_1)$. This is given by the product: $\mathcal{A} =$

$\mathcal{D}(x_2)G(x_2;x_1)\mathcal{C}(x_1)$ where $\mathcal{D}(x_2)$ is the amplitude for detection, $\mathcal{C}(x_1)$ is the amplitude for creation and $G(x_2;x_1)$ is the amplitude for the particle to propagate from one event x_1 to another event x_2. (For example, the beaten-to-death electron two slit experiment involves an electron gun to create electrons and a detector on the screen to detect them). The resulting probability $|\mathcal{A}|^2$ is expected to have an invariant, absolute, meaning.

The creation and detection processes, described by $\mathcal{C}(x_1), \mathcal{D}(x_2)$, turn out to be far more non-trivial than one would have first imagined and we will have a lot to say about them in Sect. 2.1. In this section we will take these as provided by our experimentalist friend and concentrate on $G(x_2;x_1)$.

1.2.1 Path Integral for the Non-relativistic Particle

We will start with the non-relativistic particle and obtain the amplitude $G(x_2;x_1)$ first using the path integral (in this section) and then using the Hamiltonian (in the next section). The path integral prescription says that $G(x_2;x_1)$ can be expressed as a sum over paths $\mathbf{x}(t)$ connecting the two events in the form

$$G(x_2;x_1) = \sum_{\mathbf{x}(t)} \exp iA[\mathbf{x}(t)] = \sum_{\mathbf{x}(t)} \exp i \int_{t_1}^{t_2} \frac{1}{2} m|\dot{\mathbf{x}}|^2 \, dt \qquad (1.1)$$

The paths summed over are restricted to those which satisfy the following condition: *Any given path $\mathbf{x}(t)$ cuts the spatial hypersurface $t = y^0$ at any intermediate time, $t_2 > y^0 > t_1$, at only one point.* In other words, while doing the sum over paths, we are restricting ourselves to paths of the kind shown in Fig. 1.1 (which always go 'forward in time') and will not include, for example, paths like the one shown in Fig. 1.2 (which go both forward and backward in time). The path in Fig. 1.2 cuts the constant time surface $t = y^0$ at three events suggesting that at $t = y^0$ there were three particles simultaneously present even though we started out with one particle. It is this feature which we avoid (and stick to single particle propagation) by imposing this condition on the class of paths that is included in the sum. By the same token, we will assume that the amplitude $G(x_2;x_1)$ vanishes for $x_2^0 < x_1^0$; the propagation is forward in time.

This choice of paths, in turn, implies the following 'transitivity constraint' for the amplitude:

$$G(x_2;x_1) = \int d^D \mathbf{y} \, G(x_2;y)G(y;x_1) \qquad (1.2)$$

The integration at an intermediate event $y^i = (y^0, \mathbf{y})$ (with $t_2 > y^0 > t_1$) is limited to integration over the *spatial* coordinates because each of the paths summed over cuts the intermediate spatial surface at only one point. Therefore, every path which connects the events x_2 and x_1 can be uniquely specified by the spatial location $\mathbf{y}$ at which it crosses the surface $t = y^0$. So the sum over all the paths can be divided into the sum over all the paths from x_1^i to some location $\mathbf{y}$ at $t = y^0$, followed by the sum over all the paths from y^i to x_2^i with an integration over all the locations $\mathbf{y}$ at the intermediate time $t = y^0$. This leads to Eq. (1.2).

The transitivity condition in Eq. (1.2) *is vital* for the standard probabilistic interpretation of the wave function in non-relativistic quantum

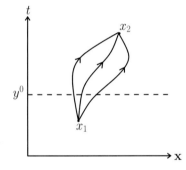

Figure 1.1: Examples of paths included in the sum in Eq. (1.1)

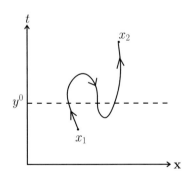

Figure 1.2: A path that goes forward and backward which is *not* included in the sum in Eq. (1.1).

mechanics. If $\psi(t_1, \mathbf{x}_1)$ is the wave function giving the amplitude to find a particle at $\mathbf{x}_1$ at time t_1, then the wave function at a later time $t = y^0$ is given by the integral:

$$\psi(y^0, \mathbf{y}) = \int d^D \mathbf{x}_1 \, G(y; x_1) \, \psi(t_1, \mathbf{x}_1) \qquad (1.3)$$

which interprets $G(y; x_1)$ as a propagator kernel allowing us to determine the solution to the Schrodinger equation at a later time $t = y^0$ from its solution at $t = t_1$. Writing the expression for $\psi(t_2, \mathbf{x}_2)$ in terms of $\psi(y^0, \mathbf{y})$ and $G(x_2; y)$ and using Eq. (1.3) to express $\psi(y^0, \mathbf{y})$ in terms of $\psi(t_1, \mathbf{x}_1)$, it is easy to see that Eq. (1.2) is *needed* for consistency. Equation (1.2) or Eq. (1.3) also implies the condition:

$$G(t, \mathbf{x}; t, \mathbf{y}) = \langle t, \mathbf{x} | t, \mathbf{y} \rangle = \delta(\mathbf{x} - \mathbf{y}) \qquad (1.4)$$

where $|t, \mathbf{x}\rangle$ is a position eigenstate at time t. Three crucial factors have gone into these seemingly innocuous results:[3]

(i) The wave function at time t can be obtained from knowing *only* the wave function at an earlier time (without, e.g., knowing its time derivative) which means that the differential equation governing ψ must be first order in time.

(ii) One can introduce eigenstates $|t, \mathbf{x}\rangle$ of the position operator $\hat{\mathbf{x}}(t)$ at time t by $\hat{\mathbf{x}}(t)|t, \mathbf{x}\rangle = \mathbf{x}|t, \mathbf{x}\rangle$ so that $\psi(t, \mathbf{x}) = \langle t, \mathbf{x} | \psi \rangle$ with Eq. (1.4) allowing the possibility of localizing a particle in space with arbitrary accuracy.

(iii) One can interpret $G(x_2; x_1)$ in terms of the position eigenstates as $\langle t_2, \mathbf{x}_2 | t_1, \mathbf{x}_1 \rangle$.

The result of the integral in Eq. (1.2) has to be independent of y^0 because the left hand side is independent of y^0. This is a *strong* restriction on the form of $G(x_2; x_1)$. The transitivity condition, plus the fact that the free particle amplitude $G(x_2; x_1)$ can only depend on $|\mathbf{x}_2 - \mathbf{x}_1|$ and $(t_2 - t_1)$ because of translational and rotational invariance, fixes the form of $G(x_2; x_1)$ to a great extent. To see this, express $G(x_2; x_1)$ in terms of its spatial Fourier transform in the form:

$$G(y; x) = \int \frac{d^D \mathbf{p}}{(2\pi)^D} \, \theta(y^0 - x^0) \, F(|\mathbf{p}|; y^0 - x^0) \, e^{i \mathbf{p} \cdot (\mathbf{y} - \mathbf{x})} \qquad (1.5)$$

and substitute into Eq. (1.2). This will lead to the condition

$$F(|\mathbf{p}|; x_2^0 - y^0) F(|\mathbf{p}|; y^0 - x_1^0) = F(|\mathbf{p}|; x_2^0 - x_1^0) \qquad (x_2^0 > y^0 > x_1^0) \quad (1.6)$$

which has a unique solution $F(|\mathbf{p}|; t) = \exp[\alpha(|\mathbf{p}|) t]$ where $\alpha(|\mathbf{p}|)$ is an arbitrary function of $|\mathbf{p}|$. Further, we note that $F(|\mathbf{p}|; y^0 - x^0)$ propagates the momentum space wave function $\phi(x^0, \mathbf{p})$ — which is the spatial Fourier transform of $\psi(x^0, \mathbf{x})$ — from $t = x^0$ to $t = y^0$. Since ϕ is the Fourier transform of ψ, this "propagation" is just multiplication by F. The probability calculated from the momentum space wave function will be well behaved for $|t| \to \infty$ only if α is pure imaginary, thereby only contributing a phase. So $\alpha = -if(|\mathbf{p}|)$ where $f(|\mathbf{p}|)$ is an arbitrary function[4] of $|\mathbf{p}|$. Thus, the spatial Fourier transform of $G(x_2; x_1)$ *must* have the form

$$\int d^D \mathbf{x} \, G(x_2; x_1) \, e^{-i \mathbf{p} \cdot \mathbf{x}} = \theta(t) \, e^{-if(|\mathbf{p}|) t} \qquad (1.7)$$

where $x^a \equiv x_2^a - x_1^a \equiv (t, \boldsymbol{x})$. That is, it must be a pure phase[5] determined

---

[3] As you will see, all of them will run into trouble in the context of a relativistic particle.

Exercise 1.1: Prove all these claims.

[4] You can also obtain the same result from the fact that $\exp(iA)$ goes to $\exp(-iA)$ under the time reversal $t_2 \Longleftrightarrow t_1$; the path integral sum must be defined such that $F \to F^*$ under $t \to -t$ which requires α to be pure imaginary.

[5] If we interpret this phase as due to the energy $\omega_{\mathbf{p}} = \mathbf{p}^2/2m$ and set $f(\mathbf{p}) = \omega_{\mathbf{p}}$ then an inverse Fourier transform of Eq. (1.7) will immediately determine $G(x_2; x_1)$ leading to our result obtained below in Eq. (1.9). But then we are getting ahead of the story and missing all the fun!.

by a single function $f(|\mathbf{p}|)$.

The sum over paths in Eq. (1.1) itself is trivial to evaluate even without us defining precisely what the sum means. (The more sophisticated definitions for the sum work — or rather, are designed to work — only because we know the answer for $G(x_2; x_1)$ from other well-founded methods!.) We first note that the sum over all $\mathbf{x}(t)$ is the same as the sum over all $\mathbf{q}(t) \equiv \mathbf{x}(t) - \mathbf{x}_c(t)$ where $\mathbf{x}_c(t)$ is the classical path for which the action is an extremum. Because of the extremum condition, $A[\mathbf{x}_c + \mathbf{q}] = A[\mathbf{x}_c] + A[\mathbf{q}]$. Substituting into Eq. (1.1) and noting that $\mathbf{q}(t)$ vanishes at the end points, we see that the sum over $\mathbf{q}(t)$ must be just a function of $(t_2 - t_1)$. (It can only depend on the time difference rather than on t_2 and t_1 individually whenever the action has no explicit time dependence; i.e., for any closed system). Thus we get

$$G(x_2; x_1) = e^{iA[\mathbf{x}_c]} \sum_{\mathbf{q}} e^{iA[\mathbf{q}(t)]} = N(t) \, \exp iA[\mathbf{x}_c] \equiv N(t) \exp\left(\frac{i}{2}\frac{m|\mathbf{x}|^2}{t}\right) \tag{1.8}$$

where $t \equiv t_2 - t_1$ and $\mathbf{x} \equiv \mathbf{x}_2 - \mathbf{x}_1$. The form of $N(t)$ is strongly constrained by the transitivity condition, Eq. (1.2) — or, equivalently, by Eq. (1.7) — which requires the $N(t)$ to have the form $(m/2\pi it)^{D/2} e^{at}$ where $a = i\varphi$, say. Thus, except for an ignorable, constant, phase factor φ (which is equivalent to adding a constant to the original Lagrangian), $N(t)$ is given by $(m/2\pi it)^{D/2}$ and we can write the full propagation amplitude for a non-relativistic particle as:[6]

$$G(x_2; x_1) = \theta(t) \left(\frac{m}{2\pi it}\right)^{D/2} \exp\left(\frac{i}{2}\frac{m|\mathbf{x}|^2}{t}\right) \tag{1.9}$$

Exercise 1.2: Prove this result for $N(t)$.

The $\theta(t)$ tells you that we are considering a particle which is created, say, at t_1 and detected at t_2 with $t_2 > t_1$. In non-relativistic mechanics all inertial observers will give an invariant meaning to the statement $t_2 > t_1$. It is also easy to see that the $G(x_2; x_1)$ in Eq. (1.9) satisfies the condition in Eq. (1.4).

[6]Obviously, a similar argument works for any action which is at most quadratic in $\dot{\mathbf{x}}$ and $\mathbf{x}$ and the sum over paths will have the form $N(t_2, t_1) \exp(iA[\mathbf{x}_c(t)])$ even when the action has an explicit time dependence. If the action has no explicit time dependence, we further know that $N(t_2, t_1) = N(t)$ which is the only non-trivial quantity left to compute. If the action is *not* quadratic in $\dot{\mathbf{x}}$ and $\mathbf{x}$, nobody knows how to do the sum over paths, except in a few rare cases, and we need to use less fancy (but more powerful) methods to quantise the system.

1.2.2 Hamiltonian Evolution: Non-relativistic Particle

Let us next review briefly how these ideas connect up with the more familiar Hamiltonian approach to NRQM. We know that the behaviour of a free non-relativistic particle is governed by a Hamiltonian $H(\mathbf{p}) = \mathbf{p}^2/2m$ and that the free particle wave functions are made out of $\exp(i\mathbf{p} \cdot \mathbf{x})$. To connect up the path integral result with such a Hamiltonian description, we only have to Fourier transform the amplitude $G(x_2; x_1)$ in Eq. (1.9) with respect to $\mathbf{x}$ getting

$$\mathcal{G}(\mathbf{p}, t) = \int G(x_2; x_1) e^{-i\mathbf{p} \cdot \mathbf{x}} \, d^D \mathbf{x} = \theta(t) \exp(-i\omega_{\mathbf{p}} t) \tag{1.10}$$

where $\omega_{\mathbf{p}} \equiv \mathbf{p}^2/2m$ is the energy of the particle of momentum $\mathbf{p}$. [This form, of course, matches with the result in Eq. (1.7).] Thus the $G(x_2; x_1)$ can be expressed, by an inverse Fourier transform, in the form[7]

[7]Note that we write $px \equiv (\omega_{\mathbf{p}} t - \mathbf{p} \cdot \mathbf{x})$ even when p^i is *not* a four-vector. The combination $(\omega_{\mathbf{p}} t - \mathbf{p} \cdot \mathbf{x})$, as well as the relations in Hamilton-Jacobi theory, $E = -\partial_t A$, $\mathbf{p} = \nabla A$ — which give you 'relativity without relativity', $p_i = -\partial_i A$ — have their origins in wave phenomena and *not* in relativity.

$$G(x_2; x_1) = \theta(t) \int \frac{d^D \mathbf{p}}{(2\pi)^D} \, e^{i(\mathbf{p} \cdot \mathbf{x} - \omega_{\mathbf{p}} t)} = \theta(t) \int \frac{d^D \mathbf{p}}{(2\pi)^D} \, e^{-ipx} \tag{1.11}$$

We will now obtain the same result from first principles in order to set the ground for the relativistic particle later on. Let us start by rapidly reviewing some basic concepts in quantum theory.

If $\hat{A}(t)$ is an operator at time t in the Heisenberg picture, then its eigenvalue equation can be written in the form: $\hat{A}(t)|t,a\rangle = a|t,a\rangle$. The eigenkets $|t,a\rangle$ are associated with the operator at time t and carry the time label t as well as the eigenvalue a. By convention, the eigenket at $t = 0$ will be denoted without the time label simply as $|a\rangle \equiv |0,a\rangle$. We will work with closed systems for which the Hamiltonian does not have explicit time dependence. In that case, the evolution equation for the operators $i\partial_t\hat{A} = [\hat{A}, \hat{H}]$ has the solution

Exercise 1.3: Prove this.

$$\hat{A}(t) = e^{i\hat{H}t}\,\hat{A}(0)\,e^{-i\hat{H}t} \tag{1.12}$$

The eigenvalue equation at time t can now be written in the form

$$\hat{A}(t)|t,a\rangle = a|t,a\rangle = e^{i\hat{H}t}\,\hat{A}(0)\left\{e^{-i\hat{H}t}|t,a\rangle\right\} \tag{1.13}$$

or equivalently,

$$\hat{A}(0)\left\{e^{-i\hat{H}t}|t,a\rangle\right\} = a\left\{e^{-i\hat{H}t}|t,a\rangle\right\} \tag{1.14}$$

This is clearly an eigenvalue equation for $\hat{A}(0)$ with eigenvalue a and hence we can identify $e^{-i\hat{H}t}|t,a\rangle$ with $|0,a\rangle$ and write

$$|t,a\rangle = e^{i\hat{H}t}|0,a\rangle; \quad \langle t,a| = \langle 0,a|e^{-i\hat{H}t} \tag{1.15}$$

Therefore the amplitude $\langle t',a'|t,a\rangle$ is given by

$$\langle t',a'|t,a\rangle = \langle 0,a'|e^{-i\hat{H}t'}e^{iHt}|0,a\rangle = \langle a'|e^{-i\hat{H}(t'-t)}|a\rangle \tag{1.16}$$

Of particular importance to us are the position and momentum operators $\hat{\mathbf{x}}(t)$ and $\hat{\mathbf{p}}(t)$ satisfying the standard commutation rules $[x^\alpha, p^\beta] = i\delta^{\alpha\beta}$. If the eigenkets of $\hat{\mathbf{x}}(0)$ are denoted by $|\mathbf{x}\rangle$, then Eq. (1.16) tells us that we can write the propagation amplitude as

$$\langle t_2, \mathbf{x}_2|t_1, \mathbf{x}_1\rangle \equiv G(x_2; x_1) = \langle \mathbf{x}_2|e^{-i\hat{H}(t_2-t_1)}|\mathbf{x}_1\rangle \tag{1.17}$$

The eigenstates of the momentum operator form a complete set obeying the relation:

$$\int \frac{d^\nu\mathbf{p}}{(2\pi)^D}\,|\mathbf{p}\rangle\langle\mathbf{p}| = 1 \tag{1.18}$$

Multiplying both sides by the eigenket $|\mathbf{k}\rangle$, it is clear that consistency requires:

$$\langle\mathbf{p}|\mathbf{k}\rangle = (2\pi)^D\delta(\mathbf{p}-\mathbf{k}) \tag{1.19}$$

Further, one can easily show that the operator $e^{-i\boldsymbol{\epsilon}\cdot\hat{\mathbf{p}}}$ generates translations in space. To see this, let us concentrate on just one dimension and consider the operators

$$\hat{U} = 1 + i\epsilon\hat{p} \quad\quad \hat{U}^{-1} = 1 - i\epsilon\hat{p} \tag{1.20}$$

where ϵ is infinitesimal. Then,

$$\hat{U}\hat{q}\hat{U}^{-1} = (1+i\epsilon\hat{p})\hat{q}(1-i\epsilon\hat{p}) = (1+i\epsilon\hat{p})(\hat{q}-i\epsilon\hat{q}\hat{p}) = \hat{q}+i\epsilon[\hat{p},\hat{q}] = \hat{q}+\epsilon \tag{1.21}$$

Consider now the operation of $\hat{q}$ on the ket $\hat{U}^{-1}|x\rangle$:

$$\hat{q}\,\hat{U}^{-1}|x\rangle = \hat{U}^{-1}(\hat{U}\,\hat{q}\,\hat{U}^{-1})|x\rangle = \hat{U}^{-1}(\hat{q}+\epsilon)|x\rangle = (x+\epsilon)\hat{U}^{-1}|x\rangle \quad (1.22)$$

which implies that $\hat{U}^{-1}|x\rangle = |x+\epsilon\rangle$. Proceeding to finite displacements from the infinitesimal displacement and taking into account all the D dimensions, we get[8]

$$|\mathbf{x}+\boldsymbol{\epsilon}\rangle = e^{-i\boldsymbol{\epsilon}\cdot\hat{\mathbf{P}}}|\mathbf{x}\rangle \quad (1.23)$$

Using this, we can write

$$|\mathbf{x}\rangle = e^{-i\mathbf{x}\cdot\hat{\mathbf{P}}}|\mathbf{0}\rangle = \int \frac{d^D\mathbf{p}}{(2\pi)^D}\, e^{-i\mathbf{p}\cdot\mathbf{x}}|\mathbf{p}\rangle\langle\mathbf{p}|\mathbf{0}\rangle = \int \frac{d^D\mathbf{p}}{(2\pi)^D}\, C_{\mathbf{p}}e^{-i\mathbf{p}\cdot\mathbf{x}}|\mathbf{p}\rangle \quad (1.24)$$

where $C_{\mathbf{p}} = \langle\mathbf{p}|\mathbf{0}\rangle$ and we have introduced a complete set of momentum basis in the second equality. It is traditional to set the phase convention such that $C_{\mathbf{p}} = 1$ so that the wave functions in coordinate space, $\langle\mathbf{x}|\psi\rangle$, and momentum space, $\langle\mathbf{p}|\psi\rangle$, are related by a simple Fourier transform:[9]

$$\langle\mathbf{x}|\psi\rangle = \int \frac{d^D\mathbf{p}}{(2\pi)^D}\langle\mathbf{x}|\mathbf{p}\rangle\langle\mathbf{p}|\psi\rangle = \int \frac{d^D\mathbf{p}}{(2\pi)^D} e^{i\mathbf{p}\cdot\mathbf{x}}\langle\mathbf{p}|\psi\rangle \quad (1.25)$$

After these preliminaries, we are in a position to compute $G(x_2; x_1)$ in standard Heisenberg picture. The propagation amplitude from $\mathbf{x}_1$ at $t = t_1$ to $\mathbf{x}_2$ at $t = t_2$ is given by (with $x_2^i - x_1^i \equiv x^i = (t, \mathbf{x})$) the expression:

$$G(x_2; x_1) = \langle\mathbf{x}_2|e^{-i\hat{H}(\mathbf{p})t}|\mathbf{x}_1\rangle = \int \frac{d^D\mathbf{p}}{(2\pi)^D}\, e^{-i\omega_{\mathbf{p}}t}\langle\mathbf{x}_2|\mathbf{p}\rangle\langle\mathbf{p}|\mathbf{x}_1\rangle \quad (1.26)$$

where we have introduced a complete set of momentum eigenstates and used the result $e^{-i\hat{H}(\mathbf{p})t}|\mathbf{p}\rangle = e^{-i\omega_{\mathbf{p}}t}|\mathbf{p}\rangle$. Using $\langle\mathbf{p}|\mathbf{x}\rangle = \exp[-i\mathbf{p}\cdot\mathbf{x}]$ in Eq. (1.26), we get

$$G(x_2; x_1) = \int \frac{d^D\mathbf{p}}{(2\pi)^D}\, e^{-i\omega_{\mathbf{p}}t+i\mathbf{p}\cdot\mathbf{x}} \quad (1.27)$$

This matches with the momentum space structure of $G(x_2; x_1)$ given by Eq. (1.11).

Taking the product of Eq. (1.18) from the left with $\langle\mathbf{x}|$ and from the right with $|\mathbf{y}\rangle$ we get the orthonormality condition for $\langle\mathbf{x}|\mathbf{y}\rangle$.

$$\langle\mathbf{x}|\mathbf{y}\rangle = \int \frac{d^D\mathbf{p}}{(2\pi)^D}\langle\mathbf{x}|\mathbf{p}\rangle\langle\mathbf{p}|\mathbf{y}\rangle = \int \frac{d^D\mathbf{p}}{(2\pi)^D} e^{i\mathbf{p}\cdot(\mathbf{x}-\mathbf{y})} = \delta(\mathbf{x}-\mathbf{y}) \quad (1.28)$$

This, in turn, requires the path integral amplitude to satisfy the condition in Eq. (1.4), which we know it does. Thus everything works out fine in this context.[10]

Incidentally, the expression in Eq. (1.26) can also be expanded in terms of the eigenkets $|E\rangle$ of energy to give an expression for $G(x_2; x_1)$ in terms of the energy eigenfunctions $\phi_E(\mathbf{x}) \equiv \langle\mathbf{x}|E\rangle$ for any system with a time independent Hamiltonian, bounded from below. In that case, we can arrange matters so that $E \geq 0$. We have

$$\begin{aligned} G(x_2; x_1) &= \langle\mathbf{x}_2|e^{-i\hat{H}(\mathbf{p})t}|\mathbf{x}_1\rangle = \int_0^\infty dE\, e^{-iEt}\langle\mathbf{x}_2|E\rangle\langle E|\mathbf{x}_1\rangle \\ &= \int_0^\infty dE\, e^{-iEt}\phi_E(\mathbf{x}_2)\phi_E^*(\mathbf{x}_1) \end{aligned} \quad (1.29)$$

---

[8]This is just a fancy way of doing the Taylor series expansion of a function to find $\psi(x+a) = \sum_n(a^n/n!)\partial_x^n\psi(x) = \exp[a\partial_x]\psi(x) \equiv \exp[iaP]\psi(x)$, using the 'generator' $P \equiv -i\partial_x$. If $\psi(x+a) = \langle x+a|\psi\rangle$ then we get $\langle x+a| = \exp[iaP]\langle x|$ which is the same as $|x+a\rangle = \exp[-iaP]|x\rangle$. The four dimensional generalization of this result is $|x+a\rangle = e^{iap}|x\rangle$ where $ap = a^0\hat{H} - \mathbf{a}\cdot\hat{\mathbf{p}}$.

[9]We are ignoring any $\mathbf{p}$-dependent phase in $C_{\mathbf{p}}$ here. This corresponds to taking $x = i(\partial/\partial p)$ rather than $x = i(\partial/\partial p) + f(p)$ to implement $[x, p] = i$.

[10]In non-relativistic quantum mechanics both position eigenkets $|\mathbf{x}\rangle$ and momentum eigenkets $|\mathbf{p}\rangle$ are equally good basis to work with. We will see later that momentum eigenkets continue to be a good basis in relativistic quantum mechanics, but defining a useful position operator for a relativistic particle becomes a nontrivial and unrewarding task.

So, in principle, you can find $G(x_2; x_1)$ for any system for which you know $\phi_E(\mathbf{x})$. This relation will be useful later on.

Now that we have discussed the calculation of $G(x_2; x_1)$ from the path integral as well as from the Hamiltonian, it is worth spending a moment to connect up the two. (The details of this approach are described in the Mathematical Supplement Sect. 1.6.1.) The basic idea is to compute the matrix element $\langle \mathbf{y} | e^{-it\hat{H}} | \mathbf{x} \rangle$ by dividing the time interval $(0, t)$ into N equal intervals of size ϵ such that $t = N\epsilon$. The matrix element is evaluated by writing it as a product of matrix elements between a complete set of position eigenstates $|\mathbf{x}_k\rangle$ with $k = 1, 2, ...(N-1)$ and integrating over the $\mathbf{x}_k$. At the end of the calculation one takes the limit $N \to \infty$, $\epsilon \to 0$ with $t = N\epsilon$ remaining finite. The result will remain finite only if the integrations and the limiting process are carried out with a suitable measure which should emerge from the calculation itself. This approach of dividing up the time into equal intervals reduces the problem to computing the matrix element $\langle \mathbf{x}_{k+1} | e^{-i\epsilon\hat{H}} | \mathbf{x}_k \rangle$. If $\hat{H}$ has the standard form $H = \hat{\mathbf{p}}^2 / 2m + V(\hat{\mathbf{x}})$, we can evaluate this matrix element by expanding the exponential in a Taylor series in ϵ and retaining only up to linear term.[11] By this procedure one can obtain

$$\langle \mathbf{x}_{k+1} | e^{-i\hat{H}\epsilon} | \mathbf{x}_k \rangle = \int \frac{d\mathbf{p}_k}{(2\pi)^D} \, e^{i\mathbf{p}_k \cdot (\mathbf{x}_{k+1} - \mathbf{x}_k)} \, e^{-i\epsilon H(\mathbf{p}_k, \mathbf{x}_k)} \qquad (1.30)$$

Interpreting $(\mathbf{x}_{k+1} - \mathbf{x}_k)/\epsilon$ as the velocity $\dot{\mathbf{x}}$ in the continuum limit, one can see that the argument of the exponential has the $\mathbf{p} \cdot \dot{\mathbf{x}} - H(\mathbf{p}, \mathbf{x}) = L$ integrated over the infinitesimal time interval. If you take the product of these factors you obtain a natural definition of the path integral in *phase space* in the form

$$\langle \mathbf{y} | e^{-i\hat{H}t} | \mathbf{x} \rangle = \int \mathcal{D}\mathbf{p} \int \mathcal{D}\mathbf{x} \, \exp i \int dt \, [\mathbf{p} \cdot \dot{\mathbf{x}} - H(\mathbf{p}, \mathbf{x})] \qquad (1.31)$$

In the case of the standard Hamiltonian with $H = \mathbf{p}^2 / 2m + V(\mathbf{x})$, we can integrate over $\mathbf{p}_k$ in Eq. (1.30) and obtain

$$\langle \mathbf{x}_{k+1} | e^{-i\hat{H}\epsilon} | \mathbf{x}_k \rangle = \left(\frac{m}{2\pi i \epsilon} \right)^{D/2} \exp \left[\frac{im}{2\epsilon} (\mathbf{x}_{k+1} - \mathbf{x}_k)^2 - i\epsilon V(\mathbf{x}_k) \right] \qquad (1.32)$$

In this case, the continuum limit will directly give the expression:

$$\langle \mathbf{y} | e^{-i\hat{H}t} | \mathbf{x} \rangle = \int \mathcal{D}\mathbf{x} \exp i \int dt \left[\frac{1}{2} m \dot{\mathbf{x}}^2 - V(\mathbf{x}) \right] \qquad (1.33)$$

In both the cases, the path integral measure, $\mathcal{D}\mathbf{x}, \mathcal{D}\mathbf{p}$ etc. are defined as a product of integrals over intermediate states with a measure which depends on ϵ, as shown in detail in mathematical supplement Sect. 1.6.1.

1.2.3 A Digression into Imaginary Time

The mathematical manipulations used in the computation of path integrals become less ambiguous if we use a technique of analytically continuing the results back and forth between real and imaginary time. This is a bit of an overkill for the non-relativistic free particle but will become a valuable mathematical tool in field theory and even for studying the quantum mechanics of a particle in external potentials. We will now introduce this procedure.

---

[11] Since $e^{A+B} \neq e^A e^B$ when A and B do not commute, we resort to this trick to evaluate the matrix element. If $H(\hat{\mathbf{p}}, \hat{\mathbf{x}})$ has a more non-trivial structure, there could be factor ordering issues, as well as questions like whether to evaluate H at $\mathbf{x}_k$ or at $\mathbf{x}_{k+1}$ or at the midpoint, etc. None of these (very interesting) issues are relevant for our purpose.

We consider all expressions to be defined in a complex t-plane with the usual time coordinate running along the real axis. The integrals etc. will be defined, though, off the real-t axis by assuming that t actually stands for $t\exp(-i\theta)$ where $0 \le \theta \le \pi/2$ and t now stands for the magnitude in the complex plane. There are two values for θ which are frequently used: If you think of θ as an infinitesimal ϵ, this is equivalent to doing the integrals etc. with $t(1 - i\epsilon)$ instead of t. Alternatively, we can choose $\theta = \pi/2$ and work with $-it$. This is done by replacing t by $-it_E$ in the relevant expressions and treating t_E as real.[12]. (The t_E is usually called the *Euclidean* time because this change makes the Lorentzian line element $-dt^2 + d\mathbf{x}^2$ with signature $(-, +, +, ...)$ go into the Euclidean line element $+dt_E^2 + d\mathbf{x}^2$.) If there are no poles in the complex t plane for the range $-\pi/2 \le \theta \le 0$, $\pi \le \theta \le 3\pi/2$ in the expressions we are interested in, then both the prescriptions — the $i\epsilon$ prescription or the $t \to -it_E$ prescription — will give the same result.

If we put $t = -it_E$, then the phase of the path integral amplitude for a particle in a potential $V(\mathbf{x})$ becomes:

$$iA = i \int dt \left[\frac{1}{2} m \left(\frac{d\mathbf{x}}{dt} \right)^2 - V(\mathbf{x}) \right] = - \int dt_E \left[\frac{1}{2} m \left(\frac{d\mathbf{x}}{dt_E} \right)^2 + V(\mathbf{x}) \right] \tag{1.34}$$

so that the sum over paths is now given by an expression which is properly convergent for a $V(\mathbf{x})$ that is bounded from below:

$$G_E(x_2; x_1) = \int \mathcal{D}\mathbf{x} \exp - \int dt_E \left[\frac{1}{2} m \left(\frac{d\mathbf{x}}{dt_E} \right)^2 + V(\mathbf{x}) \right] \tag{1.35}$$

Everything works out as before (even better as far as the integrals are concerned) and the Euclidean version of the path integral sum leads to:

$$G_E(x_2; x_1) = \theta(t_E) \left(\frac{m}{2\pi t_E} \right)^{D/2} \exp \left(-\frac{1}{2} \frac{m|\mathbf{x}|^2}{t_E} \right) \tag{1.36}$$

This analytic continuation should be thought of as a rotation in the complex t-plane with $|t|e^{-i\theta}$, rotating from $\theta = 0$ to $\theta = \pi/2$. Incidentally, $G_E(x_2; x_1)$ represents the kernel for the diffusion operator because the Schrodinger equation for a free particle becomes a diffusion equation in Euclidean time.

We will illustrate the utility of this formalism, especially in giving meaning to certain limits, with a couple of examples in the context of non-relativistic quantum mechanics. As a first example, we will show that the ground state energy and the ground state wave function of a system can be directly determined from $G_E(x_2; x_1)$ in the Euclidean sector. To do this, we will use Eq. (1.29) written in the form

$$\langle t, \mathbf{x}_2 | 0, \mathbf{x}_1 \rangle = \sum_n \phi_n(\mathbf{x}_2) \phi_n^*(\mathbf{x}_1) \exp(-iE_n t) \tag{1.37}$$

where $\phi_n(\mathbf{x}) = \langle \mathbf{x} | E_n \rangle$ is the n-th energy eigenfunction of the system under consideration. We have assumed that the energy levels are discrete for the sake of simplicity; the summation over n will be replaced by integration over the energy E if the levels form a continuum. Let us now consider the following limit:

$$W(t; \mathbf{x}_2, \mathbf{x}_1) \equiv \lim_{t \to +\infty} G(x_2; x_1) \tag{1.38}$$

[12]A relation like $t = -it_E$ cannot hold with both t and t_E being real, and this fact can be confusing. It is better to think of a complex t in the form $te^{-i\theta}$ with $\theta = 0$ giving the usual time which makes you age and $\theta = \pi/2$ giving the expressions in the Euclidean sector. Algebraically, however, this can be done by setting $t = -it_E$ and thinking of t_E as real!

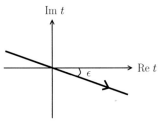

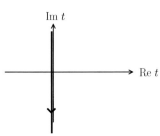

Figure 1.3: The $i\epsilon$ and Euclidean time prescription.

The t-dependence of the left hand side is supposed to indicate the leading dependence when the limit is taken. This limit cannot be directly ascertained from Eq. (1.37) because the exponent oscillates. However, we can give meaning to this limit if we first transform Eq. (1.37) to the imaginary time $\tau_1 = it_1$ and $\tau_2 = it_2$ and consider the limit $t_E = (\tau_2 - \tau_1) \to \infty$. We find that

$$W_E(t_E; \mathbf{x}_2, \mathbf{x}_1) = \lim_{t_E \to \infty} G_E(x_2; x_1) \cong \phi_0(\mathbf{x}_2)\phi_0(\mathbf{x}_1) \exp[-E_0 t_E] \quad (1.39)$$

where the zero-subscript denotes the lowest energy state. (Note that $\phi_0^* = \phi_0$ for the ground state.) Thus only the ground state contributes in this infinite time limit. Writing this as

$$\langle \infty, \mathbf{0} | - \infty, \mathbf{0} \rangle_E = |\phi_0(0)|^2 \exp - \int_{-\infty}^{\infty} dt_E \; E_0 \quad (1.40)$$

we see that the phase of $\langle \infty, \mathbf{0} | - \infty, \mathbf{0} \rangle_E$ gives the ground state energy. More explicitly, from $W_E(t_E; \mathbf{0}, \mathbf{0}) \approx (\text{constant}) \exp(-E_0 t_E)$, we have:

$$E_0 = \lim_{t_E \to \infty} \left(-\frac{1}{t_E} \ln W_E(t_E; \mathbf{0}, \mathbf{0}) \right) = \lim_{t_E \to \infty} \left(-\frac{1}{t_E} \ln \langle t_E, \mathbf{0} | 0, \mathbf{0} \rangle_E \right) \quad (1.41)$$

Using a variant of Eq. (1.39), we can also express the ground state wave function itself in terms of $G(x_2; x_1)$. From[13]

$$\frac{W_E(t_E; \mathbf{x}_2, \mathbf{x}_1)}{W_E(t_E; \mathbf{0}, \mathbf{0})} \simeq \frac{\phi_0(\mathbf{x}_2)\phi_0(\mathbf{x}_1)}{\phi_0(\mathbf{0})\phi_0(\mathbf{0})} \quad (1.42)$$

we see that

$$\frac{W_E(t_E; \mathbf{0}, \mathbf{x})}{W_E(t_E; \mathbf{0}, \mathbf{0})} \simeq \frac{\phi_0(\mathbf{x})}{\phi_0(\mathbf{0})} \quad (1.43)$$

Ignoring the proportionality constants, which can be fixed by normalizing the ground state wave function, we have the result

$$\phi_0(\mathbf{x}) \propto \lim_{t_E \to \infty} W_E(t_E; \mathbf{0}, \mathbf{x}) \propto \int \mathcal{D}q \, \exp\left(-A_E\left[\infty, 0; 0, \mathbf{x}\right]\right) \quad (1.44)$$

So if you evaluate the Euclidean path integral with the boundary conditions that at $t_E = 0$ we have $\mathbf{q} = \mathbf{x}$ and at $t_E = \infty$, $\mathbf{q} = \mathbf{0}$, then, except for unimportant constants, the path integral will reproduce the ground state wave function. This result is useful in field theoretic contexts in which one cannot explicitly solve the analog of the Schrodinger equation to find the ground state wave function but can evaluate the path integral in some suitable approximation.

The propagation amplitude can also be used to study the effect of external perturbations on the system. (We will discuss this in detail in Sect. 2.1; here, we shall just derive a simple formula which will be of use later on.) Let us suppose that the system was in the ground state in the asymptotic past ($t_1 \to -\infty$). At some time $t = -T$ we switch on an external, time dependent disturbance $J(t)$ affecting the system. Finally at $t = +T$ we switch off the perturbation. Because of the time-dependence, we no longer have stationary energy eigenstates for the system. In fact, the system is likely to have absorbed energy from the perturbation and ended up at some excited state at $t_2 \to +\infty$; so the probability for it to be found in the ground

[13]While we introduce the analytic continuation to Euclidean time, $t \to t_E$, as a mathematical trick, it will soon acquire a life of its own. One conceptual issue is that the paths you sum over in the Euclidean sector can be quite different compared to those in the Lorentzian sector. Later on, when we deal with the relativistic fields, this distinction will become more serious, because the Lorentzian space has a light cone structure which the Euclidean sector does not have. There are people who even believe quantum field theory should be actually formulated in the Euclidean sector and then analytically continued to the usual spacetime. It cannot be denied that the Euclidean formulation gives certain insights which are surprisingly difficult to obtain from other methods.

Exercise 1.4: Work this out explicitly for a harmonic oscillator and verify this claim.

state as $t_2 \to +\infty$ will be less than one. This probability can be computed from the propagation amplitude $G^J(x_2; x_1)$ in the presence of the source J. Consider the limit of:

$$
\begin{aligned}
\mathcal{P} &= \lim_{t_2 \to \infty} \lim_{t_1 \to -\infty} G^J(x_2; x_1) = \lim_{t_2 \to \infty} \lim_{t_1 \to -\infty} \langle t_2, \mathbf{x}_2 | t_1, \mathbf{x}_1 \rangle^J \qquad (1.45) \\
&= \lim_{t_2 \to \infty} \lim_{t_1 \to -\infty} \left[\int_{-\infty}^{+\infty} d^D\mathbf{x} \, d^D\mathbf{x}' \right. \\
&\qquad\qquad \left. \times \langle t_2, \mathbf{x}_2 | T, \mathbf{x} \rangle \langle T, \mathbf{x} | -T, \mathbf{x}' \rangle \langle -T, \mathbf{x}' | t_1, \mathbf{x}_1 \rangle \right]
\end{aligned}
$$

Since $J = 0$ during $t_2 > t > T$ and $-T > t > t_1$, the matrix elements in these intervals can be expressed in terms of the energy eigenstates of the original system. We can then take the limits $t_2 \to \infty$ and $t_1 \to -\infty$ by first going over to Euclidean time, taking the limits and then analytically continuing back to normal time. Using Eq. (1.39), we have:

$$
\begin{aligned}
\lim_{t_2 \to \infty} \langle t_2, \mathbf{x}_2 | T, \mathbf{x} \rangle &\cong \phi_0(\mathbf{x}_2) \phi_0(\mathbf{x}) \exp{-iE_0(t_2 - T)} \\
\lim_{t_1 \to -\infty} \langle -T, \mathbf{x}' | t_1, \mathbf{x}_1 \rangle &\cong \phi_0(\mathbf{x}') \phi_0(\mathbf{x}_1) \exp{-iE_0(-T - t_1)} \qquad (1.46)
\end{aligned}
$$

which, when substituted into Eq. (1.45), gives:

$$
\begin{aligned}
\mathcal{P} &\cong \left[\phi_0(\mathbf{x}_2) \phi_0(\mathbf{x}_1) e^{-iE_0(t_2 - t_1)} \right] \qquad\qquad\qquad\qquad\qquad\qquad (1.47) \\
&\qquad \times \int_{-\infty}^{+\infty} d^D\mathbf{x} \, d^D\mathbf{x}' (\phi_0(\mathbf{x}) e^{iE_0 T}) \langle T, \mathbf{x} | \mathbf{x}', -T \rangle (\phi_0(\mathbf{x}') e^{iE_0 T}) \\
&\cong \lim_{t \to \infty} G^{J=0}(x_2; x_1) \int_{-\infty}^{+\infty} d^D\mathbf{x} \, d^D\mathbf{x}' [\phi_0(\mathbf{x}, T)]^* \langle T, \mathbf{x} | \mathbf{x}', -T \rangle [\phi_0(\mathbf{x}', -T)]
\end{aligned}
$$

where $\phi_0(\mathbf{x}, T)$ represents the ground state wave function at time T, etc. The quantity

$$
\mathcal{Q} = \int_{-\infty}^{+\infty} d^D\mathbf{x} \, d^D\mathbf{x}' [\phi_0(\mathbf{x}, T)]^* \langle \mathbf{x}, T | \mathbf{x}', -T \rangle [\phi_0(\mathbf{x}', -T)] \qquad (1.48)
$$

represents the amplitude for the system to remain in the ground state after the source is switched off if it started out in the ground state before the source was switched on. (It is usually called the *vacuum persistence* amplitude.) From Eq. (1.47) and Eq. (1.45) we find that this amplitude is given by the limit:

Exercise 1.5: Compute $\mathcal{Q}$ for a simple harmonic oscillator of frequency ω coupled to an external time dependent source $J(t)$. Your result should match with the one we need to use later in Eq. (2.87).

$$
\mathcal{Q} = \lim_{t_2 \to \infty} \lim_{t_1 \to -\infty} \frac{G^J(x_2; x_1)}{G^{J=0}(x_2; x_1)} \qquad (1.49)
$$

This result can be further simplified by noticing that the $\mathbf{x}_2$ and $\mathbf{x}_1$ dependences cancel out in the ratio in Eq. (1.49) so that we can set $\mathbf{x}_2 = \mathbf{x}_1 = 0$. Thus the vacuum persistence amplitude can be found from the propagation amplitude by a simple limiting procedure.

Finally we mention an important application of imaginary time in statistical mechanics and condensed matter physics. The analytic continuation to imaginary values of time has close mathematical connections with the description of systems in a thermal bath. To see this, consider the mean value of some observable $\mathcal{O}(q)$ of a quantum mechanical system. If the system is in an energy eigenstate described by the wave function $\phi_n(q)$, then the expectation value of $\mathcal{O}(q)$ can be obtained by integrating $\mathcal{O}(q)|\phi_n(q)|^2$ over q. If the system is in a thermal bath at temperature β^{-1}, described by a canonical ensemble, then the mean value has to be computed by averaging over all the energy eigenstates *as well* with a weightage $\exp(-\beta E_n)$. In this case, the mean value can be expressed as

$$\langle \mathcal{O} \rangle = \frac{1}{Z} \sum_n \int dq \, \phi_n(q) \mathcal{O}(q) \phi_n^*(q) \, e^{-\beta E_n} \equiv \frac{1}{Z} \int dq \, \rho(q, q) \mathcal{O}(q) \quad (1.50)$$

where Z is the partition function and we have defined a *density matrix* $\rho(q, q')$ by

$$\rho(q, q') \equiv \sum_n \phi_n(q) \phi_n^*(q') \, e^{-\beta E_n} \quad (1.51)$$

in terms of which we can rewrite Eq. (1.50) as

$$\langle \mathcal{O} \rangle = \frac{\text{Tr}\,(\rho \mathcal{O})}{\text{Tr}\,(\rho)} \quad (1.52)$$

where the trace operation involves setting $q = q'$ and integrating over q. This result shows how $\rho(q, q')$ contains information about both thermal and quantum mechanical averaging. Comparing Eq. (1.51) with Eq. (1.37), we find that the density matrix can be immediately obtained from the Euclidean propagation amplitude:

$$\rho(q, q') = \langle \beta, q|0, q' \rangle_E \quad (1.53)$$

with the Euclidean time acting as inverse temperature. Its trace, $\rho(q, q) \propto \langle \beta, q|0, q \rangle_E$ is obtained by computing the Euclidean path integral with a periodic boundary condition on the imaginary time. The period of the imaginary time gives the inverse temperature.

Exercise 1.6: Compute $\rho(q, q')$ for a harmonic oscillator at finite temperature and study the low and high temperature limits. Can you interpret these limits?

1.2.4 Path Integral for the Jacobi Action

It is possible to use the results obtained in the Sect. 1.2.1 to give meaning to a sum over paths for an action which is *not* quadratic. Given its utility in the study of the relativistic particle in the next section, we will first develop this approach in the non-relativistic context.

In classical mechanics the usual form of the action $A[\mathbf{x}(t)] \equiv A[x^\alpha(t)]$ is taken to be a time integral [in the range (t_1, t_2)] of the Lagrangian $L = T(x^\alpha, \dot{x}^\alpha) - V(x^\alpha)$. Demanding $\delta A = 0$ with $\delta x^\alpha = 0$ at the end points leads to the equations of motion determining the trajectory $x^\alpha(t)$ satisfying the boundary conditions. This procedure not only determines the path taken by the particle in space (like e.g., the ellipse in the Kepler problem) but also the actual coordinates of the particle as a function of time t (like, for e.g., $r(t)$ and $\theta(t)$ in the Kepler problem; eliminating t between these two functions gives the equation to the path $r(\theta)$).

Suppose we are not interested in the latter information and only want to know the equation to the path taken in space by the particle. It is then possible to use a different action principle — called the *Jacobi-Maupertuis action* — the extremum condition for which leads directly to the trajectory. We will first recall this action principle from classical mechanics.

Let us consider a situation in which the kinetic energy term T of the Lagrangian is a homogeneous quadratic function of the velocities and can be written in the form

$$T = \frac{1}{2}m\left(\frac{d\ell}{d\lambda}\right)^2 = \frac{1}{2}m\left(g_{\alpha\beta}\dot{x}^\alpha\dot{x}^\beta\right) \tag{1.54}$$

where $d\ell^2 = g_{\alpha\beta}(x^\mu)dx^\alpha\,dx^\beta$ is the spatial line element in some arbitrary coordinate system and λ is the time, with $\dot{x}^\alpha = dx^\alpha/d\lambda$ etc.. Since we consider Lagrangians with no explicit time dependence, we can take $t_1 = 0, t_2 = t$ in defining the action. Using the relation between the Lagrangian and the Hamiltonian, $L = p_\alpha\dot{x}^\alpha - H$, we can write the action functional as

$$\begin{aligned} A &= \int_0^t d\lambda\left[\frac{1}{2}m\left(\frac{d\ell}{d\lambda}\right)^2 - V(x^\alpha)\right]; \qquad d\ell^2 = g_{\alpha\beta}(x^\mu)dx^\alpha\,dx^\beta \\ &= \int_0^t d\lambda\left[p_\alpha\dot{x}^\alpha - H\right] \end{aligned} \tag{1.55}$$

This form of the action motivates us to consider another action functional given by

$$A_J \equiv \int_{\mathbf{x}_1}^{\mathbf{x}_2} p_\alpha dx^\alpha = \int_{\lambda_1}^{\lambda_2} d\lambda\left(\frac{\partial L}{\partial\dot{x}^\alpha}\right)\dot{x}^\alpha \tag{1.56}$$

which differs from Eq. (1.55) by the absence of $Hd\lambda$ term. Roughly speaking, one would have expected this term not to contribute to the variation if we consider trajectories with a fixed energy E and express p_α in A_J above in terms of E. It can be easily verified that the modified action principle based on A_J in Eq. (1.56) leads to the actual *paths in space* as a solution to the variational principle $\delta A_J = 0$ when we vary all trajectories connecting the end points x_1^α and x_2^α. This trajectory will have energy E, but will contain no information about the time coordinate. In fact, the first form of the action in Eq. (1.56) makes clear that there is no time dependence in the action. Geometrically, we are only interested in various curves connecting two points in space x_1^α and x_2^α, irrespective of their parameterization. We can describe the curve with some parameter λ by giving the D functions $x^\alpha(\lambda)$ or some other set of functions obtained by changing the parameterization from $\lambda \to f(\lambda)$. The curves remain the same and the reparameterization invariance of the action expresses this fact.

It is possible to rewrite the expression for A_J in a nicer form. Using the fact that L is a homogeneous quadratic function of velocities, we have the result

$$\dot{x}^\alpha\left(\frac{\partial L}{\partial\dot{x}^\alpha}\right) = 2T = m\left(\frac{d\ell}{d\lambda}\right)^2 = 2\left[E - V(x^\alpha)\right] \tag{1.57}$$

Substituting into Eq. (1.56) we get

$$A_J = \int_{\mathbf{x}_1}^{\mathbf{x}_2} m\left(\frac{d\ell}{d\lambda}\right)d\ell = \int_{\mathbf{x}_1}^{\mathbf{x}_2}\sqrt{2m(E - V(x^\alpha))}\,d\ell \tag{1.58}$$

which is again manifestly reparameterization invariant with no reference to time.[14]

Since A_J describes a valid action principle for finding the path of a particle with energy E classically, one might wonder what happens if we try

Exercise 1.7: Vary A_J and prove these claims.

[14]If we consider an abstract space with metric $G_{\alpha\beta} \equiv 2m(E - V)g_{\alpha\beta}$, then A_J for a given value of E is just the length of the path calculated with this metric $G_{\alpha\beta}$. So the spatial paths followed by the particle are geodesics in the space with this metric. This is amusing but does not seem to lead to any deep insight.

to quantize the system by performing a sum over amplitudes $\exp(iA_J)$. We would expect it to lead to the amplitude for the particle to propagate from x_1^α to x_2^α *with energy E*. This is indeed true but since A_J is not quadratic in velocities even for a free particle, (because $d\ell$ involves a square root) it is not easy to evaluate the sum over $\exp(iA_J)$. But since we already have an alternative path integral procedure for the system, we can use it *to give meaning* to this sum, thereby defining the sum over paths for at least one non-quadratic action.

Our idea is to write the sum over all paths in the original action principle (with amplitude $\exp(iA)$) as a sum over paths with energy E followed by a sum over all E. Using the result in Eq. (1.55), we get

$$\sum_{0,\mathbf{x}_1}^{t,\mathbf{x}_2} \exp(iA) = \sum_E \sum_{\mathbf{x}_1}^{\mathbf{x}_2} e^{-iEt}\, \exp iA_J[E,\mathbf{x}(\tau)] \propto \int_0^\infty dE\, e^{-iEt} \sum_{\mathbf{x}_1}^{\mathbf{x}_2} \exp(iA_J)$$

(1.59)

In the last step we have treated the sum over E as an integral over $E > 0$ (since, for any Hamiltonian which is bounded from below, we can always achieve this by adding a suitable constant to the Hamiltonian) but there could be an extra proportionality constant which we cannot rule out. This constant will depend on the measure used to define the sum over $\exp(iA_J)$ but can be fixed by using the known form of the left hand side if required. Inverting the Fourier transform, we get:

$$\begin{aligned} \mathcal{P}(E;\mathbf{x}_2,\mathbf{x}_1) &\equiv \sum_{\mathbf{x}_1}^{\mathbf{x}_2} \exp(iA_J) = C \int_0^\infty dt\, e^{iEt} \sum_{0,\mathbf{x}_1}^{t,\mathbf{x}_2} \exp(iA) \\ &= C \int_0^\infty dt\, e^{iEt} G(x_2;x_1) \end{aligned}$$

(1.60)

where we have denoted the proportionality constant by C. This result shows that the sum over the Jacobi action involving a *square root of velocities* can be re-expressed in terms of the standard path integral; if the latter can be evaluated for a given system, then the sum over Jacobi action can be defined by this procedure.

The result also has an obvious interpretation. The $G(x_2;x_1)$ on the right hand side gives the amplitude for a particle to propagate from $\mathbf{x}_1$ to $\mathbf{x}_2$ in time t. Its Fourier transform with respect to t can be thought of as the amplitude for the particle to propagate from $\mathbf{x}_1$ to $\mathbf{x}_2$ with energy E, which is precisely what we expect the sum over the Jacobi action to give. The idea actually works even for particles in a potential if we evaluate the path integral on the right hand side by some other means like, e.g., by solving the relevant Schrodinger equation.[15]

With future applications in mind, we will display the explicit form of this result for the case of a free particle with $V = 0$. Denoting the length of the path connecting x_1^α and x_2^α by $\ell(\mathbf{x}_2,\mathbf{x}_1)$, we have:

$$\sum_{\mathbf{x}_1}^{\mathbf{x}_2} \exp i\sqrt{2mE}\,\ell(\mathbf{x}_2,\mathbf{x}_1) = C \int_0^\infty dt\, e^{iEt} \sum_{0,\mathbf{x}_1}^{t,\mathbf{x}_2} \exp \frac{im}{2}\int_0^t d\tau\, \left(g_{\alpha\beta}\dot{x}^\alpha \dot{x}^\beta\right)$$

(1.61)

This result shows that the sum over paths with a Jacobi action, involving a square root of velocities, can be re-expressed in terms of the standard path integral involving only quadratic terms in the velocities. We, of course,

[15]In this case, the $\sqrt{E-V}$ will lead to imaginary values for $E < V$, thereby describing quantum mechanical tunneling. The path integral with the Jacobi action will give exponentially suppressed amplitudes for classically forbidden processes — a result we will again allude to in the next section.

Exercise 1.8: Work out Eq. (1.61) explicitly for a D-dimensional space. You will find that the integral is trivial for $D = 1, 3$ while it involves the Bessel function for other values of D. For example, $D = 4$ — which we will consider in the next section — will give you a MacDonald function.

know the result of the path integral in the right hand side (for $g_{\alpha\beta} = \delta_{\alpha\beta}$ in Cartesian coordinates) and thus we can evaluate the sum on the left hand side.

It must be stressed that the path integral sum over the Jacobi action, given by $\mathcal{P}(E; \mathbf{x}_2, \mathbf{x}_1)$, is a very different beast compared to the path integral sum over the usual action, $G(x_2; x_1)$. The $\mathcal{P}(E; \mathbf{x}_2, \mathbf{x}_1)$ does *not* share some of the crucial properties of $G(x_2; x_1)$. Most importantly, it does not obey any kind of transitivity like that in Eq. (1.2). It can be directly verified that

$$\mathcal{P}(E; \mathbf{x}_2, \mathbf{x}_1) \neq \int d^D \mathbf{y} \, \mathcal{P}(E; \mathbf{x}_2, \mathbf{y}) \, \mathcal{P}(E; \mathbf{y}, \mathbf{x}_1) \qquad (1.62)$$

This means $\mathcal{P}(E; \mathbf{x}_2, \mathbf{x}_1)$ cannot be used to "propagate" any wave function unlike $G(x_2; x_1)$ which acts as a propagator for the solutions of Schrodinger equations in accordance with Eq. (1.3). Mathematically, this is obvious from the structure of $\mathcal{P}(E; \mathbf{x}_2, \mathbf{x}_1)$ which can be expressed formally as the matrix element:

$$\mathcal{P}(E; \mathbf{x}_2, \mathbf{x}_1) = \int_0^\infty dt \, e^{it(E+i\epsilon)} \langle \mathbf{x}_2 | e^{-it\hat{H}} | \mathbf{x}_1 \rangle = i \langle \mathbf{x}_2 | (E - \hat{H} + i\epsilon)^{-1} | \mathbf{x}_1 \rangle \tag{1.63}$$

where we have introduced an $i\epsilon$ factor, with an infinitesimal ϵ, to ensure convergence. This result can be expressed more explicitly in terms of the energy eigenfunctions using Eq. (1.29):

$$\begin{aligned}
\mathcal{P}(E; \mathbf{x}_2, \mathbf{x}_1) &= \int_0^\infty dt \int_0^\infty d\mu \, e^{-i\mu t} \phi_\mu(\mathbf{x}_2) \phi_\mu^*(\mathbf{x}_1) e^{iEt} \\
&= \int_0^\infty d\mu \, \frac{i\phi_\mu(\mathbf{x}_2) \phi_\mu^*(\mathbf{x}_1)}{(E - \mu + i\epsilon)}
\end{aligned} \tag{1.64}$$

where we have again changed E to $E + i\epsilon$ to ensure the convergence of the original integral which also avoids a pole in the real axis for the μ integration. Clearly, this expression has a very different structure compared to the one in Eq. (1.29), which accounts for lack of transitivity, etc. From either Eq. (1.63) or Eq. (1.64), it is easy to see that:

$$\int d^D \mathbf{y} \, \mathcal{P}(E; \mathbf{x}_2, \mathbf{y}) \mathcal{P}(E; \mathbf{y}, \mathbf{x}_1) = \mathcal{P}(E; \mathbf{x}_2, \mathbf{x}_1) \left[\frac{\partial \ln \mathcal{P}(E; \mathbf{x}_2, \mathbf{x}_1)}{i \partial E} \right] \quad (1.65)$$

where the factor in the square bracket gives the deviation from transitivity.

In the case of a free particle, in which energy eigenstates are labeled by the momenta $\mathbf{p}$, Eq. (1.63) has the form

$$\mathcal{P}(E; \mathbf{x}_2, \mathbf{x}_1) = \int \frac{d^D \mathbf{p}}{(2\pi)^D} \frac{i e^{i\mathbf{p} \cdot \mathbf{x}}}{E - p^2/2m + i\epsilon} \qquad (1.66)$$

In non-relativistic quantum mechanics all these are more of curiosities and we don't really care about the Jacobi action and a path integral defined by it. (It is not even clear what a Schrodinger-like "propagation" could mean when the time coordinate has been integrated out.) However these features will hit us hard when we evaluate the path integral for a relativistic particle because it will turn out to be identical in structure to the path integral defined using a Jacobi action. We shall take that up next.

1.3 Quantum Propagation Amplitude for the Relativistic Particle

The above discussion would have convinced you that, as far as the non-relativistic free particle is concerned, we can proceed from the classical theory to the quantum theory in a fairly straightforward way. We can either use the path integral and obtain the propagation amplitude or follow the more conventional procedure of setting up a Hamiltonian and quantizing the system. Encouraged by this success, we could try the same procedure with a relativistic free particle and obtain its quantum description from the classical counterpart using, say, the path integral. Here is where all sorts of strange things are expected to happen. The procedure should lead us from a single particle description to a field theoretic description. We will now see how this comes about.

1.3.1 Path Integral for the Relativistic Particle

Let us next try our hand with a relativistic particle for which all hell is expected to break loose. But since we know the classical action (as well as the Hamiltonian) for this case as well, we should be able to compute the amplitude $G(x_2; x_1)$. The standard action for a relativistic particle is given by

$$A = -m \int_{t_1}^{t_2} dt \sqrt{1 - \mathbf{v}^2} = -m \int_{x_1}^{x_2} \sqrt{\eta_{ab} dx^a dx^b} = -m \int_{\lambda_1}^{\lambda_2} d\lambda \sqrt{\eta_{ab} \dot{x}^a \dot{x}^b}$$
(1.67)

where $x^a(\lambda)$ gives a parameterized curve connecting the events x_1 and x_2 in the spacetime with the parameter λ and $\dot{x}^a \equiv dx^a/d\lambda$. In the second and third forms of the expression, the integral is evaluated for any curve connecting the two events with the limits of integration depending on the nature of the parametrization. (For example, we have chosen $x(\lambda = \lambda_1) = x_1, x(\lambda = \lambda_2) = x_2$ but the numerical value of the integral is independent of the parametrization and depends only on the curve. If we choose to use $\lambda = t$ as the parameter, then we reproduce the first expression from the last.)

It is obvious that this action has the same structure as the Jacobi action for a free particle discussed in the last section. To obtain the propagation amplitude $G(x_2; x_1)$, we need to do the path integral using the above action,

$$G(x_2; x_1) = \sum_{0,\mathbf{x}_1}^{t,\mathbf{x}_2} \exp\left[-im \int_{t_1}^{t_2} dt \sqrt{1 - \mathbf{v}^2}\right]$$

$$= \sum_{0,\mathbf{x}_1}^{t,\mathbf{x}_2} \exp\left[-im \int_0^\tau d\lambda \sqrt{\dot{t}^2 - \dot{\mathbf{x}}^2}\right] \quad (1.68)$$

which can be accomplished using the results in the last section.[16] We first take the complex conjugate of Eq. (1.61) (in order to get the overall minus sign in the action in Eq. (1.67)) and generalize the result from space to

[16]It is also possible, fortunately, to do this rigorously. (The details are given in the Mathematical Supplement in Sect. 1.6.2.) One can define the action in Eq. (1.67) in a Euclidean D-dimensional cubic lattice with lattice spacing ϵ, interpreting the amplitude as e^{-ml} where l is the length of any path connecting two given lattice points. The sum over paths can be calculated by multiplying the number of paths of a given length l by e^{-ml} and summing over all l. This is a simple problem on the lattice and the sum can be explicitly determined. The result in the continuum needs to be obtained by taking the limit $\epsilon \to 0$ using a suitable measure which has to be determined in such a way that the limit is finite. Such an analysis gives the answer which is identical to what we obtain here and has the merit of being more rigorous, while the simpler procedure used here has the advantage of having a direct physical interpretation.

spacetime, leading to

$$\sum_{\mathbf{x}_1}^{\mathbf{x}_2} \exp{-i\sqrt{2mE}}\,\ell(\mathbf{x}_2,\mathbf{x}_1) = C \int_0^\infty d\tau e^{-iE\tau} \sum_{0,\mathbf{x}_1}^{t,\mathbf{x}_2} \exp{-\frac{im}{2}\int_0^\tau d\lambda\left(g_{ab}\dot{x}^a\dot{x}^b\right)}$$

(1.69)

In order to get $-im\ell(\mathbf{x}_2,\mathbf{x}_1)$ on the left hand side we take $E = m/2$; ie., we use the above formula with the replacements

$$E = \frac{m}{2}; \qquad g_{ab} = \text{dia}\,(1,-1,-1,-1);$$

$$\ell = \int_0^\tau \sqrt{g_{ab}\dot{x}^a\dot{x}^b}\,d\lambda = \int_{t_1}^{t_2} dt\sqrt{1-\mathbf{v}^2}$$

(1.70)

The path integral over the quadratic action can be immediately borrowed from Eq. (1.9) with $D = 4$, taking due care of the fact that in the quadratic action the $\dot{t}^2$ enters with negative sign while $\dot{\mathbf{x}}^2$ enters with the usual positive sign. This gives an extra factor i to N and the answer is:

$$\langle x_2,\tau|x_1,0\rangle = \theta(\tau)i\left(\frac{m}{2\pi i\tau}\right)^2 \exp\left(-\frac{i}{2}\frac{mx^2}{\tau}\right)$$

(1.71)

The $\theta(\tau)$ is introduced for the same reason as $\theta(t)$ in Eq. (1.9) but will turn out to be irrelevant since we will integrate over it. Therefore the path integral we need to compute is given by:

$$\begin{aligned}
G(x_2;x_1) &= \sum_{0,\mathbf{x}_1}^{t,\mathbf{x}_2} \exp{-im\int_{t_1}^{t_2} dt\sqrt{1-\mathbf{v}^2}} \\
&= C\int_0^\infty d\tau \,\exp\left(-\frac{im}{2}\tau\right) i\left(\frac{m}{2\pi i\tau}\right)^2 \exp\left(-\frac{im}{2\tau}x^2\right) \\
&= (2Cm)(-i)\frac{m}{16\pi^2}\int_0^\infty \frac{ds}{s^2}e^{-ims}\exp\left(-\frac{im}{4s}x^2\right),
\end{aligned}$$

(1.72)

with $\tau = 2s$. We have thus given meaning to the sum over paths for the relativistic particle thereby obtaining $G(x_2;x_1)$. The integral expression also gives a nice interpretation for $G(x_2;x_1)$ which we will first describe before discussing this result.

The trajectory of a classical relativistic particle in *spacetime* is given by the four functions $x^i(\tau)$ where τ could be taken as the proper time shown by a clock which moves with the particle[17]. Such a description treats space and time on an equal footing with $\mathbf{x}(\tau)$ and $t(\tau)$ being *dependent* variables and τ being the *independent* variable having an observer independent, absolute, meaning. This is a natural generalization of $\mathbf{x}(t)$ in non-relativistic mechanics with (x,y,z) being dependent variables and t being the independent variable having an observer independent, absolute, status. Let us now consider an action for the relativistic particle given by

$$A[x(\tau)] = -\frac{1}{4}m\int_0^s d\tau\,\dot{x}_a\dot{x}^a$$

(1.73)

[17]This is one physically meaningful choice for timelike curves; for spacelike and null curves, the corresponding choices are proper length and affine parameter.

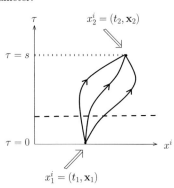

Figure 1.4: Spacetime trajectories as a function of proper time

where $\dot{x}^a \equiv (dx^a/d\tau)$ etc. This action, of course, gives the correct equations of motion $d^2x^a/d\tau^2 = 0$ but the overall constant in front of the integral — which is arbitrary as far as classical equations of motion go — is chosen based on the result obtained below in Eq. (1.75). Evaluating a path integral with this action will now lead to an amplitude of the form $\langle x_2, s | x_1, 0 \rangle$ which describes a particle propagating from an event x_1 to event x_2 when the proper time lapse is given by s. But we are interested in the amplitude $G(x_2; x_1)$ and don't care what is the amount of proper time that has elapsed. Therefore we need to also sum over (i.e., integrate) all the proper time lapses with some suitable measure. Since the rest energy of the particle m is conjugate to the proper time (which measures the lapse of time in the instantaneous co-moving Lorentz frame of a particle) it seems reasonable to choose this measure to be proportional to a phase factor e^{-ims}. Thus we have the relation

$$G(x_2; x_1) = C_m \int_{-\infty}^{\infty} ds\, e^{-ims} \langle x_2, s | x_1, 0 \rangle = C_m \int_{-\infty}^{\infty} ds\, e^{-ims} \sum_{x(\tau)} e^{iA[x(\tau)]}$$

(1.74)

where C_m is a normalization constant,[18] possibly dependent on m, which we will fix later. (We have kept the integration limits on s to be the entire real line but it will get limited to $(0, \infty)$ because of $\theta(s)$ in the path integral.) In the second equality we have used the standard path integral prescription.

Exactly as before, the sum over paths is now to be evaluated limiting ourselves to paths $x^i(\tau)$ which only go forward in the *proper time* τ (see Fig. 1.4; just as the paths in Eq. (1.8) were limited to those which go forward in the Newtonian absolute time t). However, we *have to* now allow paths like the one shown in Fig. 1.2 which go back and forth in the *coordinate time t* just as we allowed in Eq. (1.8) the paths which went back and forth in the y coordinate, say. The time coordinate $t(\tau)$ of a path now has the same status as the spatial coordinate, say $y(t)$, in the non-relativistic description. The special role played by the absolute Newtonian time t is taken over by the proper time τ in this description. This will, of course, have important consequences later on.

Let us now get back to the discussion of our main result. We will rewrite Eq. (1.72) in the form:

$$
\begin{aligned}
G(x_2; x_1) &= -(2Cm)i \left(\frac{m}{16\pi^2} \right) \int_0^{\infty} \frac{ds}{s^2} \exp\left(-ims - \frac{i}{4}\frac{mx^2}{s} \right) \\
&= -\frac{i}{16\pi^2} \int_0^{\infty} \frac{d\mu}{\mu^2} \exp\left(-i(m^2 - i\epsilon)\mu - \frac{i}{4}\frac{x^2}{\mu} \right) \quad (1.75)
\end{aligned}
$$

where we have made three modifications to arrive at the second line. First, we have rescaled the variable s to μ by $s \equiv m\mu$. Second, we have made the choice $C = 1/2m$ which, as we shall see, matches with conventional results later on and — more importantly — allows us to take the $m \to 0$ limit, if we want to study zero mass particles. Finally, we have replaced m^2 by $(m^2 - i\epsilon)$, where ϵ is an infinitesimal positive constant, in order to make the integral convergent in the upper limit.[19] We set $\epsilon = 0$ at the end of the calculations. The integral can be evaluated in terms of MacDonald functions, leading to the result:

$$G(x_2; x_1) = \frac{m}{4\pi^2 i \sqrt{x^2}} K_1(im\sqrt{x^2})$$

(1.76)

where, of course, $x^2 = t^2 - |\mathbf{x}|^2$ and hence its square-root is imaginary for spacelike intervals.

[18] Dimension alert: The amplitude $G(x_2; x_1)$ in Eq. (1.9) has the dimensions of (length)$^{-D}$ as it should. So the $G(x_2; x_1)$ in Eq. (1.74) will have the dimension (length)$^{-3}$ after integrating over s if C_m is dimensionless. People like it to have dimensions of (length)$^{-2}$ which is achieved by taking $C_m \propto (1/m)$, as we will soon do.

[19] Alternatively, you could have worked everything out in the Euclidean sector $it = t_E$, $is = s_E$ etc. It will be equivalent to the $i\epsilon$ prescription. (Try it out.)

That is it! We have found the propagation amplitude $G(x_2; x_1)$ for a relativistic particle just as we did it for a non-relativistic particle with Eq. (1.75) being the relativistic analogue of Eq. (1.9). The expression, of course, is a bit complicated algebraically but we will soon see that it has a simple expression in Fourier space. Where are the quantum fields, negative energy solutions and all the rest which we are supposed to see? To unravel all these structures we will scrutinize the expression in Eq. (1.75) more closely.

1.4   Mathematical Structure of $G(x_2; x_1)$

1.4.1   Lack of Transitivity

The first disastrous consequence arising from the nature of paths which were summed over in Eq. (1.74) is that the amplitude $G(x_2; x_1)$ does not satisfy the transitivity property:

$$G(x_2; x_1) \neq \int d^D \mathbf{y} \, G(x_2; y) G(y; x_1) \qquad (1.77)$$

So, we cannot have a Schrodinger-like wave function $\psi(x^i)$ which is propagated by $G(x_2; x_1)$ in a consistent manner, *in sharp contrast* to the situation in non-relativistic mechanics. This means that, we cannot think of the particle being 'somewhere in space' at intermediate times $x_2^0 > y^0 > x_1^0$. Mathematically, this is quite understandable because one of the paths we summed over [see Fig 1.2] suggests that the particle was at three different locations simultaneously.

1.4.2   Propagation Outside the Light Cone

Another immediate consequence is that the amplitude $G(x_2; x_1)$ does not vanish when x_2 and x_1 are separated by a spacelike interval, i.e., when x_2 lies outside the light cone originating at x_1. From the asymptotic forms of the MacDonald functions,[20] one can easily show that

$$G(x_2; x_1) \longrightarrow \begin{cases} e^{\pm imt} & (\text{for } |\mathbf{x}| = 0) \\[2mm] e^{-m|\mathbf{x}|} & (\text{for } t = 0) \end{cases} \qquad (1.78)$$

When the two events are separated by a spacelike interval, one can always find a coordinate system in which $t = 0$, in which the second form of the result applies. It shows that the amplitude is non-zero outside the light cone but decreases exponentially with a scale length $(1/m)$.

Classically a relativistic particle cannot go outside the light cone, but quantum mechanically it seems that the particle can.[21] In fact, our discussion of the Jacobi action makes it clear that the sum over paths calculated with the Jacobi action will give exponentially decaying amplitudes in classically forbidden regions due to tunneling, which is precisely what happens here. This is also obvious from the fact that the relativistic Lagrangian $L = -m\sqrt{1 - \mathbf{v}^2}$ becomes imaginary for $|\mathbf{v}| > 1$ which arises in paths which go outside the light cone. So this result is even to be expected.

The next trouble is that we can no longer localize the particle states. It is clear from Eq. (1.75) and Eq. (1.78) that $G(x_2; x_1)$ does not obey the

Exercise 1.9: Prove Eq. (1.77). It turns out that the transitivity does not hold even if we do the integral in Eq. (1.77) over 4-dimensions with $d^4 y$. The mathematical reason for this is that the amplitude we want, given by Eq. (1.74) is analogous to the propagation amplitude $\mathcal{P}(E; \mathbf{x}_2, \mathbf{x}_1)$ at fixed energy E in non-relativistic quantum mechanics, which is obvious from the fact that we evaluated the relativistic $G(x_2; x_1)$ using the Jacobi path integral. As we saw in Sect. 1.2.4, such an energy propagator does not obey transitivity even in non-relativistic quantum mechanics.

[20] This drastic difference, of course, is related to the $\sqrt{t^2 - |\mathbf{x}|^2}$ factor in Eq. (1.76). A simpler way to see this is to note that in the saddle point approximation to Eq. (1.75), the saddle point is determined by the condition $\mu = \sqrt{x^2}/2m$, which is real for $x^2 > 0$ making the exponentials in Eq. (1.75) oscillatory, while it is pure imaginary for $x^2 < 0$, leading to exponential damping.

[21] Though some text books fuss over this, it is no big deal. In non-relativistic quantum mechanics, a particle can tunnel through a barrier which is classically forbidden. So the fact that relativistic quantum theory allows certain processes which relativistic classical theory forbids, *by itself*, is no more of a surprise than the fact that tunneling amplitudes in quantum mechanics are non-zero. The real trouble arises due to issues connected to causality, as we shall see later.

normalization condition in Eq. (1.4). The fact that $\langle \mathbf{x}|\mathbf{y}\rangle \neq \delta(\mathbf{x}-\mathbf{y})$ shows that we cannot really think in terms of non-overlapping position eigenvalues representing a single particle. Suppose we try to express $G(x_2; x_1)$ as $\langle \mathbf{x_2}|e^{-iH(\mathbf{p})t}|\mathbf{x_1}\rangle$ for a relativistic, single particle Hamiltonian. Lorentz invariance plus translational invariance alone tells you that the result must have the form $\langle \mathbf{x_2}|e^{-iH(\mathbf{p})t}|\mathbf{x_1}\rangle = F(x^2)$. Suppose now that $F(x^2)$ is nonzero for two events x_1 and x_2 which have a spacelike separation — as we found in our case. Then, one can choose a Lorentz frame in which $t = 0$ and it immediately follows that

$$\lim_{t\to 0} \langle \mathbf{x_2}|e^{-iH(\mathbf{p})t}|\mathbf{x_1}\rangle = \langle \mathbf{x_2}|\mathbf{x_1}\rangle = F(-|x|^2) \neq 0 \qquad (\text{for } \mathbf{x_2} \neq \mathbf{x_1}) \quad (1.79)$$

The non-zero overlap of $\langle \mathbf{x_2}|\mathbf{x_1}\rangle$ for $\mathbf{x_2} \neq \mathbf{x_1}$ signals the impossibility of physically localized single particle states — which is a direct consequence of $F(x^2) \neq 0$ for spacelike separations. We will say more about this in the next section but it should be obvious from these facts that $G(x_2; x_1)$ is not going to permit a simple single particle Schrodinger like description.

There is also some difficulty with the simple notion of causality. We know that when x_2 and x_1 are separated by a timelike interval, the causal relation $x_2^0 > x_1^0$ has a Lorentz invariant meaning. If you create a particle at x_1 (say, by a process in which an atom emits a photon thereby creating a photon) and detect it at x_2 (say, by a process in which another atom absorbs it) and claim that the detection occurred after the creation, such a statement has a Lorentz invariant meaning *only* if x_2 and x_1 are separated by a timelike interval. All other observers will agree with you that an atom detected the photon *after* it was first created. But our discussion shows that a relativistic particle has a non-zero amplitude to reach a point x_2 which is related by a spacelike interval with respect to x_1. In one frame $x_2^0 > x_1^0$ might hold but you can always find another Lorentz frame in which $x_2^0 < x_1^0$. This suggests that some observers might end up detecting a particle before it is created.[22]

1.4.3 Three Dimensional Fourier Transform of $G(x_2; x_1)$

To come to grips with such issues and to see the physics behind $G(x_2; x_1)$, it is good to play the same trick which we played with the non-relativistic amplitude in Eq. (1.9), viz., Fourier transform it and look at it in the momentum space. Now we can do this either by Fourier transforming with respect to spatial coordinates or by Fourier transforming with respect to space *and* time. This will also get rid of MacDonald functions and express $G(x_2; x_1)$ in a human readable form.

The spatial Fourier transform is given by the integral (where we have suppressed the $i\epsilon$ factor for simplicity):

$$\int G(x_2; x_1)e^{-i\mathbf{p}\cdot\mathbf{x}}d^3\mathbf{x} = -\frac{i}{16\pi^2}\int_0^\infty \frac{d\mu}{\mu^2}e^{-im^2\mu - i(t^2/4\mu)}$$
$$\times \int d^3\mathbf{x}\, e^{i|\mathbf{x}|^2/4\mu - i\mathbf{p}\cdot\mathbf{x}} \quad (1.80)$$

The integral over $d^3\mathbf{x}$ has the value $(4\pi i\mu)^{3/2}\exp(-i|\mathbf{p}|^2\mu)$. Changing the variable of integration to ρ with $\mu = \rho^2$ we find that the integral is given by

$$\int G(x_2; x_1)e^{-i\mathbf{p}\cdot\mathbf{x}}d^3\mathbf{x} = \left(\frac{i}{\pi}\right)^{1/2}\int_0^\infty d\rho \exp\left(-i\omega_{\mathbf{p}}^2\rho^2 - \frac{it^2}{4\rho^2}\right) \quad (1.81)$$

[22] This also makes lot of people uncomfortable and may be rightly so. But no amount of advanced quantum field theory which you learn is ever going to make the amplitude $G(x_2; x_1)$ vanish when x_2 is outside the light cone of x_1. The probability for creating a particle at x_1 and detecting a particle at x_2 outside x_1's light cone is nonzero in the real world and you just have to live with it. But what quantum field theory will provide is a nicer, causal, reinterpretation of this result in terms of processes which work within the light cones. The price you pay for that is a transition from a single particle description to many (actually infinite) particle description.

Exercise 1.10: Prove the result for $I(a,b)$ which can be done using a clever trick, other than typing **Integrate** in *Mathematica*.

where $\omega_{\mathbf{p}} = +\sqrt{\mathbf{p}^2 + m^2}$ is the energy of a classical particle with momentum $\mathbf{p}$. This integral can be evaluated using the result

$$I(a,b) = \int_0^\infty dx \; e^{-ia^2 x^2 - ib^2 x^{-2}} = \frac{1}{2|a|} \left(\frac{\pi}{i}\right)^{1/2} e^{-2i|a|\,|b|} \tag{1.82}$$

to give a remarkably simple final answer:

$$\int G(x_2; x_1) e^{-i\mathbf{p}\cdot\mathbf{x}} d^3\mathbf{x} = \frac{1}{2\omega_{\mathbf{p}}} \exp(-i\omega_{\mathbf{p}}|t|) \tag{1.83}$$

We can, therefore, express the propagation amplitude as

$$
\begin{aligned}
G(x_2; x_1) &= \int \frac{d^3\mathbf{p}}{(2\pi)^3} e^{i\mathbf{p}\cdot\mathbf{x}} \left[\frac{1}{2\omega_{\mathbf{p}}} e^{-i\omega_{\mathbf{p}}|t|}\right] \\
&\equiv \int d\Omega_{\mathbf{p}} \left[\theta(t) e^{-ipx} + \theta(-t) e^{ipx}\right]
\end{aligned}
\tag{1.84}
$$

The integration measure $d\Omega_{\mathbf{p}} \equiv (d^3\mathbf{p}/(2\pi)^3)(1/2\omega_{\mathbf{p}})$ is Lorentz invariant because it arises from the measure $d^4p\,\delta(p^2 - m^2)\theta(p^0)$; the factor $\exp(\pm ipx)$ is, of course, Lorentz invariant.

Exercise 1.11: Prove that the integration measure $d\Omega_{\mathbf{p}} \equiv (d^3\mathbf{p}/(2\pi)^3)(1/2\omega_{\mathbf{p}})$ is indeed Lorentz invariant.

We see that Eq. (1.10) as well as the first expression in Eq. (1.84) can be expressed together in the form

$$
\int d^3\mathbf{x} \; G(x_2; x_1) e^{-i\mathbf{p}\cdot\mathbf{x}} =
\begin{cases}
\theta(t) e^{-i\omega_{\mathbf{p}} t} & \text{(non-relativistic)} \\[2mm]
\dfrac{1}{2\omega_{\mathbf{p}}} e^{-i\omega_{\mathbf{p}}|t|} & \text{(relativistic)}
\end{cases}
\tag{1.85}
$$

Expressed in this manner, the relativistic result looks surprisingly similar to the one obtained in the case of the non-relativistic particle in Eq. (1.11), except for two crucial differences.

First, the non-relativistic amplitude propagates modes with positive energy $\omega_{\mathbf{p}}$ forward in time and we could take it to be zero for $t < 0$. The relativistic amplitude, on the other hand, has $|t|$ leading to propagation both forward and backward in time. It propagates modes with energy $\omega_{\mathbf{p}}$ forward in time, but also propagates modes with negative energy $-\omega_{\mathbf{p}}$ backwards in time, because $\omega_{\mathbf{p}}|t| = (-\omega_{\mathbf{p}})t$ for $t < 0$. As you can guess, this will be a major talking point for us.[23]

What about the $c \to \infty$ limit when we expect to recover the non-relativistic form of $G(x_2; x_1)$ from the relativistic expression? In this limit $(1/2\omega_{\mathbf{p}}) \to (1/2m)$; so that is fine and arises from the $2mC = 1$ normalization of Eq. (1.75). In the argument of the exponentials, we set $\omega_{\mathbf{p}} \simeq mc^2 + \mathbf{p}^2/2m$. If we now pull out a factor $\exp(-imc^2 t)$ then the $t > 0$ part of the propagator will have the non-relativistic form. The $t < 0$ part of the propagator will pick up an $\exp(+2imc^2 t)$ factor. So we get $e^{imc^2 t} G_R(x_2; x_1) \approx G_{NR}(x_2; x_1)$ plus a term which oscillates rapidly due to $\exp(+2imc^2 t)$ factor when $c \to \infty$. At this stage you argue that the second term can be ignored compared to the first, thereby getting the correct NR limit with no $t < 0$ contribution.

Second, the Fourier transform of $G(x_2; x_1)$ is not a pure phase in the relativistic case due to the factor $(1/2\omega_{\mathbf{p}})$ in front. This is contrary to the result we found in Eq. (1.7) for the transitivity condition Eq. (1.2) to hold. But we know that transitivity does not hold in the relativistic case; see

[23]Incidentally, the same maths occurs in a process called *paraxial optics* when we study light propagating along the $+z$ axis, say, in spite of the fact that you start with a wave equation which is second order in z and symmetric under $z \Leftrightarrow -z$. In getting the non-relativistic quantum mechanics you are doing paraxial optics of particles in the time direction, keeping only forward propagation. For a detailed description of this curious connection, see Chapters 16,17 of T. Padmanabhan, *Sleeping Beauties in Theoretical Physics*, Springer (2015).

Eq. (1.77). Alternatively, we can use the result in Eq. (1.85) as a proof that transitivity does not hold in the relativistic case.

It is convenient at this stage to introduce *two* different Lorentz invariant amplitudes:

$$G_\pm(x_2; x_1) = \int d\Omega_{\mathbf{p}} e^{i\mathbf{p}\cdot\mathbf{x}} e^{\mp it\omega_{\mathbf{p}}} = \int \frac{d^3\mathbf{p}}{(2\pi)^3} \frac{1}{2\omega_{\mathbf{p}}} e^{i(\mathbf{p}\cdot\mathbf{x}\mp\omega_{\mathbf{p}}t)} = \int d\Omega_{\mathbf{p}} e^{\mp ipx}$$

(1.86)

In arriving at the third equality, we have flipped the sign of $\mathbf{p}$ in the integrand for $G_-(x_2; x_1)$ in order to write everything in a manifestly Lorentz invariant form. Obviously,

$$G_+(x_2; x_1) = G_-(x_1; x_2) = G_-^*(x_2; x_1)$$

(1.87)

so that only one of them is independent. The propagation amplitude in Eq. (1.84) can now be written as

$$G(x_2; x_1) = \theta(t)G_+(x_2; x_1) + \theta(-t)G_-(x_2; x_1)$$

(1.88)

We know that the left hand side of the above expression is Lorentz invariant. Therefore the $\theta(\pm t)$ factors in the right hand side should not create any problems with Lorentz invariance. When the events x_2 and x_1 are separated by a timelike interval, the notion of $x_2^0 - x_1^0 \equiv t$ being positive or negative is a Lorentz invariant statement and causes no difficulty. Only one of the two terms in the right hand side of Eq. (1.84) contributes when x_2 and x_1 are separated by a timelike interval in any Lorentz frame. The situation however is more tricky when x_2 and x_1 are separated by a spacelike interval. In that case, in a given Lorentz frame, we may have $t > 0$ and $G(x_2; x_1) = G_+(x_2; x_1)$ in this frame, while a Lorentz transformation can take us to another frame in which the same two events have $t < 0$ so that $G(x_2; x_1) = G_-(x_2; x_1)$ in this frame!

For consistency it is necessary that $G_+(x_2; x_1) = G_-(x_2; x_1)$ when x_2 and x_1 are separated by a spacelike interval, though, of course, they cannot be identically equal to each other. It is easy to verify that this is indeed the case. We have, in general,

$$\begin{aligned} G_+(x_2; x_1) - G_-(x_2; x_1) &= G_+(x_2; x_1) - G_+(x_1; x_2) \qquad\qquad (1.89)\\ &= \int d\Omega_{\mathbf{p}}[e^{-ipx} - e^{+ipx}] \qquad\qquad\qquad (1.90)\\ &= \int d\Omega_{\mathbf{p}}[e^{-i\omega_{\mathbf{p}}t + i\mathbf{p}\cdot\mathbf{x}} - e^{i\omega_{\mathbf{p}}t + i\mathbf{p}\cdot\mathbf{x}}]\\ &= \int d\Omega_{\mathbf{p}} e^{i\mathbf{p}\cdot\mathbf{x}}[e^{-i\omega_{\mathbf{p}}t} - e^{i\omega_{\mathbf{p}}t}] \end{aligned}$$

In arriving at the last equality, we have flipped the sign of $\mathbf{p}$ in the e^{ipx} term. We can now see that this expression vanishes when $x^2 < 0$. In that case, one can always chose a coordinate system such that $t = 0$ and the last line of Eq. (1.89) shows that the integrand vanishes for $t = 0$. The explicit Lorentz invariance of the expression assures us that it will vanish in all frames when $x^2 < 0$. Therefore, it does not matter which of the two terms in Eq. (1.84) is picked up when $x^2 < 0$ and everything is consistent though in a subtle way. This result will play a crucial role later on when we introduce the notion of causality in quantum field theory.

1.4.4 Four Dimensional Fourier Transform of $G(x_2; x_1)$

Let us next consider the 4-dimensional[24] Fourier transform of $G(x_2; x_1)$. This expression, which will play a crucial role in our future discussions, is given by:

$$\int G(x_2; x_1) e^{ip \cdot x} d^4 x = -\frac{i}{16\pi^2} \int_0^\infty \frac{d\mu}{\mu^2} e^{-i(m^2 - i\epsilon)\mu} \tag{1.91}$$

$$\times \int d^4 x \exp\left(-\frac{i}{4} \frac{x^2}{\mu} + ip \cdot x\right)$$

$$= \frac{i}{(p^2 - m^2 + i\epsilon)} \tag{1.92}$$

The inverse[25] Fourier transform of the right hand side should, of course, now give $G(x_2; x_1)$. We will do this in two different ways to illustrate two important technical points. First, note that we can write the inverse Fourier transform using an integral representation as

$$G(x_2; x_1) = \int \frac{d^4 p}{(2\pi)^4} \frac{ie^{-ip \cdot x}}{p^2 - m^2 + i\epsilon} = \int_0^\infty d\mu \int \frac{d^4 p}{(2\pi)^4} e^{-ipx - i\mu(m^2 - p^2 - i\epsilon)} \tag{1.93}$$

We next note that the factor $e^{-i\mu(m^2 - p^2)}$ can be thought of as a matrix element of the operator $e^{-i\mu(m^2 + \Box)}$. Hence we can write

$$G(x_2; x_1) = \int_0^\infty d\mu \langle x_2 | e^{-i\mu(\Box + m^2 - i\epsilon)} | x_1 \rangle \equiv \int_0^\infty d\mu \langle x_2, \mu | x_1, 0 \rangle \tag{1.94}$$

So our amplitude $G(x_2; x_1)$ can be thought of as the amplitude for a particle to propagate from x_1 at $\tau = 0$ to x_2 at $\tau = \mu$ followed by summing over all μ. This is identical in content to our original result Eq. (1.74).

The integrand $\langle x_2, \mu | x_1, 0 \rangle$ in Eq. (1.94) can be thought of as the propagation amplitude $\langle x_2, \mu | x_1, 0 \rangle = \langle x_2 | e^{-i\mu \mathcal{H}} | x_1 \rangle$ for a (fictitious) particle with the operator $\mathcal{H} = \Box + m^2 - i\epsilon$ sometimes called the *super-Hamiltonian* which generates translations in proper time. This (super)Hamiltonian $\mathcal{H} = -p^2 + m^2$ corresponds to a (super)Lagrangian $\mathcal{L} = -(1/4)\dot{x}^2 - m^2$. This is essentially the same as the Lagrangian in Eq. (1.73) with a rescaling $\tau \to m\tau$ and adding a constant $(-m^2)$. The above result shows that one can think of the propagation amplitude $G(x_2; x_1)$ as arising from the propagation of a particle in a 5-dimensional space with the fifth dimension being the proper time. The proper time should be integrated out to give the amplitude in physical spacetime.

The second technical aspect is related to the crucial role played by the $i\epsilon$ factor which was introduced very innocuously in Eq. (1.75). In the inverse Fourier transform, separating out $d^4 p$ as $d^3 \mathbf{p} dp^0$, we get

$$G(x_2; x_1) = \int \frac{d^4 p}{(2\pi)^4} \frac{i}{(p^2 - m^2 + i\epsilon)} e^{-ip \cdot x}$$

$$= \int \frac{d^3 \mathbf{p}}{(2\pi)^3} e^{i\mathbf{p} \cdot \mathbf{x}} \int_{-\infty}^{+\infty} \frac{dp^0}{2\pi} \frac{ie^{-ip^0 t}}{(p^0)^2 - \omega_{\mathbf{p}}^2 + i\epsilon} \tag{1.95}$$

The p^0 integral is now well defined because the zeros of the denominator

$$p^0 = \pm \left[\omega_{\mathbf{p}}^2 - i\epsilon\right]^{1/2} \Rightarrow \pm(\omega_{\mathbf{p}} - i\epsilon) \tag{1.96}$$

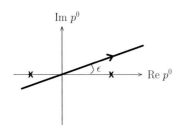

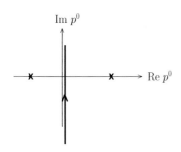

Figure 1.5: The $i\epsilon$ prescription and the Euclidean prescription in the momentum space.

are shifted off the real axis. If we did not have the $i\epsilon$ factor, the denominator would vanish at $p^0 = \pm\omega_{\mathbf{p}}$ and the integral will be ill-defined until we specify a procedure for going around the poles. The $i\epsilon$ prescription tells you that the contour has to go above the pole at $+\omega_{\mathbf{p}}$ and below the pole at $-\omega_{\mathbf{p}}$ which is equivalent to using the contour in Fig. 1.5 with integration along $p^0(1 + i\epsilon)$ keeping the poles on the real axis. This allows us to rotate the contour to the imaginary axis in the complex p^0 plane and define the expressions in the Euclidean sector by the analytic continuation. In the complex p^0 plane we think of the integral as being along $p^0 e^{i\theta}$ with $\theta = \epsilon$ giving the $i\epsilon$ prescription and $\theta = \pi/2$ giving the Euclidean prescription. The latter is equivalent to replacing p^0 by ip_E^0 and treating p_E^0 as real.

We recall from Sect. 1.2.3 that the corresponding change for time coordinate was $t \to -it_E$. So the complex phase of the Fourier transform *remains complex* when we simultaneously analytically continue in both t and p^0. That is,

$$i(p^0 t - \mathbf{p} \cdot \mathbf{x}) \to i(p_E^0 t_E - \mathbf{p} \cdot \mathbf{x}) \tag{1.97}$$

The relative sign between the two terms in the *Euclidean* expression on the right hand side might appear strange but it does not matter when we deal with integrals over functions of p^2. Using the trick of flipping the sign of spatial momenta, we have the identity:

$$
\begin{aligned}
\int d^4 p\, f(p^2) e^{-ip\cdot x} &= i \int d^4 p_E\, f(-p_E^2) e^{-i(p_E^0 t_E - \mathbf{p}\cdot\mathbf{x})} \\
&= i \int d^4 p_E\, f(-p_E^2) e^{-i(p_E^0 t_E + \mathbf{p}\cdot\mathbf{x})} \\
&= i \int d^4 p_E\, f(-p_E^2) e^{-ip_E \cdot x_E} \tag{1.98}
\end{aligned}
$$

where $p_E \cdot x_E \equiv p_E^0 t_E + \mathbf{p} \cdot \mathbf{x}$ so that everything looks sensible for functions of p^2. In the Euclidean sector our integral will become

$$\int \frac{d^4 p}{(2\pi)^4} \frac{i}{(p^2 - m^2 + i\epsilon)} e^{-ip\cdot x} = \int \frac{d^4 p_E}{(2\pi)^4} \frac{e^{-ip_E \cdot x_E}}{p_E^2 + m^2} \tag{1.99}$$

which, of course, is well defined. This is yet another illustration that if we had worked everything out in the Euclidean sector and analytically continued back to real time and real p^0, we would have got the correct result as obtained by the $i\epsilon$ prescription. Hence, one often uses the $i\epsilon$ prescription and the Euclidean definition interchangeably in field theory.

Either by contour integration or by doing the Euclidean integral and analytically continuing, one can easily show that

$$\int_{-\infty}^{+\infty} \frac{dp^0}{2\pi} \frac{i e^{-ip^0 t}}{(p^0)^2 - \omega_{\mathbf{p}}^2 + i\epsilon} = \frac{1}{2\omega_{\mathbf{p}}} e^{-i\omega_{\mathbf{p}}|t|} \tag{1.100}$$

Thus we get the final answer to be

$$
\begin{aligned}
G(x_2; x_1) &= \int \frac{d^3 \mathbf{p}}{(2\pi)^3} e^{i\mathbf{p}\cdot\mathbf{x}} \left[\frac{1}{2\omega_{\mathbf{p}}} e^{-i\omega_{\mathbf{p}}|t|} \right] \\
&\equiv \int d\Omega_{\mathbf{p}} \left[\theta(t) e^{-ipx} + \theta(-t) e^{ipx} \right] \tag{1.101}
\end{aligned}
$$

which, of course, matches with our earlier result in Eq. (1.84).

Exercise 1.12: Work out the p^0 integration along a contour which goes (a) above both poles and (b) below both poles. These expressions correspond to what are known as retarded and advanced propagators in the context of, for e.g., a harmonic oscillator.

1.4.5 The First Non-triviality: Closed Loops and $G(x; x)$

Since $G(x_2; x_1)$ does not vanish outside the light cone, one is tempted to look at the form of $G(x; x)$ which would correspond to the amplitude for a particle to propagate in a closed loop in spacetime! *This process is remarkable because it could happen even without any relativistic particle in sight.* Being a "zero particle process" (see Fig. 1.6) it contains the key essence of the new features we have stumbled upon. To analyse this, let us go back to the amplitude $\langle x_2, \mu | x_1, 0 \rangle$ given in Eq. (1.94) for a particle to propagate from x_1 to x_2 in proper time μ. We define the Fourier transform of this expression with respect to μ to obtain

$$\int_0^\infty \langle x, \mu | x, 0 \rangle \; e^{iE\mu} \, d\mu \equiv \mathcal{P}(x; E) \tag{1.102}$$

where $\mathcal{P}(x; E)$ can be thought of as the amplitude for a closed loop with energy E. (This integral could have been from $\mu = -\infty$ to $\mu = +\infty$ if we interpret $\langle x_2, \mu | x_1, 0 \rangle$ with a $\theta(\mu)$ factor.) We next integrate over all energies and define a quantity

$$\mathcal{E}(x) = \frac{1}{2} \int_0^\infty dE \; \mathcal{P}(x; E) \tag{1.103}$$

Taking the definition for $\langle x_2, \mu | x_1, 0 \rangle$ from Eq. (1.94) and integrating over E first, we get[26]

$$\mathcal{E} = \frac{1}{2} \int_0^\infty d\mu \left(\frac{-1}{i\mu} \right) \langle x, \mu | x, 0 \rangle = \frac{i}{2} \int_0^\infty \frac{d\mu}{\mu} \langle x | e^{-i\mu(\Box + m^2)} | x \rangle \tag{1.104}$$

It follows from this relation that

$$\frac{\partial \mathcal{E}}{\partial m^2} = \frac{1}{2} \, G(x; x) \tag{1.105}$$

On the other hand, from Eq. (1.84) we find that (with $\hbar$ temporarily restored)

$$\frac{1}{2} G(x; x) = \frac{1}{2} \int \frac{d^3 \mathbf{p}}{(2\pi)^3} \frac{1}{2\omega_{\mathbf{p}}} = \frac{\partial}{\partial m^2} \left[\int \frac{d^3 \mathbf{p}}{(2\pi)^3} \left(\frac{1}{2} \hbar \omega_{\mathbf{p}} \right) \right] \tag{1.106}$$

Comparing Eq. (1.105) with Eq. (1.106) we get the remarkable result

$$\mathcal{E} = \int \frac{d^3 \mathbf{p}}{(2\pi)^3} \left(\frac{1}{2} \hbar \omega_{\mathbf{p}} \right) \tag{1.107}$$

when we set the integration constant to zero. Or, more explicitly,

$$\frac{1}{2} \int_0^\infty dE \int_0^\infty d\mu \; e^{iE\mu} \langle x, \mu | x, 0 \rangle = \int \frac{d^3 \mathbf{p}}{(2\pi)^3} \left(\frac{1}{2} \hbar \omega_{\mathbf{p}} \right) \tag{1.108}$$

The left hand side can be interpreted (except for a factor $(1/2)$ which, as we said, takes care of traversing the loop in two directions) as giving the amplitude of all closed loops obtained by summing over the energy associated with each loop. We find that this is the same as summing over the zero point energies, $(1/2)\hbar\omega_{\mathbf{p}}$, of an infinite number of harmonic oscillators each labeled by the frequency $\omega_{\mathbf{p}}$. Right now there are no harmonic oscillators anywhere in sight and it is rather intriguing that we get a result like this; a physical interpretation will emerge much much later, in spite of the fact that both sides of this expression are divergent.[27]

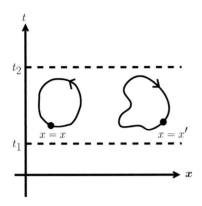

Figure 1.6: The amplitude for such closed loops is nonzero.

[26]Later on we will see that $\mathcal{E}$ can be interpreted as an effective Lagrangian of a theory when $\langle x_2, \mu | x_1, 0 \rangle = \langle x_2 | e^{-i\mu\mathcal{H}} | x_1 \rangle$ arises from more complicated operator $\mathcal{H}$ in interacting theories. The factor $(1/2)$ is added in Eq. (1.103) because each loop can be traversed in two orientations — clockwise or anticlockwise.

[27]The divergence of the expressions makes some of these operations "illegal". One can obtain the same results after "regularizing" the integrals — a procedure we will describe later on. (Whether *that* makes anything "legal" is a question you are not supposed to ask!)

1.4.6 Hamiltonian Evolution: Relativistic Particle

To understand further the structural difference in $G(x_2; x_1)$ between the relativistic and non-relativistic case, we will now try to redo our analysis in Sect. 1.2.2 for the relativistic case. The lack of transitivity in $G(x_2; x_1)$ should tell you that this is a lost cause but it is interesting to see it explicitly.

The classical Hamiltonian for a relativistic particle is given by $H(\mathbf{p}) = +\sqrt{\mathbf{p}^2 + m^2}$ which has a square-root and *is defined to be positive definite*. Squaring this is a notoriously bad idea since it allows negative energy states to creep in and is universally condemned in textbooks. So we don't want to do that. If we retain the square root and use the coordinate representation with $\mathbf{p} \to -i\nabla$, we have to deal with the operator $\sqrt{-\nabla^2 + m^2}$ which is non-local. Its action on a function $f(\mathbf{x})$ is given by

Exercise 1.13: Prove this and evaluate $\mathcal{G}(\mathbf{x})$ explicitly. It is a pretty much a useless object but you learn how to work with singular integrals in the process.

$$\sqrt{-\nabla^2 + m^2}\, f(\mathbf{x}) \equiv \int \frac{d^D\mathbf{p}}{(2\pi)^D}\, (\mathbf{p}^2 + m^2)^{1/2}\, f(\mathbf{p})e^{i\mathbf{p}\cdot\mathbf{x}}$$

$$= \int d^D\mathbf{y}\, \mathcal{G}(\mathbf{x} - \mathbf{y})\, f(\mathbf{y}) \tag{1.109}$$

where $\mathcal{G}(\mathbf{x})$ is a singular Kernel given by

$$\mathcal{G}(\mathbf{x}) \equiv \int \frac{d^D\mathbf{p}}{(2\pi)^D}\, (\mathbf{p}^2 + m^2)^{1/2}\, e^{i\mathbf{p}\cdot\mathbf{x}} \tag{1.110}$$

The simplest way to avoid this problem[28] is to work in the momentum representation in which $H(\mathbf{p})$ is diagonal and local. So we begin again by introducing a complete set of momentum eigenstates $|\mathbf{p}\rangle$. The completeness condition analogous to Eq. (1.18) will now have to be Lorentz invariant and hence should have the form

$$\int \frac{d^D\mathbf{p}}{(2\pi)^D}\, \frac{1}{2\omega_{\mathbf{p}}}\, |\mathbf{p}\rangle\langle\mathbf{p}| = \int d\Omega_{\mathbf{p}}|\mathbf{p}\rangle\langle\mathbf{p}| = 1 \tag{1.111}$$

[28] If we have to deal with a Schrodinger-like equation in coordinate space which is non-local, you might think we have already lost the battle. But note that, even in the usual non-relativistic quantum mechanics, there is nothing sacred about the coordinate representation vis-a-vis the momentum representation. If you write the Schrodinger equation for any, say, non-polynomial potential $V(\mathbf{x})$ *in the momentum representation*, you will end up getting a non-local Schrodinger equation in momentum space. Nothing should go wrong, *just because of this*.

Taking the product with $|\mathbf{k}\rangle$, we now require the Lorentz invariant orthonormality condition[29]

$$\langle\mathbf{p}|\mathbf{k}\rangle = (2\pi)^D\, (2\omega_{\mathbf{k}})\, \delta(\mathbf{p} - \mathbf{k}) \tag{1.112}$$

We next introduce position eigenstates $|\mathbf{x}\rangle$ and will try to interpret the amplitude $G(x_2; x_1)$ as $\langle\mathbf{x}_2|e^{-iH(\mathbf{p})t}|\mathbf{x}_1\rangle$. This leads to:

[29] This follows from the Lorentz invariance of 1 and $d\Omega_{\mathbf{p}}$ when expressed in the form $1 = \int d^D\mathbf{p}\,\delta(\mathbf{p} - \mathbf{k}) = \int d\Omega_{\mathbf{p}}(2\pi)^D 2\omega_{\mathbf{p}}\delta(\mathbf{p} - \mathbf{k})$

$$\langle\mathbf{x}_2|e^{-iH(\mathbf{p})t}|\mathbf{x}_1\rangle = \int d\Omega_{\mathbf{p}}\, e^{-i\omega_{\mathbf{p}}t}\langle\mathbf{x}_2|\mathbf{p}\rangle\langle\mathbf{p}|\mathbf{x}_1\rangle \tag{1.113}$$

Once again we can argue based on translational invariance (or the fact that in the momentum basis the commutation rule $[x_\alpha, p_\beta] = i\delta_{\alpha\beta}$ is implemented using $\hat{x}_\alpha = i\partial/\partial p^\alpha$) that $\langle\mathbf{p}|\mathbf{x}\rangle = C_{\mathbf{p}}\exp[-i\mathbf{p}\cdot\mathbf{x}]$. Therefore,

$$\langle\mathbf{x}_2|e^{-iH(\mathbf{p})t}|\mathbf{x}_1\rangle = \int d\Omega_{\mathbf{p}}\, e^{-i\omega_{\mathbf{p}}t}\, |C_{\mathbf{p}}|^2\, e^{i\mathbf{p}\cdot\mathbf{x}} = \int d\Omega_{\mathbf{p}}\, |C_{\mathbf{p}}|^2\, e^{-ipx} \tag{1.114}$$

We will require the amplitude $G(x_2; x_1) = \langle\mathbf{x}_2|e^{-iH(\mathbf{p})t}|\mathbf{x}_1\rangle$ to be Lorentz invariant. Since $d\Omega_{\mathbf{p}}$ and e^{-ipx} are Lorentz invariant, it follows that we must have $|C_{\mathbf{p}}|^2$ to be a constant which we can take to be unity. If we make this choice, then we will have

$$\langle\mathbf{x}_2|e^{-iH(\mathbf{p})t}|\mathbf{x}_1\rangle = G(x_2; x_1) \quad (\text{for } t_2 > t_1) \tag{1.115}$$

as can be verified by comparing with Eq. (1.84). The expression in the left hand side cannot generate the form of $G(x_2; x_1)$ for $t_2 < t_1$. This term requires e^{+ipx} which in turn involves the negative energy factor with $-\omega_\mathbf{p}$. Since we started with a Hamiltonian which is explicitly positive definite, there is no way we can get this term in this manner.

This fact already shows that our program fails; there is, however, one more difficulty which is worth pointing out. The Lorentz invariance of $G(x_2; x_1)$ and our desire to express it in the form $\langle \mathbf{x}_2 | e^{-iH(\mathbf{p})t} | \mathbf{x}_1 \rangle$ has forced us to choose $|C_\mathbf{p}|^2 = 1$. But then if we multiply Eq. (1.111) from the left by $\langle \mathbf{x} |$ and from the right by $| \mathbf{y} \rangle$ we get

$$\langle \mathbf{x} | \mathbf{y} \rangle = \int d\Omega_\mathbf{p} \, \langle \mathbf{x} | \mathbf{p} \rangle \langle \mathbf{p} | \mathbf{y} \rangle = \int \frac{d^D \mathbf{p}}{(2\pi)^D} \frac{1}{2\omega_\mathbf{p}} e^{i\mathbf{p}\cdot(\mathbf{x}-\mathbf{y})} \neq \delta(\mathbf{x} - \mathbf{y}) \quad (1.116)$$

This shows that $\langle \mathbf{x} | \mathbf{y} \rangle$ is in general non-zero for $\mathbf{x} \neq \mathbf{y}$. (In fact, from our expression in Eq. (1.76) it is clear that $\langle \mathbf{x} | \mathbf{y} \rangle$ is non-zero and decreases exponentially with a scale length of $(1/m)$.) Physically, this indicates the impossibility of localizing particles to a region smaller than $(1/m)$ and still maintain a single particle description. We have already seen in the last section that this is connected with the fact that $G(x_2; x_1)$ does not vanish for spacelike intervals.

Exercise 1.14: Construct a single particle Lorentz invariant relativistic quantum theory with a consistent interpretation. (Hint: Of course, you cannot! But a good way to convince yourself of the need for quantum *field* theory is to attempt this exercise and see for yourself how hard it is.)

[30]This is closely related to something called the *Newton-Wigner* position operator in the literature. It is known that this idea has serious problems.

There is no easy way out of these difficulties. If, for example, we have chosen $C_\mathbf{p} = \sqrt{2\omega_\mathbf{p}}$ then we could have canceled out the $(1/2\omega_\mathbf{p})$ in Eq. (1.116) thereby making $\langle \mathbf{x} | \mathbf{y} \rangle = \delta(\mathbf{x} - \mathbf{y})$. But with such a choice Eq. (1.113) would have picked up an extra $2\omega_\mathbf{p}$ factor and the resulting expression will not be Lorentz invariant.[30] Part of these technical difficulties arise from the following elementary fact: In non-relativistic quantum mechanics we have two equally good basis corresponding to either position eigenvalue $|\mathbf{x}\rangle$ or corresponding to momentum eigenvalue $|\mathbf{p}\rangle$ with integration measures $d^3\mathbf{x}$ and $d^3\mathbf{p}$. In the case of a relativistic particle we do have a Lorentz invariant integration measure $d^3\mathbf{p}/(2\omega_\mathbf{p})$ which can be used with $|\mathbf{p}\rangle$. But we do not have anything analogous which will make the integration measure $d^3\mathbf{x}$ into something which is physically meaningful and Lorentz invariant. Roughly speaking all these boil down to the fact that a free particle with momentum $\mathbf{p}$ has an associated energy $\omega_\mathbf{p}$ while a free particle with position $\mathbf{x}$ does not have "an associated time t" with it. So the momentum basis $|\mathbf{p}\rangle$ continues to be useful in quantum field theory while talking about particle interactions etc. But the position basis is harder to define and use.

The best we can do is to stick with Lorentz invariance with the choice $C_\mathbf{p} = 1$ and use *two* different Lorentz invariant amplitudes $G_\pm(x_2; x_1)$ defined earlier in Eq. (1.86). We can keep these two amplitudes separate and think of $G_+(x_2; x_1)$ arising from evolution due to the Hamiltonian $H = +\sqrt{p^2 + m^2}$, while we have to deliberately introduce another Hamiltonian with $H = -\sqrt{p^2 + m^2}$ to obtain $G_-(x_2; x_1)$. Comparing with our discussion in the case of the non-relativistic particle, we find that the existence of two energies $\pm\omega_\mathbf{p}$ requires us to define two different amplitudes.

In the path integral approach, on the other hand, we are naturally led to an expression for $G(x_2; x_1)$ which can be expressed in the form of Eq. (1.88). If we knew nothing about the path integral result, we would have made the "natural" choice of $+\omega_\mathbf{p}$ and declared the propagator to be $\theta(t)G_+(x_2; x_1)$ which is just the first term in Eq. (1.88). This is inconsistent with the path integral result; it is also *not* Lorentz invariant because $\theta(t)$ is not Lorentz invariant when x_2 and x_1 are spacelike. One could have restored Lorentz invariance with only positive energies by taking the amplitude to be just $G_+(x_2; x_1)$ (without the $\theta(t)$, for all times), but then it will not match the path integral result for $t < 0$.

The only way to interpret $G(x_2; x_1)$ is as follows: There is a non-zero amplitude for propagating either forward in time or backward in time.[31] Forward propagation occurs with positive energy $+\omega_{\mathbf{p}}$ while backward propagation occurs with negative energy $-\omega_{\mathbf{p}}$! This is the main difference between Eq. (1.84) and Eq. (1.11); in the non-relativistic case given by Eq. (1.11) there are only positive energy terms and they propagate forward in time; in the relativistic case given by Eq. (1.84) we have both positive energy terms which propagate forward in time (for $+\omega_{\mathbf{p}}$) and negative energy terms which propagate backward in time (for $-\omega_{\mathbf{p}}$). It looks as though the relativistic propagation amplitude simultaneously describes *two kinds* of particles rather than one: Those with positive energy which propagate forward and some other strange species with negative energies which propagate backwards.

We saw earlier that, because $G(x_2; x_1)$ is non-zero for spacelike separation, there can be an ambiguity when two observers try to interpret emission and absorption of particles. The existence of propagation with negative energy adds a new twist to it. To make the idea concrete, consider a two-level system with energies E_2, E_1 with $E \equiv E_2 - E_1 > 0$. We use one of these systems (call it A) to emit a particle and the second system (call it B) to absorb the particle. To do this, we first get A to the excited state and let it drop back to the ground state emitting a particle with positive energy; this particle propagates to B which is in the ground state and gets excited on absorbing positive energy. But now consider what happens when we allow for propagation of modes with negative energy. You can start with system A in the ground state, emit a particle with *negative* energy $(-E)$ and in the process get it excited to E_2. If we had kept B at an excited state, then it can absorb this particle with energy $(-E)$ and come down to the ground state. Someone who did not know about the negative energy modes would have treated this as emission of a particle by B and absorption by A, rather than the other way round. Clearly, the lack of causal ordering between emission and absorption can be reinterpreted in terms of the existence of negative energy modes. We will make this notion sharper in Sect. 2.1.

1.5 Interpreting $G(x_2; x_1)$ in Terms of a Field

All this is vaguely disturbing to someone who is accustomed to sensible physical evolution which proceeds monotonously forward in time from $t_1 < t_2 < t_3....$ and hence it would be nice if we can reinterpret $G(x_2; x_1)$ in such a nice, causal manner.[32] We have already seen that $G(x_2; x_1)$ comes from summing over paths which include the likes of the one in Figs. 1.2, 1.7. If we say that a single particle has three degrees of freedom (in $D = 3$), then we start and end (at x_1 and x_2) with three degrees of freedom in Fig. 1.7. But if a path cuts a spatial slice at an intermediate time at k points (the figure is drawn for $k = 3$), then we need to be able to describe $3k$ degrees of freedom at this intermediate time. Since k can be arbitrarily large, we conclude that if we want a description in terms of causal evolution going from t_1 to t_2 then we need to use a mathematical description involving an

[31] Note that, in the relativistic case, we cannot express the propagation amplitude $G(x_2; x_1)$ as $\langle x_2 | x_1 \rangle$. This is because, in the relativistic case, we no longer have a $|x_1\rangle = |t_1, \boldsymbol{x}_1\rangle$ which is an eigenket of some sensible position operator $\hat{\boldsymbol{x}}(t_1)$ with eigenvalue $\boldsymbol{x}_1$. We do have such an eigenket in the NRQM.

[32] Recall that the result for $G(x_2; x_1)$ in Eq. (1.75) or Eq. (1.84) is non-negotiable! All that we are allowed to do is to come up with some other system and a calculational rule which is completely causal and leads to the same expression for $G(x_2; x_1)$.

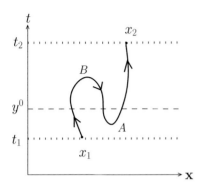

Figure 1.7: Path that represents 3 particles at $t = y^0$ though there was only one at $t = t_1$.

[34]There is an ambiguity when $t_1 = t_2$ if A and B do not commute at equal times. This turns out to be irrelevant in the present case.

infinite number of degrees of freedom.[33] Obviously, such a system must be 'hidden' in $G(x_2; x_1)$ because it knows about paths which cut the $t =$ constant surface an arbitrarily large number of times. Let us see whether we can dig it out from $G(x_2; x_1)$.

Since (i) the Fourier transform of a function has the same information as the function and since (ii) the real difference between relativistic and non-relativistic cases (see Eq. (1.85)) is in the manner in which propagation in time is treated, it makes sense to try and understand the spatial Fourier transform of $G(x_2; x_1)$ from some alternate approach. This quantity is given by (see Eq. (1.85))

$$\int d^3\mathbf{x}\, G(x_2; x_1) e^{-i\mathbf{p}\cdot\mathbf{x}} = \frac{1}{2\omega_{\mathbf{p}}} e^{-i\omega_{\mathbf{p}}|t|} \qquad (1.117)$$

but incredibly enough, the same expression arises in a completely different context. Consider a quantum mechanical harmonic oscillator with frequency $\omega_{\mathbf{p}}$ and unit mass. If $q_{\mathbf{p}}(t)$ is the dynamical variable characterizing this oscillator, then it is trivial to verify that

$$\langle 0 | T\left(q_{\mathbf{p}}(t_2) q_{\mathbf{p}}^{\dagger}(t_1)\right) | 0 \rangle = \frac{1}{2\omega_{\mathbf{p}}} e^{-i\omega_{\mathbf{p}}|t|} \qquad (1.118)$$

where $|0\rangle$ is the ground state of the oscillator and T is a time-ordering operator which arranges the variables within the bracket chronologically from right to left.[34] For example,

$$T[A(t_2)B(t_1)] \equiv \theta(t_2 - t_1)A(t_2)B(t_1) + \theta(t_1 - t_2)B(t_1)A(t_2) \qquad (1.119)$$

Usually, the dynamical variables $q_{\mathbf{p}}(t)$ of a quantum mechanical oscillator will be Hermitian and hence $q_{\mathbf{p}}^{\dagger} = q_{\mathbf{p}}$; this is because we think of the corresponding classical coordinate as real. Here we are just trying to construct a mathematical structure which will reproduce the right hand side of Eq. (1.117), so it is not necessary to impose this condition and we will keep $q_{\mathbf{p}}$ non-Hermitian. Let us briefly review how the usual results for a harmonic oscillator translate to this case before proceeding further.

For this purpose, it is useful to begin with *two* harmonic oscillators with the same frequency ω and unit mass, denoted by the dynamical variables $X(t)$ and $Y(t)$. Defining a new variable $q = (1/\sqrt{2})(X + iY)$ we can write the Lagrangian for the oscillators as

$$L = \frac{1}{2}\left(\dot{X}^2 + \dot{Y}^2\right) - \frac{1}{2}\omega^2\left(X^2 + Y^2\right) = \dot{q}\dot{q}^* - \omega^2 q q^* \qquad (1.120)$$

The usual quantum mechanical description of $X(t)$ and $Y(t)$ will proceed by introducing the creation and annihilation operators for each of them so that we can write

$$X(t) = \frac{1}{\sqrt{2\omega}}\left(\alpha e^{-i\omega t} + \alpha^{\dagger} e^{i\omega t}\right); \quad Y(t) = \frac{1}{\sqrt{2\omega}}\left(\beta e^{-i\omega t} + \beta^{\dagger} e^{i\omega t}\right) \qquad (1.121)$$

The commutation relations are $[\alpha, \alpha^{\dagger}] = 1 = [\beta, \beta^{\dagger}]$ with all other commutators vanishing. (In particular, α and $\alpha^{\dagger}$ commute with β and $\beta^{\dagger}$.) The corresponding expansion for q is given by

$$q = \frac{1}{\sqrt{2}}\frac{1}{\sqrt{2\omega}}\left[(\alpha + i\beta)e^{-i\omega t} + \left(\alpha^{\dagger} + i\beta^{\dagger}\right)e^{i\omega t}\right] \equiv \frac{1}{\sqrt{2\omega}}\left(a e^{-i\omega t} + b^{\dagger} e^{i\omega t}\right) \qquad (1.122)$$

where we have defined

$$a = \frac{1}{\sqrt{2}} (\alpha + i\beta); \qquad b = \frac{1}{\sqrt{2}} (\alpha - i\beta) \qquad (1.123)$$

It is easy to verify that

$$[a, b] = [a^\dagger, b^\dagger] = 0; \qquad [a, a^\dagger] = 1 = [b, b^\dagger] \qquad (1.124)$$

Thus we can equivalently treat a and b as the annihilation operators for two independent oscillators.[35] The variable q is made of two different oscillators and is non-Hermitian unlike X and Y which were Hermitian. Classically, this difference arises from the fact that q is a complex number with two degrees of freedom compared to X or Y which, being real, have only one degree of freedom each.

For our purpose we need to have one frequency $\omega_{\mathbf{p}}$ associated with each momentum labeled $\mathbf{p}$. Therefore we will introduce one q for each momentum label with $q_{\mathbf{p}}(t)$ now having the expansion in terms of creation and annihilation operators given by

$$q_{\mathbf{p}} = \frac{1}{\sqrt{2\omega_{\mathbf{p}}}} \left[a_{\mathbf{p}} e^{-i\omega_{\mathbf{p}} t} + b^\dagger_{-\mathbf{p}} e^{i\omega_{\mathbf{p}} t} \right]; \quad q^\dagger_{\mathbf{p}} = \frac{1}{\sqrt{2\omega_{\mathbf{p}}}} \left[a^\dagger_{\mathbf{p}} e^{+i\omega_{\mathbf{p}} t} + b_{-\mathbf{p}} e^{-i\omega_{\mathbf{p}} t} \right]$$

$$(1.125)$$

(We use $b_{-\mathbf{p}}$ rather than $b_{\mathbf{p}}$ in the second term for future convenience.) With the continuum labels $\mathbf{k}, \mathbf{p}$ etc., the standard commutator relations become:

$$[a_{\mathbf{p}}, a^\dagger_{\mathbf{k}}] = (2\pi)^3 \delta(\mathbf{p} - \mathbf{k}); \quad [b_{\mathbf{p}}, b^\dagger_{\mathbf{k}}] = (2\pi)^3 \delta(\mathbf{p} - \mathbf{k}) \qquad (1.126)$$

with all other commutators vanishing. The result in Eq. (1.118) now generalizes for any two oscillators, labeled by vectors $\mathbf{k}$ and $\mathbf{p}$, to the form:

$$\langle 0| T \left(q_{\mathbf{k}}(t_2) q^\dagger_{\mathbf{p}}(t_1) \right) |0\rangle = (2\pi)^3 \frac{\delta(\mathbf{k} - \mathbf{p})}{2\omega_{\mathbf{p}}} e^{-i\omega_{\mathbf{p}}|t|} \qquad (1.127)$$

We now see that our amplitude $G(x_2; x_1)$ can be expressed in the form

$$\begin{aligned} G(x_2; x_1) &= \int \frac{d^3\mathbf{p}}{(2\pi)^3} \int \frac{d^3\mathbf{k}}{(2\pi)^3} \langle 0| T \left[q_{\mathbf{k}}(t_2) q^\dagger_{\mathbf{p}}(t_1) \right] |0\rangle e^{i(\mathbf{k}\cdot\mathbf{x}_2 - \mathbf{p}\cdot\mathbf{x}_1)} \\ &\equiv \langle 0| T [\phi(x_2) \phi^\dagger(x_1)] |0\rangle \end{aligned} \qquad (1.128)$$

where we have introduced a new set of operators $\phi(x)$ at every event in spacetime given by

$$\phi(x) \equiv \int \frac{d^3\mathbf{p}}{(2\pi)^3} q_{\mathbf{p}}(t) \, e^{i\mathbf{p}\cdot\mathbf{x}} \equiv \int \frac{d^3\mathbf{p}}{(2\pi)^3} \frac{1}{\sqrt{2\omega_{\mathbf{p}}}} \left[a_{\mathbf{p}} e^{-ipx} + b^\dagger_{\mathbf{p}} e^{ipx} \right] \quad (1.129)$$

where we have flipped the sign of spatial momentum in the second term to get the $b^\dagger_{\mathbf{p}} e^{ipx}$ factor. (This is why we called it $b^\dagger_{-\mathbf{p}}$ in Eq. (1.125) with a minus sign.) Since G is a Lorentz invariant scalar, Eq. (1.128) suggests that $\phi(x)$ is a Lorentz invariant object. This is indeed true and can be seen by introducing two rescaled operators $A_{\mathbf{p}} = \sqrt{2\omega_{\mathbf{p}}} a_{\mathbf{p}}$, $B_{\mathbf{p}} = \sqrt{2\omega_{\mathbf{p}}} b_{\mathbf{p}}$ and writing Eq. (1.129) as

$$\phi(x) = \int d\Omega_{\mathbf{p}} \left[A_{\mathbf{p}} e^{-ipx} + B^\dagger_{\mathbf{p}} e^{ipx} \right] \qquad (1.130)$$

[35] It would have been more appropriate to define $a^\dagger$ as the annihilation operator, given the obvious relationship between daggers and annihilation. Unfortunately, the opposite notation has already been adopted!

The Lorentz invariance of $\phi(x)$ requires the commutation rules for $A_\mathbf{p}$ etc. to be Lorentz invariant. Since $[a_\mathbf{p}, a_\mathbf{k}^\dagger] = (2\pi)^3 \delta(\mathbf{p} - \mathbf{k})$, it follows that $[A_\mathbf{p}, A_\mathbf{k}^\dagger] = (2\pi)^3 (2\omega_\mathbf{p}) \delta(\mathbf{p} - \mathbf{k})$. This commutator is Lorentz invariant since $(2\omega_\mathbf{p})\delta(\mathbf{p} - \mathbf{k})$ is Lorentz invariant. So $\phi(x)$ behaves as a scalar under the Lorentz transformation. We will have more to say about Eq. (1.130) in the next section.

This result is astonishingly beautiful when you think about it. To make it explicit, what we have proved is the equality

$$\sum_{0,\mathbf{x}_1}^{t,\mathbf{x}_2} \exp\left[-im \int_{t_1}^{t_2} dt \sqrt{1 - \mathbf{v}^2}\right] = G(x_2; x_1) = \langle 0|T[\phi(x_2)\phi^\dagger(x_1)]|0\rangle$$

(1.131)

On the left side is the propagation amplitude computed using sum over paths for a *single* relativistic free particle. On the right is an expectation value in some (fictitious) harmonic oscillator ground state for the time ordered product of two field operators.[36] No two systems could have been a priori more different, but both can be used to compute the propagation amplitude which is given in the middle!

This is the result we were looking for. It suggests that if we understand the system described by these harmonic oscillators and the associated field $\phi(x)$, we can provide an alternate interpretation for the $G(x_2; x_1)$ in space-time using fully causal evolution. The last part is not yet obvious but there is cause for hope and we will see that it does work out.

1.5.1   Propagation Amplitude and Antiparticles

The decomposition in Eq. (1.130) allows us to separate $\phi(x)$ into two Lorentz invariant scalar fields with $\phi(x) = A(x) + B^\dagger(x)$ where

$$A(x) \equiv \int d\Omega_\mathbf{p} A_\mathbf{p} e^{-ipx}; \quad B(x) \equiv \int d\Omega_\mathbf{p} B_\mathbf{p} e^{-ipx}$$

(1.132)

We can then express $G_\pm(x_2; x_1)$, defined in Eq. (1.86), in terms of the expectation values of fields by comparing Eq. (1.128) with Eq. (1.88). This gives

$$\langle 0|\phi(x_2)\phi^\dagger(x_1)|0\rangle = \langle 0|A(x_2)A^\dagger(x_1)|0\rangle = \int d\Omega_\mathbf{p} e^{-ipx} = G_+(x_2; x_1)$$

$$\langle 0|\phi^\dagger(x_1)\phi(x_2)|0\rangle = \langle 0|B(x_1)B^\dagger(x_2)|0\rangle = \int d\Omega_\mathbf{p} e^{+ipx} = G_-(x_2; x_1)$$

(1.133)

so that we have

$$G(x_2; x_1) = \theta(t)\langle 0|A(x_2)A^\dagger(x_1)|0\rangle + \theta(-t)\langle 0|B(x_1)B^\dagger(x_2)|0\rangle \quad (1.134)$$

This expression is quite intriguing and let us try to understand its structure.

Let us begin with the case when x_2 and x_1 are separated by a timelike interval so that the notion of $t > 0$ or $t < 0$ is Lorentz invariant. We see from Eq. (1.134) that when $t > 0$ we have $G(x_2; x_1) = \langle 0|A(x_2)A^\dagger(x_1)|0\rangle$. This corresponds to creation of an "a-type particle", say, at x_1 followed by its annihilation at x_2; so it makes sense to think of this as a propagation of an a-type particle from x_2 to x_1.   This interpretation can be made

[36]Why does this equality in Eq. (1.131) hold? Nobody really knows, in the sense that there is no simple argument which will make this equality self evident and obvious without doing any extensive calculation. This probably means we do not yet have a really fundamental understanding of the quantum physics of a relativistic free particle, but most people would disagree and will consider this equality as merely a very convenient mathematical identity.

more specific along the following lines: Since $|0\rangle$ is the ground state of all harmonic oscillators, it follows that

$$\phi^\dagger(x)|0\rangle = A^\dagger(x)|0\rangle = \int d\Omega_{\mathbf{p}}\, e^{ipx} A_{\mathbf{p}}^\dagger|0\rangle \equiv \int d\Omega_{\mathbf{p}}\, e^{ipx}|1_{\mathbf{p}}\rangle \qquad (1.135)$$

where $|1_{\mathbf{p}}\rangle$ denotes the state in which the $\mathbf{p}$-th oscillator is in the first excited state with all other oscillators remaining in the ground state. If we make a spacetime translation by $x^i \to x^i + L^i$, this state picks up an extra factor $\exp(ipL)$ showing that this state has a four-momentum $p^i = (\omega_{\mathbf{p}}, \mathbf{p})$.[37] Therefore, it makes sense to think of $e^{ipx} A_{\mathbf{p}}^\dagger|0\rangle$ as the state $|\mathbf{p}\rangle\langle\mathbf{p}|x\rangle$ and write

$$
\begin{aligned}
\phi^\dagger(x_1)|0\rangle &= A^\dagger(x_1)|0\rangle = \int d\Omega_{\mathbf{p}}|\mathbf{p}\rangle\langle\mathbf{p}|x_1\rangle; \\
\langle 0|\phi(x_2) &= \langle 0|A(x_2) = \int d\Omega_{\mathbf{q}}\langle x_2|\mathbf{q}\rangle\langle\mathbf{q}|
\end{aligned}
\qquad (1.136)
$$

and obtain

$$\langle 0|\phi(x_2)\phi^\dagger(x_1)|0\rangle = \langle 0|A(x_2)A^\dagger(x_1)|0\rangle = \int d\Omega_{\mathbf{p}} \int d\Omega_{\mathbf{q}}\langle x_2|\mathbf{q}\rangle\langle\mathbf{q}|\mathbf{p}\rangle\langle\mathbf{p}|x_1\rangle \qquad (1.137)$$

Using the Lorentz invariant[38] normalization $\langle\mathbf{q}|\mathbf{p}\rangle = (2\pi)^3(2\omega_{\mathbf{p}})\delta(\mathbf{q} - \mathbf{p})$ we reproduce the original result

$$\langle 0|\phi(x_2)\phi^\dagger(x_1)|0\rangle = \int d\Omega_{\mathbf{p}}\langle x_2|\mathbf{p}\rangle\langle\mathbf{p}|x_1\rangle = G(x_2; x_1) \quad (t > 0) \qquad (1.138)$$

This shows that one can possibly identify the state $A_{\mathbf{p}}^\dagger|0\rangle$ with a one-particle state of momentum $\mathbf{p}$ and energy $\omega_{\mathbf{p}}$. The identity operator for the one-particle states can be expanded in terms of $|\boldsymbol{p}\rangle$ states by the usual relation

$$1 = \int \frac{d^3\mathbf{p}}{(2\pi)^3} \frac{1}{2\omega_p}|\boldsymbol{p}\rangle\langle\boldsymbol{p}| \qquad (1.139)$$

You can also easily verify that $\langle\boldsymbol{p}|\phi_0(0, \boldsymbol{x})|0\rangle = e^{-i\boldsymbol{p}\cdot\boldsymbol{x}}$. This relation looks very similar to the standard relation in NRQM giving $\langle\boldsymbol{p}|\boldsymbol{x}\rangle = \exp(-i\boldsymbol{p}\cdot\boldsymbol{x})$. It is then rather tempting to think of $\phi(0, \boldsymbol{x})$ acting on $|0\rangle$ as analogous to $|\boldsymbol{x}\rangle$ corresponding to a particle "located at" $|\boldsymbol{x}\rangle$. But, as we saw earlier, such an interpretation is fraught with difficulties.[39] In short, our field operator acting on the ground state of some (fictitious) harmonic oscillators can produce a state with a particle at x as indicated by Eq. (1.137).

Something similar happens for $x_2^0 < x_1^0$, i.e., $t < 0$, when we start with $\phi(x)|0\rangle$ in which the $B_{\mathbf{p}}^\dagger$ produces a particle. Now we get the propagation of a "b-type particle" from x_2 to x_1 which is in the direction opposite to the propagation indicated by $G(x_2; x_1)$. This can be interpreted by saying that when $t < 0$ the (backward in time) propagation of a particle from x_1 to x_2 is governed by an amplitude for the *forward* propagation of an *antiparticle* represented by B and $B^\dagger$ operators. In this interpretation, we call the b-type particle as the *antiparticle* of the a-type particle. (Note that both a-type and b-type, i.e., the particle and the antiparticle, have $\omega_{\mathbf{p}}^2 = \mathbf{p}^2 + m^2$ with the same m.) As long as x_2 and x_1 are separated by a timelike interval, the amplitude $G(x_2; x_1)$ is contributed by a particle propagating from x_1 to x_2 if $t > 0$ and is contributed by an antiparticle

[37] Treated as an eigenstate of the *standard harmonic oscillator Hamiltonian*, the state $|1_{\mathbf{p}}\rangle$ has the energy $\omega_{\mathbf{p}}[1 + (1/2)] = (3/2)\omega_{\mathbf{p}}$ and we don't know how to assign any momentum to it. We also have no clue at this stage what relevance the harmonic oscillator Hamiltonian has to the relativistic free particle we are studying. For the moment, we get around all these using the spacetime translation argument.

[38] The normalization *here* is partially a question of convention just as whether we use $a_{\mathbf{p}}$ or $A_{\mathbf{p}}$ in the expansion of the field; see Eq. (1.129) and Eq. (1.130). The latter has nicer Lorentz transformation properties and a Lorentz invariant commutator. Similarly, the states produced by $a_{\mathbf{p}}|0\rangle$ and $A_{\mathbf{p}}|0\rangle$ will be normalized differently. We choose to work with a Lorentz invariant normalization but final results should not depend on this as long as we do it consistently.

[39] Of course, there are many states in the Hilbert space which satisfy $\langle\boldsymbol{p}|\psi\rangle = \exp(-i\boldsymbol{p}\cdot\boldsymbol{x})$. Since $\langle\boldsymbol{p}|$ has non-zero matrix elements only with the one-particle state, adding two or zero particle states like $\phi^2(0, \boldsymbol{x})|0\rangle$ will still give the same $\langle\boldsymbol{p}|\psi\rangle$.

propagating from x_2 to x_1 if $t < 0$. All these notions are Lorentz invariant as long as x_2 and x_1 are separated by a timelike interval.

On the other hand, if x_2 and x_1 are separated by a spacelike interval, then the notion of $t > 0$ or $t < 0$ is not Lorentz invariant. Then, if we are in a Lorentz frame in which $t > 0$, the amplitude $G(x_2; x_1)$ is contributed by a particle propagating from x_1 to x_2; but if we transform to another frame in which the same two events have $t < 0$, then the $G(x_2; x_1)$ is contributed by an antiparticle propagating from x_2 to x_1. Thus, in the case of events separated by a spacelike interval, our interpretation of $G(x_2; x_1)$ (which, of course, is Lorentz invariant) in terms of particle or antiparticle propagation depends on the Lorentz frame.

This situation is summarized in Fig. 1.8 which shows the propagation of a particle from the origin to an event $\mathcal{P}$ for three different cases: (a) When $\mathcal{P}$ is within the future light-cone of the origin, like the event X; (b) When $\mathcal{P}$ is within the past light-cone of the origin, like the event Y; (c) When $\mathcal{P}$ is outside the light-cone of the origin (like the event Z), in which case one cannot talk of future and past in a Lorentz invariant way. The event Z could be to the future of the origin in one Lorentz frame while it could be at the past of the origin in another reference frame. The amplitude $G(x; 0)$ for the case (a) is given by $\langle 0|A(X)A^\dagger(0)|0\rangle$ which represents the creation of a particle at origin followed by its annihilation at the event X. In case (b) this amplitude is given by $\langle 0|B(0)B^\dagger(Y)|0\rangle$ which represents the creation of an antiparticle at Y followed by its subsequent annihilation at the origin. Note that the arrow in the figure refers to the propagation amplitude $G(y; 0)$ which, in this case, is a backward propagation in time. This is given by the propagation of an antiparticle forward in time. In the case (c), the amplitude could have been contributed either by creation of a particle at the origin followed by its annihilation at Z (which is shown in the figure) or by creation of an antiparticle at Z followed by its annihilation at origin depending on the Lorentz frame in which we describe the physics. The numerical value of the amplitude remains the same thanks to the crucial result obtained in Eq. (1.89) which shows that these two amplitudes are the same for spacelike separated events.

Since the field ϕ creates an antiparticle and destroys a particle (with the field $\phi^\dagger$ doing exactly the opposite), it collectively captures the full dynamics. The description uses the same $\omega_{\mathbf{p}} = \sqrt{p^2 + m^2}$ with the same m for both the particle and antiparticle; therefore, they must have the same mass. But if we add a phase α to ϕ by changing $\phi \to e^{i\alpha}\phi$, then $\phi^\dagger$ changes by $\phi^\dagger \to e^{-i\alpha}\phi^\dagger$. We will see in Sect. 3.1.5 that such phase changes are related to the charges carried by the particle, and in particular to the standard electric charge. The above behaviour shows that if the particle has a positive charge then the antiparticle must have an equal and opposite negative charge.[40]

The fact that the particle and antiparticle have opposite charges allows us to make the notion of our propagation more precise. Let us consider a physical process which increases the charge at the event x_1 by Q and decreases the charge at x_2 by Q through particle propagation. We can do this by creating a particle (which carries a charge $+Q$) at x_1, allowing it propagate to x_2 and destroying it at x_2 (thereby decreasing the charge by Q). We can also produce the same effect by creating an antiparticle (which carries a charge $-Q$) at x_2, allowing it to propagate to x_1 and destroying it at x_1 (thereby increasing the charge by Q). Which of these

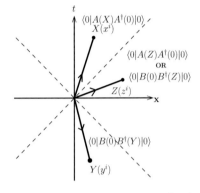

Figure 1.8: Interpretation of $G(x; y)$ in terms of propagation of particles and antiparticles.

[40]How come the path integral amplitude $G(x_2; x_1)$ we computed is the same for charged and neutral particles? Recall that we used an action for a *free* particle, so the maths knows nothing about coupling to an electromagnetic field and the particle's charge is irrelevant. Neutral and charged particles will behave the same if we ignore electromagnetic interactions and treat them as *free*. What is really curious is that one can decipher this information from the field approach. This is because we put in by hand the fact that ϕ is *not* Hermitian.

two processes we can use depends on whether $x_2^0 > x_1^0$ or $x_2^0 < x_1^0$ since we need to create a particle or antiparticle *before* we can destroy it. If $x_2^0 > x_1^0$ then we will use $A^\dagger(x_1)$ to create a particle at x_1 and $A(x_2)$ to destroy it at x_2. This process has the amplitude $\theta(t)\langle 0|A(x_2)A^\dagger(x_1)|0\rangle$. We cannot, however, use this process if we are in a Lorentz frame in which $x_2^0 < x_1^0$. But in this case we can first create an antiparticle with $B^\dagger(x_2)$ at x_2 (thereby decreasing the charge by Q at x_2 and then destroying it with $B(x_1)$ at x_1 (thereby increasing the charge by Q at x_1). This process has the amplitude $\theta(-t)\langle 0|B(x_1)B^\dagger(x_2)|0\rangle$. Thus the total amplitude for this physical process (see Fig. 1.9) is given by

$$
\begin{aligned}
G(x_2; x_1) &= \theta(t_2 - t_1)\langle 0|A(x_2)A^\dagger(x_1)|0\rangle + \theta(t_1 - t_2)\langle 0|B(x_1)B^\dagger(x_2)|0\rangle \\
&= \theta(t_2 - t_1)\langle 0|\phi(x_2)\phi^\dagger(x_1)|0\rangle + \theta(t_1 - t_2)\langle 0|\phi^\dagger(x_1)\phi(x_2)|0\rangle \\
&= \langle 0|T[\phi(x_2)\phi^\dagger(x_1)]|0\rangle \qquad (1.140)
\end{aligned}
$$

As usual, the distinction between the two terms has a Lorentz invariant meaning only when x_2 and x_1 are separated by a timelike interval. If not, which of the terms contribute to the amplitude will depend on the Lorentz frame. But in any Lorentz frame, we would have always created a particle (or antiparticle) before destroying it which takes care of an elementary notion of causality. This result is precisely what our path integral leads us to; it is a net amplitude for the physical process taking into account the existence of both particles and antiparticles.

This is also needed for consistency of causal description of the situation at an intermediate time shown in Fig. 1.7, which is what we started with. Armed with our new interpretation, we can think of a particle antiparticle pair being created at A with the antiparticle traveling forward in time to B (which is equivalent to our original particle traveling backward in time from B to A), annihilating with our original particle at B. What we detect at x_2 is the particle created at A. Obviously, each of these propagations can be within the light cone with x_1 and x_2 still being outside each other's light cone. The mechanism we set up for localized emission and detection events at x_1 and x_2 for our relativistic particle must be responsible for the pair creation as well. We, however, do not want conserved charges to appear and disappear at intermediate times like $t = y^0$. This requires the antiparticle traveling from A to B to have a charge equal and opposite to that carried by the particle traveling from A to x_2.

Most of the previous discussion will also go through if we impose an extra condition on the field $\phi(x)$ that it must be Hermitian. This is equivalent to setting $A_{\mathbf{p}} = B_{\mathbf{p}}$ for all $\mathbf{p}$. In this case, the particle happens to be its own antiparticle which can be a bit confusing when we are still trying to figure things out.[41] This also shows that our "association" of the field with the relativistic particle in order to reproduce the amplitude $G(x_2; x_1)$ is by no means unique. In the simplest context we could have done it with either a Hermitian scalar field or a non-Hermitian one.

1.5.2 Why do we *Really* Need Antiparticles?

If you were alert, you would have noticed that the field ϕ is actually made of *two* Lorentz invariant, non-Hermitian fields (which do not talk to each other) given by the two terms in Eq. (1.130). Since we can express $G(x_2; x_1)$ by adding up these two, why are we dealing[42] with $\phi(x)$ rather than with $A(x)$ and $B(x)$? This question is conceptually important because ϕ deals

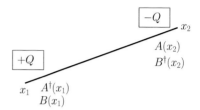

Figure 1.9: Interpretation of $G(x_2; x_1)$ in terms of particles and antiparticles.

[41] Many textbooks, which begin with a real classical scalar field (for which the particle is the same as its antiparticle) and its canonical quantization, are guilty of beginning with this potentially confusing situation!

[42] If we were using a Hermitian scalar field, you could have given it as a reason to prefer $\phi(x)$ to the non-Hermitian bits of it, $A(x)$ and $B(x)$. But since ϕ is not Hermitian either, this does not give a strong motivation.

with both particles and antiparticles on an equal footing (that is, has both negative frequency and positive frequency modes in it) while $A(x), B(x)$ are built from positive frequency modes alone. Why is it better to treat particles and antiparticles together?

The answer *is non-trivial* and has to do with the nature of commutation relations these fields satisfy amongst themselves. To begin with, it is obvious that

$$[A(x), A(y)] = 0; \quad [B(x), B(y)] = 0; \quad [A(x), B(y)] = 0 \qquad (1.141)$$

These follow from the fact that the corresponding annihilation operators commute among themselves. The only nontrivial commutator we need to evaluate is $[A(x), A^\dagger(y)]$ which should be the same as $[B(x), B^\dagger(y)]$. We see that

$$
\begin{aligned}
[A(x_2), A^\dagger(x_1)] &= \int d\Omega_{\mathbf{p}} \int d\Omega_{\mathbf{q}} e^{-ipx_2} e^{iqx_1} [A_{\mathbf{p}}, A_{\mathbf{q}}^\dagger] \\
&= \int d\Omega_{\mathbf{p}} e^{-ipx} = G_+(x_2; x_1) \qquad (1.142)
\end{aligned}
$$

This commutator does not vanish when $(x_2 - x_1)^2 < 0$, i.e., when the two events are separated by a spacelike interval. This means that one cannot, for example, specify independently the value of $A(0, \mathbf{x_1})$ and $A^\dagger(0, \mathbf{x_2})$ on a $t = 0$ spacelike hypersurface since $[A(0, \mathbf{x_1}), A^\dagger(0, \mathbf{x_2})] \neq 0$. An identical conclusion holds for $B(x)$.

In contrast, consider the corresponding commutators for $\phi(x)$. Again, $[\phi(x_1), \phi(x_2)]$ vanishes *as long as ϕ is non-Hermitian*, since in this case the A's commute with the B's. The only nontrivial commutator, $[\phi(x_2), \phi^\dagger(x_1)]$, is given by

$$[A(x_2) + B^\dagger(x_2), A^\dagger(x_1) + B(x_1)] = [A(x_2), A^\dagger(x_1)] - [B(x_1), B^\dagger(x_2)] \qquad (1.143)$$

The two terms differ only by the interchange of coordinates, $x \to -x$, and hence the commutator is given by

$$\Delta(x) \equiv [\phi(x_2), \phi^\dagger(x_1)] = G_+(x_2; x_1) - G_+(x_1; x_2); = \int d\Omega_{\mathbf{p}}[e^{-ipx} - e^{+ipx}] \qquad (1.144)$$

This is precisely the expression we evaluated in Eq. (1.89) and proved to vanish[43] when $x^2 < 0$. So we see that $\phi(x_1)$ commutes with $\phi^\dagger(x_2)$ when the events are spacelike. Therefore, unlike in the case of $A(x)$ and $B(x)$, we can measure and specify the value of $\phi(0, \mathbf{x_1})$ and $\phi^\dagger(0, \mathbf{x_2})$ on a spacelike hypersurface (without one measurement influencing the other) at $t = 0$ which is nice. From Eq. (1.143) you see clearly that the commutators involving either A or B do not vanish outside the light cone but when we combine them to form $A(x) + B^\dagger(x) = \phi(x)$, the two commutators nicely cancel each other outside the light cone. This is why we use ϕ rather than just A or B. Roughly speaking, the antiparticle contribution cancels the particle contribution outside the light cone.

If we build observables from ϕ by using bilinear, Hermitian, combinations of the form $\mathcal{O}(x) = \phi(x)D(x)\phi^\dagger(x)$ where $D(x)$ is either a c-number or a local differential operator, then a straightforward computation shows that $[\mathcal{O}(x_2), \mathcal{O}(x_1)]$ also vanishes when x_1 and x_2 are spacelike. When the fields are injected with lives of their own later on, this will turn out to be a

---

[43]The $\Delta(x)$ is, of course, non-zero inside the light cone where $x^2 > 0$ and can be expressed in terms of Bessel functions. Mercifully, we will almost never need its exact form. The commutator of the fields vanishes for spacelike separations but is non-zero otherwise. This will be rather strange if the commutators are honest-to-god analytic *functions* of the coordinates! This feat becomes possible only because the commutators are, mathematically speaking, *distributions* and not functions.

Exercise 1.15: Prove this claim.

very desirable property, viz., that the measurement of bilinear observables do not influence each other outside the light cone. *It is this version of causality which we will adopt in field theory* rather than the vanishing of the propagation amplitude $G(x_2; x_1)$ when the two events are separated by a spacelike interval.

Finally, once we have mapped the propagation of the relativistic particle to the physics of an infinite number of oscillators, it provides a strong motivation to study the dynamics of the latter system. Since each (complex) oscillator is described by a Lagrangian $L_{\mathbf{k}} = [|\dot{q}_{\mathbf{k}}|^2 - \omega_{\mathbf{k}}^2 |q_{\mathbf{k}}|^2]$, the full system is described by the action

$$A = \sum_{\mathbf{k}} \int dt [|\dot{q}_{\mathbf{k}}|^2 - \omega_{\mathbf{k}}^2 |q_{\mathbf{k}}|^2] \Rightarrow \int \frac{d^3\mathbf{k}}{(2\pi)^3} \int dt [|\dot{q}_{\mathbf{k}}|^2 - \omega_{\mathbf{k}}^2 |q_{\mathbf{k}}|^2] \quad (1.145)$$

Converting back to spacetime by expressing $q_{\mathbf{k}}(t)$ in terms of the field $\phi(x)$, this action becomes[44]

$$A = \int d^4x \left(\partial_a \phi \partial^a \phi^* - m^2 |\phi|^2 \right) \quad (1.146)$$

[44] We now think of $q_{\mathbf{k}}(t)$ and $\phi(x)$ are c-number functions, not as operators.

It is clear that this action describes a field with dynamics of its own if we decide to treat this as a physical system. By 'quantizing' this field (which is most easily done by decomposing it again into harmonic oscillators and quantizing each oscillator), we will end up getting a description of relativistic particles and antiparticles directly as the 'quanta' of the field. This, of course, is the reverse route of getting the particles from the field by what is usually called *canonical quantization* — a subject we will take up[45] in Chap. 3.

[45] Most textbooks follow the route of introducing and studying a classical field first and then quantizing it and, lo and behold, particles appear. We chose to postpone this to Chap. 3 because in that procedure it is never clear how we would have arrived at the concept of a field if we had started with a single relativistic particle and quantized it.

1.5.3 Aside: Occupation Number Basis in Quantum Mechanics

Yet another way to understand the need for using the combination $\phi = A + B^\dagger$ rather than just $\phi = A$ is to compare it with a situation in which just using $\phi = A$ *is* indeed sufficient. This example is provided by nonrelativistic quantum mechanics when we study many particle systems using the occupation number basis. Let us briefly recall this approach in some simple contexts.[46]

Think of a system made of a large number N of indistinguishable particles all located in some common external potential with energy levels $E_1, E_2, E_3 \ldots$. If you know the wave function describing the quantum state of each particle, then the full wave function of the system can be obtained by a suitable product of individual wave functions, ensuring that the symmetry (or antisymmetry) with respect to interchange of particles is taken care of depending on whether the particles are bosons or fermions; let us consider bosons for simplicity.

The quantum indistinguishability of the particles tells you that you cannot really ask which particle is in which state but only talk in terms of the number of particles in a given state. It is, therefore, much more natural to specify the state of the system by giving the *occupation number* for each energy level; that is, you use as the basis a set of states $|n_1, n_2, ...\rangle$ which corresponds to a situation in which there are n_1 particles in the energy level E_1 etc. Interactions will now change the occupation number of the

[46] Unfortunately, many textbooks use the phrase "second quantization" to describe the use of the occupation number basis in describing quantum mechanics. This is, of course, misleading. You don't quantize a many body system for the second time; you are only rephrasing what you have after the (first) quantization, in a different basis.

different energy levels and all operators can be specified by giving their matrix elements in the occupation number basis. In fact it is fairly easy to construct a "dictionary" between the matrix elements of operators in the usual, say, coordinate basis and their matrix elements in the occupation number basis. To describe the change in the occupation numbers we only need to introduce suitable annihilation and creation operators for each energy level; their action will essentially make the occupation numbers of the levels go down or up by unity.

For example, consider two non-relativistic, spin zero particles which interact via a short range two-body potential $V(\mathbf{x_1} - \mathbf{x_1})$. Suppose the two particles are initially well separated, move towards each other and scatter off to a final, well separated state. We specify the initial state of the particles by the momenta $(\mathbf{p}_1, \mathbf{p}_2)$ and the final state by the final momenta $(\mathbf{k}_1, \mathbf{k}_2)$. One can study this process by writing down the Schrodinger equation for the two particle system and solving it (possibly in some approximation) with specified boundary conditions. There is, however, a nicer way to phrase this problem by using the occupation number basis. To do this, we start with a fictitious state $|0\rangle$ of no particles and introduce a set of creation $(a_\mathbf{p}^\dagger)$ [and annihilation $(a_\mathbf{p})$] operators which can act on $|0\rangle$ and generate one-particle states with a given momentum. The corresponding *operator* (not wave function) which describes a particle at $\mathbf{x}$ at the time $t = 0$ can be taken to be

$$\psi(0, \mathbf{x}) = \int \frac{d^3\mathbf{p}}{(2\pi)^3} \, a_\mathbf{p} e^{i\mathbf{p}\cdot\mathbf{x}} \tag{1.147}$$

At a later time, the standard time evolution will introduce the phase factor $\exp(-i\omega_\mathbf{p} t)$ with $\omega_\mathbf{p} = (p^2/2m)$ leading to[47]

$$\psi(t, \mathbf{x}) = \int \frac{d^3\mathbf{p}}{(2\pi)^3} \, a_\mathbf{p} e^{-ipx}; \qquad px \equiv \omega_\mathbf{p} t - \mathbf{p} \cdot \mathbf{x} \tag{1.148}$$

[47] This operator is completely analogous to our $A(x)$; both involve only positive energy solutions and in fact if you do a Taylor expansion of $(p + m^2)^{1/2}$ in $|p^2/m^2|$, ignore the phase due to rest energy and approximate $d\Omega_\mathbf{p} = [d^3\mathbf{p}/(2\pi)^3] [1/2\omega_\mathbf{p}]$ by $d\Omega_\mathbf{p} \simeq [d^3\mathbf{p}/(2\pi)^3] [1/2m]$, you will find that $\psi(t, \mathbf{x}) \approx 2mA(x)$.

Our scattering problem can be restated in terms of the creation and annihilation operators as follows: The scattering annihilates two particles with momenta $(\mathbf{p}_1, \mathbf{p}_2)$ through the action of $a_{\mathbf{p}_1} a_{\mathbf{p}_2}$ and then creates two particles with momenta $(\mathbf{k}_1, \mathbf{k}_2)$ through the action of $a_{\mathbf{k}_1}^\dagger a_{\mathbf{k}_2}^\dagger$ with some amplitude V_1 which is just the Fourier transform of $V(\mathbf{x})$ evaluated at the momentum transfer. The whole process is described by an operator $\mathcal{H} = a_{\mathbf{k}_1}^\dagger a_{\mathbf{k}_2}^\dagger V_1 a_{\mathbf{p}_1} a_{\mathbf{p}_2}$ *which acts on the occupation number basis states.* In fact, any other interaction can be re-expressed in this language, often simplifying the calculations significantly. As we said before, there is a one-to-one correspondence between the matrix elements of the operators in the usual basis and the matrix elements evaluated in occupation number basis.

So you can see that the usual, non-relativistic, quantum mechanics can be translated into a language involving occupation numbers and changes in occupation numbers corresponding to creation and annihilation of particles. What makes quantum field theory different is therefore *not* the fact that we need to deal with situations involving a variable number of particles. The crucial difference is that just a single $\psi(x) \propto A(x)$, *propagating forward in time with positive energy, is inadequate* in the relativistic case. We need another $B(x)$ to ensure causality which — in turn — leads to the propagation of negative energy modes backward in time and the existence of antiparticles. This is what combining relativity and quantum theory leads to — which has no analogue in non-relativistic quantum mechanics even when we use a language suited for a variable number of particles.

Finally let us consider the NRQM limit of the relativistic field theory. You might have seen textbook discussions of the Klein-Gordon equation (and Dirac equation, which we will discuss in a later chapter) solved in different contexts, like, for e.g., in the hydrogen atom problem. Obviously, the scalar *field* $\phi(x)$ which we have introduced is an operator and cannot be treated as a c-number solution to some differential equation. (We are considering the Hermitian case for simplicity.) To obtain the NRQM from quantum field theory, one needs to proceed in a rather subtle way. One procedure is to *define* an NRQM "wave function" $\psi(x)$ by the relation

$$\psi(x) \equiv \langle 0|\phi(t, \boldsymbol{x})|\psi\rangle \tag{1.149}$$

It then follows that

$$
\begin{aligned}
i\partial_t \psi &= i\langle 0|\partial_t\phi(t, \boldsymbol{x})|\psi\rangle = \langle 0| \int \frac{d^3\boldsymbol{p}}{(2\pi)^3} \frac{\sqrt{\boldsymbol{p}^2 + m^2}}{2\omega_{\boldsymbol{p}}} (A_{\boldsymbol{p}} e^{-ipx} - A_{\boldsymbol{p}}^\dagger e^{ipx})|\psi\rangle \\
&= \langle 0| \int \frac{d^3\boldsymbol{p}}{(2\pi)^3} \frac{\sqrt{\boldsymbol{p}^2 + m^2}}{2\omega_{\boldsymbol{p}}} (A_{\boldsymbol{p}} e^{-ipx})|\psi\rangle \\
&= \langle 0|\sqrt{m^2 - \boldsymbol{\nabla}^2}\phi(x)|\psi\rangle \tag{1.150}
\end{aligned}
$$

where we have used the fact $\langle 0|A_{\boldsymbol{p}}^\dagger = 0$. So:

$$i\partial_t\psi(x) = \sqrt{m^2 - \boldsymbol{\nabla}^2}\psi(x) = \left(m - \frac{\boldsymbol{\nabla}^2}{2m} + \mathcal{O}\left(\frac{1}{m^2}\right)\right)\psi(x) \tag{1.151}$$

The first term on the right hand side is the rest energy mc^2 which has to be removed by changing the phase of the wave function ψ. That is, we define $\psi(x) = \Psi(x)\exp(-imt)$. Then it is clear that Ψ satisfies, to the lowest order, the free particle Schrodinger equation given by

$$i\partial_t\Psi(x) = -\frac{\boldsymbol{\nabla}^2}{2m}\Psi(x) \tag{1.152}$$

The non-triviality of the operations involved in this process is yet another reminder that the single particle description is not easy to incorporate in relativistic quantum theory.

1.6 Mathematical Supplement

1.6.1 Path Integral From Time Slicing

In the Heisenberg picture, the amplitude for a particle to propagate from $\mathbf{x} = \mathbf{x}_a$ at $t = 0$ to $\mathbf{x} = \mathbf{x}_b$ at $t = T$ is given by

$$\langle T, \mathbf{x}_b|0, \mathbf{x}_a\rangle = \langle \mathbf{x}_b|e^{-iHT}|\mathbf{x}_a\rangle \tag{1.153}$$

We now break up the time interval into N slices of each of duration ϵ and write

$$e^{-iHT} = e^{-iH\epsilon} e^{-iH\epsilon} e^{-iH\epsilon} \ldots e^{-iH\epsilon} \quad (N \text{ factors}). \tag{1.154}$$

We will next insert a complete set of intermediate states between each of these factors and sum over the relevant variables. More formally, this is done by introducing a factor unity expressed in the form

$$\mathbf{1} = \int d\mathbf{x}_k \, |\mathbf{x}_k\rangle \langle \mathbf{x}_k| \tag{1.155}$$

between each of the exponentials so that we are left with $(N-1)$ factors corresponding to $k = 1, 2, ...(N-1)$. To simplify the notation and include $\mathbf{x}_a$ and $\mathbf{x}_b$, we will *define* $\mathbf{x}_0 \equiv \mathbf{x}_a$ and $\mathbf{x}_N \equiv \mathbf{x}_b$, so that we have a product of factors involving the matrix element of $e^{-iH\epsilon}$ taken between $|\mathbf{x}_k\rangle$ and $\langle \mathbf{x}_{k+1}|$ for $k = 0, 1, ...N$.

We are ultimately interested in the limit of $\epsilon \to 0$ and $N \to \infty$ and hence we can approximate the exponentials retaining only up to first order in ϵ. Thus our problem reduces to evaluating matrix elements of the form

$$\langle \mathbf{x}_{k+1}|e^{-iH\epsilon}|\mathbf{x}_k\rangle \xrightarrow[\epsilon \to 0]{} \langle \mathbf{x}_{k+1}|1 - iH\epsilon + \cdots)|\mathbf{x}_k\rangle, \tag{1.156}$$

taking their product, and then taking their appropriate limit. This task becomes significantly simple if H has the form $A(\mathbf{p}) + B(\mathbf{x})$ like, for example, when $H = \mathbf{p}^2/2m + V(\mathbf{x})$. For any function of $\mathbf{x}$, we have the result[48]

$$\langle \mathbf{x}_{k+1}|f(\mathbf{x})|\mathbf{x}_k\rangle = f(\mathbf{x}_k)\,\delta(\mathbf{x}_k - \mathbf{x}_{k+1}) \tag{1.157}$$

which we can write as

$$\langle \mathbf{x}_{k+1}|f(\mathbf{x})|\mathbf{x}_k\rangle = f(\mathbf{x}_k) \int \frac{d\mathbf{p}_k}{(2\pi)^D} \, \exp\left[i\mathbf{p}_k \cdot (\mathbf{x}_{k+1} - \mathbf{x}_k)\right] \tag{1.158}$$

Similarly, for any term which is only a function of momenta, we can write

$$\langle \mathbf{x}_{k+1}|f(\mathbf{p})|\mathbf{x}_k\rangle = \int \frac{d\mathbf{p}_k}{(2\pi)^D} \, f(\mathbf{p}_k) \exp\left[i\mathbf{p}_k \cdot (\mathbf{x}_{k+1} - \mathbf{x}_k)\right] \tag{1.159}$$

Therefore, when H contains only terms which can be expressed in the form $A(\mathbf{p}) + B(\mathbf{x})$ the relevant matrix element becomes

$$\langle \mathbf{x}_{k+1}|H(\mathbf{p}, \mathbf{x})|\mathbf{x}_k\rangle = \int \frac{d\mathbf{p}_k}{(2\pi)^D} \, H(\mathbf{x}_k, \mathbf{p}_k) \exp\left[i\mathbf{p}_k \cdot (\mathbf{x}_{k+1} - \mathbf{x}_k)\right] \tag{1.160}$$

so that

$$\langle \mathbf{x}_{k+1}|e^{-iH\epsilon}|\mathbf{x}_k\rangle = \int \frac{d\mathbf{p}_k}{(2\pi)^D} \, \exp\left[-i\epsilon H(\mathbf{x}_k, \mathbf{p}_k)\right]$$
$$\times \, \exp\left[i\mathbf{p}_k \cdot (\mathbf{x}_{k+1} - \mathbf{x}_k)\right] \tag{1.161}$$

where we have switched back to the exponential form. Now multiplying all the factors together, we get the result

$$\langle T, \mathbf{x}_N|0, \mathbf{x}_0\rangle = \prod_k \int d\mathbf{x}_k \int \frac{d\mathbf{p}_k}{(2\pi)^D} \, \exp\left[i\Big(\mathbf{p}_k \cdot (\mathbf{x}_{k+1} - \mathbf{x}_k)\right.$$
$$\left. -\epsilon H(\mathbf{x}_k, \mathbf{p}_k)\Big)\right] \tag{1.162}$$

There is one momentum integral for each value of $k = 0, 1, 2, ...N-1$ and one coordinate integral for each $k = 1, 2, ...(N-1)$. Switching back to a continuum notation we have

$$\langle T, \mathbf{x}_b|0, \mathbf{x}_a\rangle = \int \mathcal{D}\mathbf{x}(t) \int \mathcal{D}\mathbf{p}(t) \, \exp\left[i\int_0^T dt \, (\mathbf{p} \cdot \dot{\mathbf{x}} - H(\mathbf{x}, \mathbf{p}))\right] \tag{1.163}$$

[48]To be rigorous we should evaluate f at $(1/2)[\mathbf{x}_k + \mathbf{x}_{k+1}]$ but we will get the same final result.

where the function $\mathbf{x}(t)$ takes fixed values at the end points while the functions $\mathbf{p}(t)$ remain unconstrained. The measure in the functional integral is now defined as the product of integrations in phase space

$$\mathcal{D}\mathbf{x}\,\mathcal{D}\mathbf{p} \to \int \prod_k \frac{d\mathbf{x}_k\,d\mathbf{p}_k}{(2\pi)^D} \qquad (1.164)$$

with suitable limits taken in the end. When H has a specific form $H = \mathbf{p}^2/2m + V(\mathbf{x})$, we can do the integrations over $\mathbf{p}$ using the standard result

$$\int \frac{d\mathbf{p}_k}{(2\pi)^D} \exp\left[i\left(\mathbf{p}_k\cdot(\mathbf{x}_{k+1}-\mathbf{x}_k) - \frac{\epsilon\mathbf{p}_k^2}{2m}\right)\right] = \frac{1}{C(\epsilon)}\exp\left[\frac{im}{2\epsilon}(\mathbf{x}_{k+1}-\mathbf{x}_k)^2\right] \qquad (1.165)$$

where $C(\epsilon) = (2\pi i\hbar\epsilon/m)^{1/2}$. Using this we can write the final expression involving only the $\mathbf{x}$ integration:

$$\langle T,\mathbf{x}_b|0,\mathbf{x}_a\rangle = \lim_{\epsilon\to 0}\left(\frac{1}{C(\epsilon)}\prod_k\int\frac{d\mathbf{x}_k}{C(\epsilon)}\right) \qquad (1.166)$$

$$\times\,\exp\left[i\left(\frac{m}{2}\frac{(\mathbf{x}_{k+1}-\mathbf{x}_k)^2}{\epsilon} - \epsilon V(\mathbf{x}_k)\right)\right]$$

which goes over to the original expression involving the sum over paths, with each path contributing the amplitude $\exp(iA)$:

$$\langle T,\mathbf{x}_b|0,\mathbf{x}_a\rangle = \int\mathcal{D}\mathbf{x}\exp i\int_0^T dt\left[\frac{1}{2}m\dot{\mathbf{x}}^2 - V(\mathbf{x})\right] \qquad (1.167)$$

The time slicing gives meaning to this formal functional integral in this case. As you can see, these ideas generalize naturally to more complicated systems.

1.6.2   Evaluation of the Relativistic Path Integral

In Sect. 1.3.1 we evaluated $G(x_2;x_1)$ for a relativistic particle using the Jacobi action trick. The conventional action functional for a relativistic particle is given by Eq. (1.67) and and we will now show how the path integral with this action can be *directly* evaluated by a lattice regularization procedure even though it is *not* quadratic in the dynamical variables.

We will work in the Euclidean space of D-dimensions, evaluate the path integral and analytically continue to the Lorentzian space. We need to give meaning to the sum over paths

$$\mathcal{G}_E(\mathbf{x_2},\mathbf{x_1};m) = \sum_{\text{all }\mathbf{x}(s)} \exp -m\,\ell[\mathbf{x}(s)] \qquad (1.168)$$

in the Euclidean sector, where

$$\ell(\mathbf{x_2},\mathbf{x_1}) = \int_{s_1}^{s_2} ds\sqrt{\frac{d\mathbf{x}}{ds}\cdot\frac{d\mathbf{x}}{ds}} \qquad (1.169)$$

is just the length of the curve $\mathbf{x}(s)$, parametrized by s, connecting $\mathbf{x}(s_1) = \mathbf{x}_1$ and $\mathbf{x}(s_2) = \mathbf{x}_2$. Of course, the length is independent of the parametrization and the integral defining l is manifestly invariant under the reparameterization $s \to f(s)$.

The $\mathcal{G}_E$ can be defined through the following limiting procedure: Consider a lattice of points in a D-dimensional cubic lattice with a uniform lattice spacing of ϵ. We will work out $\mathcal{G}_E$ in the lattice and will then take the limit of $\epsilon \to 0$ with a suitable measure. To obtain a finite answer, we have to use an overall normalization factor $M(\epsilon)$ in Eq. (1.168) as well as treat m (which is the only parameter in the problem) as varying with ϵ in a specific manner; i.e. we will use a function $\mu(\epsilon)$ in place of m on the lattice and will reserve the symbol m for the parameter in the continuum limit.[49] Thus the sum over paths in the continuum limit is *defined* by the limiting procedure

$$\mathcal{G}_E(\mathbf{x_2}, \mathbf{x_1}; m) = \lim_{\epsilon \to 0} \left[M(\epsilon) \mathcal{G}_E(\mathbf{x_2}, \mathbf{x_1}; \mu(\epsilon)) \right] \tag{1.170}$$

In a lattice with spacing of ϵ, Eq. (1.168) can be evaluated in a straightforward manner. Because of the translation invariance of the problem, $\mathcal{G}_E$ can only depend on $\mathbf{x_2} - \mathbf{x_1}$; so we can set $\mathbf{x_1} = 0$ and call $\mathbf{x_2} = \epsilon \mathbf{R}$ where $\mathbf{R}$ is a D-dimensional vector with integral components: $\mathbf{R} = (n_1, n_2, n_3 \cdots n_D)$. Let $C(N, \mathbf{R})$ be the number of paths of length $N\epsilon$ connecting the origin to the lattice point $\epsilon \mathbf{R}$. Since all such paths contribute a term $[\exp -\mu(\epsilon)(N\epsilon)]$ to Eq. (1.168), we get:

$$\mathcal{G}_E(\mathbf{R}; \epsilon) = \sum_{N=0}^{\infty} C(N; \mathbf{R}) \exp \left(-\mu(\epsilon) N\epsilon \right) \tag{1.171}$$

It can be shown from elementary combinatorics that the $C(N; \mathbf{R})$ satisfy the condition

$$F^N \equiv \left[e^{ik_1} + e^{ik_2} + \cdots e^{ik_D} + e^{-ik_1} + \cdots e^{-ik_D} \right]^N = \sum_{\mathbf{R}} C(N; \mathbf{R}) e^{i\mathbf{k}.\mathbf{R}} \tag{1.172}$$

Therefore,

$$\begin{aligned} \sum_{\mathbf{R}} e^{i\mathbf{k}.\mathbf{R}} \mathcal{G}_E(\mathbf{R}; \epsilon) &= \sum_{N=0}^{\infty} \sum_{\mathbf{R}} C(N; \mathbf{R}) e^{i\mathbf{k}.\mathbf{R}} \exp \left(-\mu(\epsilon) N\epsilon \right) \\ &= \sum_{N=0}^{\infty} e^{-\mu(\epsilon)\epsilon N} F^N = \sum_{N=0}^{\infty} \left[F e^{-\mu(\epsilon)\epsilon} \right]^N \\ &= \left[1 - F e^{-\mu(\epsilon)\epsilon} \right]^{-1} \end{aligned} \tag{1.173}$$

Inverting the Fourier transform, we get

$$\begin{aligned} \mathcal{G}_E(\mathbf{R}; \epsilon) &= \int \frac{d^D \mathbf{k}}{(2\pi)^D} \frac{e^{-i\mathbf{k}.\mathbf{R}}}{\left(1 - e^{-\mu(\epsilon)\epsilon} F \right)} \\ &= \int \frac{d^D \mathbf{k}}{(2\pi)^D} \frac{e^{-i\mathbf{k}.\mathbf{R}}}{\left(1 - 2e^{-\mu(\epsilon)\epsilon} \sum_{j=1}^{D} \cos k_j \right)} \end{aligned} \tag{1.174}$$

Converting to the physical length scales $\mathbf{x} = \epsilon \mathbf{R}$ and $\mathbf{p} = \epsilon^{-1} \mathbf{k}$ gives

$$\mathcal{G}_E(\mathbf{x}; \epsilon) = \int \frac{\epsilon^D d^D \mathbf{p}}{(2\pi)^D} \frac{e^{-i\mathbf{p}.\mathbf{x}}}{\left(1 - 2e^{-\mu(\epsilon)\epsilon} \sum_{j=1}^{D} \cos p_j \varepsilon \right)} \tag{1.175}$$

This is an exact result in the lattice and we now have to take the limit $\epsilon \to 0$ in a suitable manner to keep the limit finite. As $\epsilon \to 0$, the denominator

---

[49] Purely from dimensional analysis, we would expect the mass parameter $\mu(\epsilon)$ to scale inversely as lattice spacing; in fact, we will see that this is what happens.

Exercise 1.16: Prove this result. One possible way is as follows: Consider a path of length N (all lengths in units of ϵ) with r_i steps to the right and l_i steps to the left along the i-th axis ($i = 1, 2, ...D$). If $Q[N, r_i, l_i]$ is the number of such paths connecting the origin to $\mathbf{R} = (n_1, n_2, ...n_D)$, show that $(\Sigma x_i + \Sigma y_i)^N = \Sigma Q(N, r_i, l_i) x_i^{r_i} y_i^{l_i}$; i.e., the left hand side is the generating function for Q. Next show that we are interested in the case where $r_i - l_i = n_i$ is a given quantity for each i and that the number of paths with this condition can be obtained by setting $y_i = (1/x_i)$. Taking $x_i = \exp(ik_i)$ gives the result. Another (simpler) way to prove the result is to note that a path of length N which reaches $\mathbf{R}$ must have come from the paths of length $N - 1$ which have reached one of the $2D$ neighbouring sites in the previous step. This allows you to fix the Fourier transform of $C(N; \mathbf{R})$ through a recursion relation.

of the integrand becomes

$$
\begin{aligned}
1 - 2e^{-\epsilon\mu(\epsilon)}\left(D - \frac{1}{2}\epsilon^2|\mathbf{p}|^2\right) &= 1 - 2De^{-\epsilon\mu(\epsilon)} + \epsilon^2 e^{-\epsilon\mu(\epsilon)}|\mathbf{p}|^2 \quad (1.176)\\
&= \epsilon^2 e^{-\epsilon\mu(\epsilon)}\left[|\mathbf{p}|^2 + \frac{1 - 2De^{-\epsilon\mu(\epsilon)}}{\epsilon^2 e^{-\epsilon\mu(\epsilon)}}\right]
\end{aligned}
$$

so that we get, for small ϵ,

$$
\mathcal{G}_E(\mathbf{x};\epsilon) \simeq \int \frac{d^D\mathbf{p}}{(2\pi)^D}\frac{A(\epsilon)e^{-i\mathbf{p}\cdot\mathbf{x}}}{|\mathbf{p}|^2 + B(\epsilon)} \quad (1.177)
$$

where

$$
\begin{aligned}
A(\epsilon) &= \epsilon^{D-2}e^{\epsilon\mu(\epsilon)}\\
B(\epsilon) &= \frac{1}{\epsilon^2}\left[e^{\epsilon\mu(\epsilon)} - 2D\right]
\end{aligned} \quad (1.178)
$$

The continuum theory has to be defined in the limit of $\epsilon \to 0$ with some measure $M(\epsilon)$; that is, we want to choose $M(\epsilon)$ such that the limit

$$
\mathcal{G}_E(\mathbf{x};m)|_{\text{continuum}} = \lim_{\epsilon\to 0}\{M(\epsilon)\mathcal{G}_E(\mathbf{x};\epsilon)\} \quad (1.179)
$$

is finite. It is easy to see that we only need to demand

$$
\lim_{\epsilon\to 0}\left[\frac{1}{\epsilon^2}\left(e^{\epsilon\mu(\epsilon)} - 2D\right)\right] = m^2 \quad (1.180)
$$

and

$$
\lim_{\epsilon\to 0}\left[M(\epsilon)\epsilon^{D-2}e^{\epsilon\mu(\epsilon)}\right] = 1 \quad (1.181)
$$

to achieve this. The first condition implies that, near $\epsilon \approx 0$,

$$
\mu(\epsilon) \approx \frac{\ln 2D}{\epsilon} + \frac{m^2}{2D}\epsilon \approx \frac{\ln 2D}{\epsilon} \quad (1.182)
$$

The second condition Eq. (1.181), allows us to determine the measure as

$$
M(\epsilon) = \frac{1}{2D}\frac{1}{\epsilon^{D-2}} \quad (1.183)
$$

With this choice, we get

$$
\lim_{\epsilon\to 0}\mathcal{G}_E(\mathbf{x};\epsilon)M(\epsilon) = \int \frac{d^D\mathbf{p}}{(2\pi)^D}\frac{e^{-i\mathbf{p}\cdot\mathbf{x}}}{|\mathbf{p}|^2 + m^2} \quad (1.184)
$$

which is the continuum propagator we computed earlier (see Eq. (1.99)). This analysis gives a rigorous meaning to the path integral for the relativistic free particle.

The rest is easy. We write $(|\mathbf{p}|^2 + m^2)^{-1}$ as a integral over λ of $\exp[-\lambda(|\mathbf{p}|^2 + m^2)]$ and doing the $\mathbf{p}$ integration, we get

$$
\mathcal{G}_E = \int_0^\infty d\lambda\, e^{-\lambda m^2}\int \frac{d^D\mathbf{p}}{(2\pi)^D}e^{i\mathbf{p}\cdot\mathbf{x}-\lambda p^2} = \int_0^\infty d\lambda\, e^{-\lambda m^2}\left(\frac{1}{4\lambda\pi}\right)^{D/2}e^{-\frac{|\mathbf{x}|^2}{4\lambda}} \quad (1.185)
$$

When $D = 4$, this reduces to

$$\mathcal{G}_E = \frac{1}{16\pi^2} \int_0^\infty \frac{d\lambda}{\lambda^2} e^{-m^2\lambda - \frac{|\mathbf{x}|^2}{4\lambda}} \qquad (1.186)$$

To analytically continue from Euclidean to Lorentzian spacetime, we have to change the sign of one of the coordinates in $|\mathbf{x}|^2$ and obtain $|\mathbf{x}|^2 - t^2$ and set $\lambda = is$. This gives

$$\mathcal{G}_E = -\frac{i}{16\pi^2} \int_0^\infty \frac{ds}{s^2} e^{-im^2 s + \frac{i}{4s}(|\mathbf{x}|^2 - t^2)} = -\frac{i}{16\pi^2} \int_0^\infty \frac{ds}{s^2} e^{-im^2 s - \frac{i}{4s}x^2}$$

$$(1.187)$$

which matches with the expression in Eq. (1.75).

Chapter 2

Disturbing the Vacuum

2.1 Sources that Disturb the Vacuum

We started our discussion in Sect. 1.1 with the amplitude for the process $\mathcal{A} = \mathcal{D}(x_2)G(x_2;x_1)\mathcal{C}(x_1)$ involving creation, propagation and detection (usually accompanied by destruction) of a particle and concentrated on the middle part $G(x_2;x_1)$. This analysis, however, dragged us into a description in which $\phi^\dagger(x_1)|0\rangle$ could be interpreted as creating a particle from a no-particle state $|0\rangle$. Similarly, $\phi(x_2)$ acting on this state destroys this particle, taking the state back to $|0\rangle$. We now want to study more carefully the creation and destruction of the particle by external agencies and see what it can tell us. Surprisingly enough, we can learn a lot from such a study and even determine $G(x_2;x_1)$ using some basic principles. This provides a second perspective on $G(x_2;x_1)$ and — what is more — it will again lead to a description involving fields.

2.1.1 Vacuum Persistence Amplitude and $G(x_2;x_1)$

We will start by describing the notion of a *source* which is capable of creating or destroying a particle.[1] We introduce the amplitude $\langle 1_p|0_-\rangle^J \equiv iJ(p)$ which describes the creation of a one-particle state, with on-shell momentum $p = (\omega_{\mathbf{p}}, \mathbf{p})$, due to the action of a source (labeled by the superscript J) from the no-particle state $|0_-\rangle$ at very early (i.e., as $t \to -\infty$) times. The i-factor is introduced on the right hand side of $\langle 1_p|0_-\rangle^J \equiv iJ(p)$ for future convenience. The corresponding amplitude for the destruction will have an amplitude $\langle 0_+|1_p\rangle^J$ where $|0_+\rangle$ denotes the no-particle state at very late (i.e., as $t \to +\infty$) times after all the sources have ceased to operate. This state $|0_+\rangle$, in general, could be different from $|0_-\rangle$.

It is convenient to start with "weak sources", the action of which only leads to single particle, rather than multi-particle, intermediate states during the entire evolution of the system from $t = -\infty$ to $t = \infty$; i.e., we think of the source as a mild perturbation on the no-particle state. We can then easily show that $\langle 0_+|1_p\rangle^J = iJ^*(p)$. To do this, we begin with the relation $\langle 0_-|1_p\rangle^{J=0} = 0$ which represents the orthogonality of the one-particle and no-particle states in the absence of sources and expand this relation by introducing a complete set of states in the presence of the source. This will

[1] A purist will call the agency that creates as the source and the one that destroys as sink; we won't be that fussy.

give

$$0 = \langle 0_-|1_p\rangle^{J=0}$$

$$= {}^{J=0}\langle 0_-|0_+\rangle^J \; {}^J\langle 0_+|1_p\rangle^{J=0} + \sum_q {}^{J=0}\langle 0_-|1_q\rangle^J \; {}^J\langle 1_q|1_p\rangle^{J=0} + \cdots$$

$$\simeq 1 \times {}^J\langle 0_+|1_p\rangle^{J=0} + \sum_q (-iJ^*(q))\delta_{qp} \qquad (2.1)$$

In arriving at the second equality, we have ignored multi-particle intermediate states (denoted by $\cdots$) because the source is weak. For the same reason we can replace ${}^{J=0}\langle 0_-|0_+\rangle^J$ by its value for $J=0$, viz., $\langle 0_-|0_+\rangle^{J=0} = 1$ to leading order in J and replace ${}^J\langle 1_q|1_p\rangle^{J=0}$ by $\langle 1_p|1_q\rangle^{J=0} = \delta_{pq}$ (which is again the value for $J=0$) and use $\langle 0_-|1_q\rangle^J = \langle 1_q|0_-\rangle^* = -iJ^*(q)$. Equation (2.1) then leads to $\langle 0_+|1_p\rangle^J = iJ^*(p)$. (This is why we kept an extra i-factor[2] in the definition of amplitudes so that J and J^* describe creation and annihilation amplitudes.)

It will be convenient to have a corresponding spacetime description of the source. This can be achieved by introducing a real scalar function $J(x)$ and introducing the four dimensional Fourier transform

$$J(p) \equiv \int d^4x \; e^{ip\cdot x} J(x); \qquad J^*(p) = \int d^4x \; e^{-ip\cdot x} J(x) \qquad (2.2)$$

and identifying the on-shell amplitudes $-i\langle 1_p|0_-\rangle^J = J(p) = J(\omega_{\mathbf{p}}, \mathbf{p})$ as the on-shell values of the Fourier transform defined above. Note that we are considering an arbitrary $J(x)$ and a correspondingly arbitrary $J(p)$ [except for the condition $J^*(p) = J(-p)$ ensuring the reality of $J(x)$]; but the amplitude $\langle 1_p|0_-\rangle^J$ only depends on the on-shell behaviour of $J(p)$.

After these preliminaries, we will turn our attention to the amplitude $\langle 0_+|0_-\rangle^J$ for the no-particle state to remain a no-particle state even after the action of $J(x)$, once as a source and once as a sink. (The $\langle 0_+|0_-\rangle^J$ is usually called the *vacuum persistence amplitude*.) Because the source is assumed to be weak, this amplitude will differ from unity by a small quantity: $\langle 0_+|0_-\rangle^J = 1 + F[J]$. Further since each action of J changes the state only from a no-particle state to a one-particle state and vice-versa, $F[J]$ must be a quadratic functional of J to the lowest order. Therefore, the amplitude can be expressed in the form

$$\langle 0_+|0_-\rangle^J = 1 + F[J] \equiv 1 - \frac{1}{2}\int d^4x_2 \; d^4x_1 \; J(x_2)S(x_2-x_1)J(x_1) \qquad (2.3)$$

where $S(x_2-x_1)$ is a function, which can be taken to be symmetric without loss of generality,

$$S(x_2 - x_1) = S(x_1 - x_2) \qquad (2.4)$$

that needs to be determined. It can depend only on $x_2 - x_1$ because of translational invariance. (The $-1/2$ factor in Eq. (2.3) is introduced for future convenience.) To determine its form, we can use the following argument.[3]

Imagine that we divide the source $J(x)$ into two parts $J_1(x)$ and $J_2(x)$ (with $J = J_1 + J_2$) such that J_1 acts in the causal future of J_2 in some frame of reference. This is easily done by restricting the support for the functions $J_1(x), J_2(x)$ to suitable regions of spacetime. When $J_2(x)$ acts on the no-particle state it will, most of the time, do nothing, while occasionally it will

[2]The matrix element $\langle \psi|e^{-iHt}|\phi\rangle$ to the lowest order in t will have the term $-it\langle\psi|H|\phi\rangle$. On the other hand, the matrix element $\langle\phi|e^{-iHt}|\psi\rangle$ will have $-it\langle\phi|H|\psi\rangle$. Though $\langle\psi|H|\phi\rangle = \langle\phi|H|\psi\rangle^*$, we need to take into account the extra i-factor in relating the two. The origin of i-factor in $\langle 0_+|1_p\rangle^J = iJ^*(p)$ is similar.

[3]Originally due to J. Schwinger; in fact, this entire description is adapted from Schwinger's Source theory, a powerful alternative to conventional quantum field theory, almost universally ignored by particle physicists.

produce a particle with the amplitude $\langle 1_p | 0_- \rangle^{J_2}$, so that we can assume

$$\langle 0_+ | 0_- \rangle^{J_2} \simeq 1 + F[J_2] \qquad (2.5)$$

After J_2 has ceased to operate, the resulting state (either no-particle or one-particle state) will evolve undisturbed until J_1 starts operating. If it destroys the single particle state with the amplitude $\langle 0_+ | 1_p \rangle^{J_1}$, we would have reached back to the no-particle state. The entire process is described by the amplitude

$$\begin{aligned}
\langle 0_+ | 0_- \rangle^{J_1 + J_2} &= \langle 0_+ | 0_- \rangle^{J_1} \langle 0_+ | 0_- \rangle^{J_2} + \sum_p \langle 0_+ | 1_p \rangle^{J_1} \langle 1_p | 0_- \rangle^{J_2} \\
&= 1 + F[J_1] + F[J_2] + \sum_p (iJ_1^*(p))(iJ_2(p)) \qquad (2.6)
\end{aligned}$$

We now demand that the expression for $\langle 0_+ | 0_- \rangle^{J_1 + J_2}$ should not depend individually on the manner in which J is separated into J_1 and J_2 but should just be a functional of J alone, so that

$$\langle 0_+ | 0_- \rangle^{J_1 + J_2} \equiv \langle 0_+ | 0_- \rangle^J = 1 + F[J] = 1 + F[J_1 + J_2] \qquad (2.7)$$

where we have used Eq. (2.3). Substituting into Eq. (2.6), we get

$$F[J_1 + J_2] - F[J_1] - F[J_2] = \sum_p (iJ_1^*(p))(iJ_2(p)) \qquad (2.8)$$

Using the form of F in Eq. (2.3), the left hand side works out to be

$$\begin{aligned}
-\frac{1}{2} \int d^4x_2 \, d^4x_1 \, [J_1(x_1)J_2(x_2) + J_2(x_1)J_1(x_2)] \, S(x_2 - x_1) \\
= -\int d^4x_2 \, d^4x_1 \, J_2(x_2)J_1(x_1)S(x_2 - x_1) \quad (2.9)
\end{aligned}$$

where we have used the symmetry of $S(x)$. The right hand side of Eq. (2.8), on the other hand, depends only on the on-shell value of $J(p)$ and is given by, for real sources,

$$\begin{aligned}
\sum_p (iJ_1^*(p))(iJ_2(p)) &= -\int d^4x_2 \int d^4x_1 \int d\Omega_{\mathbf{p}} J_2(x_2)J_1(x_1)e^{-ip(x_2-x_1)} \\
&= -\int d^4x_2 d^4x_1 J_2(x_2) \left[\int d\Omega_{\mathbf{p}} e^{-ip(x_2-x_1)} \right] J_1(x_1)
\end{aligned}$$
$$(2.10)$$

where we have converted the sum to an integral[4] with the invariant measure $d\Omega_{\mathbf{p}}$. Comparison of Eq. (2.9) and Eq. (2.10) allows us to read off the form of $S(x)$ to be

$$S(x) = \int d\Omega_{\mathbf{p}} \, e^{-ipx} = G_+(x_2; x_1) \qquad \text{(for } t > 0\text{)} \qquad (2.11)$$

To find the form of $S(x) = S(t, \mathbf{x})$ for negative values of $t = -|t|$, we will use the symmetry property $S(-|t|, \mathbf{x}) = S(+|t|, -\mathbf{x})$ and use Eq. (2.11) for $S(|t|, -\mathbf{x})$. This gives:

$$\begin{aligned}
S(-|t|, \mathbf{x}) = S(|t|, -\mathbf{x}) &= \int d\Omega_{\mathbf{p}} e^{-i\omega_{\mathbf{p}}|t| - i\mathbf{p} \cdot \mathbf{x}} = \int d\Omega_{\mathbf{p}} e^{i\omega_{\mathbf{p}} t - i\mathbf{p} \cdot \mathbf{x}} \\
&= \int d\Omega_{\mathbf{p}} e^{ipx} = G_-(x_2; x_1) \qquad (2.12)
\end{aligned}$$

[4]The summation over momenta on the left hand side of the above equation is an on-shell summation over the 3-vector $\mathbf{p}$ and $e^{-ip(x-x')}$ is evaluated on-shell; therefore to maintain Lorentz invariance it has to be converted to an integral over $d\Omega_{\mathbf{p}}$.

In arriving at the third equality, we have used the fact that $|t| = -t$ for $t < 0$. Using the usual trick of flipping the sign of $\mathbf{p}$, we can write the full expression for $S(x)$ as

$$S(x) = \int d\Omega_{\mathbf{p}} e^{+i\mathbf{p}\cdot\mathbf{x}} e^{-i\omega_{\mathbf{p}}|t|} = G(x_2; x_1) \tag{2.13}$$

The identification of $S(x)$ with $G(x_2; x_1)$ allows us to write the amplitude $\langle 0_+ | 0_- \rangle^J$ in the form[5]

$$
\begin{aligned}
\langle 0_+ | 0_- \rangle^J &= 1 - \frac{1}{2} \int d^4x_2 d^4x_1 J(x_2) S(x_2 - x_1) J(x_1) \\
&= 1 - \frac{1}{2} \int d^4x_2 d^4x_1 J(x_2) G(x_2; x_1) J(x_1) \tag{2.14}
\end{aligned}
$$

[5]We used the causal ordering of x_2 and x_1 to determine the form of $S(x)$. Such a causal ordering has a Lorentz invariant meaning only when x_1 and x_2 are related by a timelike interval. But we know that when these two events are separated by a spacelike interval, the numerical values of the expressions in Eq. (2.11) and Eq. (2.12) are the same; see Eq. (1.89). Therefore our analysis determines $S(x)$ everywhere unambiguously.

This has an obvious and nice interpretation of a particle being created at x_1 due to the action of the source, followed by its propagation from x_1 to x_2 and its destruction by the source acting for the second time. The fact that the propagation amplitude $G(x_2; x_1)$ — originally found using the path integral for a relativistic particle — emerges from a completely different perspective (of sources perturbing the no-particle state, thereby creating and destroying particles) shows that we are on the right track in our interpretation. If we now write the amplitude $G(x_2; x_1)$ using Eq. (1.86), separating out propagation with positive and negative energies, the integral in Eq. (2.14) will also separate into two terms allowing interpretation in terms of particles and antiparticles.

So far, we have been discussing a weak source in which $J(x)$ should actually be thought of as a small perturbation. In the case of a strong source, the situation becomes more complicated. But it can be handled, essentially by exponentiating the result in Eq. (2.3) if we consider particles to be non-interacting with each other. In this case, the source can create and destroy more than one particle but each pair of creation and destruction events is treated as independent. This assumes that: (i) parts of the source which participate in the creation and destruction of particles do not influence each other and (ii) the propagation amplitude for any given particle is not affected by the presence of other particles. Then the amplitude $\langle 0_+ | 0_- \rangle^J$ is given by the product of amplitudes for individual events and we get

$$\langle 0_+ | 0_- \rangle^J = \exp -\frac{1}{2} \int d^4x_2 \, d^4x_1 J(x_2) G(x_2; x_1) J(x_1) \tag{2.15}$$

To understand this result, let us again consider a situation when $J = J_1 + J_2$ with J_1 acting in the future of J_2. On substituting $J = J_1 + J_2$ in Eq. (2.15), it separates into a product of three terms expressible as

$$\langle 0_+ | 0_- \rangle^J = \langle 0_+ | 0_- \rangle^{J_1} \exp\left[-\int d^4x d^4x' \, J_1(x) G(x; x') J_2(x') \right] \langle 0_+ | 0_- \rangle^{J_2} \tag{2.16}$$

The first and the last factors are easy to understand as the effect of individual components, while the middle factor describes the creation of particles by J_2 followed by their destruction etc. (This is just the exponential form of the result in Eq. (2.9).) Expressing the argument of the exponential in Fourier space as $\sum_p (i J_1^*(p))(i J_2(p))$ (see Eq. (2.10)) and expanding the exponential:

$$
\exp\left[\sum_p \left(iJ_{1p}^*\right)\left(iJ_{2p}\right)\right] = \prod_p \exp\left[iJ_{1p}^* iJ_{2p}\right]
$$

$$
= \prod_p \sum_{n_p=0} \frac{(iJ_{1p}^*)^{n_p}}{(n_p!)^{1/2}} \frac{(iJ_{2p})^{n_p}}{(n_p!)^{1/2}} \qquad (2.17)
$$

we find that the entire amplitude in Eq. (2.16) can be written in the form

$$
\langle 0_+|0_-\rangle^J = \sum_{\{n\}} \langle 0_+|\{n\}\rangle^{J_1} \langle\{n\}|0_-\rangle^{J_2} \qquad (2.18)
$$

Exercise 2.1: Make sure you understand the notation and the validity of interchange of products and summations in these expressions.

where we have defined

$$
\langle\{n\}|0_-\rangle^J = \langle 0_+|0_-\rangle^J \prod_p \frac{(iJ_p)^{n_p}}{(n_p!)^{1/2}}; \quad \langle 0_+|\{n\}\rangle^J = \langle 0_+|0_-\rangle^J \prod_p \frac{(iJ_p^*)^{n_p}}{(n_p!)^{1/2}}
$$

$$
(2.19)
$$

and the summation in Eq. (2.18) is over *sets* of all integers labeled by momenta. The expression in Eq. (2.18) has a straightforward interpretation: The source J_2 creates a set of particles labeled by the occupation numbers n_p in the momentum state p with amplitude $\langle\{n\}|0_-\rangle$ which are subsequently destroyed by J_1 with amplitude $\langle 0_+|\{n\}\rangle$. The Eq. (2.19) shows that each of these amplitudes is made of the source and sink acting independently to produce multiple particles indicated by the factor $J_p^{n_p}$ etc. The $(n_p!)^{1/2}$ factor in the *amplitude* gives rise to $(n_p!)$ in the probability, taking care of the indistinguishability of created particles. The standard translational invariance argument now shows that the factor $J_p^{n_p}$ gives the state $|\{n\}\rangle$ the four-momentum $P^a = \sum_p n_p p^a$ as expected. Thus, our exponentiated result in Eq. (2.15) correctly describes a set of non-interacting particles produced and destroyed by the source J.

The result Eq. (2.15) can also be obtained more formally along the following lines: Consider a strong source J divided up into a large number of small, uncorrelated and independent weak sources such that $J = \sum_\alpha J_\alpha$ where $\alpha = 1, 2, \cdots N$ with very large N. The independence of each weak component is characterized by the condition

$$
\int d^4x\, d^4x'\, J_\alpha(x) G(x; x') J_\beta(x') = 0 \qquad \text{(for } \alpha \neq \beta) \qquad (2.20)
$$

In that case, the net amplitude $\langle 0_+|0_-\rangle^J$ is given by the product over the individual amplitudes produced by each part of the source:

$$
\langle 0_+|0_-\rangle^J = \prod_\alpha \left(1 - \frac{1}{2}\int d^4x_2\, d^4x_1 J_\alpha(x_2) G(x_2; x_1) J_\alpha(x_1)\right)
$$

$$
= \exp -\frac{1}{2}\sum_\alpha \int d^4x_2\, d^4x_1 J_\alpha(x_2) G(x_2; x_1) J_\alpha(x_1)
$$

$$
= \exp -\frac{1}{2}\sum_{\alpha\beta} \int d^4x_2\, d^4x_1 J_\alpha(x_2) G(x_2; x_1) J_\beta(x_1)
$$

$$
= \exp -\frac{1}{2}\int d^4x_2\, d^4x_1 J(x_2) G(x_2; x_1) J(x_1) \qquad (2.21)
$$

The second equality follows from the fact that for small x we can write $(1 + x) \approx e^x$ in order to convert products into sums in the exponential.

The third equality follows from Eq. (2.20). Thus, even in the case of a strong source we can express the amplitude $\langle 0_+|0_-\rangle$ entirely in terms of $G(x_2;x_1)$ as long as we ignore the interaction between the particles.[6]

[6]Eventually this will translate into what is called free-field theory coupled to an external, *c*-number source, in contrast to an interacting field theory.

2.1.2 Vacuum Instability and the Interaction Energy of the Sources

The above result allows us to extract some interesting conclusions regarding the effect of external sources on the vacuum (that is, no particle) state. In general, the complex number $\langle 0_+|0_-\rangle$ will have a modulus and phase and physical consistency requires that $|\langle 0_+|0_-\rangle|^2$ should contain information about the creation of real, on-shell, particles from the vacuum by the action of the external source. This, in turn, imposes two constraints on this quantity: First, we must have $|\langle 0_+|0_-\rangle|^2 < 1$ for probabilistic interpretation. Second, it can *only* depend on the on-shell behaviour of $J(p)$.

We will now verify that these conditions are met and — in the process — obtain an expression for the mean number of particles produced by the action of the source. From Eq. (2.21) we find that $|\langle 0_+|0_-\rangle|^2$ is given by

$$|\langle 0_+|0_-\rangle|^2 = \exp-\frac{1}{2}\int d^4x_2\, d^4x_1 J(x_2)J(x_1)[G(x_2;x_1) + G^*(x_2;x_1)]$$
(2.22)

where

$$
\begin{aligned}
G(x_2;x_1) + G^*(x_2;x_1) &= \int d\Omega_{\boldsymbol{p}}\left(e^{i\mathbf{p}\cdot\mathbf{x}-i\omega_p|t|} + e^{-i\mathbf{p}\cdot\mathbf{x}+i\omega_p|t|}\right) \qquad (2.23)\\
&= \int d\Omega_{\boldsymbol{p}} e^{i\mathbf{p}\cdot\mathbf{x}}\left(e^{-i\omega_p|t|} + e^{+i\omega_p|t|}\right)\\
&= 2\int d\Omega_{\boldsymbol{p}} e^{i\mathbf{p}\cdot\mathbf{x}}\cos(\omega_p|t|)\\
&= 2\int d\Omega_{\boldsymbol{p}} e^{i\mathbf{p}\cdot\mathbf{x}}\cos(\omega_p t) = \int d\Omega_{\boldsymbol{p}}\left(e^{-ipx} + e^{ipx}\right)
\end{aligned}
$$

In arriving at the second equality, we have done the usual flipping of vector **p** in the second term; in arriving at the fourth equality we have used the fact that the cosine is an even function; and finally we have flipped the vector **p** again in the second term to get the last expression. Therefore

Exercise 2.2: Write Eq. (2.21) in momentum space and use $(z+i\epsilon)^{-1} = \mathcal{P}(z^{-1}) + i\pi\delta(z)$ (where $\mathcal{P}$ denotes the principal value) to get the result in Eq. (2.24) faster.

$$
\begin{aligned}
\frac{1}{2}&\int d^4x_2\, d^4x_1 J(x_2)J(x_1)[G(x_2;x_1) + G^*(x_2;x_1)]\\
&= \frac{1}{2}\int d^4x_2\, d^4x_1 J(x_2)J(x_1)\int d\Omega_{\boldsymbol{p}}\left(e^{-ipx} + e^{ipx}\right)\\
&= \int d^4x_2\, d^4x_1 J(x_2)J(x_1)\int d\Omega_{\boldsymbol{p}}\, e^{-ip(x_2-x_1)}\\
&= \int d\Omega_p |J(\omega_{\boldsymbol{p}},\mathbf{p})|^2 \qquad (2.24)
\end{aligned}
$$

In arriving at the second equality we have used the fact that the product $J(x_1)J(x_2)$ is symmetric in x_1 and x_2. This allows us to retain only the e^{-ipx} term and multiply the result by the factor 2. Substituting Eq. (2.24) in Eq. (2.22), we find:

$$|\langle 0_+|0_-\rangle|^2 = \exp-\int \frac{d^3\mathbf{p}}{(2\pi)^3}\frac{1}{2\omega_{\mathbf{p}}}|J(\omega_{\mathbf{p}},\mathbf{p})|^2 \qquad (2.25)$$

The result (which is gratifyingly less than unity) shows that one can interpret $(1/2\omega_{\mathbf{p}})|J(\omega_{\mathbf{p}}, \mathbf{p})|^2$ as the probability density for the creation of a particle with momentum $\mathbf{p}$. As expected, the result depends only on the on-shell behaviour $J(\omega_{\mathbf{p}}, \mathbf{p})$ of $J(p)$. Since $\omega_{\mathbf{p}} \geq m$, this probability can be non-zero only if the Fourier component $J(\omega_{\mathbf{p}}, \mathbf{p})$ is non-zero, suggesting that a static source cannot produce any particles.[7]

Going back to the notation in which the integration over $d\Omega_{\mathbf{p}}$ is replaced by a summation over momenta on-shell, we can write the same result in an intuitively understandable manner as

$$
\begin{aligned}
|\langle 0_+|0_-\rangle|^2 &= \exp - \int d\Omega_{\mathbf{p}} |J(\omega_p, \mathbf{p})|^2 = \exp - \sum_p (iJ_p)(iJ_p)^* \\
&= \prod_p \exp\left(-|\langle 1_p|0_-\rangle|^2\right)
\end{aligned}
\tag{2.26}
$$

[7]This 'obvious' result is not always true, as we will see in later chapters! It depends on the nature of the coupling.

Thus, $|\langle 0_+|0_-\rangle|^2$ is determined by the probability for production of single particles from the initial vacuum state which is the (only) input we started out with. The production of particles is essentially governed by a Poisson distribution of the form $\lambda^n e^{-\lambda}/n!$ with $\lambda = |\langle 1_p|0_-\rangle|^2$ in each mode. The total probability is given by the product over all the modes. When the argument of the exponent is small, corresponding to a weak source, this result reduces to

Exercise 2.3: Compute the probability for the source to produce n particles with momentum $\mathbf{p}$ and show that it is indeed given by a Poisson distribution.

$$
|\langle 0_+|0_-\rangle|^2 \approx 1 - \sum_p |\langle 1_p|0_-\rangle|^2
\tag{2.27}
$$

which is essentially a statement of probability conservation. Thus everything makes sense. This is the first non-trivial application of the formalism we have developed so far; it gets better.

The above analysis concentrated on the probability $|\langle 0_+|0_-\rangle|^2$ which, of course, loses the information about the phase of $\langle 0_+|0_-\rangle$. As it turns out, the phase also contains valuable information about the physical system. To unravel this, let us consider another extreme example of a source $J(x)$ which is static. Obviously, we do not expect any real particle production to take place in this context and we must have $|\langle 0_+|0_-\rangle|^2 = 1$. What we are interested in is the *phase* of the complex amplitude $\langle 0_+|0_-\rangle$. For a static source,[8] the argument of the exponent occurring in $\langle 0_+|0_-\rangle$ is given by:

[8]More precisely, we are talking about a source which varies very little over a timescale L/c where L is the spatial extent of the source; but we will be a little cavalier about this precise definition and take $J(t, \mathbf{x}) = J(\mathbf{x})$.

$$
\begin{aligned}
-\frac{1}{2} \int d^3\mathbf{x}_1 d^3\mathbf{x}_2 J(\mathbf{x}_1)J(\mathbf{x}_2) &\int_{-\infty}^{\infty} dt_1 \int_{-\infty}^{\infty} dt_2 \int d\Omega_{\mathbf{p}} e^{i\mathbf{p}\cdot\mathbf{x}} e^{-i\omega_p|t_2-t_1|} \\
= -\frac{1}{2} \int d^3\mathbf{x}_1 d^3\mathbf{x}_2 &\int d\Omega_{\mathbf{p}} J(\mathbf{x}_1)J(\mathbf{x}_2) e^{i\mathbf{p}\cdot(\mathbf{x}_2-\mathbf{x}_1)} \\
&\times \int_{-\infty}^{\infty} dt_1 \int_{-\infty}^{\infty} dt_2 e^{-i\omega_p|t_2-t_1|}
\end{aligned}
\tag{2.28}
$$

To do the time integrals, introduce the variables $T = (1/2)(t_1 + t_2), t = t_2 - t_1$ so that $dt_1 dt_2 = dT dt$. The integral over T is of course divergent but we will deal with it later on; the integral over t leads to a factor $(-2i/\omega_p)$ making the argument of the exponent to be:

$$
i \int_{-\infty}^{\infty} dT \int d^3\mathbf{x}_1 d^3\mathbf{x}_2 J(\mathbf{x}_1)J(\mathbf{x}_2) \int \frac{d^3\mathbf{p}}{(2\pi)^3} \frac{1}{2\omega_{\mathbf{p}}^2} e^{i\mathbf{p}\cdot(\mathbf{x}_2-\mathbf{x}_1)} \equiv -i \int_{-\infty}^{\infty} dT \, \mathcal{E}
\tag{2.29}
$$

where

$$\mathcal{E} \equiv \frac{1}{2} \int d^3\mathbf{x} \; d^3\mathbf{y} J(\mathbf{x}) V(\mathbf{x} - \mathbf{y}) J(\mathbf{y}) \tag{2.30}$$

with

$$V(\mathbf{x}) = -\int \frac{d^3\mathbf{p}}{(2\pi)^3} \frac{e^{i\mathbf{p}\cdot\mathbf{x}}}{\mathbf{p}^2 + m^2} = -\frac{e^{-m|\mathbf{x}|}}{4\pi|\mathbf{x}|} \tag{2.31}$$

So we can express Eq. (2.28) as:

$$\langle 0_+|0_-\rangle = \exp\left(-i \int_{-\infty}^{\infty} dT \, \mathcal{E}\right) \tag{2.32}$$

Thus all that the static source has done is to increase the phase of the amplitude $\langle 0_+|0_-\rangle$ at a steady *rate* of $\mathcal{E}$ for infinite duration. This suggests interpreting $\mathcal{E}$ as the energy due to the presence of the static source $J(\mathbf{x})$. (This also agrees with the interpretation based on Eq. (1.40) generalized to the field theory context.)[9]

The expression in Eq. (2.31) shows that $\mathcal{E}$ is negative, indicating that the sources attract each other. Further, the explicit form of $V(\mathbf{x})$ allows us to interpret the resulting force of attraction as due to a Yukawa potential of mass m. It is as though the exchange of particles of mass m between the two sources governed by the amplitude $G(x_2; x_1)$ appears in the Fourier space as an effective potential of interaction between the sources which are capable of exchanging these particles. This is remarkable when you remember that the amplitude $G(x_2; x_1)$ in real space shows no hint of such varied dynamical properties.[10] All that remains is to bring out the fields which are hiding inside these expressions. This task, as you will see, turns out to be somewhat easier now.

2.2   From the Source to the Field

We have found earlier in Sect. 1.5 that $G(x_2; x_1)$ can be reinterpreted in terms of a field $\phi(x)$, which in turn, is made of infinite number of harmonic oscillators each labeled by a momentum vector $\mathbf{p}$. In that interpretation, $G(x_2; x_1)$ was related to the expectation value of the operator $T[\phi(x_2)\phi^\dagger(x_1)]$ in the ground state of all oscillators. Somewhat intriguingly, we also found that the first excited state of the oscillator labeled by $\mathbf{p}$ corresponds to a one-particle state with momentum $\mathbf{p}$.

In Sect. 2.1 we approached the problem from a different perspective by explicitly introducing sources $J(x)$ which can create or destroy a particle. Consistency of the formalism demands that we should be able to recover the field $\phi(x)$, as well as the harmonic oscillators it is made of, from the expression for $G^J(x_2; x_1)$. (This is exactly in the same spirit as our 'discovering' the fields from the expression for $G(x_2; x_1)$ in Sect. 1.5). As you might have guessed, both these goals can indeed be achieved, thereby connecting the discussion in Sect. 2.1 with that in Sect. 1.5.

We will first discuss, in the next section, how one can obtain the field $\phi(x)$ and its action functional from the expression for $G^J(x_2; x_1)$. In Sect. 2.2.3, we will show how to relate this expression to individual harmonic oscillators (which turns out to be a lot easier).

[9]The integral in Eq. (2.31) is essentially the Fourier transform of the propagation amplitude in the static case — obtained by setting $p^0 = 0$ in the Fourier transform of the propagator. So we find that the spatial Fourier transform of the propagation amplitude gives the effective interaction potential of the theory. This turns out to be a fairly general feature.

[10]The fact that "like" sources attract while interacting through a scalar field — which is what our $G(x_2; x_1)$ was reinterpreted as in Sect. 1.5 — is related to its Lorentz transformation properties. We will see later (see Sect. 3.6.2) that if the sources were vector fields $J_a(x)$ interacting via a corresponding vector field, then like sources will repel each other (as in the case of electromagnetism). If the sources were second rank symmetric tensor fields $T_{ab}(x)$ then like sources will again attract; the corresponding field in that case is called *gravity*, but we don't understand it.

2.2.1 Source to Field: Via Functional Fourier Transform

To motivate the connection between $\langle 0_+|0_-\rangle^J$ and the fields in spacetime, we start with the expression for the vacuum persistence amplitude:

$$\langle 0_+|0_-\rangle^J = \exp -\frac{1}{2}\int d^4x d^4y J(x)G(x;y)J(y) \qquad (2.33)$$

which depends both on the nature of the source $J(x)$ as well as the propagation amplitude $G(x;y)$, and ask how we can extract the information contained in $G(x;y)$ without having to bother about the source $J(x)$. To do this we will rewrite Eq. (2.33) in a suggestive form as

$$\langle 0_+|0_-\rangle^J \to \mathcal{G}(J) \equiv \exp -\frac{1}{2}\sum_{x,y} J_x M_{xy} J_y = \exp -\frac{1}{2}(J^T M J) \qquad (2.34)$$

where $M(x-y) \equiv G(x;y)$ is treated as a symmetric matrix with elements denoted by M_{xy} and $J(y)$ is treated as a column vector with elements denoted by J_y. This is a discretised representation of the spacetime continuum in which the integrals over d^4x and d^4y (suitably discretized in a dimensionless form) become summation over the discrete indices. It is now obvious that we can express M_{xy} as the second derivative of $\langle 0_+|0_-\rangle^J$ evaluated at $J=0$:

$$M_{xy} = G(x;y) = -\left.\frac{\partial^2 \langle 0_+|0_-\rangle^J}{\partial J_x \partial J_y}\right|_{J=0} \qquad (2.35)$$

On the other hand, we know from earlier analysis (see Eq. (1.128)) that $G(x;y) = \langle 0|T[\phi(x)\phi(y)]|0\rangle$ where we have taken the field ϕ to be Hermitian for simplicity. This gives the identification

$$-\left.\frac{\partial^2 \langle 0_+|0_-\rangle^J}{\partial J_x \partial J_y}\right|_{J=0} = G(x;y) = \langle 0|T[\phi(x)\phi(y)]|0\rangle \qquad (2.36)$$

Such a result is reminiscent of generating functions related to probability distributions. The right hand side of Eq. (2.36) is the expectation ('mean') value of the time-ordered $\phi(x)\phi(y)$ which could be compared to the correlation function of a stochastic variable. The $\langle 0_+|0_-\rangle^J$ in the left hand side then becomes analogous to the generating function for the probability distribution. Recall that the generating function $\mathcal{G}(\lambda_i)$ (given in terms of a set of variables $\lambda_1, \lambda_2...\lambda_N$) and the corresponding probability distribution $\mathcal{P}(q_i)$ (given in terms of some variables $q_1, q_2...q_N$) are related[11] by:

$$\mathcal{P}(q_i) = \int \left(\prod_{j=1}^{N} d\lambda_j e^{-i\lambda_j q_j}\right) \mathcal{G}(\lambda_j); \quad \mathcal{G}(\lambda_i) = \int \left(\prod_{j=1}^{N} dq_j e^{i\lambda_j q_j}\right) \mathcal{P}(q_j) \qquad (2.37)$$

so that

$$\left.\frac{\partial^2 \mathcal{G}}{\partial \lambda_i \partial \lambda_j}\right|_{\lambda=0} = -\int \left(\prod_{k=1}^{N} dq_k\right) \mathcal{P}(q_k)\, q_i q_j \equiv -\langle q_i q_j\rangle \qquad (2.38)$$

If the generating function is a Gaussian, $\mathcal{G}(\lambda_i) = \exp -(1/2)\sum_{ij}\lambda_i M_{ij}\lambda_j$, then, we will have $-M_{ij} = [\partial^2 \mathcal{G}/\partial \lambda_i \partial \lambda_j]|_{\lambda=0}$ leading to

$$\langle q_i q_j\rangle = M_{ij} \qquad (2.39)$$

[11] Usually one uses the Laplace transform for greater convergence but with a suitable contour in the complex plane, one can also work with the Fourier transforms; in any case, we are only doing some formal manipulation here.

This is completely analogous to what we have in Eq. (2.36) suggesting that we can think of $\langle 0_+|0_-\rangle^J$ as the analogue of a generating function and look for the corresponding probability distribution $\mathcal{P}(q_i)$. Since Eq. (2.36) tells us that the generating function leads to the mean value of $\phi(x)\phi(y)$, it makes sense to think of $\phi(x)$ as the analogue of the q_i. So we should compute the Fourier transform of $\langle 0_+|0_-\rangle^J$ with respect to J by introducing a conjugate variable ϕ_x which is the discretised version of some field $\phi(x)$ in spacetime.

For evaluating this Fourier transform we need a preliminary result. In the case of a *finite* N-dimensional matrix we know that

$$
\begin{aligned}
\mathcal{P}(\phi) &\equiv \int d^D J \exp\left(-\frac{1}{2}J^T M J - i\phi^T J\right) \qquad\qquad (2.40)\\
&= (2\pi)^{D/2}(\det M)^{-1/2}\exp\left(-\frac{1}{2}\phi^T M^{-1}\phi\right) \equiv \mathcal{P}(0)e^{-\frac{1}{2}\phi^T M^{-1}\phi}
\end{aligned}
$$

where M^{-1} is the inverse of the matrix M. We are interested in the case of an infinite dimensional matrix where the indices are continuous variables. In that case we can think of the integration $d^D J$ as a functional integral $\mathcal{D}J$ over the functions $J(x)$. The prefactor $(2\pi)^{D/2}(\det M)^{-1/2}$ can become ill-defined for an infinite continuous matrix, but it is independent of ϕ; so, for the moment[12] we can just call it $\mathcal{P}(0)$. What remains is to give meaning to the expression

$$
\phi^T M^{-1}\phi = \sum_{x,y}\phi_x(M^{-1})_{xy}\phi_y \Rightarrow \int d^4x\, d^4y\, \phi(x)D(x-y)\phi(y) \qquad (2.41)
$$

so that we can write the final result of the functional Fourier transform as:

$$
\begin{aligned}
\mathcal{P}(\phi) &= \mathcal{P}(0)\exp -\frac{1}{2}(\phi_T M^{-1}\phi)\\
&\Rightarrow \mathcal{P}(0)\exp -\frac{1}{2}\int d^4x\, d^4y\, \phi(y)D(x-y)\phi(x) \qquad (2.42)
\end{aligned}
$$

So, in the continuum limit, we need to find some operator $D(x-y)$ which will play the role of the inverse matrix. This requires giving a meaning to the relation $M^{-1}M = 1$ for a matrix with continuous indices by the relation:

$$
\sum_z (M^{-1})_{xz}M_{zy} = \delta_{xy} \Rightarrow \int dz\, D(x-z)M(z-y) = \delta(x-y) \qquad (2.43)
$$

To find the form of $D(x-y)$ from this integral relation, we express the left hand side in Fourier space obtaining

$$
\int \frac{d^4p}{(2\pi)^4}D(p)M(p)e^{-ip(x-y)} = \delta(x-y) \qquad (2.44)
$$

which is satisfied if $D(p) = 1/M(p)$. From Eq. (1.92), we know that $M(p) = i(p^2 - m^2 + i\epsilon)^{-1}$ and hence $D(p) = -i(p^2 - m^2 + i\epsilon)$. In position space, the expression for $D(x-y)$ is given by the Fourier transform:

$$
\begin{aligned}
D(x-y) &= -i\int \frac{d^4p}{(2\pi)^4}\left(p^2 - m^2 + i\epsilon\right)e^{-ip(x-y)} \qquad (2.45)\\
&= i\int \frac{d^4p}{(2\pi)^4}\left(\Box_x + m^2 - i\epsilon\right)e^{-ip(x-y)}\\
&= i(\Box_x + m^2 - i\epsilon)\,\delta(x-y).
\end{aligned}
$$

Exercise 2.4: Prove Eq. (2.40) by diagonalizing the matrix M.

[12]The expression $\mathcal{P}(0) \propto (\det M)^{-1/2}$ actually contains interesting physics which we will soon come back to.

Obviously this is a somewhat singular operator but it leads to meaningful results when substituted into Eq. (2.41). We get

$$\int d^4x \int d^4y \; \phi(x)D(x-y)\phi(y)$$

$$= \; i\int d^4x \int d^4y \; \phi(x)(\Box_x + m^2 - i\epsilon)\delta(x-y)\phi(y)$$

$$= \; i\int d^4x \; \phi(x)(\Box_x + m^2 - i\epsilon)\int d^4y \; \delta(x-y)\phi(y)$$

$$= \; i\int d^4x \; \phi(x)(\Box_x + m^2 - i\epsilon)\phi(x) \qquad (2.46)$$

Substituting back into Eq. (2.42), we get the final result to be

$$\mathcal{P}(\phi) = \mathcal{P}(0)\exp-\frac{i}{2}\int d^4x \; \phi(x)(\Box + m^2 - i\epsilon)\phi(x) \qquad (2.47)$$

This is the first example of functional integration and functional Fourier transforms which we have come across and — as you can guess — many more will follow. All of them require (i) some kind of discretisation followed by (ii) using a result for a finite dimensional integral and finally (iii) taking a limit to the continuum. Most of the time this will involve finding the inverse of either an operator or a symmetric function in two variables acting as a kernel to a quadratic form. The procedure we use in all these cases is the same. The inverses of the operators are defined in momentum space as just reciprocals (of algebraic expressions) and can be Fourier transformed to give suitable results in real space. In this particular example, the relation

$$-i(p^2 - m^2 + i\epsilon)\frac{i}{(p^2 - m^2 + i\epsilon)} = 1 \qquad (2.48)$$

in momentum space translates into the result

$$i(\Box_x + m^2 - i\epsilon)G(x;y) = \delta(x-y) \qquad (2.49)$$

in real space. One can, of course, directly verify in real space that Eq. (2.49) does hold.

The term $\phi(x)\Box\phi(x) = \phi(x)[\partial_a\partial^a\phi(x)]$ occurring in Eq. (2.47) can be expressed in a different form using the result

Exercise 2.5: Prove this result using the expression in Eq. (1.128) for $G(x;y)$.

$$\int_{\mathcal{V}} d^4x \; \phi\Box\phi \;=\; \int_{\mathcal{V}} d^4x \; \partial_a(\phi\partial^a\phi) - \int_{\mathcal{V}} d^4x \; \partial_a\phi\partial^a\phi$$

$$= \; \int_{\partial\mathcal{V}} d^3x \; n_a \, \phi\partial^a\phi - \int_{\mathcal{V}} d^4x \; \partial_a\phi\partial^a\phi \qquad (2.50)$$

where the integration is over a 4-volume $\mathcal{V}$ with a boundary $\partial\mathcal{V}$ which has a normal n_a. If we now assume that, for all the field configurations we are interested in, we can ignore the boundary term,[13] then one can replace $(\partial_a\phi\partial^a\phi)$ by $(-\phi\Box\phi)$. So, an equivalent form for $\mathcal{P}(\phi)$ is given by

$$\mathcal{P}(\phi) = \mathcal{P}(0)\exp\frac{i}{2}\int d^4x \left[\partial_a\phi \; \partial^a\phi - (m^2 - i\epsilon)\phi^2\right] \equiv \mathcal{P}(0)e^{iA[\phi]} \qquad (2.51)$$

where A is given by

$$A = \frac{1}{2}\int d^4x \left(\partial_a\phi\partial^a\phi - m^2\phi^2\right) \qquad (2.52)$$

[13]The cavalier attitude towards boundary terms is a disease you might contract if not immunized early on in your career. Most of the time you can get away with it; but not always, so it is good to be cautious. In this case, the equivalence (in the momentum space) arises from two ways of treating $p^2|\phi(p)|^2$ either as $\phi^*(p)p^2\phi(p)$ and getting $-\phi(x)\Box\phi(x)$ or writing it as $[p\phi(p)][p^*\phi^*(p)]$ leading to $[-i\partial_a\phi][i\partial^a\phi] = \partial_a\phi\partial^a\phi$.

which is essentially the action functional for the field $\phi(x)$ we constructed from the oscillators — see Eq. (1.146) and note that we are now dealing with a *real* $\phi(x)$. (It is understood that m^2 should be replaced by $(m^2 - i\epsilon)$ when required; we will not display it explicitly all the time.) Our entire analysis can therefore be summarized by the equation:

$$\int \mathcal{D}J \, \langle 0_+|0_-\rangle^J \exp -i \left[\int J(x)\phi(x)d^4x \right] = \mathcal{P}(0)e^{iA[\phi]} \qquad (2.53)$$

This relation tells you that the (functional) Fourier transform of the vacuum persistence amplitude leads to the action functional for the field. If we think of $\langle 0_+|0_-\rangle^J$ as a generating function, then $e^{iA[\phi]}$ is the corresponding probability amplitude distribution.[14]

[14]This analogy gets better in the Euclidean space. The factor e^{-A_E} does make sense as a probability for a field configuration in many contexts.

Clearly all the information about our propagation amplitude $G(x_2; x_1)$ is contained in A and, in fact, we can extract it fairly easily. This is because A can be thought of as an action functional for a real scalar field which — in turn — can be decomposed, in Fourier space, into a bunch of harmonic oscillators. Then the action decomposes into a (infinite) sum of actions for individual oscillators as in Eq. (1.145). The dynamics of the system (either classical or quantum) is equivalent to that of an infinite number of harmonic oscillators, each labeled by a vector $\mathbf{p}$ and frequency $\omega_{\mathbf{p}}$. To study the quantum theory, one can introduce the standard creation and annihilation operators for each oscillator. It is now obvious that we will be led to exactly the same field operator which we introduced earlier in Eq. (1.129). The ground state will now be the ground state of all the oscillators and the amplitude $G(x_2; x_1)$ can again be constructed as the ground state expectation value of the time ordered product as done in Eq. (1.128).

In this (more conventional) approach, one will be interested in obtaining $\langle 0_+|0_-\rangle^J$ starting from a theory specified by a given action functional. This is given by the inverse Fourier transform of Eq. (2.53):

$$[\mathcal{P}(0)]^{-1}\langle 0_+|0_-\rangle^J = \int \mathcal{D}\phi \exp \left[iA[\phi] + i \int J(x)\phi(x)d^4x \right] \equiv Z[J] \quad (2.54)$$

where the last equality defines $Z[J]$. Given the fact that $\langle 0_+|0_-\rangle^{J=0} = 1$, we also have the relation[15]

[15]The $\mathcal{P}[(0)]$ stands for $\mathcal{P}[\phi = 0]$ while $Z[(0)]$ is $Z[J = 0]$. But since both are just constants, this notation is unlikely to create any confusion.

$$[\mathcal{P}(0)]^{-1} = Z(0) = \int \mathcal{D}\phi \, e^{iA[\phi]} \qquad (2.55)$$

which essentially gives $(\det M)^{-1/2}$ as a functional integral. (We will come back to this term soon.). From the explicit expression of $\langle 0_+|0_-\rangle^J$ given by Eq. (2.33), it is also obvious that

$$Z(J) = Z(0) \exp -\frac{1}{2} \int d^4x_2 \, d^4x_1 J(x_2)G(x_2; x_1)J(x_1) \qquad (2.56)$$

This result can be expressed in a slightly different form which makes clear some of the previous operations related to functional integrals. In the Eq. (2.54) — which expresses the vacuum persistence amplitude in the presence of a source J in terms of a functional Fourier transform — we will ignore a surface term in A and write $\exp(iA)$ in the form

$$e^{iA} = \exp -\frac{i}{2} \int d^4x \, \phi(x)(\Box + m^2 - i\epsilon)\phi(x) = \exp -\frac{1}{2} \int d^4x \, \phi(x)D\phi(x)$$
$$(2.57)$$

where D is the operator $D \equiv i(\square + m^2 - i\epsilon)$. We also know from Eq. (2.49) that $G(x_2; x_1)$ can be thought of as a coordinate representation of the inverse D^{-1} of this operator. Comparing these results, we see that Eq. (2.56) can be expressed[16] in the form

$$
\begin{aligned}
Z(J) &= \int \mathcal{D}\phi \exp\left[-\frac{1}{2}\int d^4x\ \phi(x)D\phi(x) + i\int J(x)\phi(x)d^4x\right] \\
&= Z(0)\exp\left[-\frac{1}{2}\int d^4x_2\, d^4x_1\, J(x_2)D^{-1}J(x_1)\right] \quad (2.58)
\end{aligned}
$$

[16]Mnemonic: We have $-(1/2)D\phi^2 + iJ\phi = -(1/2)D[(\phi - iJ/D)^2 + J^2/D^2]$ on completing the square. The Gaussian integral then leads to $\exp[-(1/2)D^{-1}J^2]$.

with $D^{-1} \to G(x_2; x_1)$ in coordinate representation. In our approach, we *first* obtained the right hand side of this equation from physical considerations *and then* related it to the left hand side by a functional Fourier transform thereby obtaining Eq. (2.54). Alternatively, we could have started with the expression for A, calculated the functional Fourier transform of $\exp iA$ in the left hand side and thus obtained the right hand side — which is what usual field theory text books do.

In such an approach, we can obtain an explicit expression for $G(x_2; x_1)$ in terms of the *functional* derivatives[17] of $Z(J)$ which generalizes the derivatives with respect to J_x etc which we started out in the discretised version in Eq. (2.35) and Eq. (2.36). On calculating the functional derivatives of the right hand side of Eq. (2.58) we have the result':

$$
\frac{1}{Z(0)}\left(i\frac{\delta}{\delta J(x_1)}\right)\left(i\frac{\delta}{\delta J(x_2)}\right)Z[J]\Big|_{J=0} = G(x_2; x_1) \quad (2.59)
$$

[17]Mathematical supplement Sect. 2.3.1 describes the basic mathematics behind functional differentiation and other related operations. This is a good time to read it up if you are in unfamiliar territory. Fortunately, if you do the most obvious thing you will be usually right as far as functional operations are concerned!

which is in direct analogy with Eq. (2.36). On the other hand, we can also explicitly evaluate the functional derivatives on the left hand side of Eq. (2.58), bringing down two factors of ϕ. Further, we know that the $G(x_2; x_1)$ arising from the right hand side is expressible as the vacuum expectation value of the time ordered product (see Eq. (1.128)) of the scalar fields. This leads to the relation:

$$
\frac{1}{Z(0)}\left(i\frac{\delta}{\delta J(x_2)}\right)\left(i\frac{\delta}{\delta J(x_1)}\right)Z[J]\Big|_{J=0}
$$
$$
= \int \mathcal{D}\phi\, \phi(x_2)\phi(x_1)\exp\left[-\frac{1}{2}\int d^4x\ \phi(x)D\phi(x)\right]
$$
$$
= \langle 0|T[\phi(x_2)\phi(x_1)]|0\rangle \quad (2.60)
$$

This relation will play a key role in our future discussions.[18] The fact that you reproduce a ground state expectation value in Eq. (2.60) shows that these path integrals are similar to the ones discussed in Sect. 1.2.3. We usually assume that the time integral in d^4x is from $-\infty$ to ∞ with t interpreted as $t(1 - i\epsilon)$. Therefore the functional integrals automatically reproduce ground state expectation values.

[18]Note that we need a time ordering operator in the expectation value $\langle 0|T[\phi(x_2)\phi(x_1)]|0\rangle$ expressed in the Heisenberg picture while we don't need to include it explicitly in the path integral expression. The reason should be clear from the way we define the path integral — by time slicing, as discussed in Sect. 1.6.1. Depending on whether $x_1^0 > x_2^0$ or the other way around, the time slicing will put the two operators in the time ordered manner when we evaluate the path integral. Make sure you understand how this happens because the time ordering — which is vital in our discussion — is hidden in the path integral.

2.2.2 Functional Integral Determinant: A First Look at Infinity

It is now time to look closely at the expression for $Z(0)$ which we promised we will come back to. From the discussion around Eq. (2.40), we see that

$$
\mathcal{P}(0) \propto (\det M)^{-1/2} \propto (\det D)^{1/2} = Z[0]^{-1} \Rightarrow Z(0) \propto (\det D)^{-1/2} \quad (2.61)
$$

where we have used Eq. (2.55) and the fact that D and M are inverses of each other. This result expresses a relation which will also play a crucial role in our later discussions:

$$Z(0) = \int \mathcal{D}\phi \, \exp\left[-\frac{1}{2}\int d^4x \, \phi(x)D\phi(x)\right] \propto (\text{Det } D)^{-1/2} \qquad (2.62)$$

It should be obvious from our discussions leading up to Eq. (2.40) that this result is independent of the nature of the operator D. In later chapters we will have to evaluate this for different kinds of operators and we will now develop a few tricks for evaluating the determinant of an infinite matrix or operator.

Let us work in the Euclidean sector to define the path integral in Eq. (2.62). It is then physically meaningful to write $Z(0)$ as $\exp(-A_{\text{eff}})$ where A_{eff} is called the *effective action*.[19] In many situations — like the present one — it is possible to write A_{eff} as a spacetime integral over an effective Lagrangian L_{eff}. In this particular case one can provide a direct interpretation of L_{eff} as vacuum energy density along the following lines: When the Euclidean time integrals go from $-\infty$ to $+\infty$ we know that (see Sect. 1.2.3) the path integral will essentially give $\exp(-TE_0)$ where T is the formally infinite time interval and E_0 is the ground state energy. If the energy *density* in space is $\mathcal{H}$, then we would expect the exponential factor to be $\exp(-TV\mathcal{H})$ where V is the volume of the space. In other words we expect the result

$$Z(0) \propto \exp\left(-\int d^4x_E \, \mathcal{H}\right) = \exp\left(-\int d^4x_E \, L_{\text{eff}}^E\right) \qquad (2.63)$$

where L_{eff}^E is the effective Lagrangian in the Euclidean sector and $\mathcal{H}$ should be some kind of energy density; the last equality arises from the fact that, in the Euclidean sector, $\mathcal{H} = L_{\text{eff}}^E$ in the present context. If we analytically continue to the Lorentzian spacetime by $t_E = it, d^4x_E = id^4x$, this relation becomes:

$$Z(0) \propto \exp\left(-i\int d^4x \, \mathcal{H}\right) = \exp\left(-i\int d^4x \, L_{\text{eff}}^E\right) \equiv \exp\left(i\int d^4x \, L_{\text{eff}}^M\right) \tag{2.64}$$

where L_{eff}^M is the effective Lagrangian in the Lorentzian ('Minkowskian') spacetime. The first proportionality clearly shows that $\mathcal{H}$ leads to the standard energy dependent phase while the last equality defined the effective Lagrangian in the Lorentzian spacetime. From $\mathcal{H} = L_{\text{eff}}^E = -L_{\text{eff}}^M$ we see that the effective Lagrangians in Lorentzian and Euclidean sectors differ by a sign.

Let us work this out explicitly and see what we get. We will work in the Euclidean sector and drop the superscript E from L_{eff}^E for simplicity of notation. We begin by taking the logarithm of Eq. (2.62) and ignoring any unimportant additive constant:

$$\ln Z[0] = -\int d^4x_E \, L_{\text{eff}} = \ln(\det D)^{-\frac{1}{2}} = -\frac{1}{2}\text{Tr } \ln D \qquad (2.65)$$

where we have used the standard relation $\ln \det D = \text{Tr } \ln D$. There are several tricks available to evaluate the logarithm and the trace, which we will now describe.

[19] If you think of $\exp(-A_E[\phi])$ as analogous to a probability, the path integral obtained by summing over all ϕ is analogous to the partition function in statistical mechanics. (In fact, the notation Z is a reminder of this.) We know that the physically useful quantity is the free energy F, related to the partition function by $Z = \exp(-\beta F)$. The effective action A_{eff} bears the same relation to our Z as free energy does to partition function in statistical mechanics.

Exercise 2.6: Prove this.

The simplest procedure is to introduce a (fictitious) Hilbert space with state vectors $|x\rangle$ on which D can operate, and write:

$$-\frac{1}{2}\text{Tr}\,\ln\,D = -\frac{1}{2}\int d^4x_E < x|\ln D|x> = -\int d^4x_E\,L_{\text{eff}} \qquad (2.66)$$

so that $L_{\text{eff}} = (1/2) < x|\ln D|x >$. Going over to the momentum representation in which D is diagonal, and working in n *spatial* dimensions for future convenience, this result becomes:

$$
\begin{aligned}
L_{\text{eff}} &= \frac{1}{2} < x|\ln\,D|x > = \frac{1}{2}\int \frac{d^{n+1}p}{(2\pi)^{n+1}}\,\langle x|p\rangle \ln D(p)\langle p|x\rangle \\
&= \frac{1}{2}\int \frac{d^{n+1}p}{(2\pi)^{n+1}}\,\ln D(p) \qquad (2.67)
\end{aligned}
$$

In our case, the form of the operator D in the Euclidean space is

$$D = (-\partial_a\partial^a + m^2)_E = -\left(\frac{d^2}{dt_E^2} + \nabla^2\right) + m^2 = -\Box_E + m^2 \qquad (2.68)$$

so that $D(p) = p^2 + m^2$, giving L_{eff} to be:

$$L_{\text{eff}} = \frac{1}{2}\int \frac{d^{n+1}p}{(2\pi)^{n+1}}\,\ln(p^2 + m^2) \qquad (2.69)$$

The expression is obviously divergent[20] and we will spend considerable effort in later chapters to interpret such divergent expressions. In the present case, we can transform this (divergent) expression to another (divergent) expression which is more transparent and reveal the fact that L_{eff} is indeed an energy density.

For this purpose, we first differentiate L_{eff} with respect to m^2 and then separate the $d^{n+1}p$ as a n-dimensional integral d^np (which will correspond to real space) and an integration over dp^0 (which is the analytic continuation of the zeroth coordinate to give the Euclidean sector from the Lorentzian sector) and perform the dp^0 integration. These lead to the result (with our standard notation $\omega_{\mathbf{p}}^2 = \mathbf{p}^2 + m^2$):

$$
\begin{aligned}
\frac{\partial L_{\text{eff}}}{\partial m^2} &= \frac{1}{2}\int \frac{d^{n+1}p}{(2\pi)^{n+1}}\frac{1}{(p^2+m^2)} = \frac{1}{2}\int \frac{d^n\mathbf{p}}{(2\pi)^n}\int_{-\infty}^{\infty}\frac{dp^0}{(2\pi)}\frac{1}{((p^0)^2+\omega_{\mathbf{p}}^2)} \\
&= \frac{1}{2}\int \frac{d^n\mathbf{p}}{(2\pi)^n}\frac{1}{2\omega_{\mathbf{p}}} \qquad (2.70)
\end{aligned}
$$

which is gratifyingly Lorentz invariant, though divergent[21]. If we now integrate both sides with respect to m^2 between the limits m_1^2 and m_2^2 we will get the difference between the L_{eff} for two theories with masses m_1 and m_2. We will instead perform one more illegal operation and just write the formal indefinite integral of both sides with respect to m^2. Then we find that L_{eff} has the nice, expected form

$$L_{\text{eff}} = \frac{1}{2}\int \frac{d^n\mathbf{p}}{(2\pi)^n}\sqrt{\mathbf{p}^2+m^2} = \int \frac{d^n\mathbf{p}}{(2\pi)^n}\left(\frac{1}{2}\hbar\omega_{\mathbf{p}}\right) \qquad (2.71)$$

The $Z(0)$ factor is now related to an energy density which is the sum of the zero point energies of an infinite number of harmonic oscillators each having

[20] .... not to mention the fact that we are taking the logarithm of a dimensionful quantity. It is assumed that finally we need to only interpret the difference between two L_{eff}'s, say, with two different masses, which will take care of these issues.

[21] This is the same expression as Eq. (1.105); the right hand side is just $(1/2)\int d\Omega_{\mathbf{p}}$ — the Lorentz invariant, infinite, on-shell volume in momentum space.

a frequency $\omega_{\mathbf{k}}$. Once again we see the fields and their associated harmonic oscillators making their presence felt in our mathematical expressions — this time through a functional determinant.

In more complicated situations, the operator D can depend on other fields or external sources. In that case the ground state energy will be modified compared to the expression obtained above and the difference in the energy can have observable consequences. In the case discussed above, L_{eff} is real, and hence — when we analytically continue to the Lorentzian sector — it will contribute a pure phase to $\langle 0_+ | 0_- \rangle$ of the form:

$$Z(0) \propto \exp\left(-i \int d^4 x\, \mathcal{H}\right); \quad \mathcal{H} = L_{\text{eff}} = \int \frac{d^3 k}{(2\pi)^3} \left(\frac{1}{2}\hbar\omega_{\mathbf{k}}\right) \quad (2.72)$$

which is similar to the situation we encountered in Sect.2.1.2 (see Eq. (2.32)). In some other contexts L_{eff} can pick up an imaginary part which will lead to the vacuum persistence probability becoming less than unity, which will be interpreted as the production of particles by other fields or external sources.

Most part of the above discussion can be repeated for such a more general D which we will encounter in later chapters. With these applications in mind, we will now derive some other useful formulas for the effective Lagrangian and discuss their interpretation.

One standard procedure is to use an integral representation[22] for $\ln D$ and write:

$$
\begin{aligned}
L_{\text{eff}} &= \frac{1}{2} <x| \ln D |x> = -\frac{1}{2} \int_0^\infty \frac{ds}{s} <x| \exp -sD |x> \\
&\equiv -\frac{1}{2} \int_0^\infty \frac{ds}{s} K(x,x;s)
\end{aligned}
\quad (2.73)
$$

where $K(x,x;s)$ is the coincidence limit of the propagation amplitude for a (fictitious) particle from y^i to x^i under the action of a (fictitious) "Hamiltonian" D: That is,

$$K(x,y;s) = <x|e^{-sD}|y> \quad (2.74)$$

Thus our prescription for computing $Z(0)$ is as follows. We first consider a quantum *mechanical* problem for a quantum particle under the influence of a Hamiltonian D and evaluate the path integral propagator $K(x,y;s)$ for this particle. (The coordinates x^i_E describe a 4-dimensional Euclidean space with s denoting time. So this is quantum *mechanics* in four dimensional *space*.) We then integrate[23] the coincidence limit of this propagator $K(x,x;s)$ over (ds/s) to find L_{eff} which, in principle, can have a surviving dependence on x^i. The functional determinant is then given by the expression in Eq. (2.66). We thus have the result

$$
\begin{aligned}
Z(0) &= \int \mathcal{D}\phi \, \exp -A_E(\phi) = \int \mathcal{D}\phi \, \exp -\frac{1}{2} \int d^4 x_E \phi \, D\phi \quad (2.75) \\
&= \exp - \int d^4 x_E \, L_{\text{eff}} = \exp - \int d^4 x_E \left[-\frac{1}{2} \int_0^\infty \frac{ds}{s} \langle x| e^{-sD} |x\rangle \right]
\end{aligned}
$$

We will use this result extensively in the next chapter.[24]

If we are working in the Lorentzian ('Minkowskian') sector, Eq. (2.62) still holds but we define the L_{eff}^M with an extra i-factor as

$$\int \mathcal{D}\phi \exp i \int (-\phi D\phi) d^4 x \propto (\text{Det } D)^{-1/2} = \exp i \int d^4 x \, L_{\text{eff}}^M \quad (2.76)$$

[22]There is a singularity at $s = 0$ in this integral; in all practical calculations we will use this expression to evaluate $\ln(F/F_0)$ for some F_0 when the expression becomes well defined. As we said before, it is also civil to keep arguments of logarithms dimensionless.

[23]This is the same analysis as the one we did to arrive at Eq. (1.104) which provides the physical meaning of the effective Lagrangian. In the conventional descriptions, the $(1/s)$ factor in the effective Lagrangian, Eq. (2.73) appears to be a bit of a mystery. But we saw that if we first define an amplitude $\mathcal{P}(x;E)$ for the closed loop to have energy E and then integrate over all E, the $(1/s)$ factor arises quite naturally. This is the proper interpretation of L_{eff} in terms of closed loops of virtual particles. More importantly, now we *know* the oscillators behind $G(x;x)$.

[24]It is amusing that the functional integral over the *field configurations* is related to the coincidence limit of a path integral involving *particle trajectories*. The coincidence limit of the latter path integral gives the propagation amplitude for the (fictitious?) particle to start from an event x^i and end up at the same event x^i after the lapse of proper time s corresponding to a closed curve in spacetime.

with

$$L_{\text{eff}}^M = \frac{i}{2}\langle x|\ln D|x\rangle = -\frac{i}{2}\int_0^\infty \frac{ds}{s}\langle x|e^{-isD}|x\rangle \qquad (2.77)$$

In this form,[25] it is obvious that D acts as an effective Schrodinger Hamiltonian. (In the specific case of a massive scalar field, the operator in the Euclidean sector is given by Eq. (2.68) which is a Schrodinger Hamiltonian for a quantum mechanical particle of mass $(1/2)$ in 4-space dimensions in a constant potential $(-(1/2)m^2)$. With future applications in mind, we will switch to $(n+1)$ Euclidean space.) The coincidence limit of the propagation amplitude for this quantum mechanical particle can be immediately written down and substituted into Eq. (2.73) and L_{eff} can be evaluated.

We also often need a general approach that allows handling the divergences in the integrals defining L_{eff} in a more systematic manner. This is obtained by again introducing a complete set of momentum eigenstates in the $d \equiv (1+n)$ Euclidean space and interpreting the action of $-\square$ in momentum space as leading to the $+p^2$ factor. Then, we get:

$$L_{\text{eff}} = -\frac{1}{2}\int_0^\infty \frac{ds}{s}\langle x|e^{-s(-\square+m^2)}|x\rangle = -\frac{1}{2}\int_0^\infty \frac{ds}{s}\int\frac{d^d p}{(2\pi)^d}e^{-s(p^2+m^2)} \qquad (2.78)$$

The p-integration can be easily performed leading to the result

$$L_{\text{eff}} = -\frac{1}{2}\int_0^\infty \frac{ds}{s(4\pi s)^{d/2}}e^{-m^2 s} = -\frac{1}{2}\int_0^\infty \frac{d\lambda}{\lambda(2\pi\lambda)^{d/2}}e^{-\frac{1}{2}m^2\lambda} \qquad (2.79)$$

where $\lambda = 2s$. (On the other hand, if we had done the s integral first, interpreting the result as the logarithm, we would have got the result we used earlier in Eq. (2.69)). This integral is divergent for $d > 0$ at the lower limit and needs to be 'regularized' by some prescription.[26]

The simplest procedure will be to introduce a cut-off in the integral at the lower limit and study how the integral behaves as the cut-off tends to zero, in order to isolate the nature of the divergences. For example, the integral for L_{eff} in $n = 3, d = 4$ dimensions is given by

$$L_{\text{eff}} = -\frac{1}{8\pi^2}\int_0^\infty \frac{d\lambda}{\lambda^3}e^{-\frac{1}{2}m^2\lambda} \qquad (2.80)$$

This integral is quadratically divergent near $\lambda = 0$. If we introduce a cut-off at $\lambda = (1/M^2)$ where M is a large mass (energy) scale[27], then the integral is rendered finite and can be evaluated by a standard integration by parts. We then get:

$$
\begin{aligned}
L_{\text{eff}} &= -\frac{1}{8\pi^2}\int_{M^{-2}}^\infty \frac{d\lambda}{\lambda^3}e^{-\frac{1}{2}m^2\lambda} \\
&= -\frac{1}{16\pi^2}\left[M^4 - \frac{m^2 M^2}{2}\right] + \frac{1}{64\pi^2}m^4\ln\frac{m^2}{2M^2} + \frac{1}{64\pi^2}\gamma_E m^4
\end{aligned}
\qquad (2.81)
$$

where γ_E is Euler's constant, defined by

$$\gamma_E \equiv -\int_0^\infty e^{-x}\ln x\, dx \qquad (2.82)$$

To separate out the terms which diverge when $M \to \infty$ from the finite terms, it is useful to introduce another arbitrary but *finite* energy scale

[25]Note that $\langle x, s|y, 0\rangle \equiv \langle x|e^{-isD}|y\rangle$ in the Lorentzian spacetime becomes i times the corresponding Euclidean expression $\langle x, s|y, 0\rangle_E$. That is, $\langle x, s|y, 0\rangle = i\langle x_E, s_E|y_E, 0\rangle$ if we analytically continue by $t = -it_E, s = -is_E$. This is clear from, for e.g., Eq. (1.71), which displays an explicit extra i factor. Hence $L_{\text{eff}}^M \equiv (i/2)\langle x|\ln D|x\rangle = -(1/2)\langle x_E|\ln D_E|x_E\rangle = -L_{\text{eff}}^E$ as it should.

[26]In the specific context of a free field with a constant m^2, there is not much point in worrying about this expression. But we will see later on that we will get the same integral when we study, e.g., the effective potential for a self-interacting scalar field. There, it is vital to know the m^2 dependence of the integral. This is why we are discussing it at some length here.

[27]Note that the $\lambda \to 0$ limit corresponds to short distances and high energies. Such a divergence is called a *UV divergence.*

μ and write $m/M = (m/\mu)(\mu/M)$ in the logarithmic term in Eq. (2.81). Then we can write it as:

$$L_{\text{eff}} = -\frac{1}{16\pi^2}\left[M^4 - \frac{m^2 M^2}{2} + \frac{1}{4}m^4 \ln\frac{M^2}{\mu^2}\right]$$
$$+ \frac{1}{64\pi^2}m^4 \ln\frac{m^2}{2\mu^2} + \frac{1}{64\pi^2}\gamma_E m^4 \qquad (2.83)$$

When we take the $M \to \infty$ limit, the terms in the square bracket diverge while the remaining two terms — proportional to m^4 and $m^4 \ln m^2$ — stay finite. Among the divergent terms, the first term, proportional to M^4, can be ignored since it is just an infinite constant independent of m. So there are two divergences, one which is quadratic (scaling as $M^2 m^2$) and the other which is logarithmic (scaling as $m^4 \log M^2$). In any given theory we need to interpret these divergences or eliminate them in a consistent manner before we can take the limit $M \to \infty$. Even after we do it, the surviving, finite, terms will depend on an arbitrary scale μ we introduced and we need to ensure that no physical effect depends on μ. We shall see in later chapters how this can indeed be done in different contexts.

Another possibility[28] — which we will use extensively in the later chapters — is to evaluate the last integral in $d \equiv n+1$ "dimensions", treating d as just a parameter and analytically continue the result to $d = 4$. In d-dimensions, we have:

$$L_{\text{eff}} = -\frac{1}{2}\int_0^\infty \frac{d\lambda}{\lambda(2\pi\lambda)^{d/2}} e^{-\frac{1}{2}m^2\lambda} = -\frac{1}{2}\frac{\Gamma(-d/2)}{(4\pi)^{d/2}}(m^2)^{d/2} \qquad (2.84)$$

The last expression is an analytic continuation of the integral after a formal evaluation in terms of Γ functions. As it stands, it clearly shows the divergent nature of the integral for $d = 4$ since $\Gamma(-2)$ is divergent. In this case, one can isolate the divergences by writing $2\epsilon \equiv 4 - d$ and taking the $\epsilon \to 0, d \to 4$ limit. It is again convenient to introduce an arbitrary but finite energy scale μ and write $(m^2)^{d/2} = \mu^d(m^2/\mu^2)^{d/2}$. Then the limit gives: [29]

$$L_{\text{eff}} = -\frac{1}{2}\frac{\Gamma(-d/2)}{(4\pi)^{d/2}}\mu^d\left(\frac{m^2}{\mu^2}\right)^{d/2}$$
$$= -\frac{1}{2}\frac{1}{\frac{d}{2}\left(\frac{d}{2}-1\right)}\frac{\Gamma\left(2-\frac{d}{2}\right)}{(4\pi)^{d/2}}\mu^d\left(\frac{m^2}{\mu^2}\right)^{d/2}$$
$$\simeq -\frac{1}{4}\frac{m^4}{(4\pi)^2}\left(\frac{2}{\epsilon} - \gamma_E - \ln\frac{m^2}{4\pi\mu^2} + \frac{3}{2}\right) \qquad (2.85)$$

where γ_E is the Euler's constant defined earlier in Eq. (2.82). We now need to take the limit of $\epsilon \to 0$ in this expression. We see that the finite terms in Eq. (2.85) again scales as m^4 and $m^4 \ln(m^2/4\pi\mu^2)$ which matches with the result obtained in Eq. (2.83). On the other hand, the divergence in Eq. (2.85) has a m^4/ϵ structure which is different from the quadratic and logarithmic divergences seen in Eq. (2.83). That is, the divergent terms in Eq. (2.83) had a m^4 and m^2 dependence while the divergent term Eq. (2.85) has only a m^4 dependence. This is a general feature of dimensional regularization about which we will comment further in a later section.

[28]This procedure is called *dimensional regularization* and is usually more useful in computation, though its physical meaning is more abstract compared to that of evaluating the expression with a cut-off. You never know what you are subtracting out when you use dimensional regularization!

[29]You need to use the fact that $\Gamma(\epsilon) \approx -\epsilon^{-1} + \gamma_E + \mathcal{O}(\epsilon)$ in this limit. Expand both $(m^2/4\pi)^{d/2}$ and the $(d/2)(d/2 - 1)$ factors correct to linear order in ϵ, using $(m^2/4\pi)^{2-\epsilon} \approx m^4[1 - \epsilon\ln(m^2/4\pi)]$ etc.

We thus have several equivalent ways of expressing L_{eff} which can be summarized by the equation

$$
\begin{aligned}
L_{\text{eff}} &= \frac{1}{2} \int \frac{d^n k}{(2\pi)^n} \sqrt{\mathbf{k}^2 + m^2} = \frac{1}{2} \int \frac{d^{n+1} p}{(2\pi)^{n+1}} \ln(p^2 + m^2) \\
&= -\frac{1}{2} \int_0^\infty \frac{d\lambda}{\lambda (2\pi\lambda)^{\frac{1}{2}(n+1)}} e^{-\frac{1}{2} m^2 \lambda} = -\frac{1}{2} \frac{\Gamma(-d/2)}{(4\pi)^{d/2}} (m^2)^{d/2} \quad (2.86)
\end{aligned}
$$

where $d \equiv n + 1$. Each of these expressions are useful in different contexts and allow different interpretations, some of which we will explore in the later chapters.

2.2.3 Source to the Field: via Harmonic Oscillators

It is also obvious that we should also be able to interpret the amplitude $\langle 0_+ | 0_- \rangle^J$ in Eq. (2.21) completely in terms of the harmonic oscillators introduced earlier directly. In particular, if these ideas have to be consistent, then we should be able to identify the no-particle states $|0_\pm\rangle$ with the ground state of our (so far hypothetical!) oscillators and the one-particle state $|1_p\rangle$ generated by the action of $J(x)$ with the first excited state of the **p**-th oscillator. We will briefly describe, for the sake of completeness, how this can be done.

The key idea is to again write $G(x_2; x_1)$ in terms of spatial Fourier components using Eq. (1.117). Doing this in Eq. (2.21) allows us to write it as

$$
\begin{aligned}
\langle 0_+ | 0_- \rangle^J &= \exp -\frac{1}{2} \int d^4 x_2 \, d^4 x_1 \, J(x_2) G(x_2; x_1) J(x_1) \\
&= \prod_{\mathbf{p}} \exp \left[-\frac{1}{2} \int_{-\infty}^\infty dt \int_{-\infty}^\infty dt' \, J_{\mathbf{p}}^*(t) \left(\frac{e^{-i\omega_{\mathbf{p}} |t - t'|}}{2\omega_{\mathbf{p}}} \right) J_{\mathbf{p}}(t') \right] \\
&\equiv \prod_{\mathbf{p}} e^{-R_{\mathbf{p}}} \quad (2.87)
\end{aligned}
$$

The amplitude splits up into a product of amplitudes — one for each Fourier mode $J_{\mathbf{p}}$. What is more, the individual amplitude $e^{-R_{\mathbf{p}}}$ arises in a completely different context in usual quantum mechanics. To see this, let us again consider an oscillator with frequency $\omega_{\mathbf{p}}$ and let a time-dependent external force $J_{\mathbf{p}}(t)$ act on it by adding an interaction term[30]

$$
A_I^{(\mathbf{p})} = \frac{1}{2} \int dt \left[J_{\mathbf{p}}(t) q_{\mathbf{p}}^*(t) + J_{\mathbf{p}}^*(t) q_{\mathbf{p}}(t) \right] \quad (2.88)
$$

to the individual term in the action given by Eq. (1.145). The effect of such an external, time-dependent force is to primarily cause transitions between the levels of the oscillator. In particular, if we start the oscillator in the ground state $|0, -\infty\rangle$ at very early times ($t \to -\infty$, when we assume that $J_{\mathbf{p}}(t)$ is absent), then the oscillator might end up in an excited state at late times ($t \to +\infty$, when again we assume the external force has ceased to act). The amplitude that the oscillator is found in the ground state $|0, +\infty\rangle$ at late times can be worked out using usual quantum mechanics. *This is given by exactly the same expression we found above.* That is,

$$
\begin{aligned}
\langle 0, +\infty | 0, -\infty \rangle &= e^{-R_{\mathbf{p}}} \quad (2.89) \\
&= \exp \left[-\frac{1}{2} \int_{-\infty}^\infty dt \int_{-\infty}^\infty dt' J_{\mathbf{p}}^*(t) \left(\frac{e^{-i\omega_{\mathbf{p}} |t - t'|}}{2\omega_{\mathbf{p}}} \right) J_{\mathbf{p}}(t') \right]
\end{aligned}
$$

[30] Usually $q_{\mathbf{p}}(t)$ will be real/Hermitian for an oscillator and hence we take $J_{\mathbf{p}}(t)$ to be real/Hermitian as well. But we saw in Sect. 1.5 that the oscillators which are used to interpret $G(x_2; x_1)$ are not Hermitian; hence we explicitly include two terms in the action A_I.

Exercise 2.7: Compute this probability amplitude for an oscillator in NRQM (which is most easily done using path integral techniques).

Further, for weak perturbations, it is easy to see from the form of the linear coupling in Eq. (2.88) that the oscillator makes the transition to the first excited state. These facts allow us to identify the state $|0_-\rangle$ with the ground state of all oscillators at early times and $|0_+\rangle$ with the corresponding ground state of the oscillators at late times. Similarly we can identify the state $|1_\mathbf{p}\rangle$ with the first excited state of the $\mathbf{p}$-th oscillator.

It is now clear that the action of source $J(x)$ on the no-particle state can be represented through an interaction term in Eq. (2.88) for each oscillator. If we add up the interaction terms for all the oscillators we get the total action — which can be expressed entirely in terms of spacetime variables $J(x)$ and $\phi(x)$ by an inverse Fourier transform. An elementary calculation gives

$$
\begin{aligned}
A_I &= \sum_\mathbf{p} A_I^{(\mathbf{p})} \Rightarrow \int \frac{d^3\mathbf{p}}{(2\pi)^3} \int dt\, \mathrm{Re}\left[J_\mathbf{p}(t) q_\mathbf{p}^*(t)\right] \\
&= \frac{1}{2} \int d^4x \left[J(x)\phi^\dagger(x) + J^\dagger(x)\phi(x)\right]
\end{aligned}
\tag{2.90}
$$

So, once again, the physics can be described in terms of a field $\phi(x)$ now interacting with an externally specified c-number source.

2.3   Mathematical Supplement

2.3.1   Aspects of Functional Calculus

A functional $F(\phi)$ is a mapping from the space of functions to real or complex numbers. That is, $F(\phi)$ is a rule which allows you to compute a real or complex number for every function $\phi(x)$ which itself could be defined, say, in 4-dimensional spacetime. The functional derivative $(\delta F[\phi]/\delta\phi(x))$ tells you how much the value of the functional (i.e, the value of the computed number) changes if you change $\phi(x)$ by a small amount at x. It can be defined through the relation[31]

$$
\delta F[\phi] = \int d^4x\, \frac{\delta F[\phi]}{\delta\phi(x)}\, \delta\phi(x)
\tag{2.91}
$$

Most of the time we will encounter functional derivatives when we vary an integral involving the function ϕ and the above definition will be the most useful one. One can also define the functional derivatives exactly in analogy with ordinary derivatives. If we change the function by $\delta\phi(x) = \epsilon\delta(x-y)$, then the above equation tells you that

$$
\delta F[\phi] = F\left[\phi + \epsilon\delta(x-y)\right] - F[\phi] = \int d^4x\, \frac{\delta F[\phi]}{\delta\phi(x)}\, \epsilon\delta(x-y) = \epsilon\frac{\delta F}{\delta\phi(y)}
\tag{2.92}
$$

This allows us to define the functional derivative as

$$
\frac{\delta F[\phi]}{\delta\phi(y)} = \lim_{\epsilon\to 0} \frac{F\left[\phi + \epsilon\delta(x-y)\right] - F[\phi]}{\epsilon}
\tag{2.93}
$$

(To avoid ambiguities, we assume that the limit $\epsilon \to 0$ is taken before any other possible limiting operation.) Note that the right hand side is independent of x in spite of the appearance.

---

[31] This also means the dimension of $(\delta F/\delta\phi)$ is *not* the dimension of (F/ϕ) — unlike in the case of partial derivatives — because of the d^4x factor.

From the above definition one can immediately prove that the product rule of differentiation works for functional derivatives. The chain rule also works, leading to

$$\frac{\delta}{\delta\phi(y)}F\left[G[\phi]\right] = \int d^4x \, \frac{\delta F[G]}{\delta G(x)} \frac{\delta G[\phi]}{\delta\phi(y)} \tag{2.94}$$

From the definition in Eq. (2.93), it is easy to obtain the functional derivatives of several simple functionals. For example, we have:

$$F[\phi] = \int d^4x \, (\phi(x))^n \, ; \qquad \frac{\delta F[\phi]}{\delta\phi(y)} = n \, (\phi(y))^{n-1} \tag{2.95}$$

This result can be generalized to any function $g[\phi(x)]$ which has a power series expansion, thereby leading to the result

$$\frac{\delta}{\delta\phi(y)} \int d^4x \, g(\phi(x)) = g'(\phi(y)) \tag{2.96}$$

where prime denotes differentiation with respect to the argument. Next, if x is one dimensional, we can prove:

$$F[\phi] = \int dx \left(\frac{d\phi(x)}{dx}\right)^n ; \qquad \frac{\delta F[\phi]}{\delta\phi(y)} = -n \, \frac{d}{dx} \left(\frac{d\phi}{dx}\right)^{n-1}\Bigg|_y \tag{2.97}$$

which again generalizes to any function $h[\partial_a\phi]$ in the form

$$\frac{\delta}{\delta\phi(y)} \int dx \, h[\partial_a\phi] = -\partial_a \, \frac{\partial h}{\partial\left(\partial_a\phi\right)}\Bigg|_y \tag{2.98}$$

Very often we will have expressions in which F is a functional of ϕ as well as an ordinary function of another variable. Here is one example:

$$F[\phi; y] = \int d^4x' \, K(y, x') \, \phi(x'); \qquad \frac{\delta F[\phi; y]}{\delta\phi(x)} = K(y, x) \tag{2.99}$$

Finally, note that sometimes F is intrinsically a function of x through its functional dependence on ϕ like e.g. when $F(\phi, x) = \nabla\phi(x)$, or, even more simply, $F[\phi; x] = \phi(x)$; then the functional derivatives are:

$$F[\phi; x] = \phi(x); \qquad \frac{\delta F[\phi; x]}{\delta\phi(y)} = \delta(x - y) \tag{2.100}$$

and

$$F(\phi, x) = \nabla\phi(x); \qquad \frac{\delta F_x[\phi]}{\delta\phi(y)} = \nabla_x \, \delta(x - y) \tag{2.101}$$

This is different from the partial derivative $\partial F/\partial\phi$ which is taken usually at constant $\nabla\phi$, leading to $(\partial\nabla\phi/\partial\phi) = 0$. The results in the text can be easily obtained from the above basic results of functional differentiation.

Chapter 3

From Fields to Particles

In the previous chapters we obtained the propagation amplitude $G(x_2; x_1)$ by two different methods. First we computed it by summing over paths in the path integral for a free relativistic particle. Second, we obtained $G(x_2; x_1)$ by studying how an external c-number source creates and destroys particles from the vacuum state. Both approaches, gratifyingly, led to the same expression for $G(x_2; x_1)$ but it was also clear — in both approaches — that one cannot really interpret this quantity in terms of a single, relativistic particle propagating forward in time. These approaches strongly suggested the interpretation of $G(x_2; x_1)$ in terms of a system with an infinite number of degrees of freedom, loosely called a field.

In the first approach, we were led (see Eq. (1.131)) to the result:

$$\sum_{0,\mathbf{x}_1}^{t,\mathbf{x}_2} \exp\left[-im \int_{t_1}^{t_2} dt \sqrt{1 - \mathbf{v}^2}\right] = G(x_2; x_1) = \langle 0|T[\phi(x_2)\phi(x_1)]|0\rangle \quad (3.1)$$

expressing $G(x_2; x_1)$ in terms of the vacuum expectation value of the time ordered product of two ϕs. (For simplicity, we have now assumed that ϕ is Hermitian.) In the second approach, we were again led (see Eq. (2.54)) to the relation

$$\int \mathcal{D}J \, Z(J) \exp\left[-i \int J(x)\phi(x)dx\right] = \exp\left[i \int d^4x \, \frac{1}{2} \left(\partial_a \phi \partial^a \phi - m^2 \phi^2\right)\right] \tag{3.2}$$

which expresses the functional Fourier transform of the vacuum persistence amplitude $[Z(J)/Z(0)] = \langle 0_+|0_-\rangle^J$ in the presence of a source $J(x)$, in terms of a functional of a scalar field given by:

$$A = \frac{1}{2} \int d^4x \, \left(\partial_a \phi \partial^a \phi - m^2 \phi^2\right) \tag{3.3}$$

We also saw that this functional can be thought of as the action functional for the scalar field (see Eq. (1.146)) and could have been obtained as the sum of the action functionals for an infinite number of harmonic oscillators which were used to arrive at the interpretation of Eq. (3.1). Given the fact that $Z(J) \propto \langle 0_+|0_-\rangle^J$ in the left hand side of Eq. (3.2) contains $G(x_2; x_1)$, this relation again links it to the dynamics of a field ϕ. More directly, we found that

$$G(x_2; x_1) = \langle 0|T[\phi(x_2)\phi(x_1)]|0\rangle = \int \mathcal{D}\phi \, \phi(x_2)\phi(x_1) \exp\left(iA[\phi]\right) \tag{3.4}$$

thus expressing the propagation amplitude *entirely* in terms of a path integral average of the fields.

In this chapter we shall reverse the process and obtain the relativistic particle itself as an excitation of a quantum field. In the process, we will re-derive Eq. (3.1) and other relevant results, starting from the dynamics of a field described by the action functionals like the one in Eq. (3.3). Part of this development will be quite straightforward since we have already obtained all these results through one particular route and we only have to reverse the process and make the necessary connections. So we will omit the obvious algebra and instead concentrate on new conceptual issues which crop up.

3.1   Classical Field Theory

Let us assume that someone has given you the functional in Eq. (3.3) and has told you that this should be thought of as the action functional for a real scalar field $\phi(t, \mathbf{x})$. You are asked to develop its quantum dynamics and you may very well ask 'why'. The reason is that one can eventually get relativistic particles and their propagators out of such a study and *more importantly, field theory provides a simple procedure to model interactions between the particles in a Lorentz invariant, quantum mechanical, manner.* This is far from obvious at this stage but you will be able to see it before the end of this chapter.

So we want to study the (classical and) quantum dynamics of fields and the easiest procedure is to start with action principles. We will recall the standard action principle in the case of classical mechanics and then make obvious generalizations[1] to proceed from point mechanics to relativistically invariant field theory.

[1]Some textbooks attempt to "introduce" field theory after some mumbo-jumbo about mattresses or springs connected to balls etc. which, if anything, confuses the issue. The classical dynamics of a field based on an action functional is a *trivial* extension of classical mechanics based on the action principle, unless you make it a point to complicate it.

3.1.1   Action Principle in Classical Mechanics

The starting point in classical mechanics is an action functional defined as an integral (over time) of a Lagrangian:

$$A = \int_{t_1, q_1}^{t_2, q_2} dt \, L(\dot{q}, q) \tag{3.5}$$

The Lagrangian depends on the function $q(t)$ and its time derivative $\dot{q}(t)$ and the action is defined for all functions $q(t)$ which satisfy the boundary conditions $q(t_1) = q_1$, $q(t_2) = q_2$. For each of these functions, the action A will be a pure number; thus the action can be thought of as a *functional* of $q(t)$. Very often, the limits of integration on the integral will not be explicitly indicated or will be reduced to just t_1 and t_2 for notational convenience.

Let us now consider the change in the action when the form of the function $q(t)$ is changed from $q(t)$ to $q(t) + \delta q(t)$. The variation gives

$$
\begin{aligned}
\delta A &= \int_{t_1}^{t_2} dt \left[\frac{\partial L}{\partial q} \delta q + \frac{\partial L}{\partial \dot{q}} \delta \dot{q} \right] \\
&= \int_{t_1}^{t_2} dt \left[\frac{\partial L}{\partial q} - \frac{d}{dt} \left(\frac{\partial L}{\partial \dot{q}} \right) \right] \delta q + \int_{t_1}^{t_2} dt \frac{d}{dt} \left(\frac{\partial L}{\partial \dot{q}} \delta q \right) \\
&= \int_{t_1}^{t_2} dt \left[\frac{\partial L}{\partial q} - \frac{dp}{dt} \right] \delta q + p \delta q \Big|_{t_1}^{t_2}
\end{aligned}
\tag{3.6}
$$

In arriving at the second equality we have used $\delta \dot{q} = (d/dt)\delta q$ and have carried out an integration by parts. In the third equality we have defined the *canonical momentum* by $p \equiv (\partial L/\partial \dot{q})$. Let us first consider those variations δq which preserve the boundary conditions, so that $\delta q = 0$ at $t = t_1$ and $t = t_2$. In that case, the $p\delta q$ term vanishes at the end points. If we now demand that $\delta A = 0$ for arbitrary choices of δq in the range $t_1 < t < t_2$, we arrive at the equation of motion

$$
\left[\frac{\partial L}{\partial q} - \frac{dp}{dt} \right] = 0
\tag{3.7}
$$

It is obvious that two Lagrangians L_1 and $L_1 + (df(q,t)/dt)$, where $f(q,t)$ is an arbitrary function, will lead to the same equations of motion.[2]

The Hamiltonian for the system is defined by $H \equiv p\dot{q} - L$ with the understanding that H is treated as function of p and q (rather than a function of $\dot{q}$ and q). By differentiating H with respect to time and using Eq. (3.7) we see that $(dH/dt) = 0$.

It is also useful to introduce another type of variation which allows us to determine the canonical momentum in terms of the action itself. To do this, we shall treat the action as a function of the upper limits of integration (which we denote simply as q, t rather than as q_2, t_2) but evaluated for a particular solution $q_c(t)$ which satisfies the equation of motion in Eq. (3.7). This makes the action a function of the upper limits of integration; i.e., $A(q,t) = A[q,t;q_c(t)]$. We can then consider the variation in the action when the value of q at the upper limit of integration is changed by δq. In this case, the first term in the third line of Eq. (3.6) vanishes and we get $\delta A = p\delta q$, so that

$$
p = \frac{\partial A}{\partial q}
\tag{3.8}
$$

From the relations

$$
\frac{dA}{dt} = L = \frac{\partial A}{\partial t} + \frac{\partial A}{\partial q} \dot{q} = \frac{\partial A}{\partial l} + p\dot{q}
\tag{3.9}
$$

we find that

$$
H = p\dot{q} - L = -\frac{\partial A}{\partial t}
\tag{3.10}
$$

This description forms the basis for the Hamilton-Jacobi equation in classical mechanics. In this equation we can express $H(p,q)$ in terms of the action by substituting for p by $\partial A/\partial q$ thereby obtaining a partial differential equation for $A(q,t)$ called the *Hamilton-Jacobi* equation:

$$
\frac{\partial A}{\partial t} + H\left(\frac{\partial A}{\partial q}, q \right) = 0
\tag{3.11}
$$

all of which must be familiar to you from your classical mechanics course.

[2]Note that f can depend on q and t but *not* on $\dot{q}$.

Exercise 3.1: Make sure you understand why $\partial A/\partial t$ is not just L but $-H$. That is, why there is an extra $p\dot{q}$ term and a sign flip. Also note that these relations can be written in the "Lorentz invariant" form $p_i = -\partial_i A$ though classical mechanics knows nothing about special relativity. Figure out why.

3.1.2   From Classical Mechanics to Classical Field Theory

Let us now proceed from classical mechanics to classical field theory. As you will see, everything works out as a direct generalization of the ideas from point mechanics as highlighted in Table. 3.1 for ready reference. Let us review some significant new features.

In classical mechanics, the action is expressed as an integral of the Lagrangian over a time coordinate with the measure dt. In relativity, we want to deal with space and time at an equal footing and hence will generalize this to an integral over the *spacetime* coordinates with a Lorentz invariant measure d^4x in any inertial Cartesian system. The Lagrangian L now becomes a scalar so that its integral over d^4x leads to another scalar.

Further, in classical mechanics, the Lagrangian for a closed system depends on the dynamical variable $q(t)$ and its first time derivative $\dot{q}(t) \equiv \partial_0 q$. In relativity, one cannot treat the time coordinate preferentially in a Lorentz invariant manner; the dynamical variable describing a field, $\phi(x^i)$, will depend on both time and space and the Lagrangian will depend on the derivatives of the dynamical variable with respect to both time and space, $\partial_i \phi$. Hence, the action for the field has the generic form

$$A = \int_{\mathcal{V}} d^4x \, L(\partial_a \phi, \phi) \tag{3.12}$$

The integration is over a 4-dimensional region $\mathcal{V}$ in spacetime, the boundary of which will be a 3-dimensional surface,[3] denoted by $\partial \mathcal{V}$.

At this stage, it is convenient to introduce several fields (or field components) into the Lagrangian and deal with all of them at one go.[4] We will denote by $\phi_N(t, \mathbf{x})$ a set of K fields where $N = 1, 2, ...K$. In fact, we can even use the same notation to denote different components of the same field when we are working with a vector field or tensorial fields. In the expressions below we will assume that we sum over all values of N whenever the index is repeated in any given term. The action now becomes

$$A = \int_{\mathcal{V}} d^4x \, L(\partial_a \phi_N, \phi_N) \tag{3.13}$$

To obtain the dynamics of the field, we need to vary the dynamical variable ϕ_N. Performing the variation, we get (in a manner very similar to the corresponding calculation in classical mechanics):

$$\begin{aligned}
\delta A &= \int_{\mathcal{V}} d^4x \, \left(\frac{\partial L}{\partial \phi_N} \delta \phi_N + \frac{\partial L}{\partial(\partial_a \phi_N)} \delta(\partial_a \phi_N) \right) \\
&= \int_{\mathcal{V}} d^4x \, \left[\frac{\partial L}{\partial \phi_N} - \partial_a \left(\frac{\partial L}{\partial(\partial_a \phi_N)} \right) \right] \delta \phi_N \\
&\quad + \int_{\mathcal{V}} d^4x \, \partial_a \left[\frac{\partial L}{\partial(\partial_a \phi_N)} \delta \phi_N \right] \tag{3.14}
\end{aligned}$$

In obtaining the second equality, we have used the fact that $\delta(\partial_a \phi_N) = \partial_a(\delta \phi_N)$ and have performed an integration by parts. The last term in the second line is an integral over a four-divergence, $\partial_a[\pi_N^a \delta \phi_N]$ where $\pi_N^a \equiv [\partial L / \partial(\partial_a \phi_N)]$. This quantity π_N^a generalizes the expression $[\partial L / \partial(\partial_0 q)] = [\partial L / \partial \dot{q}]$ from classical mechanics and can be thought of as the analogue of the canonical momentum. In fact, the 0-th component of this quantity is

[3]This generalizes the notion in classical mechanics in which integration over time is in some interval $t_1 \leq t \leq t_2$. We stress that, in this action, the dynamical variable is the field $\phi(t, \boldsymbol{x})$ and $x^i = (t, \boldsymbol{x})$ are just parameters.

[4]This is identical in spirit to working with a Lagrangian $L(q_i, \dot{q}_i)$ where $i = 1, 2, ...K$ denotes a system with K generalized coordinates.

Property	Mechanics	Field theory
Independent variable	t	$(t, \boldsymbol{x})$
Dependent variable	$q_j(t)$	$\phi_N(t, \boldsymbol{x})$
Definition of Action	$A = \int dt\, L$	$A = \int d^4x\, L$
Form of Lagrangian	$L = L(\partial_0 q_j, q_j)$	$L = L(\partial_i \phi_N, \phi_N)$
Domain of integration	$t \in (t_1, t_2)$ 1-dimensional interval	$x^i \in \mathcal{V}$ 4-dimensional region
Boundary of integration	two points; $t = t_1, t_2$	3D surface $\partial\mathcal{V}$
Canonical Momentum	$p_j = \dfrac{\partial L}{\partial(\partial_0 q_j)}$	$\pi_N^j = \dfrac{\partial L}{\partial(\partial_j \phi_N)}$
General form of the variation	$\delta A = \int_{t_1}^{t_2} dt\, \mathcal{E}[q_j]\delta q_j$ $+ \int_{t_1}^{t_2} dt\, \partial_0(p_j \delta q_j)$	$\delta A = \int_{\mathcal{V}} d^4x\, \mathcal{E}[\phi_N]\delta\phi_N$ $+ \int_{\mathcal{V}} d^4x\, \partial_j(\pi_N^j \delta\phi_N)$
Form of $\mathcal{E}$	$\mathcal{E}[q_j] = \dfrac{\partial L}{\partial q_j} - \partial_0 p_j$	$\mathcal{E}[\phi_N] = \dfrac{\partial L}{\partial\phi_N} - \partial_j \pi_N^j$
Boundary condition to get equations of motion	$\delta q_j = 0$ at the boundary	$\delta\phi_N = 0$ at the boundary
Equations of motion	$\mathcal{E}[q_j] = \dfrac{\partial L}{\partial q_j} - \partial_0 p_j = 0$	$\mathcal{E}[\phi_N] = \dfrac{\partial L}{\partial\phi_N} - \partial_j \pi_N^j = 0$
Form of δA when $\mathcal{E} = 0$ gives momentum	$\delta A = (p_j \delta q_j)\big|_{t_1}^{t_2}$	$\delta A = \int_{\partial\mathcal{V}} d^3\sigma_j(\pi_N^j \delta\phi_N)$
Energy	$E = p_j \partial_0 q_j - L$	$T_b^a = [\pi_N^a \partial_b\phi_N - \delta_b^a L]$

Table 3.1: Comparison of action principles in classical mechanics and field theory

indeed $\pi_N^0 = [\partial L/\partial\dot\phi_N]$, as in classical mechanics. We now need to use the 4-dimensional divergence theorem,

$$\int_{\mathcal{V}} d^4x\ \partial_i v^i = \int_{\partial\mathcal{V}} d^3\sigma_i\ v^i \tag{3.15}$$

where $\mathcal{V}$ is a region of 4-dimensional space bounded by a 3-surface $\partial\mathcal{V}$ and $d^3\sigma_i$ is an element of the 3-surface. You might recall the Fig. 3.1 from your special relativity course which explains the usual context in which we use this result: We take the boundaries of a 4-dimensional region $\mathcal{V}$ to be made of the following components: (i) Two 3-dimensional surfaces at $t = t_1$ and $t = t_2$ both of which are spacelike; the coordinates on these surfaces being the regular spatial coordinates (x, y, z) or (r, θ, ϕ). (ii) One timelike surface at a large spatial distance ($r = R \to \infty$) at all time t in the interval $t_1 < t < t_2$; the coordinates on this 3-dimensional surface could be (t, θ, ϕ). In the right hand side of Eq. (3.15) the integral has to be taken over the surfaces in (i) and (ii). If the vector field v^j vanishes at large spatial distances, then the integral over the surface in (ii) vanishes for $R \to \infty$. For the integral over the surfaces in (i), the volume element can be parametrized as $d\sigma_0 = d^3x$. It follows that

$$\int_{\mathcal{V}} d^4x\ \partial_i v^i = \int_{t_2} d^3\boldsymbol{x}\ v^0 - \int_{t_1} d^3\boldsymbol{x}\ v^0$$

with the minus sign arising from the fact that the normal has to be always treated as outwardly directed. If $\partial_i v^i = 0$ then the integral of v^0 over all space is conserved in time.

In our case, we can convert the total divergence term in the action into a surface term

$$\delta A_{\text{sur}} \equiv \int_{\mathcal{V}} d^4x\ \partial_a(\pi_N^a \delta\phi_N) = \int_{\partial\mathcal{V}} d\sigma_a\ \pi_N^a \delta\phi_N \to \int_{t=\text{cons}} d^3x\ \pi_N^0 \delta\phi_N \tag{3.16}$$

where the last expression is valid if we take the boundary to be the spacelike surfaces defined by $t =$ constant and assume that the surface at spatial infinity does not contribute.[5] We see that, we can obtain sensible dynamical equations for the field ϕ_N by demanding $\delta A = 0$ if we consider variations $\delta\phi_N$ which vanish everywhere on the boundary $\partial\mathcal{V}$. (This is similar to demanding $\delta q = 0$ at $t = t_1$ and $t = t_2$ in classical mechanics.) For such variations, the demand $\delta A = 0$ leads to the field equations

$$\partial_a\left(\frac{\partial L}{\partial(\partial_a \phi_N)}\right) = \partial_a \pi_N^a = \frac{\partial L}{\partial\phi_N} \tag{3.17}$$

Given the form of the Lagrangian, this equation determines the classical dynamics of the field.

We can also consider the change in the action when the field configuration is changed on the boundary $\partial\mathcal{V}$ assuming that the equations of motion are satisfied. In classical mechanics this leads to the relation $p = (\partial A/\partial q)$ where the action is treated as a function of its end points. In our case, Eq. (3.16) can be used to determine different components of π_N^a by choosing different surfaces. In particular, if we take the boundary to be $t =$ constant, we get $\pi_N^0 = (\delta A/\delta\phi_N)$ on the boundary, where $(\delta A/\delta\phi_N)$ is the functional derivative and is defined through the second equality in

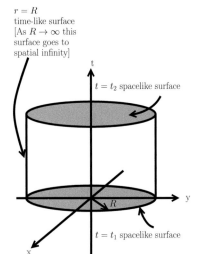

$r = R$
time-like surface
[As $R \to \infty$ this surface goes to spatial infinity]

t

$t = t_2$ spacelike surface

y

R

$t = t_1$ spacelike surface

x

Figure 3.1: Divergence theorem in the spacetime.

[5]In classical mechanics, the corresponding analysis leads to $p\delta q$ at the end points $t = t_1$ and $t = t_2$. Since the integration is over one dimension, the "boundary" in classical mechanics is just two points. In the relativistic case, the integration is over four dimensions leading to a boundary term which is a 3-dimensional integral.

Eq. (3.16). This provides an alternative justification for interpreting π_N^a as the canonical momentum.

Another useful quantity which we will need is the energy momentum tensor of the field. In classical mechanics, if the Lagrangian has no explicit dependence on time t, then one can prove that the energy defined by $E = (p\dot{q}) - L$ is conserved. By analogy, when the relativistic Lagrangian has no explicit dependence on the spacetime coordinate x^i, we expect to obtain a suitable conservation law. In this case, we expect $\dot{q}$ to be replaced by $\partial_i \phi_N$ and p to be replaced by π_N^a. This suggests considering a generalization of $E = (p\dot{q}) - L$ to the second rank tensor

$$T^a_{\ i} \equiv [\pi_N^a(\partial_i \phi_N) - \delta^a_i L] \tag{3.18}$$

Again, we see that the component $T^0_{\ 0} = \pi_N^0 \dot{\phi}_N - L$ is identical in structure to E in classical mechanics, making $T_{00}(= T^0_{\ 0})$ the energy density. To check its conservation, we calculate $\partial_a T^a_{\ i}$ treating L as an implicit function of x^a through ϕ_N and $\partial_i \phi_N$. We get:

$$
\begin{aligned}
\partial_a T^a_{\ i} &= (\partial_i \phi_N)(\partial_a \pi_N^a) + \pi_N^a \partial_a \partial_i \phi_N - \frac{\partial L}{\partial \phi_N} \partial_i \phi_N - \pi_N^a \partial_i \partial_a \phi_N \\
&= (\partial_i \phi_N)\left[\partial_a \pi_N^a - \frac{\partial L}{\partial \phi_N}\right] = 0
\end{aligned}
\tag{3.19}
$$

In arriving at the second equality, we have used $\partial_i \partial_a \phi_N = \partial_a \partial_i \phi_N$ to cancel out a couple of terms and the last equality follows from the equations of motion, Eq. (3.17). It is obvious that the quantity $T^a_{\ i}$ is conserved when the equations of motion are satisfied. (It is also obvious that if we accept the expression for $T^a_{\ i}$ given in Eq. (3.18), then demanding $\partial_a T^a_{\ i} = 0$ will lead to the equations of motion for the scalar field.) Integrating the conservation law $\partial_a T^a_{\ i} = 0$ over a four-volume and using the Gauss theorem in Cartesian coordinates, we find that the quantity

$$P^i = \int d\sigma_k T^{ki} = \int d^3\boldsymbol{x}\, T^{0i} \tag{3.20}$$

is a constant which does not vary with time. We will identify P^i with the total four-momentum of the field.[6] [One difficulty with our definition of the energy momentum tensor through Eq. (3.18) is that — in general — it will not be symmetric; there are alternative definitions which will tackle this problem, which we will describe later.]

In addition to symmetries related to spacetime, the action can have some symmetries related to the transformation of fields, which can also lead to conservation laws. A simple example is provided by a set of K fields ϕ_N with $N = 1, 2, ...K$ with the Lagrangian being invariant under some infinitesimal transformation $\phi_N \to \phi_N + \delta\phi_N$. (That is, the Lagrangian does not change under such a transformation *without* our using the field equations.) This allows us to write

$$0 = \delta L = \frac{\partial L}{\partial \phi_N}\delta\phi_N + \pi_N^j \partial_j(\delta\phi_N) = \left(\frac{\partial L}{\partial \phi_N} - \partial_j \pi_N^j\right)\delta\phi_N + \partial_j(\pi_N^j \delta\phi_N) \tag{3.21}$$

If we now further assume that the field equations hold, then the first term on the right hand side vanishes and we get a conserved current

$$J^i \propto \pi_N^i \delta\phi_N \propto \sum_N \frac{\partial \mathcal{L}}{\partial(\partial_i \phi_N)}\frac{\delta\phi_N}{\delta\alpha} \tag{3.22}$$

[6]The absence of x^i in the Lagrangian is equivalent to the 4-dimensional translational invariance of the Lagrangian. We see that the symmetry of 4-dimensional translational invariance leads to the conservation of both energy and momentum at one go. In classical mechanics, time translation invariance leads to energy conservation and spatial translation invariance leads to momentum conservation, separately. But since Lorentz transformation mixes space and time coordinates, the conservation law in relativity is for the four-momentum.

where the second relation is applicable if the changes $\delta\phi_N$ were induced by the change of some parameter α so that $\delta\phi_N = (\partial\phi_N/\partial\alpha)\delta\alpha$. This result will be needed later.

3.1.3   Real Scalar Field

All these work for any action functional. If we specialize to the action in Eq. (3.3), then the Lagrangian, in explicit $(1+3)$ form, becomes

$$L = \frac{1}{2}[\partial_a\phi\partial^a\phi - m^2\phi^2] = \frac{1}{2}\dot{\phi}^2 - \frac{1}{2}(\nabla\phi)^2 - \frac{1}{2}m^2\phi^2 \tag{3.23}$$

[7]This equation is called the Klein-Gordon equation though it was first written down by Schrodinger, possibly because he anyway has another equation named after him! We will come across several such naming conventions as we go along.

The field equation[7] reduces to

$$\left(\partial^a\partial_a + m^2\right)\phi = 0; \qquad \left(\frac{\partial^2}{\partial t^2} - \nabla^2 + m^2\right)\phi = 0 \tag{3.24}$$

The general solution to the Klein-Gordon equation can be easily determined by Fourier transforming the equation in the spatial variables. We find that

$$\phi(t, \boldsymbol{x}) = \int \frac{d^3\boldsymbol{k}}{(2\pi)^3} \frac{1}{2\omega_{\mathbf{k}}} e^{i\boldsymbol{k}\cdot\boldsymbol{x}} \left(A(\boldsymbol{k})e^{-i\omega_{\boldsymbol{k}}t} + B(-\boldsymbol{k})e^{i\omega_{\boldsymbol{k}}t}\right) \tag{3.25}$$

[8]We write $B(-\boldsymbol{k})$ rather than the natural $B(\boldsymbol{k})$ in Eq. (3.25) for future convenience.

where $\omega_{\boldsymbol{k}} \equiv \sqrt{\boldsymbol{k}^2 + m^2}$ and $A(\boldsymbol{k})$ and $B(\boldsymbol{k})$ are arbitrary scalar functions satisfying $A^*(\boldsymbol{k}) = B(\boldsymbol{k})$ to ensure that ϕ is real.[8] It is clear that the solution represents a superposition of waves with wave vector $\boldsymbol{k}$ and that the frequency of the wave is given by the dispersion relation $\omega_{\boldsymbol{k}}^2 = \boldsymbol{k}^2 + m^2$ corresponding to a four-vector k^i with $k^i k_i = m^2$. Flipping the sign of $\mathbf{k}$ in the second term of Eq. (3.25), we can write it in manifestly Lorentz invariant form:

$$\phi(\mathbf{x}) = \int d\Omega_{\mathbf{k}} \left[A(\mathbf{k})e^{-ikx} + A^*(\mathbf{k})e^{ikx}\right] \tag{3.26}$$

The existence of two arbitrary real functions in the solution (in the form of one complex function $A(\mathbf{k})$) is related to the fact that Klein-Gordon equation is second order in time; to find $\phi(t, \mathbf{x})$ you need to know both $\phi(t_0, \mathbf{x})$ and $\dot{\phi}(t_0, \mathbf{x})$.

Exercise 3.2: Find the solution $\phi(t, \mathbf{x})$ to the Klein-Gordon equation explicitly in terms of the initial conditions $\phi(t_0, \mathbf{x})$ and $\dot{\phi}(t_0, \mathbf{x})$ at an earlier time t_0.

The canonical momentum corresponding to the field ϕ is given by $\pi^a = \partial^a\phi$, and in particular

$$\pi(t, \mathbf{x}) \equiv \frac{\partial L}{\partial\dot{\phi}(t, \mathbf{x})} = \dot{\phi}(t, \mathbf{x}) \tag{3.27}$$

The energy momentum tensor T_{ab} and the energy density $\mathcal{H} = T_{00}$ are given by

$$T^a_b = \partial^a\phi\partial_b\phi - \delta^a_b L; \qquad \mathcal{H} = \frac{1}{2}\left[\dot{\phi}^2 + (\nabla\phi)^2\right] + \frac{1}{2}m^2\phi^2 \tag{3.28}$$

The Hamiltonian density $\mathcal{H}$ can also be found from the expression $\pi\dot{\phi} - L$. This leads to the Hamiltonian when we integrate $\mathcal{H}$ over all space:

$$H = \int d^3\boldsymbol{x}\,\mathcal{H} = \int d^3\boldsymbol{x}\left[\frac{1}{2}\pi^2 + \frac{1}{2}(\nabla\phi)^2 + \frac{1}{2}m^2\phi^2\right] \tag{3.29}$$

The energy momentum tensor in Eq. (3.28) *happens* to be symmetric but this is a special feature of the Lagrangian for a scalar field. The energy momentum tensor as defined in Eq. (3.18), T_{ai}, will involve $\pi_a^A \partial_i \phi_N$ and hence — in general — need not be symmetric; as we said before this is a serious difficulty with the definition in Eq. (3.18). (The problem arises from the fact that angular momentum will not be conserved if T_{ab} is not symmetric; but we will not pause to prove this.) There are tricks to get around this difficulty in an ad-hoc manner but the only proper way to address this question is to ask for the physical meaning of T_{ab}. Since we expect the energy momentum tensor to be the source of gravity, the correct definition of the energy momentum tensor must involve aspects of gravity. We will now briefly describe how such a definition emerges; as a bonus we will also understand how to do field theory in arbitrary curvilinear coordinates and in curved spacetime.

Let us consider the action in Eq. (3.3) and ask how one would rewrite it in an arbitrary curvilinear coordinate system. We know that when one transforms from the Cartesian coordinates $x^\alpha \equiv (x, y, z)$ in 3-dimensional space to, say, spherical polar coordinates $\bar{x}^\alpha \equiv (r, \theta, \phi)$, we need to use a general metric $g_{\alpha\beta}$ rather than $\delta_{\alpha\beta}$ because the line interval changes from $d\ell^2 = \delta_{\alpha\beta} dx^\alpha dx^\beta$ to $d\ell^2 = g_{\alpha\beta} d\bar{x}^\alpha d\bar{x}^\beta$. The dot products between vectors etc. will now involve $g_{\alpha\beta} v^\alpha u^\beta$ instead of $\delta_{\alpha\beta} v^\alpha u^\beta$ in Cartesian coordinates. Moreover, transforming from x^α to $\bar{x}^\alpha$ will also change the volume element of integration d^3x to $d^3\bar{x}\sqrt{g}$ where g is the determinant of $g_{\alpha\beta}$.

Exactly the same sort of changes occur when we go from 3-dimensions to 4-dimensions. If we decide to use some set of curvilinear coordinates $\bar{x}^i$ instead of the standard Cartesian inertial coordinates x^i, the expression for the spacetime interval will change from $ds^2 = \eta_{ik} dx^i dx^k$ to $ds^2 = g_{ik} d\bar{x}^i d\bar{x}^k$ and the volume element[9] will become $d^4x\sqrt{-g}$. The raising and lowering of indices will be now done with g_{ab} rather than η_{ab} with g^{ab} defined as the inverse of the matrix g_{ab}.

With these modifications the action in Eq. (3.3) can be expressed in any curvilinear coordinate system in the form

$$A = \int d^4x \sqrt{-g} \left[\frac{1}{2} g^{ab} \partial_a \phi \partial_b \phi - V(\phi) \right] \tag{3.30}$$

where we have promoted the term $(1/2)m^2\phi^2$ to an arbitrary potential $V(\phi)$ for future convenience. Using the principle of equivalence — which may be broadly interpreted as saying that physics in curvilinear coordinates should be the same as physics in a gravitational field in a sufficiently small region of spacetime — one can make a convincing case[10] that the above expression should also hold in arbitrary curved spacetime described by a line interval $ds^2 = g_{ik} dx^i dx^k$.

With these modifications, we have obtained a *generally covariant* version of the action principle for the scalar field, given by Eq. (3.30), which retains its form under arbitrary coordinate transformations. This action now has two fields in it, $\phi(x)$ and $g_{ab}(x)$. Varying $\phi(x)$ will give the generalization of the field equations in Eq. (3.24), valid in a background gravitational field described by the metric tensor g_{ab}:

$$\frac{1}{\sqrt{-g}} \partial_a \left(\sqrt{-g}\, g^{ab} \partial_b \phi \right) = -V'(\phi) \tag{3.31}$$

where the operator $(|g|)^{-1/2} \partial_a [|g|^{1/2}\, g^{ab} \partial_b]$ is the expression for Laplacian in curvilinear coordinates or curved spacetime.[11] On the other hand, the

[9]We write $\sqrt{-g}$ rather than $\sqrt{g}$ because we know that the determinant will be negative definite in spacetime due to the flip of sign between space and time coordinates.

[10]This is, of course, not a course on GR so we will not spend too much time on these aspects; fortunately, this is all you need to know to get by for the moment. At the first available opportunity, do learn GR if you are not already familiar with it!

[11]It is the same maths as the one involved in writing ∇^2 in spherical polar coordinates, except for some sign changes and promotion from $D = 3$ to $D = 4$.

variation of the action with respect to the metric tensor will have the generic structure

$$\delta A \equiv \frac{1}{2} \int d^4x \sqrt{-g} \, T_{ab} \, \delta g^{ab} \tag{3.32}$$

where the right hand side defines some second rank symmetric tensor T_{ab}. When we study gravity we will start with a total action $A_{tot} = A_g(g) + A_m(\phi, g)$ which is the sum of the action for the gravitational field A_g depending on the metric and the action for the matter source depending on the matter variables (like the scalar field ϕ) and the metric, as in Eq. (3.30). The variation of the total action will give the field equations of gravity with the variation of the matter action with respect to g_{ab} providing the source. This justifies using the symbol T_{ab} and identifying the tensor which appears in Eq. (3.32) as the energy momentum tensor of matter.

It is also easy to see that the conservation law for this tensor T_{ab} arises from the fact that our action functional is now a scalar with respect to arbitrary coordinate transformations $x^a \to \bar{x}^a$. The infinitesimal coordinate transformation would correspond to $\bar{x}^a = x^a + \xi^a(x)$ with the functions $\xi^a(x)$ treated as a first order infinitesimal quantity that vanishes on the boundary $\partial \mathcal{V}$ of the spacetime region $\mathcal{V}$ of interest. The metric tensor in the new coordinates will be

$$g^{ab} = \eta^{ij} \, \partial_i \bar{x}^a \, \partial_j \bar{x}^b = \eta^{ij} \left(\delta_i^a + \partial_i \xi^a \right) \left(\delta_j^b + \partial_j \xi^b \right) \approx \eta^{ab} + \partial^a \xi^b + \partial^b \xi^a \tag{3.33}$$

so that $\delta g^{ab} = \partial^a \xi^b + \partial^b \xi^a \equiv \partial^{(a} \xi^{b)}$. For such a variation of the metric arising from an infinitesimal coordinate transformation, $\delta A = 0$ in the left hand side of Eq. (3.32) when ξ^a vanishes on the boundary.[12] Using $\delta g^{ab} = \partial^{(a} \xi^{b)}$ on the right hand side, we get

$$\begin{aligned} 0 &= \frac{1}{2} \int_{\mathcal{V}} d^4x \sqrt{-g} \, T_{ab} \, \partial^{(a} \xi^{b)} = \int_{\mathcal{V}} d^4x \sqrt{-g} \, T^{ab} \, \partial_a \xi_b \\ &= \int_{\mathcal{V}} d^4x \sqrt{-g} \left[\partial_a (T^{ab} \xi_b) - \xi_b \partial_a T^{ab} \right] \end{aligned} \tag{3.34}$$

In arriving at the second equality we have used the symmetry of T_{ab} and to get the third equality we have done an integration by parts. This leads to a surface term containing $T^{ab}\xi_b$ on the surface $\partial \mathcal{V}$ which vanishes because $\xi_b = 0$ on $\partial \mathcal{V}$. We then get the conservation law $\partial_a T^{ab} = 0$, reinforcing the physical interpretation of T_{ab} as energy momentum tensor of matter.

Usually, we only have to deal with matter Lagrangians in which the dependence on g^{ab} is either linear or quadratic (apart from the $\sqrt{-g}$ factor) and in this case one can write down a general formula for T_{ab}. If the matter action has the form:

$$A = \int d^4x \, \sqrt{-g} \, L = \int d^4x \, \sqrt{-g} \left[-V + U_{ij} g^{ij} + W_{ijkl} g^{ij} g^{kl} \right] \tag{3.35}$$

where V, U_{ij}, W_{ijkl} are functionals of the matter field variables and are independent of the metric, then we can carry out the variation of the action A with respect to g^{ij} in a straight forward manner using the result

$$\delta \sqrt{-g} = \frac{1}{2} \sqrt{-g} \, g^{ab} \delta g_{ab} = -\frac{1}{2} \sqrt{-g} \, g_{ab} \delta g^{ab} \tag{3.36}$$

and obtain:[13]

$$T_{ab} = -\eta_{ab} L + 2U_{ab} + 4W_{abkl} \eta^{kl} \tag{3.37}$$

[12] Sometimes textbooks — even good ones — claim that 'since A is a scalar $\delta A = 0$' which is misleading. Since $\delta L = -\xi^a \partial_a L$ and $\delta \sqrt{g} = -\partial_a(\sqrt{g} \xi^a)$ we actually have $\delta[L\sqrt{(g)}] = -\partial_a[\sqrt{g} L \xi^a]$ which is a total divergence that vanishes if (and only if) ξ^a vanishes on the boundary.

Exercise 3.3: Prove Eq. (3.36) by varying $\det M = \exp(\mathrm{Tr} \ln M)$.

[13] In general the prescription gives the T_{ab} in a general curved spacetime. For us, who pretend to live in flat spacetime, the introduction of g^{ab} is just a trick to get T_{ab} and we can set $g^{ab} = \eta^{ab}$ at the end.

where we have set $g^{ab} = \eta^{ab}$ at the end of the calculation. In the case of the scalar field, $W_{ijkl} = 0, U_{ab} = (1/2)\partial_a\phi\,\partial_b\phi$, thus giving:

$$T_{ab} = \partial_a\phi\,\partial_b\phi - \eta_{ab}\,L \tag{3.38}$$

which tells you that we get back the same expression[14] in Eq. (3.28).

There is another natural conserved quantity for a system described by the equation Eq. (3.24) which will acquire some significance later on. If ϕ_1 and ϕ_2 are any two solutions to Eq. (3.24) (which could, in general, be complex even though we are dealing with a real scalar field), then it is easy to show that the current

$$J_a \propto i\left(\phi_1^*\partial_a\phi_2 - \phi_2\partial_a\phi_1^*\right) \equiv i\phi_1^* \overleftrightarrow{\partial_a} \phi_2 \tag{3.39}$$

is conserved in the sense that $\partial_a J^a = 0$. (The constant i in front of the definition is included to ensure J_a is real.) If we write this conservation law in the form $(\partial\rho/\partial t) + \nabla\cdot\mathbf{J} = 0$ with $J^a = (\rho, \mathbf{J})$, then we can obtain a conserved charge by integrating the density ρ over all space. In this particular context, one can use this fact to define a conserved *scalar product* between any two solutions. This is denoted by the symbol (ϕ_1, ϕ_2) and defined by

$$(\phi_1, \phi_2) \equiv \frac{i}{2}\int d^3x\,\phi_1^* \overleftrightarrow{\partial_t} \phi_2 = \frac{i}{2}\int d^3x\,\left[\phi_1^*\dot\phi_2 - \phi_2\dot\phi_1^*\right] \tag{3.40}$$

One can verify explicitly that (i) $\partial(\phi_1,\phi_2)/\partial t = 0$ and that (ii) ρ need *not* be positive definite for a general solution.

Exercise 3.4: Prove these statements.

3.1.4 Complex Scalar Field

The above analysis generalizes in a straightforward manner for a complex scalar field ϕ — which, in fact, can be thought of as being made of two real scalar fields with $\phi = (\phi_1 + i\phi_2)/\sqrt{2}$ — with the number of degrees of freedom doubling up. The action is now given by

$$A = \int d^4x \left(\partial_a\phi^*\partial^a\phi - m^2\phi^*\phi\right) \tag{3.41}$$

where we treat ϕ and ϕ^* as independent variables while varying the action.[15] This will lead to the Klein-Gordon equations with the same mass m for both ϕ and ϕ^*. The canonical momentum $\pi_{(\phi)}^a$ conjugate to ϕ is $\partial^a\phi^*$, and the one conjugate to ϕ^* is $\partial^a\phi$.

There is an important new feature which emerges from the action in Eq. (3.41). The Lagrangian is now invariant under the transformation

$$\phi(x) \to e^{-i\theta}\phi(x); \qquad \phi^*(x) \to e^{i\theta}\phi^*(x) \tag{3.42}$$

where θ is a constant. This symmetry will lead to a conserved current which actually can be interpreted as the electromagnetic current associated with the complex scalar field. To find the conserved current we note that the above symmetry transformation reduces to $\delta\phi = -i\theta\phi$, $\delta\phi^* = i\theta\phi^*$ in the infinitesimal form, to first order in θ. Using these in Eq. (3.22), taking $\phi_N = (\phi, \phi^*), \pi_N = (\partial^a\phi^*, \partial^a\phi)$, we get the conserved current to be

$$J_m \propto i\left(\phi\partial_m\phi^* - \phi^*\partial_m\phi\right) = -i\phi^* \overleftrightarrow{\partial_m} \phi \tag{3.43}$$

[14] But the effort is not wasted because we will need this procedure to find the correct T_{ab} of the electromagnetic field later on. Besides, this is a physically meaningful way to obtain T_{ab}.

[15] If you do not want to think of ϕ^* as being independent of ϕ, go back to ϕ_1 and ϕ_2 and vary them; it amounts to the same thing.

with the corresponding charge being

$$Q \propto \int d^3x\, j^0 = i \int d^3x\, \phi^* \overleftrightarrow{\partial_0} \phi \qquad (3.44)$$

This is essentially the same conserved scalar product introduced earlier in Eq. (3.40). We will see in the next section that this conserved charge plays an important role.

The non-relativistic limit of this system and, in particular, of the Lagrangian, is also of some interest. It can be obtained by writing[16]

$$\phi = \frac{1}{\sqrt{2m}}\, e^{-imt}\, \psi \qquad (3.45)$$

and expressing the Lagrangian in terms of ψ. We note that $|\nabla\phi|^2 = (|\nabla\psi|^2/2m)$ and, to the leading order,

$$|\dot{\phi}|^2 = \frac{1}{2m}\left(\dot{\psi} - im\psi\right)\left(\dot{\psi}^* + im\psi^*\right) \approx \frac{1}{2}m|\psi|^2 + \frac{i}{2}\left(\psi^*\partial_0\psi - \psi\partial_0\psi^*\right) \qquad (3.46)$$

Hence, in the action in Eq. (3.41), the $m^2|\phi|^2 = (1/2)m|\psi|^2$ term is *cancelled out* by the first term in the above expression! Further, doing an integration by parts over the time variable, we can combine the last two terms in Eq. (3.46). This reduces the Lagrangian in Eq. (3.41) to the form

$$L = i\psi^* \partial_0 \psi - \frac{1}{2m}|\nabla\psi|^2 \qquad (3.47)$$

The variation of this Lagrangian will lead to the Schrodinger equation $i\dot{\psi} = -(1/2m)\nabla^2\psi$ for a free particle of mass m. It is amusing to see that the mass term $m^2|\phi|^2$ does not survive in the non-relativistic limit and that the $(1/2m)$ factor in the Schrodinger equation has a completely different origin in the action.

3.1.5   Vector Potential as a Gauge Field

Once you have a complex scalar field, there is a very natural way of "discovering" the existence of an electromagnetic field. This approach forms the corner-stone of gauge field theories, one of the most successful paradigms in particle physics.

To see how this comes about, let us recall that the Lagrangian for the complex scalar field in Eq. (3.41) is invariant under the transformation $\phi \to e^{-i\theta}\phi \equiv e^{-iq\alpha}\phi$ where we have put $\theta = q\alpha$ with q and α being real constants, for future convenience. On the other hand, if $\alpha = \alpha(x)$ is a function of spacetime coordinates (with q constant), the action is not invariant because the derivatives $\partial_i\phi$ are *not* invariant and will pick up $\partial_i\alpha$ terms.

We now want to modify the action for the complex scalar field such that it is invariant even under the local transformation with $\alpha = \alpha(x)$. One way to do this is to replace the ordinary partial derivative $\partial_i\phi$ by another quantity $D_i\phi$ [called the *gauge covariant derivative*] involving another vector field A_i [called the *gauge field*] and arrange matters such that this invariance is maintained.

Let us postulate an ansatz for the gauge covariant derivative to be $D_i = \partial_i + iqA_i(x)$. We now demand that, when $\phi \to \phi' = e^{-iq\alpha(x)}\phi$,

[16]We have subtracted out the rest energy $mc^2 = m$ by separating the phase e^{-imt}. The normalization is a relic of $(1/\sqrt{2\omega_k})$ which becomes $(1/\sqrt{2m})$ when $c \to \infty$.

Exercise 3.5: This is a bit nontrivial; convince yourself that we are retaining terms of $\mathcal{O}(c^3)$ and $\mathcal{O}(c)$ and ignoring terms of $\mathcal{O}(1/c)$.

Exercise 3.6: (a) Convince yourself that this is not an artificiality due to our normalization in Eq. (3.45) and will arise for arbitrary normalization. (b) Write $\psi = \sqrt{\rho}\exp(i\theta)$ and re-express the Lagrangian in terms of ρ and θ. Show that when we quantise this system, we will be led to $[N, \theta] = i$ where N is the integral of ρ over all space. In many condensed matter applications this leads to a conjugate relationship between the number of particles in a condensate and the phase of the wave function.

Exercise 3.7: Prove this.

the A_i transforms to A'_i such that $D'_i \phi' = e^{-iq\alpha(x)} D_i \phi$. [That is, $D_i \phi$ transforms just like ϕ.] A simple calculation shows that this is achieved if A_i transforms as

$$A'_i = A_i + \partial_i \alpha \qquad (3.48)$$

That is, if we modify the action for complex scalar field to the form[17]

$$
\begin{aligned}
S &= \int d^4 x \left[D_i \phi \, [D^i \phi]^* - m^2 |\phi|^2 \right] \\
&= \int d^4 x \left[(\partial_i + iq A_i(x)) \phi \, (\partial^i - iq A^i(x)) \phi^* - m^2 |\phi|^2 \right]
\end{aligned} \qquad (3.49)
$$

then the action remains invariant under the simultaneous transformations:

$$\phi \to \phi' = e^{-iq\alpha(x)} \phi; \quad A'_i = A_i + \partial_i \alpha \qquad (3.50)$$

[17] We denote the action by S (rather than A) for notational clarity vis-a-vis A_i.

Thus, we can construct an action for ϕ which has *local gauge invariance* if we couple it to a vector field A_i and postulate that both fields change as in Eq. (3.50).

The gauge covariant derivative has the structure $D_j = \partial_j + iq A_j(x)$ (where $A^i = (A^0, \mathbf{A})$), so that $-iD_\alpha = -i\partial_\alpha + q A_\alpha(x) = -i\partial_\alpha - q A^\alpha$. So this is equivalent to replacing $\mathbf{p}$ by $\mathbf{p} - q\mathbf{A}$. This is precisely what happens when we couple a particle of electric charge q to an electromagnetic vector potential $\mathbf{A}$. It makes sense to identify q with the charge of the quanta of the scalar field (which is also consistent with the fact that for the transformation $\phi \to \phi' = e^{-iq\alpha} \phi$ with constant α, the conserved charge will scale with q) and A_j with the standard electromagnetic vector potential. We have discovered electromagnetism.

A couple of points are worth remembering about this Lagrangian. First, note that while the Lagrangian has both $(D_m \phi)^*$ and $(D_m \phi)$ in which the complex conjugation changes $D_m = \partial_m + iq A_m$ to $D_m^* = \partial_m - iq A_m$. The field equations resulting from, say, varying ϕ^* in the Lagrangian, have the form

$$D_m D^m \phi + m^2 \phi = -\left[(i\partial - qA)^2 - m^2 \right] \phi = 0 \qquad (3.51)$$

involving only D_m without the complex conjugate. The variation of ϕ, of course, leads to the complex conjugate version of the above equation with $D_m^* D^{m*}$. Second, one can rewrite the Lagrangian in the form $\phi^* M \phi$ by ignoring a total divergence term just as in the case of the real scalar field. In this case, we use the identity

$$D_j^* \phi^* D^j \phi = \partial_j \left(\phi^* D^j \phi \right) - \phi^* D_j D^j \phi \qquad (3.52)$$

and ignore the first term (which is a total divergence) to obtain

$$L = (D_m \phi)^* D^m \phi - m^2 \phi^* \phi \Rightarrow \phi^* \left[(i\partial - qA)^2 - m^2 \right] \phi \qquad (3.53)$$

which will turn out to be quite useful in future.

The action in Eq. (3.49) now depends on two fields, A_i as well as ϕ. The variation of the action with respect to ϕ and ϕ^* leads to the field equation in Eq. (3.51) and its complex conjugate. But now we can also vary the action with respect to A^i. This is identical in spirit to our varying the metric tensor in the action for the scalar field in Eq. (3.32) to define the energy momentum tensor T_{ab}. Just as we discovered the source of gravity by this procedure, varying A^i in the action for the complex scalar field

Exercise 3.8: Show that $[D_m, D_l]\phi = iq F_{ml} \phi$ where $D_m = \partial_m + iq A_m$ and $F_{lm} = \partial_l A_m - \partial_m A_l$. So gauge covariant derivatives do not commute.

will lead us to the source for electromagnetism, viz., the current vector J^i through

$$\delta S \equiv -\int d^4x \; J_i \delta A^i \qquad (3.54)$$

This is a very general definition for the current J^i to which the vector field A^i couples.[18] Further, just as we obtained the conservation law for T_{ab} by considering the δg_{ab} arising from coordinate transformations in Eq. (3.34), we can obtain the conservation law $\partial_i J^i = 0$ for the current from the gauge invariance of the action A. Under a gauge transformation, $\delta A_i = \partial_i \alpha$ and $\delta S = 0$. Using these in Eq. (3.54), doing an integration by parts we get

$$0 = \int d^4x \; J^i \partial_i \alpha = \int d^4x \left[\partial_i(J^i \alpha) - \alpha \partial_i J^i \right] \qquad (3.55)$$

As usual, the surface contribution resulting from the first term can be made to vanish by a suitable choice of α, thereby leading to the conservation law $\partial_i J^i = 0$.

Using the specific form of the action in Eq. (3.49), and varying A^i, we find the current to be

$$J_i = -iq \left[\phi(D_i\phi)^* - \phi^*(D_i\phi) \right] = -iq \left[\phi\partial_i\phi^* - \phi^*\partial_i\phi \right] - 2q^2|\phi|^2 A_i \quad (3.56)$$

which reduces to current obtained earlier in Eq. (3.43) (except for an overall factor) when $A^i = 0$. In the presence of A^i, we get an extra term $2q^2|\phi|^2 A_i$ which might appear a bit strange. However, it is not only needed for gauge invariance but also plays a crucial role in superconductivity and Higgs phenomena.

There is another, equivalent, way of identifying the current in Eq. (3.56). The action in Eq. (3.49) does not change under the *global* transformation $\phi \to e^{-iq\alpha}\phi$, $\phi^* \to e^{iq\alpha}\phi^*$ even if we do *not* change A_i. For this transformation, $\delta\phi/\delta\alpha = -iq\phi$ and $\delta\phi^*/\delta\alpha = iq\phi^*$. Therefore, Eq. (3.22) tells us that there is a conserved current given by

$$J_m = \sum_N \frac{\partial \mathcal{L}}{\partial(\partial_m \phi_N)} \frac{\delta\phi_N}{\delta\alpha} = -iq(\phi\partial_m\phi^* - \phi^*\partial_m\phi) - 2q^2 A_m \phi^* \phi \quad (3.57)$$

which matches with Eq. (3.56).

More generally, consider any Lagrangian $L(\phi, \partial\phi)$ that remains invariant when you change the dynamical variables by $\delta_\epsilon\phi \equiv \epsilon f(\phi, \partial\phi)$ where ϵ is an infinitesimal *constant* parameter. Given this symmetry, let us ask what happens if we make ϵ a function of coordinates, $\epsilon \to \epsilon(x)$, and look at the form of δL for $\delta_\epsilon\phi \to \epsilon(x)f(\phi, \partial\phi)$. Since $\delta L = 0$ for constant ϵ, we must have δL scaling linearly with $\partial_a\epsilon$ now. Therefore δL must have the form $\delta L = J^a\partial_a\epsilon$ for some J^a. Let us now consider these variations around an extremum of the action; i..e when the equations of motion hold. Then δS has to be zero for *any variation* including the one, $\delta_\epsilon\phi \to \epsilon(x)f(\phi, \partial\phi)$, we are studying. Writing $J^a\partial_a\epsilon = \partial_a(J^a\epsilon) - \epsilon\partial_a J^a$, and noting that the first term — being a total divergence — does not contribute to δS, we find that $\delta S = 0$ implies $\partial_a J^a = 0$. So, whenever you upgrade a global symmetry (constant ϵ) to a local symmetry (ϵ dependent on x) you get a current J^a which will be conserved on-shell; i.e. when the equations of motion hold.[19]

Finally, you need to be aware of the following point which sometimes causes a bit of confusion. The expanded form of the Lagrangian is now given by

$$L = (D_m\phi)^* D^m\phi - m^2\phi^*\phi \qquad (3.58)$$
$$= \partial_m\phi\partial^m\phi^* + iqA_m(\phi\partial^m\phi^* - \phi^*\partial^m\phi) + q^2|\phi|^2 A_m A^m - m^2\phi^*\phi$$

in which there is a coupling term of the form $-A^m K_m$ where $K_m = J_m + q^2|\phi|^2 A_m$. So the current K_m we identify from the coupling term $-A^m K_m$ in the Lagrangian is *not* quite the genuine conserved current J^m which acts as the source for the vector field. The A_m independent parts of K_m and J_m are identical but the terms involving A_m differ by a factor 2. This is because the Lagrangian in Eq. (3.58) has a quadratic term in A_m which, under variation [used to define J_m through Eq. (3.54)], will give an extra factor 2. The current that is gauge invariant and conserved is J_m and not K_m.

3.1.6 Electromagnetic Field

The action resulting from Eq. (3.58) contains the scalar field, its derivatives and the vector field but *not* the derivatives of the vector field. This cannot be the whole story because an external field like $A_i(x)$ can exchange energy and momentum with the scalar field ϕ (thereby violating conservation of energy etc.) without any dynamics for A_i itself. What we lack is a term in the action containing the derivatives of A_i in order to complete the dynamics. Only then will we have a closed system, with conserved energy, momentum, etc. with two interacting fields ϕ and A_i. We will address this issue next.

The term in the action leading to the dynamics of the field A_i will be expressible as an integral over the four-volume d^4x of some scalar Lagrangian (density) $L = L(A^i, \partial_j A^i)$, which could be a function of A^i and its first derivative.[20] We have seen earlier that the equations of motion for the scalar field and the coupling of A_i and ϕ respect the gauge transformation: $A_i' = A_i + \partial_i f$. Therefore, it makes sense to demand that the action for the field should also be invariant under the gauge transformation. Clearly, the "kinetic energy" term for A_i must have the structure $M^{ijkl}\partial_i A_j \partial_k A_l$ where M^{ijkl} is a suitable fourth rank tensor which has to be built out of the only two covariant tensors: η^{ij} and the completely antisymmetric[21] tensor in 4-dimensions ϵ^{ijkl}. One obvious choice, of course, is $M^{ijkl} = \epsilon^{ijkl}$. To determine the terms that can be constructed from η^{ij}, we note that there are essentially three different ways of pairing the indices in $\partial_i A_j \partial_k A_l$ with a product of two η's. One can contract: (1) i with j and k with l; (2) i with k and j with l; (3) i with l and j with k. Adding up all these possible terms with arbitrary coefficients leads to a Lagrangian of the form

$$L = c_1\epsilon^{ijkl}\partial_i A_j \partial_k A_l + c_2(\partial_i A^i)^2 + c_3(\partial_i A_j \partial^i A^j) + c_4(\partial_i A_j \partial^j A^i) \quad (3.59)$$

It is convenient at this stage to express $\partial_i A_j$ as $(1/2)[F_{ij} + S_{ij}]$ where S_{ij} is the symmetric tensor $S_{ij} \equiv \partial_i A_j + \partial_j A_i$ which complements the information contained in the antisymmetric part $F_{ij} = \partial_i A_j - \partial_j A_i$. We will call F_{ij} the electromagnetic field tensor, which is obviously gauge invariant.[22] The last three terms in Eq. (3.59) can be expressed in terms of these two tensors and, using the fact that $F_{ij}S^{ij} = 0$, one can easily work out the Lagrangian

[20]In the case of electromagnetism, experiments show that electromagnetic fields obey the *principle of superposition*, viz. that the field due to two independently specified currents is the sum of the fields produced by each of them in the absence of the other. For this to be true, the differential equations governing the dynamics have to be linear in the field; alternatively, the Lagrangian can be at most quadratic in the field variable. We will assume this is the case in determining the form of the action.

[21]In case you haven't seen this before, revise the relevant part of tensor analysis. We define ϵ^{ijkl} as being completely antisymmetric in all indices with $\epsilon^{0123} = 1$ (so that $\epsilon_{0123} = -1$) in Cartesian coordinates. You should be able to verify that such a structure is Lorentz invariant.

[22]Notation: We take $A^i = (A^0, \mathbf{A}) = (\phi, \mathbf{A})$ with the Cartesian components of $\mathbf{A}$ denoted by a superscript as A^α. Then $A_i = (\phi, -\mathbf{A})$ and $F_{0\alpha} = \partial_0 A_\alpha - \partial_\alpha A_0$. This corresponds to the relation $\mathbf{E} = -\dot{\mathbf{A}} - \nabla\phi$.

to be of the form

$$
\begin{aligned}
4L &= 4c_1\epsilon^{ijkl}\partial_i A_j \partial_k A_l + c_2(S_k^k)^2 \tag{3.60}\\
&\quad + c_3(F_{ik}F^{ik} + S_{ik}S^{ik}) + c_4(S_{ik}S^{ik} - F_{ik}F^{ik})\\
&= 4c_1\epsilon^{ijkl}\partial_i A_j \partial_k A_l + c_2(S_k^k)^2 + (c_3 - c_4)F_{ik}F^{ik} + (c_3 + c_4)S_{ik}S^{ik}
\end{aligned}
$$

In the first term with c_1, we can replace $\partial_i A_j \partial_k A_l$ by $F_{ij}F_{kl}$ since the ϵ_{ijkl} assures that only the antisymmetric part contributes. Therefore, this term is clearly gauge invariant. But we can also write this term as

$$
\epsilon^{ijkl}\partial_i A_j \partial_k A_l = \partial_i[\epsilon^{ijkl}A_j\partial_k A_l] + \epsilon^{ijkl}A_j\partial_i\partial_k A_l = \partial_i[\epsilon^{ijkl}A_j\partial_k A_l] \tag{3.61}
$$

where the second equality arises from the fact that $\partial_i \partial_k A_l$ is symmetric in i and k but ϵ_{ijkl} is completely antisymmetric, making the contraction vanish. The surviving term in Eq. (3.61) is a four-divergence which — when integrated over all space, with the usual assumption that all fields vanish at spatial infinity — will contribute only at the two boundaries at $t = (t_1, t_2)$. So we need to deal[23] with the surface terms of the kind:

$$
\begin{aligned}
\int d^4x \partial_i[\epsilon^{ijkl}A_j\partial_k A_l] &= \int_{t=cons} d^3x[\epsilon^{0jkl}A_j\partial_k A_l]\\
&= \int_{t=cons} d^3x[\epsilon^{0\alpha\beta\mu}A_\alpha\partial_\beta A_\mu] \tag{3.62}
\end{aligned}
$$

[23] As an aside, we mention that the integrand in Eq. (3.62) is essentially $\mathbf{A} \cdot \mathbf{B}$; it is usually called the *Chern-Simons term*.

In the variational principle we use to get the equations of motion, we will consider variations with A_μ fixed at the $t = t_1, t_2$ surfaces. If A_μ is fixed everywhere on this surface, its *spatial* derivatives are also fixed and hence the term in Eq. (3.62) will not vary. Hence this term does not make a contribution.[24]

Among the remaining three terms in Eq. (3.60), the term with $(c_3 - c_4)$ also remains invariant under a gauge transformation but the other two terms do not. Hence, if we want the Lagrangian to be gauge invariant we must have $c_3 = -c_4 = a_1$, say, and $c_2 = 0$. Thus, only the scalar $F_{ik}F^{ik}$ survives as a possible choice for the Lagrangian of the electromagnetic field and the action will be proportional to the integral of this term over d^4x. It is conventional to write this part of the action as

[24] It is important to realize that *only the spatial derivatives* of the field variables remain fixed at the $t = $ constant boundary surface and *not* the time derivatives. In this particular case, we do not have surviving time derivative terms on the boundary and hence ignoring this term is acceptable. When this does not happen, as in the case of action for gravity, you need to work much harder, or be a little dishonest to throw away boundary terms.

$$
S_f = -\frac{1}{4}\int F_{ik}F^{ik}d^4x = \frac{1}{2}\int \left(\mathbf{E}^2 - \mathbf{B}^2\right) d^4x. \tag{3.63}
$$

The magnitude of the constant in front is arbitrary and merely decides the units used for measuring the electromagnetic field. We have taken this prefactor to be a dimensionless numerical factor $(1/4)$, thereby making the field A_i have the dimensions of inverse length in natural units[25]. (From $D_i = \partial_i + iqA_i$, we see that q is dimensionless in natural units. In normal units, q stands for $q/(c\hbar)^{1/2}$.) The sign is chosen so that the term $(\partial\mathbf{A}/\partial t)^2$ has a positive coefficient in S_f. This is needed to ensure that the energy of the plane wave solutions should be positive. The second equality in Eq. (3.63) allows us to identify the Lagrangian for the field as the integral over $d^3\mathbf{x}$ of the quantity $\mathcal{L} \equiv (1/2)(\mathbf{E}^2 - \mathbf{B}^2)$. This is the action for the electromagnetic field which should be familiar to you from a course on electrodynamics.

[25] Another system of units uses a factor $1/16\pi$ instead $1/4$, differing by a 4π factor. We will try to be consistent.

Exercise 3.9: The electromagnetic Lagrangian $F^{ik}F_{ik}$ is clearly invariant under the gauge transformation $A_i \to A_i + \partial_i\alpha$. Show that this symmetry does not lead to a new conserved current. How come we don't get a conserved current from this symmetry?

The above analysis simplifies significantly, if we work with the action expressed in the Fourier space in terms of the Fourier transform of $A_j(x)$

which we will denote by $A_j(k)$. The partial differentiation with respect to coordinates becomes multiplication by k_i in Fourier space. The most general quadratic action in Fourier space must have the form $M^{ij}(k,\eta)A_iA_j$ where M^{ij} is built from k^m and η^{ab}. Since the Lagrangian is quadratic in first derivatives of $A_i(x)$, the expression $M^{ij}(k,\eta)A_iA_j$ must be quadratic in k_i. Hence, $M_{ij}A^iA^j$ must have the form $[\alpha k_i k_j + (\beta k^2 + \gamma)\eta_{ij}]A^iA^j$, where α, β, and γ are constants. The gauge transformation has the form $A_j(k) \rightarrow A_j(k) + k_j f(k)$ in the Fourier space. Demanding that $M_{ij}A^iA^j$ should be invariant under such a transformation — except for the addition of a term that is independent of $A_i(k)$ — we find that $\alpha = -\beta, \gamma = 0$. Therefore, the action in the Fourier space will be

$$S_f \propto \int \frac{d^4k}{(2\pi)^4} A_i[k^ik^j - k^2\eta^{ij}]A_j \propto \int \frac{d^4k}{(2\pi)^4}[k_jA_i - k_iA_j]^2 \quad (3.64)$$

which reduces to Eq. (3.63) in real space.[26]

The field equations for the gauge field can now be obtained by varying $L = -(1/4)F^2 + L_m(\phi, A_m)$ with respect to A_m. Using Eq. (3.54) and ignoring the total divergence term in

$$-(1/4)\delta(F_{lm}F^{lm}) = -(1/2)F_{lm}(2\partial^l\delta A^m) = (\partial^l F_{lm})\delta A^m - \partial^l(F_{lm}\delta A^m) \quad (3.65)$$

we get the field equation to be

$$\partial^l F_{lm} = J_m \quad (3.66)$$

The antisymmetry of F^{lm} again ensures the conservation of J^m. In the case of the source being the complex scalar field, we get

$$\partial_l F^{lm} = J^m = -iq[\phi\partial^m\phi^* - \phi^*\partial^m\phi] - 2q^2|\phi|^2A^m \quad (3.67)$$

Incidentally, you are just one step away from discovering a more general class of (non-Abelian) gauge fields by proceeding along similar lines.[27] Consider an $N-$component field $\phi_A(x)$ with $A = 1, 2, ...N$ which is a vector in some $N-$dimensional linear vector space and a "rotation" in the internal symmetry space is generated through a matrix transformation of the type

$$\phi' = U\phi; \qquad U = \exp(-i\tau_A\alpha^A) \quad (3.68)$$

where ϕ is treated as a column vector, U is a unitary matrix (i.e., $U^\dagger U = 1$), α^A is a set of N parameters and τ_A are N matrices which satisfy the commutation rules $[\tau_A, \tau_B] = iC^J_{AB}\tau_J$ where C^J_{AB} are constants. In the standard context, one considers a theory based on a gauge group, say, $SU(N)$, in which case the C^J_{AB} will be the structure constants of the group. Consider now a field theory for ϕ_A based on a Lagrangian of the form

$$L = -\frac{1}{2}\left[\partial_i\phi^\dagger \, \partial^i\phi + \mu^2\phi^\dagger \phi\right] \quad (3.69)$$

where $\phi^\dagger$ is the Hermitian conjugate. This Lagrangian is clearly invariant under the transformations $\phi' = U\phi$ with *constant* parameters α^A.

Consider now the "local rotations" with $\alpha^A = \alpha^A(x)$ depending on spacetime coordinates. You can easily see that above Lagrangian is no longer invariant under such transformations. As before, introduce an N component *gauge field* $A^K_i(x)$ with $i = 0, 1, 2, 3$ being a spacetime index

[26]Originally, the complex scalar field had a single parameter m which we interpreted as its mass through the relation $\omega^2 = p^2 + m^2$. Writing the phase of the transformation as $e^{iq\alpha}$ and introducing a q we have attributed a second parameter to the complex scalar field. So, can we have different complex scalar fields with arbitrary values for (i) mass m, (ii) charge q? The answer to (i) is "yes" while the answer to (ii) is "no". The electric charges of all the particles which exist in unbound state (which excludes quarks) seem to be in multiples of a quantum of charge. Nobody really knows why; so we don't usually talk about it.

[27]We will not study non-Abelian gauge theories in this book except for this brief discussion.

Exercise 3.10: Prove this.

and $K = 1, 2, 3, ...N$ being an internal space index. Let the $A_i \equiv \tau_J A_i^J$ denote a set of matrices corresponding to the gauge field. You can then show that our Lagrangian can be made invariant under the local rotations if: (i) we replace partial derivatives by *gauge covariant derivatives*

$$D_i = \partial_i + i A_i \tag{3.70}$$

where the first term is multiplied by the unit matrix and (ii) we assume that the gauge field transforms according to the rule:

$$A_i' = U A_i U^{-1} - i U \partial_i U^{-1} \tag{3.71}$$

This is a natural generalization of the Abelian gauge field (which is a fancy name for the electromagnetic field). Continuing as before, you can now introduce a field tensor by

$$F_{ik} = \partial_i A_k - \partial_k A_i + i [A_i, A_k] \tag{3.72}$$

in which each term is interpreted as an $N \times N$ matrix, which is gauge covariant. That is, when A^i changes as in Eq. (3.71), the F_{ik} changes to $U F_{ik} U^{-1}$. This allows us to construct a Lagrangian for the gauge field which is gauge covariant

$$L = -\frac{1}{4} \operatorname{Tr} \left(F_{ik} F^{ik} \right) \tag{3.73}$$

which is the generalization of the electromagnetic Lagrangian.

Exercise 3.11: Prove these claims about Eq. (3.70) – Eq. (3.73) and determine the field equations of the theory.

Let us get back to the electromagnetic field. We can compute the canonical momenta $\pi^{i(j)}$ corresponding to A_j (where the bracket around j denotes which component of the field we are considering) and try to determine the energy momentum tensor by the procedure laid down in Eq. (3.18). Since

$$\pi^{i(j)} = \frac{\partial L}{\partial (\partial_i A_j)} = -\frac{1}{4} F^{ij} \times 4 = -F^{ij} \tag{3.74}$$

we get

$$T_{ab} = -\eta_{ab} L + \pi_a^{(j)} \partial_b A_j = -\eta_{ab} L - F_{aj} \partial_b A^j \tag{3.75}$$

which is *not* symmetric. This is the difficulty we mentioned earlier about using Eq. (3.18). On the other hand, we can easily determine[28] the energy momentum tensor by introducing a g^{ab} in to the electromagnetic action and varying with respect to it. Since $F^2 = g^{ab} g^{cd} F_{ac} F_{bd}$, we now find, on using Eq. (3.35) with $V = U_{ab} = 0, W_{abcd} = -(1/4) F_{ac} F_{bd}$ that:

$$T_{ab} = -\eta_{ab} L - F_{ak} F_b{}^k \tag{3.76}$$

which is clearly symmetric and gauge invariant.

[28]There is a subtlety here. Because $A^i = A_j g^{ij}$ etc. we cannot assume both A_j and A^j are independent of the metric, while varying the action with respect to g_{ab}. Here we assume the A_i and consequently F_{ij} are independent of the metric. This is related, in turn, to the fact that A is actually a one-form and F is a two-form; but if you haven't heard of such things, just accept the result.

The fact that electromagnetic interactions are gauge invariant has some important consequences when we attempt to quantize it. As a preamble to this topic which we will take up in Sect. 3.6 we will describe some of these features at the classical level where they are easier to understand.

Exercise 3.12: Compute $T_{00}, T_{0\alpha}$ and $T_{\alpha\beta}$ in terms of the electric and magnetic fields and identify the resulting expressions with more familiar constructs.

To begin with, gauge invariance demands that A_i should be coupled to a current J^i which is conserved. This is because we want to keep the coupling term to be gauge invariant under the transformation $A_i \rightarrow A_i + \partial_i \alpha$. We have already seen how the current J^i defined through Eq. (3.54) is conserved when the action is gauge invariant.

Second, note that when the Lagrangian in Eq. (3.63) is expressed in terms of the vector potential, we get

$$L = \frac{1}{2}\left[\left(\nabla\phi + \dot{\mathbf{A}}\right)^2 - (\nabla\times\mathbf{A})^2\right] \tag{3.77}$$

which shows that the momentum associated with $\mathbf{A}$ is $(\partial L/\partial\dot{\mathbf{A}}) = -\mathbf{E}$ while the canonical momentum associated with ϕ vanishes.[29] This, in turn, implies that the equation of motion obtained by varying ϕ in the electromagnetic Lagrangian will have the form $(\delta L/\delta\phi) = 0$ which cannot contain any second derivatives with respect to time. Such an equation (called a *constraint equation*) puts a constraint on the initial data which we give for evolving the system forward in time. The existence of such constraint equations usually creates mathematical difficulties in quantizing the system which we will come across when we study the quantum theory of electromagnetic field.

> [29]One can also see this from Eq. (3.74); the momentum conjugate to A^j is $\pi^{0(j)} = -F^{0j}$ which is clearly zero for $j = 0$.

Third, gauge invariance tells you that a description in terms of A_i is actually redundant in the sense that A_is can be modified by gauge transformations without affecting the physical consequences. This raises the question as to what are the true degrees of freedom contained in the set of four functions A_i. This question can be answered differently depending on what exactly we want to do with the electromagnetic field. For our purpose we are interested in determining the *propagating* degrees of freedom contained in the set A_i. These are the degrees of freedom which propagate in spacetime at the speed of light in the case of electromagnetic interactions. It turns out that these degrees of freedom are contained in the transverse part of $\mathbf{A}$; that is, if we write[30] $\mathbf{A} = \mathbf{A}^\perp + \mathbf{A}^\parallel$ where $\nabla\cdot\mathbf{A}^\perp = 0 = \nabla\times\mathbf{A}^\parallel$, then the propagating degrees of freedom are contained in $\mathbf{A}^\perp$. Hence a natural gauge choice which isolates the propagating degrees of freedom will be the two conditions $A_0 = 0$ and $\mathbf{A}^\parallel = 0$. The second condition is, of course, equivalent to imposing $\nabla\cdot\mathbf{A} = 0$. (This is called the *radiation gauge* and we will use it to quantize the electromagnetic field in Sect. 3.6.)

> [30]This separation is possible for any vector field $V^\alpha(t, x^\mu)$. This "theorem" is rather trivial because the spatial Fourier transform $V^\alpha(t, k^\mu)$ can always be written as a sum of a term parallel to $\mathbf{k}$ [given by $(k^\alpha k^\beta/k^2)V_\beta(t, k^\mu)$] and a term perpendicular to $\mathbf{k}$ [given by the rest $V^\alpha(t, k^\mu) - (k^\alpha k^\beta/k^2)V_\beta(t, k^\mu)$]. Obviously, these two terms will Fourier transform back to curl-free and divergence-free vectors.

The simplest way to see that such a gauge condition can indeed be imposed is to proceed through two successive gauge transformations. Let us first consider a gauge transformation from A_i to $\bar{A}_i \equiv A_i + \partial_i\alpha$ such that $\bar{A}_0 = 0$. This is easily achieved by choosing α to be

$$\alpha(t,\mathbf{x}) = -\int dt\, A_0(t,\mathbf{x}) + F(\mathbf{x}) \tag{3.78}$$

Such a transformation will also change the original $\mathbf{A}$ to some other $\bar{\mathbf{A}}(t,\mathbf{x})$. This function, however, satisfies the condition

$$\frac{\partial}{\partial t}(\nabla\cdot\bar{\mathbf{A}}) = \nabla\cdot\left(\frac{\partial\bar{\mathbf{A}}}{\partial t}\right) = -\nabla\cdot\mathbf{E} = 0 \tag{3.79}$$

since we are dealing with source-free electromagnetic fields. This result shows that $\nabla\cdot\bar{\mathbf{A}}$ is independent of time. We now make another gauge transformation with a gauge function f changing $\bar{\mathbf{A}}$ to $\mathbf{A}' = \bar{\mathbf{A}} + \nabla f$ such that $\nabla\cdot\mathbf{A}' = 0$. This leads to the condition $\nabla^2 f = -\nabla\cdot\bar{\mathbf{A}}$ with the solution:

$$f(\mathbf{x}) = \frac{1}{4\pi}\int d^3\mathbf{y}\,\frac{1}{|\mathbf{x}-\mathbf{y}|}(\nabla\cdot\mathbf{A})(\mathbf{y}) \tag{3.80}$$

Since $\nabla\cdot\bar{\mathbf{A}}$ is independent of t, the gauge function f is also independent of t and hence this transformation does not change the value $\bar{A}_0 = 0$ which we had already achieved. Thus with these two transformations, we have successfully gone to a gauge with $A_0 = 0, \nabla \cdot \mathbf{A} = 0$.

Of the two conditions which define the Coulomb gauge, viz., $\nabla \cdot \mathbf{A} = 0$ and $\phi = A_0 = 0$, the second condition cannot be imposed in the presence of external sources because it will lead to trouble with $\nabla \cdot \mathbf{E} = 4\pi\rho$ condition. This Coulomb's law, however, has no dynamic content because it has no time derivative. Therefore one would expect that the Hamiltonian, for the electromagnetic field interacting with the source, should decouple into one involving transverse components of $\mathbf{A}$ and the one involving the Coulomb part. To see how this comes about, let us decompose the electric field into a transverse (divergence-free) $\mathbf{E}^{\perp}$ and longitudinal (curl-free) $\mathbf{E}^{\parallel}$ components given by $\mathbf{E}^{\perp} = -\partial_0 \mathbf{A}$, $\mathbf{E}^{\parallel} = -\nabla A_0$ (with $\nabla \cdot \mathbf{A} = 0$). In the free part of the electromagnetic Lagrangian, the E^2 term will pick up the squares of these two parts; the cross term

$$\mathbf{E}^{\parallel} \cdot \mathbf{E}^{\perp} = -(\nabla A_0) \cdot \mathbf{E}^{\perp} = -\nabla \cdot (A_0 \mathbf{E}^{\perp}) + A_0 \nabla \cdot \mathbf{E}^{\perp} = -\nabla \cdot (A_0 \mathbf{E}^{\perp}) \quad (3.81)$$

is a total divergence and can be discarded. The canonical momentum corresponding to $\mathbf{A}$ (which, of course, is transverse since $\nabla \cdot \mathbf{A} = 0$) is given by $\boldsymbol{\pi}^{\perp} = \mathbf{E}^{\perp}$. Using this, you can compute the corresponding Hamiltonian to be

$$\mathcal{H}_{\mathrm{em}} = \frac{1}{2} \left(\mathbf{E}^{\perp 2} + \mathbf{B}^2 \right) - \frac{1}{2} \mathbf{E}^{\parallel 2} = \mathcal{H}_0 + \mathcal{H}^{\parallel} \quad (3.82)$$

The part $\mathcal{H}^{\parallel}$ combines in a natural manner with the interaction Hamiltonian $J_i A^i$ which couples the vector potential to the external current. Writing

$$\mathcal{H} = \left(\mathcal{H}_0 + \mathcal{H}^{\parallel} \right) + \mathcal{H}_{\mathrm{int}} = \mathcal{H}_0 + \mathcal{H}'_{\mathrm{int}} \quad (3.83)$$

where

$$\mathcal{H}'_{\mathrm{int}} = \mathcal{H}^{\parallel} + \mathcal{H}_{\mathrm{int}} = \mathcal{H}^{\parallel} - L_{\mathrm{int}} \quad (3.84)$$

we can express the Hamiltonian arising from integrating the interaction term as

$$\begin{aligned} H'_{\mathrm{int}} &= \int d^3 \boldsymbol{x} \left(-\frac{1}{2} \nabla A_0 \cdot \nabla A_0 + J_0 A^0 - \mathbf{J} \cdot \mathbf{A} \right) \\ &= \int d^3 \boldsymbol{x} \left(\frac{1}{2} A_0 \nabla^2 A_0 + J_0 A^0 - \mathbf{J} \cdot \mathbf{A} \right) \end{aligned} \quad (3.85)$$

where we have done one integration by parts and discarded the surface term. Using the Poisson equation $\nabla^2 A_0 = -J_0$, the first two terms combine to give the Coulomb interaction energy:

$$H_{\mathrm{coul}} = \int d^3 \boldsymbol{x} \frac{1}{2} J_0 A^0 = \frac{1}{2} \int d^3 \boldsymbol{x}\, d^3 \boldsymbol{x}'\, J_0(\mathbf{x}, t) \frac{1}{4\pi|\mathbf{x} - \mathbf{x}'|} J_0(\mathbf{x}', t) \quad (3.86)$$

leaving the remaining interaction term as[31]

$$H'_{\mathrm{int}} = -\int d^3 x\, \mathbf{J}(\mathbf{x}, t) \cdot \mathbf{A}(\mathbf{x}, t) \quad (3.87)$$

which only couples $\mathbf{A}$ to the transverse part of $\mathbf{J}$. Therefore, after removing the non-dynamical interaction energy term H_{coul}, we are essentially left with the transverse part of $(E^2 + B^2)$ and $\mathbf{J} \cdot \mathbf{A}$ which are the physical degrees of freedom one would deal with in the quantum theory.

[31] In general, you cannot use the equations of motion in the Hamiltonian to change its form. Here we could do this only because the equation $\nabla^2 A_0 = -J_0$ is not a dynamical equation — i.e.., it has no time derivatives — but a constraint equation which relates A_0 at some time t to J_0 at the same time t. Since there are no propagating degrees of freedom, we can eliminate A_0 from the Hamiltonian by this process.

3.2 Aside: Spontaneous Symmetry Breaking

A real scalar field described by the action in Eq. (3.30) with $V(\phi) = (1/2)m^2\phi^2$ describes excitations with the dispersion relation $\omega^2 = \mathbf{k}^2 + m^2$ which can be thought of as relativistic free particles with mass m. We have already seen that such a simple system has an equivalent description in terms of $G(x_2; x_1)$ obtained either from a path integral or from studying the action of sources and sinks on the vacuum. The scalar field theory comes alive only when we go beyond the free particles and introduce interactions, requiring a $V(\phi)$ which is more general than a simple quadratic.

In classical field theory one could have introduced any $V(\phi)$ (except possibly for the condition that it should be bounded from below) but we will see later that meaningful quantum field theories exist only for very special choices of $V(\phi)$. One such example, of considerable importance, is a $V(\phi)$ which has both quadratic and quartic terms with the structure $V(\phi) = \alpha\phi^2 + \beta\phi^4$. We definitely need $\beta > 0$ for the potential to be bounded from below at large ϕ. But we now have the possibility of choosing α with either sign. When $\alpha > 0$ and $\beta > 0$, one can think of the theory as describing particles with mass $m = \sqrt{2\alpha}$ which are interacting through the potential $\beta\phi^4$. We will study this extensively in Chapter 4 as a prototype of an interacting field theory.

But when $\alpha < 0$ and $\beta > 0$, we are led to a completely different kind of theory which exhibits one of the most beautiful and unifying phenomena in physics called *spontaneous symmetry breaking*, albeit in the simplest form. Obviously we cannot now think of the $\alpha\phi^2$ term as describing the mass of the scalar field since α has the wrong sign. In this case, by adding a suitable constant to the potential, we can re-express it in the form

$$V(\phi) = \frac{1}{4}\lambda \left[\phi^2 - v^2\right]^2 \tag{3.88}$$

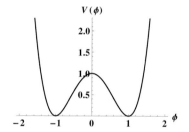

$$V(\phi)$$

Figure 3.2: The shape of the potential $V(\phi)$ which produces spontaneous symmetry breaking.

which is shown in Fig. 3.2. The potential has two degenerate[32] minima at $\phi = \pm v$ and a maximum at $\phi = 0$. This implies that the ground state with, say, $\langle\phi\rangle = \pm v$ will break the symmetry under $\phi \to -\phi$ present in the action. Thus, physical phenomena arising in this system will not exhibit the full symmetry of the underlying Hamiltonian because the ground state breaks the symmetry; this situation is called *spontaneous symmetry breaking*.

Condensed matter physics is full of beautiful examples in which this occurs. Consider, for example, a ferromagnet which exhibits non-zero magnetization $\mathbf{M}$ below a critical temperature T_c thereby breaking the rotational invariance. Observations show that the magnetization varies with temperature in the form $M \propto (T - T_c)^\mu$ where $0 < \mu < 1$. Such a *non-analytic* dependence on $(T - T_c)$ has puzzled physicists until Ginzburg and Landau came up with an elegant way of describing such phenomena. They suggested that the effective potential energy $V(\mathbf{M}, T)$ (more precisely the free energy of the system) describing the magnetization has the form $V(\mathbf{M}, T) = a(T)M^2 + b(T)M^4 + \dots$ with the conditions: (i) $a(T_c) = 0$ while (ii) $b(T)$ is smooth and non-zero around T_c. Expanding $a(T)$ in a Taylor series around T_c and writing $a(T) \approx \alpha(T - T_c)$ with $\alpha > 0$, one sees that the nature of $V(\mathbf{M}, T)$ changes drastically when we go from $T > T_c$ to $T < T_c$. In the high temperature phase, the minimum of the potential is at $M = 0$ and the system exhibits no spontaneous magnetization. But when $T < T_c$, the potential takes the shape in Fig.3.2 and the minimum

[32] The behaviour of such a system is quite different in quantum *field theory* compared to quantum *mechanics*. In quantum mechanics the tunneling probability between the two minima is nonzero which breaks the degeneracy and leads to a sensible ground state. In quantum field theory, the tunneling amplitude between the minima vanishes because an infinite number of modes have to tunnel; mathematically, the tunneling amplitude will be of the form $\exp(-\Gamma)$ where Γ will involve an integration over all space of a constant factor, making the amplitude vanish. Thus, in field theory, we have genuine degeneracy of the two vacua.

of the potential is now at $M^2 = (\alpha/2b)(T_c - T)$ thereby exhibiting the non-analytic dependence $M \propto (T - T_c)^\mu$ with $\mu = 0.5$.

In fact, this phenomenological model can also be used to describe the response of the system to externally imposed magnetic field $\mathbf{H}$, thereby explaining the behaviour of the correlation length and magnetic susceptibility of the system. In this context, one could use a phenomenological Hamiltonian of the form[33]

$$H = \frac{1}{2} (\partial^\alpha \mathbf{M}) \cdot (\partial_\alpha \mathbf{M}) + V(M) - \mathbf{H} \cdot \mathbf{M} \tag{3.89}$$

Taking $V \approx a(T)M^2 \approx \alpha(T - T_c)M^2$ as the leading behaviour for $T > T_c$ and minimizing this Hamiltonian, we get the equation

$$-\nabla^2 \mathbf{M} + a\mathbf{M} = \mathbf{H} \tag{3.90}$$

which leads to the Yukawa potential solution for the induced magnetic field $\mathbf{M}(\mathbf{x})$ in terms of the applied magnetic field $\mathbf{H}(\mathbf{y})$ in the form

$$\mathbf{M}(\mathbf{x}) = \int d^3\mathbf{y} \; \mathbf{H}(\mathbf{y}) \frac{1}{4\pi} \frac{1}{|\mathbf{x} - \mathbf{y}|} e^{-\sqrt{a}|\mathbf{x} - \mathbf{y}|} \tag{3.91}$$

The effect of the applied magnetic field is felt essentially over a distance (called the *correlation length*) of the order of $\xi = (1/\sqrt{a}) \propto (T - T_c)^{-1/2}$. The correlation length is finite at $T > T_c$ but diverges as $T \to T_c$ indicating the development of long range order in the system. Further, for a constant applied field $\mathbf{H}$ we can integrate over $\mathbf{y}$ in Eq. (3.91) and get the resulting magnetization to be $\mathbf{M} = \chi\mathbf{H}$, where the magnetic susceptibility $\chi \propto \xi^2 \propto (T - T_c)^{-1}$. Thus, the simple model of spontaneous symmetry breaking also predicts that the magnetic susceptibility diverges as $T \to T_c$, all thanks to the behaviour $a(T) \propto (T - T_c)$ which, in turn, is the root cause of spontaneous symmetry breaking.

Back to field theory. Since the ground state is at non-zero ϕ, it makes sense to shift the field ϕ in order to study excitations around the true ground state. If we take the ground state to be at $\phi = v$ and define $\psi \equiv \phi - v$, then the potential takes the form

$$V(\phi) = \frac{\lambda}{4} \left[\phi^2 - v^2 \right]^2 = \lambda v^2 \psi^2 + \left[\frac{\lambda}{4} \psi^4 + \lambda v \psi^3 \right] \tag{3.92}$$

We now see that the theory describes excitations with a well defined mass $m_\psi = v\sqrt{2\lambda}$ for the ψ field with a self-interaction described by the terms $(\lambda/4)\psi^4 + \lambda v\psi^3$. This will provide a description of a sensible quantum field theory in terms of the shifted field ψ. In this particular case, the situation is fairly simple since there is only one degree of freedom to deal with in the ϕ or ψ field.

Let us next consider a *complex* scalar field, having a potential with quadratic and quartic terms, which exhibits spontaneous symmetry breaking. This happens if we replace the $m^2|\phi|^2$ term in Eq. (3.41) by

$$V(|\phi|) = \frac{\lambda}{4} \left[|\phi|^2 - v^2 \right]^2 \tag{3.93}$$

The situation now is, however, qualitatively different from the case of the real scalar field because of the existence of *two* degrees of freedom ϕ_1 and ϕ_2 in the complex scalar field $\phi \equiv \phi_1 + i\phi_2$. The potential now has a minimum

[33]As you can guess, the inter-relationship between field theory and condensed matter physics is very strong and the concepts in each clarify and enrich the understanding of the other. The third term in Eq. (3.89) is analogous to adding the external source term $J(x)\phi(x)$ in the case of the scalar field theory; you should be able to see the parallels in what follows.

Exercise 3.13: (a) Show that, in the (1+1) spacetime, there exist static, finite energy field configurations $\phi(x)$ which are solutions to the field equation with the double-well potential. These solutions, with energy density concentrated in a finite region of space, are examples of *solitons*. (b) Show that such solutions cannot exist in $(1 + D)$ spacetime with $D \geq 2$. This is called *Derrick's theorem*. [Hint: Study the scaling behaviour of the total kinetic energy and the potential energy for a static solution under $\mathbf{x} \to \lambda\mathbf{x}$ and use the fact that the total energy should be a minimum with respect to λ at $\lambda = 1$ for any valid solution.]

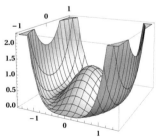

Figure 3.3: The shape of the potential $V(\phi)$ that leads to a zero mass particle.

on the circle $\phi_1^2 + \phi_2^2 = v^2$ in the $\phi_1 - \phi_2$ plane (see Fig. 3.3). The ϕ field can roll along the circle in the angular direction without changing the potential energy while any radial oscillations about the minimum will cost potential energy. Since the mass of the field gives the natural frequency of oscillations, it follows that we will have one massless degree of freedom (corresponding to the mode that rolls along the valley of the potential) and one massive degree of freedom (corresponding to the radial oscillations).

This is easily verified by writing $\phi(x) = \rho(x)\exp[i\theta(x)/v]$ and working with the two fields $\rho(x)$ and $\theta(x)$. The kinetic term in the action now becomes

$$\partial_a \phi \, \partial^a \phi^* = \partial_a \rho \, \partial^a \rho + \frac{\rho^2}{v^2} \partial_a \theta \, \partial^a \theta \qquad (3.94)$$

while the potential energy term becomes, in terms of the field $\psi \equiv (\rho - v)$ defined with respect to the minimum,

$$V = \frac{\lambda}{4}\Big[\psi(2v + \psi)\Big]^2 = \lambda v^2 \psi^2 + \frac{\lambda}{4}\psi^4 + v\lambda\psi^3 \equiv \lambda v^2 \psi^2 + U(\psi) \quad (3.95)$$

where the first term is the mass term for ψ and the rest represent the self-interaction of the field. Putting them together, we can express the Lagrangian in the form

$$L = \partial_a \psi \, \partial^a \psi - m_\psi^2 \psi^2 - U(\psi) + \partial_a \theta \, \partial^a \theta + \left[\frac{\psi^2}{v^2} + 2\frac{\psi}{v}\right]\partial_a \theta \, \partial^a \theta \quad (3.96)$$

with $m_\psi = v\sqrt{\lambda}$. This form of the Lagrangian shows that our system decomposes to one ψ field with mass m_ψ (and self-interaction described by $U(\psi)$) one field θ with zero mass, with the remaining terms describing the interaction between these two fields. The emergence of a zero mass field in such systems with spontaneous symmetry breaking is a universal feature and we will provide a general proof of this result when we discuss the quantum structure of this theory, at the end of Sect. 3.5.

Finally, let us consider what happens when we couple the charged scalar field to the gauge field. In the absence of the gauge field, when we wrote $\phi(x)$ as $\rho(x)\exp[i\theta(x)/v]$ we found that that $\theta(x)$ became a massless scalar field. But when we have a gauge field present, we can eliminate the $\theta(x)$ by a gauge transformation! This allows us to go from the 3 real fields (ρ, θ, A_j) to two new (real) fields $(\rho, 0, B_j)$ with $B_j \equiv A_j + (1/qv)\partial_j\theta$. Since $D_j\phi$ is gauge invariant, it now becomes $\partial_j\rho + iqB_j\rho$; similarly, since F_{ij} is gauge invariant it can be expressed in terms of B_j. With these modifications, the relevant Lagrangian will now become:

$$\begin{aligned}
L &= -\frac{1}{4}F^2 + |D\phi|^2 - \frac{\lambda}{4}[|\phi|^2 - v^2]^2 \\
&= -\frac{1}{4}F^2 + |\partial_m\rho + iqB_m\rho|^2 - \frac{\lambda}{4}(\rho^2 - v^2)^2 \\
&= -\frac{1}{4}F^2 + (\partial\rho)^2 + q^2\rho^2 B^2 - \frac{\lambda}{4}(\rho^2 - v^2)^2 \qquad (3.97)
\end{aligned}$$

As usual we now introduce a shifted field ψ by $\psi = \rho - v$ to get:

$$L = \left[-\frac{1}{4}F^2 + q^2v^2 B_j B^j\right] + \left[(\partial\psi)^2 - \lambda v^2\psi^2 - U(\psi)\right] + 2vq^2\psi B^2 + q^2\psi^2 B^2 \qquad (3.98)$$

The terms in the first square bracket describe a *massive* vector field B_j with mass $m_B = \sqrt{2}qv$; the second square bracket represents a massive scalar field ψ with mass $m_\psi = v\sqrt{\lambda}$ and a self-interaction described by $U(\psi)$ as before. The remaining terms represent interactions between the two fields. *The massless mode has disappeared!* Rather, the one degree of freedom in θ and two degrees of freedom in the massless gauge field A_j have combined into a massive vector field B_j with three degrees of freedom.[34]

[34]The phenomenon, which is of considerable importance in condensed matter physics and particle physics, was discovered, in one form or another, by (in alphabetical order) Anderson, Brout, Englert, Guralnik, Hagen, Higgs and Kibble; hence it is usually called the *Higgs mechanism*.

An important example in condensed matter physics in which the above result plays a role is *superconductivity*. Under certain circumstances, one can describe the electron pairs (called *Cooper pairs*) in terms of a bosonic complex scalar field ϕ (carrying a charge $q = 2e = -2|e|$) with a nonzero vacuum expectation value. In that case, assuming ϕ does not vary significantly over space, the current in the right hand side of Eq. (3.67) will have spatial components $\mathbf{J} = -2q^2v^2\mathbf{A}$. The Maxwell equation $\nabla \times \mathbf{B} = \mathbf{J} = -2q^2v^2\mathbf{A}$ reduces, on taking a curl and using $\nabla \cdot \mathbf{B} = 0$, to the form

$$\nabla^2\mathbf{B} = 2q^2v^2\mathbf{B} \equiv \ell^{-2}\mathbf{B} \tag{3.99}$$

with solutions which decay exponentially with a characteristic length scale $\ell = 1/(\sqrt{2}\,qv)$ which is essentially the mass scale of the vector boson. This fact underlies the so called *Meissiner effect* in superconductivity, which leads to a superconductor expelling a magnetic field from its interior.

3.3   Quantizing the Real Scalar Field

Given the action functional in classical mechanics, one can obtain the quantum theory by elevating the coordinates and the canonical momenta to the operator status and imposing the equal time commutation rules (ETCR) given by $[q_\alpha(t), p_\beta(t)] = i\delta_{\alpha\beta}$ with all other commutators vanishing.[35] The time evolution of the dynamical variables is determined through the evolution equations $i\dot{q}_\alpha = -[H, q_\alpha]$, $i\dot{p}_\alpha = -[H, p_\alpha]$ where $H(p,q)$ is the Hamiltonian operator. Alternatively, one can choose to work in the Schrodinger picture in the coordinate representation and take $p_\alpha = -i\partial_\alpha$, and express $H(p,q) \to H(-i\partial, q)$ as a differential operator acting on the wave functions $\psi(t,q)$, with the time evolution determined by $i\dot{\psi} = H\psi$. Finally one can also obtain a quantum theory from a classical theory (in principle) by computing the path integral amplitude by summing over paths connecting fixed end points, with the amplitude for a given path given by $\exp(iA)$.

[35]Recall that, in the Heisenberg picture in which we are working, these commutation relations hold only when the operators are evaluated at the same time; $[q_\alpha(t), q_\alpha(t')]$ need not vanish when $t \neq t'$.

Exercise 3.14: How do you know that the commutation rules are consistent with the Hamiltonian evolution? That is, if you impose the commutation rules at $t = t_0$ and evolve the system using a Hamiltonian, you better make sure the commutation rules hold at $t > t_0$. Is this a worry?

Move on from mechanics to field theory. In the case of a real scalar field, the dynamical variable which is varied in the action is $\phi(t, \mathbf{x})$ and the canonically conjugate momenta is $\pi(t, \mathbf{x}) = \dot{\phi}$ with the Hamiltonian given by Eq. (3.29). If we try to quantize this system using the Schrodinger picture — which is what you are likely to have spent most time on in a QM course — we immediately run into a mathematical complexity. The coordinate q, which was just a *real number* in Schrodinger picture quantum mechanics, will now have to be replaced by a *function* giving the field configuration $\phi(\mathbf{x})$ at a given time t. The wave*function* $\psi(q, t)$ giving the probability amplitude for the dynamical variable to have value q at time t will now have to be replaced by a *functional* $\Psi[\phi(\mathbf{x}), t]$ giving the probability amplitude for the field to be specified in space by the function $\phi(\mathbf{x})$ at time t. The Hamiltonian, which was a partial differential operator in Schrodinger quantum mechanics, will now become a functional differential operator and the equivalent of the Schrodinger equation will become a functional differential equation. The technology for dealing with these is not yet at a sufficiently high state of development — which prevents us from handling realistic quantum field theory problems using the Schrodinger picture.[36]

Similar difficulties arise in the case of using path integrals to quantize the fields. What we now need to compute is an amplitude for transition from a given field configuration $\phi_1(\mathbf{x})$ at time $t = t_1$ to a field configuration $\phi_2(\mathbf{x})$ at time $t = t_2$ which is given by the path integral

$$\langle t_2, \phi_2(\mathbf{x}) | t_1, \phi_1(\mathbf{x}) \rangle = \int_{t=t_1, \phi=\phi_1(\mathbf{x})}^{t=t_2, \phi=\phi_2(\mathbf{x})} \mathcal{D}\phi \, \exp i \int d^4x \, L(\partial_a \phi, \phi) \quad (3.100)$$

The sum over paths (or rather the functional integrals) have to be now computed with the specific boundary conditions, as indicated. Once we know $\langle t_2, \phi_2(\mathbf{x}) | t_1, \phi_1(\mathbf{x}) \rangle$, we can evaluate the wave functional at a given time from its form at an earlier time by the standard rule

$$\Psi[\phi_2(\mathbf{x}), t_2] = \int \mathcal{D}\phi_1 \, \langle t_2, \phi_2(\mathbf{x}) | t_1, \phi_1(\mathbf{x}) \rangle \Psi[\phi_1(\mathbf{x}), t_1] \quad (3.101)$$

It is obvious that this path integral quantization procedure is at least as complicated technically as using the Schrodinger picture. To begin with, we cannot compute it except for a quadratic action functional (which essentially corresponds to non-interacting fields). Second, even after such a computation, we have to work continuously with functional integrals and wave functionals to make any sense of the theory. Finally this approach (as well as the one based on the Schrodinger picture) makes it difficult to maintain manifest Lorentz invariance because it treats the time coordinate in a preferential manner.[37]

Fortunately, the Heisenberg picture works fine in quantum field theory without any of these difficulties and you having to learn new mathematical physics. In the case of a real scalar field, the dynamical variable which is varied in the action is $\phi(t, \mathbf{x})$ and the canonically conjugate momentum is $\pi(t, \mathbf{x}) = \dot{\phi}$ which suggests that we should try to quantize this system by postulating the commutation rules

$$[\phi(t, \mathbf{x}), \pi(t, \mathbf{y})] = i\delta(\mathbf{x} - \mathbf{y}); \quad [\phi(t, \mathbf{x}), \phi(t, \mathbf{y})] = [\pi(t, \mathbf{x}), \pi(t, \mathbf{y})] = 0$$
$$(3.102)$$

This will indeed lead to the quantum theory of a real scalar field but there is a much simpler procedure based on what we have already learnt in the last chapter. We will follow that route.

We had seen that the action functional for the scalar field in Eq. (3.3) is the same as that for an infinite number of harmonic oscillators and can be expressed in the form

$$A = \frac{1}{2} \int d^4x \, \left(\partial_a \phi \partial^a \phi - m^2 \phi^2 \right) = \int \frac{d^3\mathbf{k}}{(2\pi)^3} \frac{1}{2} [|\dot{q}_\mathbf{k}|^2 - \omega_\mathbf{k}^2 |q_\mathbf{k}|^2] \quad (3.103)$$

where

$$\phi(x) = \int \frac{d^3\mathbf{p}}{(2\pi)^3} q_\mathbf{p}(t) \, e^{i\mathbf{p}\cdot\mathbf{x}} \quad (3.104)$$

So, if we quantize each of the oscillators by standard quantum mechanical rules, we would have quantized the field itself. This reduces the quantum

[36] We will give some simple examples of Schrodinger picture quantum field theory later on because of its intuitive appeal but one cannot do much with it.

[37] We will, of course, use functional integrals extensively in developing quantum field theory in later chapters. But these are functional integrals evaluated from $t_1 = -\infty$ to $t_2 = +\infty$ with either the $i\epsilon$ or the Euclidean prescription. As we saw in Sect. 1.2.3, such path integrals do not require specific boundary configurations $\phi_1(\mathbf{x})$ and $\phi_2(\mathbf{x})$ — or rather, they can be set to zero — and lead to the *ground state expectation values* of various operators. So we will use path integrals as a convenient tool for calculating vacuum expectation values etc. and not as a procedure for quantization — which is what we are discussing here. These two roles of path integrals are somewhat different.

field theory problem to the quantum mechanics of harmonic oscillators which you are familiar with.

In this approach, the dynamical variables for each oscillator should become $q_{\mathbf{k}}$ but there is a minor issue that the $q_{\mathbf{k}}$'s are not real if ϕ is real and satisfy the condition $q_{-\mathbf{k}} = q_{\mathbf{k}}^*$. This constraint is easily taken care[38] of by writing $q_{\mathbf{k}}(t) = a_{\mathbf{k}}(t) + a_{-\mathbf{k}}^*(t)$ without any constraint on the variables $a_{\mathbf{k}}(t)$. The standard harmonic oscillator time evolution, arising from the Heisenberg equations of motion now gives

$$q_{\mathbf{p}} = \frac{1}{\sqrt{2\omega_{\mathbf{p}}}} \left[a_{\mathbf{p}} e^{-i\omega_{\mathbf{p}} t} + a_{-\mathbf{p}}^\dagger e^{i\omega_{\mathbf{p}} t} \right] \tag{3.105}$$

[This has the same form as the classical evolution in Eq. (3.25).] Quantizing each harmonic oscillator is now trivial and one can think of $a_{\mathbf{k}}^\dagger$ and $a_{\mathbf{k}}$ as the standard creation and annihilation operators. In the continuum case, the creation and annihilation operators satisfy the commutation rule

$$[a_{\mathbf{p}}, a_{\mathbf{k}}^\dagger] = (2\pi)^3 \delta(\mathbf{p} - \mathbf{k}) \tag{3.106}$$

with all other commutators vanishing. The evolution of the field ϕ itself follows from Fourier transforming Eq. (3.105) with an $\exp[i\mathbf{p} \cdot \mathbf{x}]$ factor. Using the usual trick of flipping the sign of $\mathbf{p}$ in the second term, we get back essentially the result we found in the previous chapter (see Eq. (1.129)): So our final answer is:

$$\phi(x) = \int \frac{d^3\mathbf{p}}{(2\pi)^3} \frac{1}{\sqrt{2\omega_{\mathbf{p}}}} \left[a_{\mathbf{p}} e^{-ipx} + a_{\mathbf{p}}^\dagger e^{ipx} \right] \tag{3.107}$$

We will now explicitly verify that this does lead to the correct commutation rule for the fields. Since we need to verify the commutation rules only at a given time, we can always choose it to be $t = 0$. To work out the commutators, we start with the expressions for ϕ and π at $t = 0$, given by

$$\phi(0, \mathbf{x}) = \int \frac{d^3\mathbf{p}}{(2\pi)^3} \frac{1}{\sqrt{2\omega_{\mathbf{p}}}} \left(a_{\mathbf{p}} e^{i\mathbf{p} \cdot \mathbf{x}} + a_{\mathbf{p}}^\dagger e^{-i\mathbf{p} \cdot \mathbf{x}} \right) \tag{3.108}$$

$$\pi(0, \mathbf{x}) = \int \frac{d^3\mathbf{p}}{(2\pi)^3} (-i) \sqrt{\frac{\omega_{\mathbf{p}}}{2}} \left(a_{\mathbf{p}} e^{i\mathbf{p} \cdot \mathbf{x}} - a_{\mathbf{p}}^\dagger e^{-i\mathbf{p} \cdot \mathbf{x}} \right) \tag{3.109}$$

It is convenient to rearrange these two expressions by the usual trick of flipping the sign of momentum in the second term and writing

$$\phi(0, \mathbf{x}) = \int \frac{d^3\mathbf{p}}{(2\pi)^3} \frac{1}{\sqrt{2\omega_{\mathbf{p}}}} \left(a_{\mathbf{p}} + a_{-\mathbf{p}}^\dagger \right) e^{i\mathbf{p} \cdot \mathbf{x}} \tag{3.110}$$

$$\pi(0, \mathbf{x}) = \int \frac{d^3\mathbf{p}}{(2\pi)^3} (-i) \sqrt{\frac{\omega_{\mathbf{p}}}{2}} \left(a_{\mathbf{p}} - a_{-\mathbf{p}}^\dagger \right) e^{i\mathbf{p} \cdot \mathbf{x}} \tag{3.111}$$

Evaluating the commutator and using Eq. (3.106), we find that[39]

$$[\phi(\mathbf{x}), \pi(\mathbf{x}')] = \int \frac{d^3\mathbf{p}\, d^3\mathbf{p}'}{(2\pi)^6} \frac{-i}{2} \sqrt{\frac{\omega_{\mathbf{p}'}}{\omega_{\mathbf{p}}}} \tag{3.112}$$

$$\times \left([a_{-\mathbf{p}}^\dagger, a_{\mathbf{p}'}] - [a_{\mathbf{p}}, a_{-\mathbf{p}'}^\dagger] \right) e^{i(\mathbf{p} \cdot \mathbf{x} + \mathbf{p}' \cdot \mathbf{x}')} = i\delta^{(3)}(\mathbf{x} - \mathbf{x}')$$

[38] More formally, this can be done by separating $q_{\mathbf{k}}$ into real and imaginary parts with $q_{\mathbf{k}} = X_{\mathbf{k}} + iY_{\mathbf{k}}$ with the conditions $X_{-\mathbf{k}} = X_{\mathbf{k}}$, $Y_{-\mathbf{k}} = -Y_{\mathbf{k}}$ and noting that the dynamical variables $X_{\mathbf{k}}, Y_{\mathbf{k}}$ are independent only for half the $\mathbf{k}$ values. In going from $q_{\mathbf{k}}$ to $X_{\mathbf{k}}, Y_{\mathbf{k}}$ you are doubling the number of degrees of freedom, but the fact that only half of them are independent restores the balance. Therefore we can introduce one single real variable $Q_{\mathbf{k}}$ in place of $X_{\mathbf{k}}, Y_{\mathbf{k}}$ but now retaining all possible values of $\mathbf{k}$. This means the action in Eq. (3.103) can be expressed entirely in terms of one real harmonic oscillator coordinate $Q_{\mathbf{k}}$. We won't bother to spell out the details of this approach since the final result is the same.

[39] We could have done this starting from Eq. (3.102) and then found that the a and $a^\dagger$ satisfy the usual commutation rules for a creation and an annihilation operator. We took the opposite route since we have actually done all the hard work in the last chapter. We will see later that the identification and quantization of oscillators work even when the approach based on commutators requires special treatment.

Similarly, you can work out the other commutators and you will find that they all vanish. Thus we have successfully quantized the real scalar field and obtained the time evolution of the dynamical variable in Eq. (3.107).

For the sake of completeness, we will briefly mention how the same result comes about if we had taken the more formal approach based on Eq. (3.102). Here one would associate with the system the Hamiltonian

$$H = \frac{1}{2} \int d^3\mathbf{x} \left[\pi^2 + (\nabla \phi)^2 + m^2 \phi^2 \right] \qquad (3.113)$$

and use the standard result in quantum mechanics $-i\partial_t \hat{O} = [H, \hat{O}]$ which describes the time evolution of any operator $\hat{O}$ to determine $\dot{\phi}$ and $\dot{\pi}$. This calculation gives (on using the commutation rules in Eq. (3.102)) the results

$$
\begin{aligned}
-i\dot{\phi}(t,\mathbf{y}) &= \frac{1}{2}\int d^3\mathbf{x}\left[\pi^2(t,\mathbf{x}),\phi(t,\mathbf{y})\right] = \int d^3\mathbf{x}\left[\pi(t,\mathbf{x}),\phi(t,\mathbf{y})\right]\pi(t,\mathbf{x}) \\
&= -i\int d^3\mathbf{x}\,\delta(\mathbf{x}-\mathbf{y})\,\pi(t,\mathbf{x}) = -i\pi(t,\mathbf{y}) \qquad (3.114)
\end{aligned}
$$

and

$$
\begin{aligned}
-i\partial_t \pi(t,\mathbf{y}) &= \frac{1}{2}\int d^3\mathbf{x}\left\{ m^2\left[\phi^2(\mathbf{x}),\pi(\mathbf{y})\right] + \left[\partial_\alpha \phi_{\mathbf{x}} \partial_\alpha \phi_{\mathbf{x}}, \pi_{\mathbf{y}}\right]\right\} \\
&= im^2\phi(\mathbf{y}) + \int d^3\mathbf{x}\left[\partial_\alpha \phi_{\mathbf{x}}, \pi_{\mathbf{y}}\right]\partial_\alpha \phi \\
&= im^2\phi(\mathbf{y}) + \int i\partial_\alpha \delta(\mathbf{x}-\mathbf{y})(\partial_\alpha \phi_{\mathbf{x}})d^3\mathbf{x} \\
&= im^2\phi(\mathbf{y}) - i\nabla^2 \phi_{\mathbf{y}} = i\left(-\nabla^2 + m^2\right)\phi(\mathbf{y},t) \qquad (3.115)
\end{aligned}
$$

Combining these two we get the operator equation for ϕ which is just the Klein-Gordon equation:

$$\ddot{\phi} - \nabla^2 \phi + m^2 \phi = 0 \qquad (3.116)$$

Obviously, normalized solutions to this equation can be expressed in the standard form as

$$\phi(\mathbf{x}) = \int d\Omega_{\mathbf{p}}\left[A_{\mathbf{p}}e^{-ipx} + A_{\mathbf{p}}^\dagger e^{ipx}\right] \qquad (3.117)$$

which is exactly our result in Eq. (3.107) with $a_{\mathbf{p}} = A_{\mathbf{p}}/\sqrt{2\omega_{\mathbf{p}}}$ etc. In fact, if we solve equations Eq. (3.110) and Eq. (3.111) for $a_{\mathbf{p}}$ and $a_{-\mathbf{p}}^\dagger$ in terms of $\phi(0,\mathbf{x})$ and $\pi(0,\mathbf{x})$ and use Eq. (3.102), we can obtain Eq. (3.106). [See Problem 3.] This is a more formal procedure for quantizing the system treating ϕ as a dynamical variable and π as the conjugate momentum leading to the results.

This is a good time to take stock of what we have achieved. We are interested in defining a quantum theory for the field $\phi(t,\mathbf{x})$ such that it obeys the commutation rules in Eq. (3.102). We find that this is most easily done in terms of a set of creation and annihilation operators $a_{\mathbf{p}}$ and $a_{\mathbf{p}}^\dagger$, for each mode labeled by a 3-vector $\mathbf{p}$, which satisfy the standard commutation rule. This allows us to introduce states in the Hilbert space $|\{n_{\mathbf{p}}\}\rangle$ labeled by a set of integers $\{n_{\mathbf{p}}\}$ with the creation and annihilation operators for each mode increasing or decreasing the corresponding integer by one unit[40]

[40] By and large, if you understand harmonic oscillators, you can understand noninteracting fields; they are essentially the same except that there are an infinite number of oscillators.

etc. The state $|0\rangle$ with all integers set to zero will be the ground state of all harmonic oscillators which we will expect to be the vacuum state. (We will have quite a bit to say about this state soon.) Acting on $|0\rangle$ with the creation operator, we can generate the excited states of the system.

To investigate further the nature of these states $|\{n_{\mathbf{p}}\}\rangle$ we can work out the Hamiltonian for the system in terms of $a_{\mathbf{k}}$ and $a_{\mathbf{k}}^{\dagger}$. A direct way of obtaining the Hamiltonian operator is to substitute Eq. (3.110) and Eq. (3.111) into the expression for the Hamiltonian in Eq. (3.29). Again, since H is independent of time, we can evaluate it at $t = 0$. This gives

$$
\begin{aligned}
H &= \int d^3\mathbf{x} \int \frac{d^3\mathbf{p}\,d^3\mathbf{p}'}{(2\pi)^6}\, e^{i(\mathbf{p}+\mathbf{p}')\cdot\mathbf{x}} \left\{ -\frac{\sqrt{\omega_{\mathbf{p}}\omega_{\mathbf{p}'}}}{4}(a_{\mathbf{p}} - a_{-\mathbf{p}}^{\dagger})(a_{\mathbf{p}'} - a_{-\mathbf{p}}^{\dagger}) \right. \\
&\qquad\left. + \frac{-\mathbf{p}\cdot\mathbf{p}' + m^2}{4\sqrt{\omega_{\mathbf{p}}\omega_{\mathbf{p}'}}}(a_{\mathbf{p}} + a_{-\mathbf{p}}^{\dagger})(a_{\mathbf{p}'} + a_{-\mathbf{p}}^{\dagger}) \right\} \\
&= \int \frac{d^3\mathbf{p}}{(2\pi)^3}\, \omega_{\mathbf{p}} \left(a_{\mathbf{p}}^{\dagger}a_{\mathbf{p}} + \frac{1}{2}\left[a_{\mathbf{p}}, a_{\mathbf{p}}^{\dagger}\right] \right)
\end{aligned}
\tag{3.118}
$$

In the first term, each mode contributes $n_{\mathbf{p}}\omega_{\mathbf{p}}$ to the energy density which is consistent with interpreting the state $|\{n_{\mathbf{p}}\}\rangle$ as having $n_{\mathbf{p}}$ particles with energy $\omega_{\mathbf{p}}$. The second term is more nontrivial: It can be thought of as the limit of Eq. (3.106) as $\mathbf{p} \to \mathbf{k}$ which will lead to a $\delta(0)$ type of divergence in 3-dimensions. As usual, we interpret this as arising out of a volume integration with

$$
\lim_{\mathbf{p}\to\mathbf{k}} (2\pi)^3 \delta(\mathbf{p}\to\mathbf{k}) = \lim_{\mathbf{p}\to\mathbf{k}} \int d^3\mathbf{x}\, \exp[i\mathbf{x}\cdot(\mathbf{p}\to\mathbf{k})] = \lim_{V\to\infty} V \tag{3.119}
$$

so that the second term can be thought of as giving an energy density which exists even in the vacuum state:

$$
\frac{E_{\text{vac}}}{V} = \frac{1}{2} \int \frac{d^3\mathbf{p}}{(2\pi)^3}\, \omega_{\mathbf{p}} \tag{3.120}
$$

which is quartically divergent in the upper limit. Physically, this term arises due to the infinite number of harmonic oscillators each having a ground state energy $(1/2)\omega_{\mathbf{k}}$. We have encountered the same expression earlier when we computed $Z(0)$ in Sect. 2.2.1 (see Eq. (2.71)) and also when we computed the energy due to the closed loop of particles in Sect. 1.4 [see Eq. (1.107)].

The usual procedure to deal with this term is to simply throw this away and hope for the best. A formal way of implementing this procedure is to postulate that physical quantities (like the Hamiltonian) should be defined by a procedure called *normal ordering*. The normal ordering of an expression containing several a s and $a^{\dagger}$s is to simply move all the $a^{\dagger}$s to the left of a so that the vacuum expectation value of any normal ordered operator is automatically zero. One possible way of "justifying" this procedure is to note that the ordering of physical variables in the classical theory is arbitrary, and when one constructs the quantum theory, one has to make a choice regarding the way these variables are ordered, since they may not commute. The normal ordering is supposed to give a physically meaningful ordering of operators so that divergent vacuum expectation values are avoided.[41] We use the symbol $: O :$ to denote a normal ordered operator O. The expectation value of this (normal ordered)

---

[41] If you are uncomfortable with this entire discussion, then you are doing fine. Nobody really knows how to handle this divergence and it is rather surprising that this procedure actually 'works' even to the extent it does. We will see later that an experimentally observed result, called the *Casimir effect*, can be interpreted in terms of the zero point energies. This alone would have been enough to argue that normal ordering is a dubious procedure but for the fact that there exist more complicated ways of obtaining Casimir effect without using the zero point energy.

Hamiltonian, denoted by $: H :$, in a state $|\{n_{\mathbf{p}}\}\rangle$ is given by

$$\langle\{n_{\mathbf{p}}\}| : H : |\{n_{\mathbf{p}}\}\rangle = \int \frac{d^3\mathbf{p}}{(2\pi)^3} \omega_{\mathbf{p}} [n_{\mathbf{p}}] \qquad (3.121)$$

By a similar procedure we can also express the total momentum operator in Eq. (3.20) in terms of the creation and annihilation operators. We now find

$$\mathbf{P} = -\int d^3x \; \pi(t, \mathbf{x}) \, \nabla\phi(t, \mathbf{x}) = \int \frac{d^3p}{(2\pi)^3} \, \mathbf{p} \, a_{\mathbf{p}}^\dagger a_{\mathbf{p}} \qquad (3.122)$$

There is no zero point contribution in this case because the contributions from the vectors $\mathbf{p}$ and $-\mathbf{p}$ cancel.[42] This shows that the expectation value $\mathbf{P}$ in the state labeled by $\{n_{\mathbf{p}}\}$ is given by

$$\langle\{n_{\mathbf{p}}\}|P|\{n_{\mathbf{p}}\}\rangle = \int \frac{d^3\mathbf{p}}{(2\pi)^3} \, \mathbf{p} \, [n_{\mathbf{p}}] \qquad (3.123)$$

It is obvious that (except for the zero point energy which we have dropped) we can think of this state $|\{n_{\mathbf{p}}\}\rangle$ as containing $n_{\mathbf{p}}$ particles with momentum $\mathbf{p}$ and energy $\omega_{\mathbf{p}}$. This agrees with the conclusion we reached in the last chapter that the first excited state of any given oscillator labeled by $\mathbf{p}$ can be thought of as a one-particle state describing a particle with momentum $\mathbf{p}$ and energy $\omega_{\mathbf{p}}$.

An alternative way of obtaining the same result is to simply note that the action for our system is made of a sum of actions for harmonic oscillators which do not interact with each other. Then, it is obvious that the Hamiltonian can be expressed in the form

$$H = \int \frac{d^3\mathbf{p}}{(2\pi)^3} \omega_{\mathbf{p}} [a_{\mathbf{p}}^\dagger a_{\mathbf{p}}] \qquad (3.124)$$

where we have dropped the zero point energy term.

We mentioned earlier that it is not easy to do quantum field theory in the Schrodinger picture and this is definitely true in general. But in the case of a free scalar field we are only dealing with a bunch of harmonic oscillators and these can, of course, be handled in the Schrodinger picture. For each harmonic oscillator we can write down the time independent Schrodinger equation with the eigenfunction $\phi_{n_{\mathbf{k}}}(q_{\mathbf{k}})$ and the eigenvalue determined by $n_{\mathbf{k}}$. The full wave functional of the system is just the product of all the $\phi_{n_{\mathbf{k}}}(q_{\mathbf{k}})$ over $\mathbf{k}$.

Just to see what is involved, let us consider the *ground state wave functional* of the system which is given by the product of all the ground state wave functions of the oscillators. We have

$$\Psi[\{q_{\mathbf{k}}\}, 0] \propto \prod_{\mathbf{k}} \phi_0(q_{\mathbf{k}}) \propto \exp\left[-\frac{1}{2}\int \frac{d^3\mathbf{k}}{(2\pi)^3} \, \omega_{\mathbf{k}}|q_{\mathbf{k}}|^2\right] \qquad (3.125)$$

The argument of the exponential can be expressed in terms of the field ϕ itself using the inverse Fourier transform in the form

$$\frac{1}{2}\int \frac{d^3\mathbf{k}}{(2\pi)^3}\omega_{\mathbf{k}}|q_{\mathbf{k}}|^2 = \frac{1}{2}\int d^3\mathbf{x}\int d^3\mathbf{y} \; \phi(\mathbf{x})\mathcal{G}(\mathbf{x} - \mathbf{y})\phi(\mathbf{y}); \qquad (3.126)$$

$$\mathcal{G}(\mathbf{x}) = \int \frac{d^3\mathbf{k}}{(2\pi)^3}(k^2 + m^2)^{1/2}e^{i\mathbf{k}\cdot\mathbf{x}} \qquad (3.127)$$

[42]This is fortunate because any non-zero momentum vector for the vacuum would have even violated the rotational symmetry of the vacuum and given a preferential direction in space!

so that the ground state wave functional of the system is given by

$$\Psi[\phi(\mathbf{x}), 0] \propto \exp\left[-\frac{1}{2}\int d^3\mathbf{x}\int d^3\mathbf{y}\ \phi(\mathbf{x})\mathcal{G}(\mathbf{x}-\mathbf{y})\phi(\mathbf{y})\right] \qquad (3.128)$$

Exercise 3.15: Evaluate $\mathcal{G}(\mathbf{x})$ paying particular attention to its singular structure. Just a sec.; isn't it the same one we saw in Eq. (1.110) ? Is this a coincidence ?

The form of $\mathcal{G}(\mathbf{x})$ is rather complicated for $m \neq 0$ but you can write the vacuum functional in a simpler fashion when $m = 0$ by making use of the fact that, for a massless scalar field, $\omega_{\mathbf{k}} = |\mathbf{k}|$. The trick now is to multiply the numerator and denominator of the integrand in Eq. (3.126) by $|\mathbf{k}|$ and note that $\mathbf{k}q_{\mathbf{k}}$ is essentially the Fourier transform of $\nabla\phi$. This gives

$$\frac{1}{2}\int\frac{d^3\mathbf{k}}{(2\pi)^3}|\mathbf{k}|\,|q_{\mathbf{k}}|^2 = \int\frac{d^3\mathbf{k}}{(2\pi)^3}\frac{k^2|q_{\mathbf{k}}|^2}{2|\mathbf{k}|} = \frac{1}{4\pi^2}\int d^3\mathbf{x}\ d^3\mathbf{y}\frac{\nabla_{\mathbf{x}}\phi(\mathbf{x})\cdot\nabla_{\mathbf{y}}\phi(\mathbf{y})}{|\mathbf{x}-\mathbf{y}|^2} \qquad (3.129)$$

so that the ground state wave functional (also called the *vacuum functional*) for a massless scalar field has the simple closed form expression given by

$$\Psi[\phi(\mathbf{x}), 0] = N\exp\left[-\frac{1}{4\pi^2}\int d^3\mathbf{x}\ d^3\mathbf{y}\frac{\nabla_{\mathbf{x}}\phi(\mathbf{x})\cdot\nabla_{\mathbf{y}}\phi(\mathbf{y})}{|\mathbf{x}-\mathbf{y}|^2}\right] \qquad (3.130)$$

[43] We have ignored the overall normalization factor and the time dependence $\exp(-iEt)$ in writing the wave functional (for very good reasons; both are divergent).

This wave functional is proportional to the amplitude that you will find a non-trivial field configuration $\phi(\mathbf{x})$ in the vacuum state; that is, for any given functional form $\phi(\mathbf{x})$ you can evaluate the integral in $\Psi[\phi(\mathbf{x}), 0]$ and obtain a number which gives the probability amplitude.[43] The ratio of $\Psi[\phi_1(\mathbf{x}), 0]/\Psi[\phi_2(\mathbf{x}), 0]$ for any two functions $\phi_1(\mathbf{x}), \phi_2(\mathbf{x})$ correctly gives the relative probability for you to find one field configuration compared to another *in the vacuum state*. This is a concrete demonstration of the non-triviality of the vacuum state in quantum field theory. If you look for non-zero field configurations in the vacuum state, you will find them.[44]

[44] The way we have described the entire quantization procedure, this should not come as a surprise to you. The field is a collection of oscillators and the vacuum state is the simultaneous ground state of all the oscillators. The probability amplitude for a non-zero displacement of the dynamical variable $q_{\mathbf{k}}$ is finite in the ground state of the harmonic oscillator in quantum mechanics; it follows that the amplitude for a non-zero field configuration in the vacuum state *must* necessarily be non-zero in field theory. There is no way you can avoid it.

3.4 Davies-Unruh Effect: What is a Particle?

In the earlier sections we started with the notion of a relativistic free particle, used the path integral formalism to obtain the quantum propagator and reinterpreted it in terms of a field, say, a scalar field. In the last section we took the opposite point of view: We started with a scalar field, treated it as a dynamical system governed by a Hamiltonian, quantized it using the commutation rules in Eq. (3.102) and discovered that the states in the Hilbert space can be described in terms of occupation numbers $\{n_{\mathbf{k}}\}$ allowing a particle interpretation. In this approach, we start with a vacuum state containing no particles and we can create a one-particle state, say, by acting on the vacuum with a creation operator. The propagator for the relativistic particle, for example, can now be obtained in terms of the vacuum expectation value of the time-ordered product of two field operators. At the level of propagators, we seem to have established the connection with the original approach starting from a relativistic particle.

Conceptually, however, the procedure of describing relativistic particles as excitations of a field has certain subtleties which need to be recognized. The first one is the existence of vacuum fluctuations in the no-particle state which, obviously, cannot be interpreted in terms of real relativistic particles. We had already mentioned it while describing the wave functional for the vacuum state given by Eq. (3.128) and will discuss it further later on when we study the Casimir effect in Sect. 3.6.3.

There is a second subtlety which goes right to the foundation of defining the concept of the particle, which we will discuss in this section. It turns out that while the notion of a vacuum state and that of a particle are Lorentz invariant, these are not *generally covariant* notions. That is, while two observers who are moving with respect to each other with a uniform velocity will agree on the definition of a vacuum state and the notion of a particle, two observers who are moving with respect to each other non-inertially may not agree[45] on their definitions of vacuum state or particles!

For example, the vacuum state — as we have defined in the previous section — for an inertial observer will not appear to be a vacuum state for another observer who is moving with a uniform acceleration. The accelerated observer, it turns out, will see the vacuum state of the inertial observer to be populated by a Planckian spectrum of particles with a temperature $T = (\hbar/ck_B)(g/2\pi)$ where g is the acceleration. This peculiar result has far reaching implications when one tries to combine the principles of quantum theory with gravity. We will now describe how this comes about.

Let us begin with the expansion of the scalar field in terms of creation and annihilation operators as given in Eq. (3.117). Given the commutation rules in Eq. (3.102), we found that $A_\mathbf{k}$ and $A_\mathbf{k}^\dagger$ can be interpreted as creation and annihilation operators. However, this description in terms of creation and annihilation operators is not unique. Consider, for example, a new set of creation and annihilation operators $B_\mathbf{k}$ and $B_\mathbf{k}^\dagger$, related to $A_\mathbf{k}$ and $A_\mathbf{k}^\dagger$ by the linear transformations

$$A_\mathbf{k} = \alpha B_\mathbf{k} + \beta B_\mathbf{k}^\dagger, \qquad A_\mathbf{k}^\dagger = \alpha^* B_\mathbf{k}^\dagger + \beta^* B_\mathbf{k} \qquad (3.131)$$

It is easy to verify that $B_\mathbf{k}$ and $B_\mathbf{k}^\dagger$ obey the standard commutation rules if we choose α and β with the condition $|\alpha|^2 - |\beta|^2 = 1$. Substituting Eq. (3.131) into Eq. (3.117) we find that $\phi(x)$ can be expressed in the form

$$\phi(x) = \int d\Omega_\mathbf{k} \left\{ B_\mathbf{k} f_\mathbf{k}(x) + B_\mathbf{k}^\dagger f_\mathbf{k}^*(x) \right\}, \qquad (3.132)$$

where

$$f_\mathbf{k}(x) = \alpha e^{-ikx} + \beta^* e^{ikx} \qquad (3.133)$$

One can now define a new vacuum state ("B-vacuum") state by the condition $B_\mathbf{k}|0\rangle_B = 0$ and the corresponding one-particle state, etc. by acting on $|0\rangle_B$ by $B_\mathbf{k}^\dagger$, etc. It is easy to see that this vacuum state is not equivalent to the original vacuum state (let us call it the "A-vacuum") defined by $A_\mathbf{k}|0\rangle_A = 0$. In fact, the number of A-particles in the B-vacuum is given by

$$_B\langle 0|A_k^\dagger A_k|0\rangle_B = |\beta|^2 \qquad (3.134)$$

At first sight this result seems to throw a spanner in the works! How can we decide on our vacuum state and concept of particle if the linear transformations in Eq. (3.133) (called the *Bogolioubov transformations*) change the notion of vacua? One could have worked either with the modes e^{-ikx}, e^{+ikx} or with $f_k(x), f_k^*(x)$ and these two quantization procedures will lead to different vacuum states and particle concepts. We need some additional criteria to choose the vacuum state.

In standard quantum field theory, one tacitly assumes such an additional criterion which is the following: We demand that the mode functions we work with must have "positive frequency" with respect to the Hamiltonian that we use. That is, the mode functions must be eigenfunctions

[45]It is rather surprising that most of the standard QFT textbooks do not even alert you about this fact!

[46] Actually one can be a little bit more generous and allow for mode functions which are arbitrary superpositions of positive frequency mode functions. But for our illustrative purposes, this criterion is good enough. Also note that we do not put any condition on the *spatial* dependence of the modes. One could use the standard plane waves $\exp(i\mathbf{k} \cdot \mathbf{x})$ or one could expand the field in terms of, say, the spherical harmonics. The notion of a vacuum state is only sensitive to the time dependence. Of course, a one-particle state describing a particle with momentum $\mathbf{k} = (0, 0, k_z)$ will be a superposition of different $Y_{lm}(\theta, \phi)$-s when expressed in spherical coordinates. This need not scare anyone who has studied scattering theory in NRQM.

Exercise 3.16: Prove these statements.

of the operator $H = i\partial/\partial t$ with a positive eigenvalue $\omega_{\mathbf{k}}$ so that the time dependence of the modes is of the form $\exp(-i\omega_{\mathbf{k}}t)$. This excludes using modes of the form $f_{\mathbf{k}}(x)$ in Eq. (3.133) which is a linear combination of positive and negative frequency modes.[46] It is easy to verify that the notion of a positive frequency mode is Lorentz invariant; a positive frequency mode will appear to be a positive frequency mode to all Lorentz observers.

If we introduce this extra restriction and define our vacuum state, it will appear to be a vacuum state to all other inertial observers. In fact most textbooks will not even bother to tell you that this criterion has been tacitly assumed because one usually deals with only Lorentz invariant field theory.

Things get more interesting when we allow for transformations of coordinates from inertial to non-inertial frames of reference. For example, consider a transformation from the standard inertial coordinates (t, x) to a non-inertial coordinate system (τ, ξ) by the equations

$$x = \xi \cosh g\tau; \qquad t = \xi \sinh g\tau \qquad (3.135)$$

(see Fig. 3.4). This transformation introduces a coordinate system which is appropriate for an observer who is moving with a uniform acceleration g along the x−axis and is called the *Rindler transformation*. World lines of observers who are at rest in the new coordinate system $\xi =$constant, corresponds to hyperbolas $x^2 - t^2 = \xi^2$ in the inertial frame. You can easily convince yourself that: (a) These are trajectories of uniformly accelerated observers. (b) The τ represents the proper time of the clock carried by the uniformly accelerated observer. The line element in terms of the new coordinates is easily seen to be

$$ds^2 = dt^2 - d\mathbf{x}^2 = g^2\xi^2 d\tau^2 - d\xi^2 - dy^2 - dx^2 \qquad (3.136)$$

(This form of the metric is called the *Rindler metric*.) Note that the transformations in Eq. (3.135) only cover the region with $|x| > |t|$ which is the region in which uniformly accelerated trajectories exist. The trajectory corresponding to $\xi = 0$ maps to $x = \pm t$ which are the null rays propagating through the origin. These null surfaces act as a horizon for the uniformly accelerated observer. That is, an observer with $\xi > 0$ will not have access to the region beyond $x = t$. This horizon divides the x-axis into two parts: $x > 0$ and $x < 0$ which are causally separated as far as the accelerated observer is concerned.

The key feature of the Rindler metric in Eq. (3.136) which is of relevance to us is its static nature; that is, the metric does not depend on the new time coordinate τ. This implies that when we study quantum field theory in these coordinates, one can again obtain mode functions of the form $e^{-i\omega\tau} f_\omega(\xi, y, z)$ which are positive frequency modes with respect to the proper time τ of the accelerated observer. But given the rather complicated nature of the transformation in Eq. (3.135), it is obvious that these modes will be a superposition of positive and negative frequency modes of the inertial frame. This is indeed true.

So even after imposing our condition that one should work with positive frequency mode functions, we run into an ambiguity — positive frequency modes of the inertial frame are not positive frequency modes of the accelerated (Rindler) frame! This, in turn, means that the vacuum state defined using the modes $e^{-i\omega\tau} f_\omega(\xi, y, z)$ will be different from the inertial vacuum

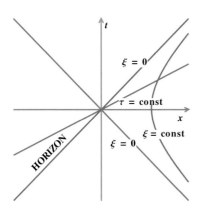

Figure 3.4: The Rindler coordinates appropriate for a uniformly accelerated observer. The ξ=constant curves are hyperbolas while τ= constant curves are straight lines through the origin. The $x = t$ surface of the Minkowski space (which maps to ξ=0) acts as a horizon for the uniformly accelerated observers in the right wedge.

state. The accelerated observer will not consider the inertial vacuum to be a no-particle state. In short, the notion of a vacuum state and the concept of particles (as excitations of the field) are Lorentz invariant concepts but they are not invariant notions if we allow transformations to non-inertial coordinates.

We will now show that the inertial vacuum functional leads to a thermal density matrix when viewed in an accelerated frame. To obtain this result, let us begin with the result we have obtained earlier in Sect. 1.2.3 which expresses the quantum mechanical path integral in terms of the stationary states of the system:

$$\int \mathcal{D}q \exp\left(iA[t_2, q_2; t_1, q_1]\right) = \langle q_2|q_1 \rangle = \sum_n \phi_n(q_2)\phi_n^*(q_1)e^{-iE_n(t_2-t_1)}$$

(3.137)

We now analytically continue to Euclidean time and set $t_1^E = 0, q_2 = 0, q_1 = q$ and take the limit $t_2^E \to \infty$. In the infinite time limit, the right hand side will be dominated by the lowest energy eigenvalue which will correspond to the ground state ϕ_0. We can always add a constant to the Hamiltonian to make the ground state energy zero. In that case, the only term which survives in the right hand side will be $\phi_0(q_2)\phi_0^*(q_1) = \phi_0(0)\phi_0(q) \propto \phi_0(q)$. Thus the ground state wave function can be expressed as a Euclidean path integral with very specific boundary conditions (see Eq. (1.44)):

$$\phi_0(q) \propto \int \mathcal{D}q \, \exp(-A_E[\infty, 0; 0, q])$$

(3.138)

This result will directly generalize to field theory if we think of q_1 and q_2 as field configurations $q_1(\mathbf{x}), q_2(\mathbf{x})$ and $\phi_0(q_1)$ as the ground state wave functional $\Psi_{\mathrm{gs}}[\phi(\mathbf{x})]$. Then we get

$$\Psi_{\mathrm{gs}}[\phi(\mathbf{x})] \propto \int \mathcal{D}\phi \, e^{-A_E(\infty, 0; 0, \phi(\mathbf{x}))}$$

(3.139)

The path integral on the right hand side can be evaluated in the Euclidean sector obtained (i) either by analytically continuing the inertial time coordinate t to t_E or (ii) analytically continuing the Rindler time coordinate τ to τ_E. The transformations in Eq. (3.135) go over to

$$x_E = \xi \cos g\tau_E; \qquad t_E = \xi \sin g\tau_E$$

(3.140)

when we analytically continue to the Euclidean sector. The corresponding line interval

$$-ds_E^2 = dt_E^2 + d\mathbf{x}^2 = \xi^2 d(g\tau_E)^2 + d\xi^2 + dy^2 + dz^2$$

(3.141)

clearly shows that we are just going over from planar coordinates to polar coordinates in the Euclidean $(g\tau_E, \xi)$ coordinates with $\theta \equiv g\tau_E$ having a period of 2π.

This situation is shown in Fig. 3.5. The field configuration $\phi(\mathbf{x})$ on the $t_E = 0$ surface can be thought of as given by two functions $\phi_L(\mathbf{x}), \phi_R(\mathbf{x})$ in the left and right halves $x < 0$ and $x > 0$ respectively. Therefore, the ground state wave functional $\Psi_{\mathrm{gs}}[\phi(\mathbf{x})]$ now becomes a functional of these two functions: $\Psi_{\mathrm{gs}}[\phi(\mathbf{x})] = \Psi_{\mathrm{gs}}[\phi_L(\mathbf{x}), \phi_R(\mathbf{x})]$. The path integral expression in Eq. (3.139) now reads

$$\Psi_{\mathrm{gs}}[\phi_L(\mathbf{x}), \phi_R(\mathbf{x})] \propto \int_{t_E=0; \phi=(\phi_L, \phi_R)}^{t_E=\infty; \phi=(0,0)} \mathcal{D}\phi \, e^{-A}$$

(3.142)

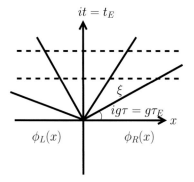

Figure 3.5: Analytic extension to the imaginary time in two different time coordinates in the presence of a horizon. When one uses the path integral to determine the ground state wave functional on the $t_E = 0$ surface, one needs to integrate over the field configurations in the upper half ($t_E > 0$) with a boundary condition on the field configuration on $t_E = 0$. This can be done either by using a series of hypersurfaces parallel to the horizontal axis (shown by broken lines) or by using a series of hypersurfaces corresponding to the radial lines. Comparing the two results, one can show that the ground state in one coordinate system appears as a thermal state in the other.

But from Fig. 3.5 it is obvious that this path integral could also be evaluated in the polar coordinates by varying the angle $\theta = g\tau_E$ from 0 to π. While the evolution in t_E will take the field configuration from $t_E = 0$ to $t_E \to \infty$, the same time evolution gets mapped in terms of τ_E into evolving the "angular" coordinate τ_E from 0 to π/g. (This is clear from Fig. 3.5.) It is obvious that the entire upper half-plane $t > 0$ is covered in two completely different ways in terms of the evolution in t_E compared to the evolution in τ_E. In (t_E, x) coordinates, we vary x in the range $(-\infty, \infty)$ for each t_E and vary t_E in the range $(0, \infty)$. In (τ_E, ξ) coordinates, we vary ξ in the range $(0, \infty)$ for each τ_E and vary τ_E in the range $(0, \pi/g)$. When $\theta = 0$, the field configuration corresponds to $\phi = \phi_R$ and when $\theta = \pi$ the field configuration corresponds to $\phi = \phi_L$. Therefore Eq. (3.142) can also be expressed as

$$\Psi_{\mathrm{gs}}[\phi_L(\mathbf{x}), \phi_R(\mathbf{x})] \propto \int_{g\tau_E=0;\phi=\phi_R}^{g\tau_E=\pi;\phi=\phi_L} \mathcal{D}\phi\, e^{-A} \qquad (3.143)$$

. Let H_R be some Rindler Hamiltonian which is the (Euclidean version of the) Hamiltonian that describes the evolution in terms of the proper time coordinate of the accelerated observer. Then, in the Heisenberg picture, 'rotating' from $g\tau_E = 0$ to $g\tau_E = \pi$ is a time evolution governed by H_R. So the path integral Eq. (3.143) can also be represented as a matrix element of the Rindler Hamiltonian, H_R giving us the result:

$$\Psi_{\mathrm{gs}}[\phi_L(\mathbf{x}), \phi_R(\mathbf{x})] \propto \int_{g\tau_E=0;\phi=\phi_R}^{g\tau_E=\pi;\phi=\phi_L} \mathcal{D}\phi\, e^{-A} = \langle \phi_L | e^{-(\pi/g)H_R} | \phi_R \rangle \quad (3.144)$$

If we denote the proportionality constant by C, then the normalization condition

$$
\begin{aligned}
1 &= \int \mathcal{D}\phi_L\, \mathcal{D}\phi_R \left| \Psi_{\mathrm{gs}}[\phi_L(\mathbf{x}), \phi_R(\mathbf{x})] \right|^2 \\
&= C^2 \int \mathcal{D}\phi_L\, \mathcal{D}\phi_R \, \langle \phi_L | e^{-\pi H_R/g} | \phi_R \rangle \, \langle \phi_R | e^{-\pi H_R/g} | \phi_L \rangle \\
&= C^2 \, \mathrm{Tr}\left(e^{-2\pi H_R/g} \right) \qquad (3.145)
\end{aligned}
$$

fixes the proportionality constant C, allowing us to write the normalized vacuum functional in the form:

$$\Psi_{\mathrm{gs}}[\phi_L(\mathbf{x}), \phi_R(\mathbf{x})] = \frac{\langle \phi_L | e^{-\pi H_R/g} | \phi_R \rangle}{\left[\mathrm{Tr}(e^{-2\pi H_R/g}) \right]^{1/2}} \qquad (3.146)$$

From this result, we can show that for operators $\mathcal{O}$ made out of variables having support in $x > 0$, the vacuum expectation values $\langle \mathrm{vac}| \mathcal{O}(\phi_R)|\mathrm{vac}\rangle$ become thermal expectation values. This arises from straightforward algebra of inserting a complete set of states appropriately:

$$
\begin{aligned}
\langle \mathrm{vac}| \mathcal{O}(\phi_R)|\mathrm{vac}\rangle &= \sum_{\phi_L} \sum_{\phi_R^{(1)}, \phi_R^{(2)}} \Psi_{\mathrm{gs}}[\phi_L, \phi_R^{(1)}] \langle \phi_R^{(1)} | \mathcal{O}(\phi_R) | \phi_R^{(2)} \rangle \Psi_{\mathrm{gs}}[\phi_R^{(2)}, \phi_L] \\
&= \sum_{\phi_L} \sum_{\phi_R^{(1)}, \phi_R^{(2)}} \frac{\langle \phi_L | e^{-(\pi/g)H_R} | \phi_R^{(1)} \rangle \langle \phi_R^{(1)} | \mathcal{O} | \phi_R^{(2)} \rangle \langle \phi_R^{(2)} | e^{-(\pi/g)H_R} | \phi_L \rangle}{Tr(e^{-2\pi H_R/g})} \\
&= \frac{Tr(e^{-2\pi H_R/g}\mathcal{O})}{Tr(e^{-2\pi H_R/g})} \qquad (3.147)
\end{aligned}
$$

Thus, tracing over the field configuration ϕ_L in the region $x < 0$ (behind the horizon) leads to a thermal density matrix $\rho \propto \exp[-(2\pi/g)H]$ for the observables in the region $x > 0$. In particular, the expectation value of the number operator will be a thermal spectrum at the temperature $T = g/2\pi$ in natural units. So, an accelerated observer will consider the vacuum state of the inertial frame to be a thermal state with a temperature proportional to his acceleration. This result (called the *Davies-Unruh effect*) contains the essence of more complicated results in quantum field theory in curved spacetimes, like for example the association of a temperature with black holes.[47]

This result shows the limitations of concepts developed within the context of Lorentz invariant quantum field theory; they can run into conceptual non-trivialities in the presence of gravity described by a curved spacetime.

3.5 Quantizing the Complex Scalar Field

Let us get back to QFT in the inertial frame and consider a complex scalar field. This is completely straightforward and follows exactly the same route as that of the real scalar field except for the doubling of degrees of freedom because $\phi \neq \phi^\dagger$. This has the consequence that the expansion of the field in terms of the creation and annihilation operators now takes the form

$$\phi(x) = \int \frac{d^3\mathbf{p}}{(2\pi)^3} \frac{1}{\sqrt{2\omega_\mathbf{p}}} \left(a_\mathbf{p} e^{-ipx} + b_\mathbf{p}^\dagger e^{ipx}\right) \tag{3.148}$$

and

$$\phi^\dagger(x) = \int \frac{d^3\mathbf{p}}{(2\pi)^3} \frac{1}{\sqrt{2\omega_\mathbf{p}}} \left(a_\mathbf{p}^\dagger e^{ipx} + b_\mathbf{p} e^{-ipx}\right) \tag{3.149}$$

which is precisely what we found earlier in Chapter 1 in Eq. (1.129). The creation and annihilation operators now satisfy the commutation rules

$$\left[a_\mathbf{p}, a_\mathbf{q}^\dagger\right] = \left[b_\mathbf{p}, b_\mathbf{q}^\dagger\right] = (2\pi)^3\, \delta(\mathbf{p} - \mathbf{q}) \tag{3.150}$$

with all other commutators vanishing. Obviously we now have double the number of creation and annihilation operators compared to a real scalar field, because when $\phi \neq \phi^\dagger$ the antiparticle is distinct from the particle while for a real scalar field with $\phi = \phi^\dagger$, the antiparticle is the same as the particle.

One can also obtain the same results by treating $\phi(x)$ and $\phi^\dagger(x)$ as two independent dynamical variables and imposing the canonical commutation rules between them and their conjugate momenta. It should be noted, however, that for the Lagrangian given by $L = \partial_a \phi^\dagger \partial^a \phi - m^2 \phi^\dagger \phi$, the canonical momentum for ϕ is $\pi^a_{(\phi)} = (\partial L/\partial(\partial_a \phi)) = \partial^a \phi^\dagger$ and vice-versa. (This is in contrast to the real scalar field for which the canonical momentum corresponding to ϕ is just $\dot{\phi}$.) The algebra proceeds exactly as in the case of the real scalar field and one finds that both ϕ and $\phi^\dagger$ satisfy the Klein-Gordon equation, allowing the expansions in Eq. (3.148) and Eq. (3.149).

Having quantized the scalar field (real and complex), we can work out all sorts of vacuum expectation values involving the fields by using its decomposition in terms of the creation and annihilation operators. But we have already done all these in the last chapter and hence will only recall a couple of key results for future reference. First, we know that the

[47]Mathematically, this result arises because analytic continuation to Euclidean time is a Lorentz invariant procedure but *not* a generally covariant procedure. This is obvious from the fact that the Euclidean metric expressed in terms of τ_E exhibits periodicity in the imaginary time with a period $\beta = (2\pi/g)$. We have seen in Sect. 1.2.3 that the period of the imaginary time can be interpreted as the inverse temperature which is precisely what happens in this case.

vacuum expectation value of the time ordered product is given by (see e.g., Eq. (1.128)):

$$\langle 0|T[\phi(x_2)\phi^\dagger(x_1)]|0\rangle = G(x_2;x_1) = \int \frac{d^4p}{(2\pi)^4} \frac{i}{(p^2 - m^2 + i\epsilon)} e^{-ip\cdot x}$$

$$= \int d\Omega_{\mathbf{p}} \left[\theta(t)e^{-ipx} + \theta(-t)e^{ipx}\right] \qquad (3.151)$$

This quantity (usually called the *Feynman propagator*) will play a key role in our future discussions. The two individual pieces in this expression are given by (see Eq. (1.133)):

$$\langle 0|\phi(x_2)\phi^\dagger(x_1)|0\rangle = \int d\Omega_{\mathbf{p}} e^{-ipx} = G_+(x_2;x_1)$$

$$\langle 0|\phi^\dagger(x_1)\phi(x_2)|0\rangle = \int d\Omega_{\mathbf{p}} e^{+ipx} = G_-(x_2;x_1) \qquad (3.152)$$

Further, the commutator of the fields at arbitrary times is given by (see Eq. (1.89)):

$$[\phi(x_2), \phi^\dagger(x_1)] = G_+(x_2;x_1) - G_+(x_1;x_2) = \int d\Omega_{\mathbf{p}}[e^{-ipx} - e^{+ipx}]$$

$$= \int d\Omega_{\mathbf{p}} e^{i\mathbf{p}\cdot\mathbf{x}} \left[e^{-i\omega_{\mathbf{p}}t} - e^{i\omega_{\mathbf{p}}t}\right] \qquad (3.153)$$

[48] In contrast, note that $[\phi(x),\phi(y)] = [\phi^\dagger(x),\phi^\dagger(y)] = 0$ for any pair of events for the complex scalar field.

This commutator vanishes outside the light cone i.e., when x_1 and x_2 are related by a spacelike interval.[48] As we explained in Sect. 1.5.1, this is crucial for the validity of causality in field theory. The observables of the theory are now built from field operators — usually quadratic functionals of the field operators like the Hamiltonian etc. — and causality is interpreted as the measurement of any such observable at a spacetime location not affecting observables which are causally disconnected from that event. We no longer think in terms of a single particle description or localized position eigenstates of particles.

The main difference between the real and complex scalar field lies in the fact that now we have an antiparticle which is distinct from the particle so that we need two sets of creation and annihilation operators in Eq. (3.148) and Eq. (3.149). The existence of a distinct antiparticle created by $b_{\mathbf{p}}^\dagger$ is reflected in all the physical quantities which we compute. Following exactly the same steps as in the case of real scalar fields, we can now compute the (normal ordered) Hamiltonian and the momentum, obtaining

Exercise 3.17: Verify all these.

$$: H := \int \frac{d^3\mathbf{p}}{(2\pi)^3} \omega_{\mathbf{p}} \left(a_{\mathbf{p}}^\dagger a_{\mathbf{p}} + b_{\mathbf{p}}^\dagger b_{\mathbf{p}}\right) \qquad (3.154)$$

$$P^\alpha = \int \frac{d^3\mathbf{p}}{(2\pi)^3} p^\alpha \left(a_{\mathbf{p}}^\dagger a_{\mathbf{p}} + b_{\mathbf{p}}^\dagger b_{\mathbf{p}}\right) \qquad (3.155)$$

which shows that both the particle and the antiparticle contribute additively to the energy and momentum.

To see the physical distinction between the particle and the antiparticle, we will compute the operator corresponding to the conserved charge of the complex scalar field introduced in Eq. (3.44). The invariance of the action for complex scalar field under the transformations in Eq. (3.42) leads to a conserved current J^a, the zeroth component of which is given by (see Eq. (3.43)):

$$J^0 = \sum_A \pi_{(A)} \delta\phi_{(A)} = -iq\,\phi\,\partial_t\phi^\dagger + iq\,\phi^\dagger\partial_t\phi = iq(\phi^\dagger\partial_t\phi - \phi\partial_t\phi^\dagger) \quad (3.156)$$

The integral of J^0 over all space gives the charge Q. Substituting the expansion in Eq. (3.148) and Eq. (3.149) into Eq. (3.44), we get

$$
\begin{aligned}
Q &= i\int d^3\mathbf{x}\,\phi^\dagger \overleftrightarrow{\partial_0}\, \phi = i\int d^3\mathbf{x}\,\frac{d^3\mathbf{q}}{(2\pi)^3}\frac{1}{\sqrt{2\omega_\mathbf{q}}}\frac{d^3\mathbf{p}}{(2\pi)^3}\frac{1}{\sqrt{2\omega_\mathbf{p}}}\\
&\quad \times\left[\left(a_\mathbf{q}^\dagger e^{iqx} + b_\mathbf{q}e^{-iqx}\right)\partial_0\left(a_\mathbf{p}e^{-ipx} + b_\mathbf{p}^\dagger e^{ipx}\right)\right.\\
&\qquad\left. -\left(\left(a_\mathbf{p}e^{-ipx} + b_\mathbf{p}^\dagger e^{ipx}\right)\partial_0\left(a_\mathbf{q}^\dagger e^{iqx} + b_\mathbf{q}e^{-iqx}\right)\right)\right]\\
&= \int d^3\mathbf{x}\,\frac{d^3\mathbf{q}}{(2\pi)^3}\frac{1}{\sqrt{2\omega_\mathbf{q}}}\frac{d^3\mathbf{p}}{(2\pi)^3}\frac{1}{\sqrt{2\omega_\mathbf{p}}}\\
&\quad \times\left[\left(a_\mathbf{q}^\dagger e^{iqx} + b_\mathbf{q}e^{-iqx}\right)\omega_\mathbf{p}\left(a_\mathbf{p}e^{-ipx} - b_\mathbf{p}^\dagger e^{ipx}\right)\right.\\
&\qquad\left. +\omega_\mathbf{q}\left(a_\mathbf{q}^\dagger e^{iqx} - b_\mathbf{q}e^{-iqx}\right)\left(a_\mathbf{p}e^{-ipx} + b_\mathbf{p}^\dagger e^{ipx}\right)\right] \quad (3.157)
\end{aligned}
$$

Different terms respond differently to the integration over $d^3\mathbf{x}$. We get $\delta(\mathbf{p} - \mathbf{q})$ from the terms having $a_\mathbf{q}^\dagger a_\mathbf{p}$ and $b_\mathbf{q}b_\mathbf{p}^\dagger$ while the terms with $b_\mathbf{q}a_\mathbf{p}$ and $a_\mathbf{q}^\dagger b_\mathbf{p}^\dagger$ lead to $\delta(\mathbf{p} + \mathbf{q})$. In either case, $|\mathbf{p}| = |\mathbf{q}|$ leading to $\omega_\mathbf{q} = \omega_\mathbf{p}$. This fact leads to cancellation of terms with $b_\mathbf{q}a_\mathbf{p}$ and $a_\mathbf{q}^\dagger b_\mathbf{p}^\dagger$ leaving behind only the terms with $a_\mathbf{q}^\dagger a_\mathbf{p}$ and $b_\mathbf{q}b_\mathbf{p}^\dagger$. In these terms, the time dependence goes away[49] when we use $\omega_\mathbf{q} = \omega_\mathbf{p}$. So, we finally get:

$$
\begin{aligned}
Q &= \int \frac{d^3\mathbf{p}}{(2\pi)^3}\frac{1}{\sqrt{2\omega_\mathbf{p}}}\frac{d^3\mathbf{q}}{(2\pi)^3}\frac{1}{\sqrt{2\omega_\mathbf{q}}}(2\pi)^3\delta(\mathbf{p} - \mathbf{q})2\omega_\mathbf{p}\left(a_\mathbf{q}^\dagger a_\mathbf{p} - b_\mathbf{q}b_\mathbf{p}^\dagger\right)\\
&= \int \frac{d^3\mathbf{p}}{(2\pi)^3}\left(a_\mathbf{p}^\dagger a_\mathbf{p} - b_\mathbf{p}b_\mathbf{p}^\dagger\right) \quad (3.158)
\end{aligned}
$$

We are again in a bit of trouble because the vacuum expectation value of the second term will not vanish and will lead to an infinite amount of charge due to antiparticles. This can be avoided by the usual remedy of normal ordering Q to obtain[50]

$$: Q := \int \frac{d^3p}{(2\pi)^3}\left(a_\mathbf{p}^\dagger a_\mathbf{p} - b_\mathbf{p}^\dagger b_\mathbf{p}\right) \quad (3.159)$$

This expression shows that the net charge is the *difference* of the charges contributed by the particles and antiparticles. So, if the particles carry $+1$ unit of charge, then the antiparticles should carry -1 unit of charge. This provides a simple and important procedure for distinguishing between processes involving the particle from those with the antiparticle.

We conclude this section with a discussion of a couple of more features related to the existence of a conserved current and charge. From the definition of the charge as an integral over J^0 and Eq. (3.156), we can compute

[49] Since we know that charge is conserved, we could have done all the calculation at $t = 0$, but once in a while it is good not to be clever and see explicitly how cancellations occur.

[50] The normal ordering is somewhat easier to justify in this case because the classical expression for the charge involves the product of ϕ^* and $\dot\phi$; in quantum theory this could have been written either as $\phi^\dagger\dot\phi$, or as $\dot\phi\phi^\dagger$ or as some symmetric combination. Since $\phi^\dagger$ does not commute with ϕ, each of these choices will lead to a different expression for the total charge and it makes sense (may be?) to settle the issue by demanding that the vacuum should have zero charge.

the commutator $[Q, \phi_{(A)}(0, \mathbf{x})]$ to be

$$[Q, \phi_{(A)}(\mathbf{x})] = \int d^3\mathbf{y}\, \delta\phi_{(B)}(\mathbf{y})[\pi_{(B)}(\mathbf{y}), \phi_{(A)}(\mathbf{x})] = i[\delta\phi_{(A)}(\mathbf{x})] \quad (3.160)$$

This result shows that one can think of the charge Q as the generator of the infinitesimal transformation $\phi_{(A)} \to \phi_{(A)} + \delta\phi_{(A)}$. The well known special case of this is the translation in time and space generated by the 4-momentum.

Since a symmetry transformation generates a conserved current and an associated charge, some of the crucial features of spontaneous symmetry breaking can be related to the existence of the charge. For example, we saw in Sect. 3.1.4 that the presence of degenerate vacua led to a zero mass particle in the theory. This result happens to be very general and independent of the detailed structure of the Lagrangian in the relativistic field theory. We will describe briefly how one can prove such a result in fairly general terms.

Let us consider a system with a conserved charge Q so that $[Q, H] = 0$. The ground state $|0\rangle$ of the system is taken to have zero energy so that $H|0\rangle = 0$. Since $[Q, H]|0\rangle = 0 = HQ|0\rangle$ it follows that $Q|0\rangle$ also has zero energy. When this state $Q|0\rangle$ differs from $|0\rangle$, we have a situation with degenerate vacua. We want to show that under this circumstance there will exist zero mass particles in the theory; that is, we have to show that there exist states with some momentum $\mathbf{k}$ and energy E such that $E \to 0$ when $|\mathbf{k}| \to 0$, which is the relativistic characterization of a zero mass state. To do this, consider the state defined by

$$|s\rangle = \mathcal{R}|0\rangle \equiv \int d^3\mathbf{x}\, e^{i\mathbf{k}\cdot\mathbf{x}} J_0(\mathbf{x})|0\rangle \quad (3.161)$$

where $J_0(\mathbf{x})$ is the charge density. It is easy to see that this state is an eigenstate of the momentum operator P^α with eigenvalue k_α. We have

$$P^\alpha|s\rangle = \int d^3\mathbf{x}\, e^{i\mathbf{k}\cdot\mathbf{x}}[P^\alpha, J_0]|0\rangle = i \int d^3\mathbf{x}\, e^{i\mathbf{k}\cdot\mathbf{x}}(\partial_\alpha J^0)|0\rangle \quad (3.162)$$

where we have used the facts that $P^\alpha|0\rangle = 0$ and $[P^\alpha, J_0] = -i\partial^\alpha J_0 = i\partial_\alpha J^0$. Doing an integration by parts and ignoring the surface term, we get

$$P^\alpha|s\rangle = -i \int d^3\mathbf{x}(\partial_\alpha e^{i\mathbf{k}\cdot\mathbf{x}})J^0|0\rangle = k_\alpha \int d^3\mathbf{x}\, e^{i\mathbf{k}\cdot\mathbf{x}} J^0|0\rangle = k_\alpha|s\rangle \quad (3.163)$$

Consider now what happens to $|s\rangle$ when we take the limit of $|\mathbf{k}| \to 0$. We see that, in this limit, $|s\rangle \to Q|0\rangle$ which we know is a state with zero energy. In other words, the state $|s\rangle$ has the property that when $|\mathbf{k}| \to 0$, its energy vanishes. This shows that there are indeed zero mass excitations in the theory.

3.6 Quantizing the Electromagnetic Field

We will next take up the example of the electromagnetic field which — historically — started it all. Our aim is to quantize the classical electromagnetic field and obtain the photons as its quanta. This actually turns out to be much more difficult than one would have first imagined because of the gauge invariance. So we will first describe the difficulties qualitatively and then take up the quantization.

We saw that quantizing a field involves identifying the relevant dynamical variables (which are the ones that you vary in the action to get the classical field equations), calculating the corresponding canonical momenta and then imposing the commutation rules between the two. The Lagrangian for the electromagnetic field is proportional to $F^{ij}F_{ij}$ where $F_{ij} = \partial_i A_j - \partial_j A_i$ and we vary A_i to obtain the classical field equations. This identifies A_i as the dynamical variables. To determine the canonical momenta corresponding to A_i, you need to compute $\partial L/\partial(\partial_0 A_i)$ which turns out to be proportional to F^{0i}. The trouble begins here.

The canonical momentum conjugate to A_0 is F^{00}, which, of course, vanishes identically.[51] So we cannot upgrade all A_is to operator status and impose the commutation rules. Of course, this is related to the fact that the physical observables of the theory are F^{ik} and not A^i, and, because of gauge invariance, there is redundancy in the latter variables.

To get around this difficulty, we need to impose some gauge condition and work with true dynamical variables. In principle, one can impose any one of the many possible gauge conditions, but most of them will lead to fairly complicated description of the quantum theory. Two natural conditions are the following: (i) $\phi = 0$, $\nabla \cdot \mathbf{A} = 0$; (ii) $\partial_j A^j = 0$. The first condition (which is what we will work with) has the disadvantage that it is not manifestly Lorentz invariant and so you need to check that final results are Lorentz invariant explicitly by hand. Further, when we try to impose the condition

$$\left[A^\alpha(t, \mathbf{x}), \dot{A}^\beta(t, \mathbf{y})\right] = -i\eta^{\alpha\beta}\delta(\mathbf{x} - \mathbf{y}) = i\delta^{\alpha\beta}\,\delta(\mathbf{x} - \mathbf{y}) \tag{3.164}$$

we run into trouble if we take the divergence of both sides with respect to x^α. We find that

$$\left[\partial_\alpha A^\alpha(t, \mathbf{x}), \dot{A}^\beta(t, \mathbf{y})\right] = -i\partial^\beta\delta(\mathbf{x} - \mathbf{y}) \neq 0 \tag{3.165}$$

violating the $\nabla \cdot \mathbf{A} = 0$ condition. The second gauge[52] is Lorentz invariant but if we impose the commutation relation

$$\left[A^i(t, \mathbf{x}), \dot{A}^j(t, \mathbf{y})\right] = -i\eta^{ij}\delta(\mathbf{x} - \mathbf{y}) \tag{3.166}$$

The time component again creates a problem because the right hand side now has a flipped sign for the time component vis-a-vis the space component. This leads to the existence of negative norm states and related difficulties. All these *can* be handled but not as trivially as in the case of scalar fields. We will use the first gauge and get around the difficulty mentioned above by a different route.

3.6.1 Quantization in the Radiation Gauge

Let us consider a free electromagnetic field, with the vector potential $A^i = (\phi, \mathbf{A})$ satisfying the gauge conditions: $\phi = 0, \nabla \cdot \mathbf{A} = 0$. As usual we expand $\mathbf{A}(t, \mathbf{x})$ in Fourier space as:

$$\mathbf{A}(t, \mathbf{x}) = \int \frac{d^3\mathbf{k}}{(2\pi)^3}\mathbf{q}_\mathbf{k}(t)\exp i\mathbf{k}.\mathbf{x} = \int \frac{d^3\mathbf{k}}{(2\pi)^3}\left(\boldsymbol{a}_\mathbf{k}(t) + \boldsymbol{a}^*_{-\mathbf{k}}(t)\right)e^{i\mathbf{k}.\mathbf{x}} \tag{3.167}$$

[51] If you write the Lagrangian as $(E^2 - B^2)$ and note that $\mathbf{E} = -\nabla\phi - (\partial\mathbf{A}/\partial t)$, $\mathbf{B} = \nabla \times \mathbf{A}$, it is obvious that the Lagrangian has $\dot{\mathbf{A}}$ but no $\dot{\phi}$. So clearly the momentum conjugate to ϕ vanishes.

[52] Almost universally called the *Lorentz gauge* because it was discovered by *Lorenz*; yet another example of dubious naming conventions used in this subject. This gauge was first used by L. V. Lorenz in 1867 when the more famous H.A. Lorentz was just 14 years old and wasn't particularly concerned about gauges. We shall give the credit to Lorenz and use the correct spelling. The original reference is L.Lorenz, *On the Identity of the Vibrations of Light with Electrical Currents*, Philos. Mag. **34**, 287-301 (1867).

which has exactly the same structure as in the case of the scalar field, given by Eq. (3.105). The condition $\nabla \cdot \mathbf{A} = 0$, translates into

$$\mathbf{k} \cdot \mathbf{q_k} = \mathbf{k} \cdot \boldsymbol{a_k} = \mathbf{k} \cdot \boldsymbol{a_k^*} = 0; \qquad (3.168)$$

i.e., for every value of $\mathbf{k}$, the vector $\boldsymbol{a_k}$ is perpendicular to $\mathbf{k}$. This allows the vectors $\mathbf{q_k}, \boldsymbol{a_k}$ to have two components $\boldsymbol{a_{k\lambda}}$ ($\lambda = 1, 2$) and $q_{\mathbf{k}\lambda}(\lambda = 1, 2)$ in the plane perpendicular to $\mathbf{k}$ with $\{\boldsymbol{a_{k1}}, \boldsymbol{a_{k2}}, \mathbf{k}\}$ forming an orthogonal system. The electric and magnetic fields corresponding to this vector potential $A^i = (0, \mathbf{A})$ are

$$\mathbf{E} = -\dot{\mathbf{A}} = -\int \frac{d^3\mathbf{k}}{(2\pi)^3} \dot{\mathbf{q}}_k e^{i\mathbf{k}\cdot\mathbf{x}}; \quad \mathbf{B} = \nabla \times \mathbf{A} = i \int \frac{d^3\mathbf{k}}{(2\pi)^3} \left(\mathbf{k} \times \mathbf{q_k}\right) e^{i\mathbf{k}\cdot\mathbf{x}}.$$
$$(3.169)$$

Substituting this into the action in Eq. (3.63) and using

$$(\mathbf{k} \times \mathbf{q_k}) \cdot (\mathbf{k} \times \mathbf{q_k^*}) = \mathbf{q_k^*} \cdot [(\mathbf{k} \times \mathbf{q_k}) \times \mathbf{k}] = (\mathbf{q_k^*} \cdot \mathbf{q_k}) \, k^2 \qquad (3.170)$$

we can again express the action as that of a sum of oscillators:

$$\mathcal{A} = \frac{1}{2} \int d^3\mathbf{x} \left(E^2 - B^2\right) = \frac{1}{2} \sum_{\lambda=1,2} \int \frac{d^3\mathbf{k}}{(2\pi)^3} \left(|\dot{q}_{\mathbf{k}\lambda}|^2 - k^2|q_{\mathbf{k}\lambda}|^2\right). \quad (3.171)$$

So, once again, we find that the action for the electromagnetic field can be expressed as a sum of the actions for harmonic oscillators with each oscillator labeled by the wave vector $\mathbf{k}$ and a polarization. The polarization part is the only extra complication compared to the scalar field and the rest of the mathematics proceeds exactly as before. The dispersion relation now corresponds to $\omega_{\mathbf{k}} = |\mathbf{k}|$ showing that the quanta are massless, as we expect for photons.

The wave equation satisfied by $\mathbf{A}\,(t, \mathbf{x})$ implies that $\mathbf{q_k}\,(t)$ satisfies the harmonic oscillator equation $\ddot{\mathbf{q}}_k + k^2\mathbf{q_k} = 0$, allowing us to take the time evolution to be $\boldsymbol{a_k} \propto \exp\left(-ikt\right)$. With the usual normalization, this gives the mode expansion for the vector potential in the standard form

$$\mathbf{A}\,(t, \mathbf{x}) = \sum_{\lambda=1,2} \int \frac{d^3\mathbf{k}}{(2\pi)^3} \frac{1}{\sqrt{2\omega_{\mathbf{k}}}} \left(\boldsymbol{a_{k\lambda}} e^{-ikx} + \boldsymbol{a_{k\lambda}^\dagger} e^{ikx}\right) \qquad (3.172)$$

It is convenient in manipulations to separate out the vector nature from the creation and annihilation operators and write $\boldsymbol{a_{k\lambda}} = \boldsymbol{\epsilon}(\mathbf{k}, \lambda) a_{\mathbf{k}\lambda}$ where $a_{\mathbf{k}\lambda}$ is the annihilation operator for the photon with momentum $\mathbf{k}$ and polarization λ while $\boldsymbol{\epsilon}(\mathbf{k}, \lambda)$ is just a c-number vector carrying the vector nature of $\mathbf{A}$. Since $\boldsymbol{\epsilon}(\mathbf{k}, 1), \boldsymbol{\epsilon}(\mathbf{k}, 2)$ and $\mathbf{k}/|\mathbf{k}|$ form an orthonormal basis, they satisfy the constraint:[53]

$$\sum_{\lambda=1}^{2} \epsilon^{\alpha}(\mathbf{k}, \lambda) \, \epsilon^{\beta}(\mathbf{k}, \lambda) + \frac{k^{\alpha}k^{\beta}}{\mathbf{k}^2} = \delta^{\alpha\beta} \qquad (3.173)$$

The creation and annihilation operators satisfy the standard commutation rules, now with an extra polarization index:

$$\left[a_{\mathbf{k}'\lambda'}, a_{\mathbf{k}\lambda}^\dagger\right] = \delta(\mathbf{k}' - \mathbf{k})\delta_{\lambda\lambda'} \,, \qquad [a_{\mathbf{k}'\lambda'}, a_{\mathbf{k}\lambda}] = \left[a_{\mathbf{k}'\lambda'}^\dagger, a_{\mathbf{k}\lambda}^\dagger\right] = 0 \quad (3.174)$$

[53]Notice that the operator $a_{\boldsymbol{k}\lambda}^\dagger$ creates a photon with a polarization state labelled by λ and momentum labelled by $\boldsymbol{k}$ which will define a direction (of propagation) in the 3-dimensional space. The polarization vectors $\epsilon^\alpha(\boldsymbol{k}, \lambda)$ necessarily have to depend on the momentum vector $\mathbf{k}$ of the photon. For example, if we try to pick 3 four-vectors with components $\epsilon_a^1 = (0, 1, 0, 0)$, $\epsilon_a^2 = (0, 0, 1, 0)$ and $\epsilon_a^3 = (0, 0, 0, 1)$ as the bases, the Lorentz transformations will mix them up with the "time-like polarization" vector $\epsilon_a^0 = (1, 0, 0, 0)$.

Everything else proceeds as before. The Hamiltonian governing the free electromagnetic field can be expressed as a sum of Hamiltonians for harmonic oscillators, with each oscillator labeled by a wave vector $\mathbf{k}$ and polarization index α. The quantum states of the system are labeled by by a set of integers $|\{n_{\mathbf{k}\alpha}\}\rangle$, one for each oscillator labeled by $\mathbf{k}\alpha$. The expectation value of H in such a state will be

$$E = \sum_{\alpha=1,2} \int \frac{d^3\mathbf{k}}{(2\pi)^3} \hbar\omega_{\mathbf{k}} \left(n_{\mathbf{k}\alpha} + \frac{1}{2} \right) \tag{3.175}$$

where the second term is the the energy of the system when all the oscillators are in the ground state — which can again be removed by normal ordering.

You might have thought all this is pretty obvious and standard. But, if you think about it, the second term in Eq. (3.172) involving $\boldsymbol{a}_{k\lambda}^{\dagger}$ is precisely the term about which we made a song and dance in Eq. (1.130) while discussing the scalar field. *It is related to propagating negative energy antiphotons backward in time!* The quantization of the electromagnetic field based on the separation in Eq. (3.172) describes photons and antiphotons, carrying positive frequency and negative frequency modes and propagating forward and backward in the language we have used earlier.[54] This term creates an antiphoton from the vacuum, but since the antiphoton is the same as the photon, you think of it as a creation of photons — which every light bulb does. This is the simplest example of a particle popping out of nowhere thereby making the standard description in terms of a Schrodinger equation completely inadequate. The more esoteric processes — like, for example, $e^+ - e^-$ annihilation producing some exotic new particles is fundamentally no different from a light bulb emitting a photon through the action of the second term which has $\boldsymbol{a}_{k\lambda}^{\dagger}$ in it.

Let us next take a look at the commutation relation between the fields. Once again, we have quantized a field by mapping the relevant dynamical variables to a bunch of independent harmonic oscillators and then quantizing the oscillators. While discussing the scalar field, we said that this procedure is equivalent to imposing the canonical commutation relations between the dynamical variables and the canonical momenta; indeed — in the case of the scalar field — one can *derive* the commutation relations between the creation and annihilation operators from the canonical commutation relations between the dynamical variables and the canonical momenta. But in the case of the electromagnetic field in the radiation gauge, the situation is somewhat different. We know from Eq. (3.164) and Eq. (3.165) that we *cannot* impose the standard commutation relations and maintain consistency with the radiation gauge condition. So, one cannot start from Eq. (3.164) and get the commutation rules for the creation and annihilation operators in a straightforward manner.

Fortunately, our procedure — of first identifying the relevant oscillators and then just quantizing them — bypasses this problem. Having quantized the system by this procedure, we can now go back and *compute* the relevant commutators between the field and its momentum and see what happens. The canonical momenta[55] corresponding to A^{α}:

$$\pi^{\alpha} = \frac{\partial L}{\partial(\partial_0 A_{\alpha})} = -\partial_0 A^{\alpha} = E^{\alpha} \tag{3.176}$$

[54]While every textbook makes some noise about these "new features" when it quantizes the scalar field, these are not emphasized in the context of the familiar electromagnetic field. You talk about π^+ and π^-, electrons and positrons and their strange relationships but the same thing happens with photons and antiphotons contained in Eq. (3.172). It is the same maths and same physics (except for $m = 0$).

[55]Note the placement of α and the sign flips due to our signature; the components of $\mathbf{A}, \mathbf{E}$ etc. have, by definition, a superscript index.

has the explicit expansion:

$$\mathbf{E}(t,\mathbf{x}) = \int \frac{d^3\mathbf{k}}{(2\pi)^3\sqrt{2\omega_\mathbf{k}}} \sum_{\lambda=1}^{2} i\omega_\mathbf{k}\boldsymbol{\epsilon}(\mathbf{k},\lambda)\left(a_{\mathbf{k}\lambda}e^{-ikx} - a_{\mathbf{k}\lambda}^\dagger e^{ikx}\right) \quad (3.177)$$

We need to compute $[A^\alpha(t,\mathbf{x}), E^\beta(t,\mathbf{x}')]$ which is given by:

$$\left[A^\alpha(t,\mathbf{x}), E^\beta(t,\mathbf{x}')\right] \qquad\qquad\qquad\qquad\qquad (3.178)$$

$$= \int \frac{d^3\mathbf{k}}{(2\pi)^3\sqrt{2\omega_\mathbf{k}}} \sum_{\lambda=1}^{2} \int \frac{d^3\mathbf{k}'}{(2\pi)^3\sqrt{2\omega_{\mathbf{k}'}}} \sum_{\lambda'=1}^{2} \epsilon^\alpha(\mathbf{k},\lambda)\epsilon^\beta(\mathbf{k}',\lambda')$$

$$\times \left(\left[a_{\mathbf{k}\lambda}, a_{\mathbf{k}'\lambda'}^\dagger\right]e^{-i(kx-k'x')}(-i\omega_{\mathbf{k}'}) + \left[a_{\mathbf{k}\lambda}^\dagger, a_{\mathbf{k}'\lambda'}\right]e^{i(kx-k'x')}(+i\omega_{\mathbf{k}'})\right)$$

$$= \int \frac{d^3\mathbf{k}}{(2\pi)^3 2\omega_\mathbf{k}}(-i\omega_\mathbf{k})\left(e^{i\mathbf{k}\cdot(\mathbf{x}-\mathbf{x}')} + e^{-i\mathbf{k}\cdot(\mathbf{x}-\mathbf{x}')}\right)\sum_{\lambda=1}^{2}\epsilon^\alpha(\mathbf{k},\lambda)\epsilon^\beta(\mathbf{k},\lambda)$$

We now do the usual flip of $\mathbf{k}$ to $-\mathbf{k}$ in the second term and use Eq. (3.173) to obtain:

$$\left[A^\alpha(t,\mathbf{x}), E^\beta(t,\mathbf{x}')\right] = -i\int \frac{d^3\mathbf{k}}{(2\pi)^3}e^{i\mathbf{k}\cdot(\mathbf{x}-\mathbf{x}')}\left(\delta^{\alpha\beta} - \frac{k^\alpha k^\beta}{\mathbf{k}^2}\right)$$

$$\equiv -i\delta_\perp^{\alpha\beta}(\mathbf{x}-\mathbf{x}') \qquad\qquad (3.179)$$

where the last equality defines the *transverse delta function*:

$$\delta_{\alpha\beta}^\perp(\mathbf{x}-\mathbf{x}') \equiv \int \frac{d^3\mathbf{k}}{(2\pi)^3}e^{i\mathbf{k}\cdot(\mathbf{x}-\mathbf{x}')}\left(\delta_{\alpha\beta} - \frac{k_\alpha k_\beta}{\mathbf{k}^2}\right) \qquad (3.180)$$

Its key property is that $\partial_\alpha\delta_\perp^{\alpha\beta}(\mathbf{x}) = 0$ (which is obvious in the momentum space representation) thereby making the commutation relations consistent with $\nabla\cdot\mathbf{A} = 0$. One can write down the explicit form of $\delta_\perp^{\alpha\beta}(\mathbf{x})$ in coordinate space by noting that it can be written, formally, as

$$\delta_{\alpha\beta}^\perp(\mathbf{x}) = \left(\delta_{\alpha\beta} - \partial_\alpha\frac{1}{\nabla^2}\partial_\beta\right)\delta(\mathbf{x}) \equiv (P_\perp)^{\alpha\beta}\,\delta(\mathbf{x}) \qquad (3.181)$$

Exercise 3.18: Prove this.

By defining the inverse of the Laplacian as an integral operator, one can easily show that:

$$\delta_{\alpha\beta}^\perp(\mathbf{x}) = \delta_{\alpha\beta}\,\delta(\mathbf{x}) + \frac{1}{4\pi}\partial_\alpha\partial_\beta\frac{1}{|\mathbf{x}|} \qquad (3.182)$$

[56]The way we approached the problem, this should not worry you; finding the oscillators and quantizing them is the safest procedure, when it can be done. If, instead, you start from the commutator for fields and canonical momenta, you need to first argue — somewhat unconvincingly — that we should introduce the transverse delta function for the commutators and then proceed with the quantization.

The analysis shows that the commutator $[A^\alpha(t,\mathbf{x}), E^\beta(t,\mathbf{x}')]$ does *not* have the value $[i\delta^{\alpha\beta}\delta(\mathbf{x}-\mathbf{x}')$ which we might have naively thought it should have. In the radiation gauge, if we identify the correct harmonic oscillator variables and quantise them, everything works fine and this commutator has[56] a value given by Eq. (3.179).

The price we have paid is the loss of manifest Lorentz invariance and, of course, loss of gauge invariance. The first problem can be avoided by using a Lorentz invariant gauge like the Lorenz gauge. The second problem is going to stay with us because we have to always fix a gauge — naively or in a more sophisticated manner — to isolate the physical variables but we expect the meaningful results to be gauge invariant. We will say more about this in the next section.

Let us next consider the commutator between the fields at arbitrary events. A straightforward calculation gives:

$$
\begin{aligned}
\left[A^{\alpha}(x), A^{\beta}(y)\right] &= \int \frac{d^3\mathbf{k}}{(2\pi)^3} \frac{1}{2\omega_{\mathbf{k}}} \left(\sum_{\lambda=1}^{2} \epsilon^{\alpha}(\mathbf{k}, \lambda) \epsilon^{\beta}(\mathbf{k}, \lambda)\right) \\
&\quad \times \left(e^{-ik(x-y)} - e^{+ik(x-y)}\right) \\
&= \int \frac{d^3\mathbf{k}}{(2\pi)^3} \frac{1}{2\omega_{\mathbf{k}}} \left(\delta^{\alpha\beta} - \frac{k^{\alpha}k^{\beta}}{\mathbf{k}^2}\right) \left(e^{-ik(x-y)} - e^{+ik(x-y)}\right)
\end{aligned}
\tag{3.183}
$$

This can again be expressed in the coordinate space using the non-local projection operator and in terms of the function $\Delta(x) \equiv G_{+}(x) - G_{+}(-x)$ we introduced in Eq. (1.89) but now evaluated for $m = 0$; it is usual to define

$$
D(x) \equiv -i\Delta(x)|_{m=0} = -\frac{1}{2\pi}\theta(t)\delta(x^2)
\tag{3.184}
$$

in the massless case where the last equality is easy to obtain from any of the integral representations for Δ. We find

Exercise 3.19: Prove Eq. (3.184).

$$
\begin{aligned}
\left[A^{\alpha}(x), A^{\beta}(y)\right] &= \left(\delta^{\alpha\beta} - \frac{\partial^{\alpha}\partial^{\beta}}{\nabla^2}\right) \int \frac{d^3\mathbf{k}}{(2\pi)^3} \frac{1}{2\omega_{\mathbf{k}}} \left(e^{-ik(x-y)} - e^{+ik(x-y)}\right) \\
&= (P_{\perp})^{\alpha\beta} [iD(x-y)]
\end{aligned}
\tag{3.185}
$$

A straightforward evaluation of the integrals reveals, at first sight, another problem: this commutator now does not vanish outside the light cone — something we have been claiming is rather sacred! But this is sacred only for observable quantities and A^i is not an observable. (In fact, if we use some other gauge we will get a different result for this commutator; in the Lorenz gauge it *does* happen to vanish outside the light cone.) If the commutator between the electric fields, for example, does not vanish outside the light cone, you will be in real trouble. This, as you might have guessed from the fact that we are still in business, does not happen. The commutator of the electric fields is given by:

Exercise 3.20: Show that $\left[A^{\alpha}(x), A^{\beta}(y)\right]$ is given by $i\delta^{\alpha\beta}D(x - y) + \partial^{\alpha}\partial^{\beta}H(x - y)$ where $H(x) = -(i/8\pi)(1/r)[v\,\mathrm{sgn}(v) - u\,\mathrm{sgn}(u)]$ with $v \equiv t + r$ and $u \equiv t - r$. Hence argue that the commutator does not vanish outside the light cone.

$$
\begin{aligned}
\left[E^{\alpha}(x), E^{\beta}(y)\right] &= \partial_x^0\partial_y^0 \left[A^{\alpha}(x), A^{\beta}(y)\right] = i\partial^0\partial^0 (P_{\perp})^{\alpha\beta} D(x-y) \\
&= i\delta^{\alpha\beta}\partial^0\partial^0 D(x-y) \\
&\quad - \partial^{\alpha}\partial^{\beta} \int \frac{d^3\mathbf{k}}{(2\pi)^3} \frac{1}{2\omega_{\mathbf{k}}} \frac{\omega_{\mathbf{k}}^2}{\mathbf{k}^2} \left(e^{-ik(x-y)} - e^{+ik(x-y)}\right) \\
&= i\left(\delta^{\alpha\beta}\partial^0\partial^0 - \partial^{\alpha}\partial^{\beta}\right) D(x-y)
\end{aligned}
\tag{3.186}
$$

which vanishes for spacelike separation. Further, this commutator has the same expression in all gauges, as one would have expected.

3.6.2 Gauge Fixing and Covariant Quantization

In the case of scalar fields we saw that the propagation amplitude $G(x_2; x_1)$ (or propagator, for short) can be related to the vacuum expectation value of the time ordered product of the fields. In a similar manner, one can obtain the propagator for the photon from the vacuum expectation value of the time ordered product of the vector potential. We would, however, like to have a photon propagator which is Lorentz invariant and this cannot be obtained using the radiation gauge discussed above because it is not Lorentz invariant. While we will not work out in detail the issue of quantizing the electromagnetic field in Lorentz invariant gauges, it will be useful to describe some important features which will be needed in later chapters. We will now take up some of these topics.

We have seen in the last chapter (see Sect. 2.2.1; especially Eq. (2.49) and Eq. (2.48)) that the propagator arises as the inverse of a differential operator D which occurs in the Lagrangian in the form $\phi D \phi$. In the case of the electromagnetic field, the Lagrangian can be rewritten, by removing a total divergence, in the form

$$
-\frac{1}{4} F_{ik} F^{ik} = -\frac{1}{2} \left(\partial_i A_k \right) \left(\partial^i A^k - \partial^k A^i \right) \tag{3.187}
$$
$$
= -\frac{1}{2} \partial_i \left[A_k \partial^i A^k - A_k \partial^k A^i \right] + \frac{1}{2} A_k \left(\delta_i^k \Box - \partial^k \partial_i \right) A^i
$$

Therefore the action can be expressed in the form

$$
\mathcal{A} = -\frac{1}{4} \int d^4 x \, F_{ik} F^{ik} \Longrightarrow \frac{1}{2} \int d^4 x \, A_k \left(\delta_i^k \, \Box - \partial^k \partial_i \right) A_i
$$
$$
= -\frac{1}{2} \int \frac{d^4 p}{(2\pi)^4} \, A_k(p) \left[\delta_i^k p^2 - p^k p_i \right] A^i(p) \tag{3.188}
$$

In arriving at the first step we have ignored the total divergence and the last equality is obtained by writing everything in the momentum space. To find the propagator in the momentum space we need to find the inverse of the matrix $M_i^k = \delta_i^k p^2 - p^k p_i$. But since $M_i^k p^i = 0$ it is obvious that this matrix has an eigenvector p^i with zero eigenvalue. Therefore, its inverse does not exist.

The reason is that the action is invariant under the gauge transformation $A_i \to A_i + \partial_i \Lambda$ which translates to $A_i(p) \to A_i(p) - ip_i \Lambda(p)$ in the momentum space. The gauge invariance requires the extra term proportional to p_i *not* to contribute to the action in momentum space. This is why the matrix M_i^k *necessarily* has to satisfy the constraint $M_i^k p^i = 0$; so it *must* have a zero eigenvalue if the action is gauge invariant. In other words, we cannot find the inverse if we work with a gauge invariant action.

The simplest way out is to fix the gauge in some suitable form and hope that final results will be gauge invariant.[57] There are several ways of fixing the gauge and we will discuss a couple of them and a formal procedure to do this at the level of the action.

The first procedure is a cheap trick. If we add a mass term to the photon by adding a term $m^2 A_j A^j$ term to the Lagrangian, we immediately break the gauge invariance. In this case the action for the vector field interacting with, say, a current J^m will be

$$
\mathcal{A} = \int d^4 x \, L = \int d^4 x \left[\frac{1}{2} A_m \left[\left(\partial^2 + m^2 \right) \eta^{mn} - \partial^m \partial^n \right] A_n + A_m J^m \right] \tag{3.189}
$$

Now the relevant propagator D_{nl} will satisfy the equation

$$
\left[\left(\partial^2 + m^2 \right) \eta^{mn} - \partial^m \partial^n \right] D_{nl}(x) = \delta_l^m \delta(x) \tag{3.190}
$$

This is exactly analogous to Eq. (2.49) for the scalar field with a couple of extra indices. The solution is trivial in the Fourier space and we get

$$
D_{nl}(k) = \frac{-\eta_{nl} + (k_n k_l / m^2)}{k^2 - m^2 + i\epsilon} \tag{3.191}
$$

[57] This is the party line and, of course, this procedure has been enormously successful in practice. But the fact that the cleanest description of nature seems to require gauge redundant variables — like A_j in electrodynamics and it gets worse in gauge theories — is rather strange when you think about it. Mathematically it has something to do with the representations of the Lorentz group and elementary combinatorics; we will say something about it in a later chapter. Physically this issue is related to whether A_j is a local observable in quantum theory (and all indications are that it is not) and if not whether we can formulate the entire theory using only gauge invariant variables like F_{ik}. There have been attempts but nobody has succeeded in coming up with a useful, simple, formalism which does not use gauge dependent variables. One can indeed rewrite all the expressions involving the vector potentials A_j in terms of the field tensor F_{ab} *after* choosing a specific gauge. But this relation will be non-local and the resulting theory will appear to be non-local though we know that it is actually local — and you need to first choose a gauge to do this inversion. This is one of the reasons we do not try to express everything in terms of F_{ab}, and learn to love gauge fields.

which is a well defined propagator for a massive vector field. In the last step, we have also added the $i\epsilon$ factor with the usual rule that $m^2 \to m^2 - i\epsilon$.

The propagator for the photon needs to be obtained by taking the $m \to 0$ limit of our expressions, which makes Eq. (3.191) blow up in your face. The idea is to leave m non-zero, do calculations with it and take the limit right at the end after computing the relevant amplitude. This trick usually works as long as the vector field is coupled to a conserved current. When the current J^m is conserved, the equation $\partial_m J^m = 0$ translates to $k_m J^m(k) = 0$ in momentum space. While computing amplitudes, you usually have to contract the propagator with the current, obtaining $D_{nm}(k) J^m(k)$. In these expressions the $k_m k_n / m^2$ does not contribute because $k_m J^m(k) = 0$ and we get the same result as with no mass term. We can take the $m \to 0$ limit trivially.

As an application of this trick, let us consider the functional Fourier transform of the massive vector field with respect to an external source $J^m(x)$ exactly in analogy with what we did for scalar fields in Sect. 2.2.1. We will now get the result

$$\int \mathcal{D}A_j \exp i \int d^4x \left\{ \frac{1}{2} A_k \left(\delta^k_i \left(\Box + m^2 \right) - \partial^k \partial_i \right) A^i + A_m J^m \right\}$$
$$= \exp iW(J) \tag{3.192}$$

where

$$W(J) = -\frac{1}{2} \int \frac{d^4k}{(2\pi)^4} J^m(k)^* \left[\frac{-\eta_{mn} + (k_m k_n / m^2)}{k^2 - m^2 + i\epsilon} \right] J^n(k) \tag{3.193}$$

For a conserved current, the terms in the numerator with $(1/m^2)$ do not contribute, and we get

$$W(J) = \frac{1}{2} \int \frac{d^4k}{(2\pi)^4} J^m(k)^* \frac{1}{k^2 - m^2 + i\epsilon} J_m(k) \tag{3.194}$$

in which we can take the limit of m going to zero in the denominator without any difficulty and obtain:

$$W(J) = \frac{1}{2} \int \frac{d^4k}{(2\pi)^4} \frac{J^*_m(k) J^m(k)}{k^2 + i\epsilon} \tag{3.195}$$

If you compare this with the scalar field case you see a crucial sign difference in $W(J)$; so, if you rework the energy for a static source, it will come out to be positive.[58] Like charges repel in electromagnetism.

While the above example shows that you can sometimes fix the gauge by adding a mass term to the photon, this is no good if you want to write down a *covariant propagator* which is finite. As we saw above, the propagator itself was divergent in the $m \to 0$ limit. One way out of this is to add an explicit gauge fixing term into the Lagrangian in which we do not have to take any singular limits. A simple way to impose the gauge condition $\partial_m A^m = $ constant, say, is to modify the Maxwell Lagrangian by adding an arbitrary function $Q(\partial A)$ of $\partial_m A^m \equiv \partial A$. Let P be the derivative of Q with respect to its argument; ie., $P(\partial A) = Q'(\partial A)$. The modified Lagrangian

$$L = -\frac{1}{4} F_{ij} F^{ij} - Q(\partial A) \tag{3.196}$$

[58] The sign difference arises from the $-\eta_{nl}$ term in Eq. (3.191) which leads to a term $D_{00} = -(k^2 + i\epsilon)^{-1}$ which has an extra minus sign compared to the scalar field case.

then leads to the field equation:

$$\Box A^j - \partial^j(\partial A) + \partial^j P = 0 \tag{3.197}$$

On taking the divergence we get the condition $\Box P = 0$ to which one can consistently take the solution to be $P = 0$. This requires Q and hence ∂A to be a constant which is our gauge condition. The usual procedure is to take Q to be a quadratic function such as $Q = (1/2)\zeta(\partial A)^2$, where ζ is a constant.[59] In this case, we modify the Lagrangian to the form:

[59]Though any non-trivial Q will fix the gauge, quadratic terms make the path integrals simpler.

$$L = -\frac{1}{4}F_{mn}F^{mn} - \frac{1}{2}\zeta\left(\partial_s A^s\right)^2 \tag{3.198}$$

The field equations now get modified to

$$\Box A^m - (1-\zeta)\,\partial^m\left(\partial_s A^s\right) = 0 \tag{3.199}$$

If we take the divergence of this equation with respect to x^m, we get the constraint $\zeta\Box\left(\partial_s A^s\right) = 0$ for which one can consistently choose the solution $\partial_m A^m = 0$ thereby imposing the Lorenz gauge condition on the vector potential. With this condition, the field equation reduces to $\Box A^m = 0$ which, of course, has a well defined propagator etc. One can also obtain the same result by choosing $\zeta = 1$ in Eq. (3.198) so that the second term in Eq. (3.199) vanishes.

The differential operator D_{mn}^{-1} in the Lagrangian $(A^m D_{mn}^{-1} A^n)$ now takes the form, in momentum space,

$$D_{mn}^{-1}(k) = -\eta_{mn}k^2 + (1-\zeta)\,k_m k_n \tag{3.200}$$

The propagator in momentum space is the inverse of this matrix. Assuming that the matrix has the form

$$(D)^{ns}(k) = A(k^2)\eta^{ns} + B(k^2)k^n k^s \tag{3.201}$$

and imposing the condition $D_{mn}(k)(D^{-1})^{ns}(k) = \delta_m^s$, we get the two constraints

$$-k^2 A(k^2) = 1; \qquad \zeta k^2 B(k^2) = (1-\zeta)\,A(k^2) \tag{3.202}$$

One can immediately see that if $\zeta = 0$, these equations are not compatible showing again that, without the gauge fixing term, the differential operator has no inverse. When $\zeta \neq 0$, we can solve for A and B and obtain the propagator:

$$D^{mn}(k) = \frac{-\eta^{mn}}{k^2 + i\epsilon} + \frac{\zeta - 1}{\zeta}\frac{k^m k^n}{(k^2 + i\epsilon)^2} \tag{3.203}$$

The choice $\zeta = 1$ clearly simplifies the structure of the propagator and reduces it to the form we encountered earlier in obtaining Eq. (3.195). The gauge choice with $\zeta \to \infty$ is called the *Landau gauge*, and in this case the propagator becomes

$$D^{mn}(k) = \frac{k^m k^n - \eta^{mn}k^2}{(k^2 + i\epsilon)^2} \tag{3.204}$$

One can show, with a fair amount of effort and after surmounting some additional complications, that these propagators can actually be expressed as the vacuum expectation value of the time ordered product of the A_js when we quantize the field in the relevant gauge.

The way we have introduced the gauge fixing term in Eq. (3.198) allows taking the limit of $\zeta \to 0$ easily. From the structure of Eq. (3.203), it is obvious that the parameter $\lambda \equiv (1/\zeta)$ expresses the propagator in an equivalent but simpler form as

$$
\begin{aligned}
D_{mn}(k) &= -\frac{1}{k^2 + i\epsilon}\left(\eta_{mn} - \frac{k_m k_n}{k^2}\right) - \lambda \frac{k_m k_n}{k^2} \\
&= -\frac{i}{k^2}\left[\eta_{mn} - (1-\lambda)\frac{k_m k_n}{k^2}\right]
\end{aligned} \tag{3.205}
$$

The first line of this equation shows that the part of $D^{mn}(k)$ — which is independent of the gauge fixing term — is orthogonal to k^m because of which we do not get a sensible inverse in the absence of gauge fixing terms. The structure of this propagator will play a crucial role in our discussion of QED.

There is a very cute way[60] of introducing the gauge fixing terms we saw above by using path integrals. While this is not essential for our discussion, we will describe it because of its cleverness as well as the fact that this procedure plays a crucial role in quantizing non-Abelian gauge theories.

The starting point for this approach is the formal path integral for the vector field[61] written in the form

$$
Z = \int \mathcal{D}A_j \exp[iS(A_j)] \tag{3.206}
$$

The first thing you note is that this functional integral, as it stands, is ill-defined if we try to integrate over all functions $A_j(x)$. If we consider two vector potentials $A_m^{(\alpha)}$ and A_m related by a gauge transformation $A_m^{(\alpha)} \equiv A_m - \partial_m \alpha$, then the action as well as the measure will be the same for both because both are gauge invariant. So if you sum over both $A_m^{(\alpha)}$ and A_m, you will get a divergent result.[62] This should be no more mysterious than integrating over x and y of a function which is actually independent of x. Nothing you do can make this divergence *go away* but the idea is to *separate it out* in a sensible form as an infinite constant in Z such that the remaining part of the functional integral is well defined.

So we want to reduce Z to a form $Z = NF$ where N is an infinite constant *independent* of A_j and F is a well defined functional of A_j. Obviously, this will require restricting the gauge in some manner but the trick is to do it in such a way that the integration over the gauge transformation function $\alpha(x)$ can be separated out.

Let us say that we want to impose a gauge condition $G(A) \equiv \partial_m A^m - \omega(x) = 0$ where ω is some function. This can be done by introducing a Dirac delta functional $\delta(G)$ inside the functional integral. But that will be cheating, since it changes the value of the integral. A legitimate way of introducing the delta functional is to write the factor unity in the form

$$
\mathbb{I} = \int \mathcal{D}\alpha(x)\delta(G(A_m^{(\alpha)}))\,\mathrm{Det}\left(\frac{\delta G(A_m^{(\alpha)})}{\delta \alpha}\right) \tag{3.207}
$$

This is just an integral over $\mathcal{D}G\delta(G)$ [which is unity by definition] rewritten with α as the variable of integration with the Jacobian determinant taking care of the transformation from G to α. Since $A_m^{(\alpha)} = A_m + \partial_m \alpha$, we have

[60]This is called the Fadeev-Popov procedure

[61]Notation alert: We use S (rather than $\mathcal{A}$) for the action, for typographical clarity.

[62]There is a subtlety here. The rigorous definition of path integrals from time slicing leads to integration in phase space over both coordinates and momenta (see Sect. 1.6.1). This generalizes to, for example, the scalar field in a natural fashion. The phase space for the electromagnetic field, however, is constrained because of gauge invariance. So it is not obvious that one can define the theory using a path integral over A_j at all. We won't discuss this because there are ways of getting around it formally leading to the same results.

$G(A_m^{(\alpha)}) = \partial_m A^m + \partial^2 \alpha - \omega$ and $(\delta G/\delta \alpha) = \partial^2$; the determinant is just Det $[\partial^2]$ which is independent of A_j. Introducing this factor of unity into the path integral and pulling out the determinant because it is independent of A_j, we get the result

$$
\begin{aligned}
Z &= \int \mathcal{D}\alpha \int \mathcal{D}A_j \, e^{iS[A]} \, \delta(G(A_j^{(\alpha)})) \, \text{Det}[\partial^2] \\
&= \int \mathcal{D}\alpha \, \text{Det}[\partial^2] \int \mathcal{D}A_j \, e^{iS[A]} \delta(G(A_j^{(\alpha)}))
\end{aligned}
\tag{3.208}
$$

But since both the measure and the action are gauge invariant, we know that $\mathcal{D}A_j = \mathcal{D}A_j^{(\alpha)}$ and $S[A] = S[A_j^{(\alpha)}]$, which allows us to write

$$
\begin{aligned}
Z &= \int \mathcal{D}\alpha \, \text{Det}[\partial^2] \int \mathcal{D}A_j^{(\alpha)} \, e^{iS[A^{(\alpha)}]} \delta(G(A_j^{(\alpha)})) \\
&= \int \mathcal{D}\alpha \, \text{Det}[\partial^2] \int \mathcal{D}A_j \, e^{iS[A]} \delta(G)
\end{aligned}
\tag{3.209}
$$

In arriving at the second equality we have just replaced integration over $A_j^{(\alpha)}$ by integration over A_j. We have now isolated the infinite integration measure over the gauge function α in the first factor and all the dependence on A_j is contained in the second functional integral over A_j. The Dirac delta function in it restricts integration to configurations for which $G = \partial_m A^m - \omega(x) = 0$.

There is one final trick which we can use to eliminate the ω dependence of the result. Consider a functional integral over ω of some arbitrary function $F(\omega)$ leading to some finite constant C:

$$
\int \mathcal{D}\omega \, F(\omega) = C; \qquad 1 = \frac{1}{C} \int \mathcal{D}\omega \, F(\omega)
\tag{3.210}
$$

We now introduce this factor of unity into the path integral and write

$$
\int \mathcal{D}A_j \, e^{iS[A]} \delta \left[\partial A - \omega \right] = \frac{1}{C} \int \mathcal{D}\omega \int \mathcal{D}A_j \, e^{iS[A]} \delta \left[\partial A - \omega \right] F(\omega) \tag{3.211}
$$

Interchanging the order of functional integrations, we find that we have the result

$$
\int \mathcal{D}A_j \, e^{iS[A]} \delta \left[\partial A - \omega \right] = \frac{1}{C} \int \mathcal{D}A \, e^{iS[A]} F(\partial A)
\tag{3.212}
$$

which is expressed entirely in terms of the vector potential A_j. If you take $F \propto \exp -i \int Q(\partial A) d^4 x$ then we are essentially adding $-Q(\partial A)$ as a gauge fixing term to the Lagrangian, obtaining the result in Eq. (3.196). But since we like quadratic Lagrangians, we choose $Q \propto (\partial A)^2$ by taking

$$
F(\omega) = \exp \left(-i \int d^4 x \, \frac{\zeta}{2} \omega^2 \right)
\tag{3.213}
$$

in which case the final result will be

$$
\begin{aligned}
Z &= \left\{ \frac{1}{C} \det \left(\partial^2 \right) \int \mathcal{D}\alpha \right\} \\
&\quad \times \int \mathcal{D}A_j \, \exp \left(-i \int d^4 x \, \left[\frac{1}{4} F^{mn} F_{mn} + \frac{\zeta}{2} (\partial_m A^m)^2 \right] \right)
\end{aligned}
\tag{3.214}
$$

Except for the infinite constant in the curly bracket, we are now working with an action with an extra gauge fixing term which we introduced earlier in Eq. (3.198) somewhat arbitrarily.

The above procedure also works in the case of Coulomb gauge thereby justifying the quantization scheme, which we have originally adopted, in the language of the path integral. While integrating over $\mathcal{D}A^j = \mathcal{D}A^0\mathcal{D}\mathbf{A}$ we can impose the $\nabla \cdot \mathbf{A} = 0$ condition by introducing a factor

$$1 = \Delta \int [\mathcal{D}\omega]\delta[\nabla \cdot \mathbf{A}^\omega], \quad A_m^\omega \equiv A_m + \partial_m\omega \qquad (3.215)$$

where Δ is the determinant which is independent of A_i, into the path integral, thereby making $\mathbf{A}$ purely transverse. Going through the same procedure as above, this will lead to a path integral of the form

$$Z[J] = \int \mathcal{D}A^0\,\mathcal{D}\mathbf{A}\,\,\delta[\nabla \cdot \mathbf{A}]\,e^{iS[A]+i\int d^4x\,J^m A_m} \qquad (3.216)$$

when the field is coupled to a source. We have already seen in Sect. 3.1.5 that in the Coulomb gauge, the A^0 decouples from $\mathbf{A}$ in the action and can be integrated out separately. This will essentially add a phase containing the Coulomb energy of interaction which can be ignored. The rest of the integration is over the transverse degree of freedom which will lead to a term like

$$Z[J] \propto \exp\left[\frac{1}{2}\int d^4x\,d^4y\,\,J_\kappa(x)G^{\kappa\lambda}(x,y)\,J_\lambda(y)\right] \qquad (3.217)$$

where $G^{\kappa\lambda}$ is the *transverse* propagator coupling the transverse degrees of freedom of the source. This result ties up with the previous analysis done in the canonical formalism in Sect. 3.6.1 and relates the transverse propagator to the interaction between the currents.

3.6.3 Casimir Effect

It is obvious that the success of our program, for quantizing a field and interpreting the states in the Hilbert space in terms of particles, is closely tied to decomposing the field into an infinite number of harmonic oscillators. The facts: (i) the energy spectrum of the harmonic oscillator is equally spaced and (ii) its potential is quadratic, are the features which allow us to introduce quantum states labeled by a set of integers with the energy of the state changing by $\hbar\omega_{\mathbf{k}}$ when $n_{\mathbf{k}} \to n_{\mathbf{k}} + 1$. But if we take these oscillators seriously, we get into trouble with their zero point energy $(1/2)\hbar\omega_{\mathbf{k}}$ which adds up to infinity. The conventional wisdom — which we have been faithfully advocating — is to simply throw this away by introducing normal ordering of operators like the Hamiltonian.

There is however one effect (called the *Casimir effect*, which is observed in the lab) for which the *most natural explanation* is provided in terms of the zero point energy of the oscillators. This is probably one of the most intriguing results in field theory and is worth understanding.

Consider two large, plane, perfect conductors of area L^2 kept at a distance a apart in otherwise empty space. This is like two capacitor plates with $L \gg a$ (which allows us to ignore the edge effects) but we have not put any charges on them. Experiments show that these two plates exert a force of attraction on each other which varies as a^{-4}. You need to explain why two perfect conductors kept in the vacuum should exert such a force on each other!

The simplest explanation is the following: In the absence of the plates we are in the vacuum state of the electromagnetic field (made of an infinite number of oscillators) with corresponding mode functions varying in space as $\exp(i\mathbf{k}\cdot\mathbf{x})$ with all possible values for $\mathbf{k}$. When you introduce two plates, you have to impose the perfect conductor boundary condition for the electromagnetic field at the location of the plate. If the plates are located at $z = 0$ and $z = a$, this necessarily limits the z-component of the wave vector $\mathbf{k}$ to be discrete with the allowed values being $k_z = n\pi/a$ with integral n. The form of the total zero point energy for the oscillators will now be different (though, of course, still divergent) from the original form of the total zero point energy. The difference between the zero point energies calculated with and without the plates will involve subtracting one infinity from other. There are several ways of giving meaning to such an operation, thereby extracting a finite result for the difference. This difference turns out to be negative and scales as $(-a^{-3})$; this leads to an attractive force because reducing a decreases this energy. Thus, whether zero point energies exist or not, their *differences* seem to exist and seem to be finite!

Given the conceptual importance of this result, we will provide a fairly detailed discussion of its derivation. To understand the key issues which are involved, we will first consider a toy model of a scalar field in the (1+1) dimension and then study electromagnetic fields in (3+1).

Consider the quantum theory for a massless scalar field in the (1+1) dimension when the field is constrained to vanish[63] at $x = 0$ and $x = L$. The allowed wave vectors for the modes are now given by $k_n = (n\pi/L)$. The total zero-point energy of the vacuum in the presence of the plates is therefore given by

$$E = \sum_{n=0}^{\infty} \frac{1}{2}\frac{n\pi}{L} = \frac{\pi}{2L}\sum_{n=0}^{\infty} n \tag{3.218}$$

To give meaning to this expression, we need to evaluate the sum over all positive integers. It turns out that this sum[64] is equal to $-(1/12)$ so that $E = -(\pi/24L)$. If one imagines two point particles at $x = 0$ and $x = L$ (which are the analogs of two perfect conductors in the case of electromagnetism), this energy will lead to an attractive force $(\partial E/\partial L) = (\pi/24L^2)$ between the particles. Let us now ask how we can obtain this result.

There are rigorous ways of defining sums of certain divergent series which could be used for this purpose. But before discussing this approach, let us consider a more physical way of approaching the question. We first note that the zero-point energy in the absence of plates is given by the expression

$$E_0 = \int_{-\infty}^{\infty} \frac{L\,dk}{2\pi}\frac{1}{2}|k| = \frac{2L}{4\pi}\left(\frac{\pi^2}{L^2}\right)\int_0^{\infty} dn\, n = \frac{\pi}{2L}\int_0^{\infty} dn\, n \tag{3.219}$$

where we have used the substitution $k = (\pi n/L)$ with n being a continuous variable. What we are really interested in is the energy difference given by

$$\Delta E = \frac{\pi}{2L}\left[\sum_{n=0}^{\infty} n - \int_0^{\infty} dn\, n\right] \tag{3.220}$$

---

[63]The vanishing boundary condition is expected to mimic what happens in the case of electromagnetic field on the surfaces of the perfect conductors.

[64]This can be quite shocking to anyone innocent of the practices in high energy physics. One might have expected that the sum, if not infinite, should at least be decent enough to be a positive integer rather than a negative fraction! Such a conclusion, coming from physicists, would be quite suspect; but it was first obtained by a very respectable mathematician, Euler. In fact, most of our discussion will revolve around attributing meaning to divergent expressions in a systematic manner.

To evaluate this difference, let us introduce a high-n cut-off by a function $e^{-\lambda n}$ into these expressions and take the limit of $\lambda \to 0$ at the end of the calculation. That is, we think of the energy difference as given by[65]

$$\Delta E = \frac{\pi}{2L} \lim_{\lambda \to 0} \left[\sum_{n=0}^{\infty} n e^{-\lambda n} - \int_0^{\infty} dn \, n e^{-\lambda n} \right] \tag{3.221}$$

For finite λ, the energy difference can be evaluated as

$$\Delta E = -\frac{\pi}{2L} \frac{d}{d\lambda} \left[\sum_{n=0}^{\infty} e^{-\lambda n} - \frac{1}{\lambda} \right] = -\frac{\pi}{2L} \frac{d}{d\lambda} \left[\frac{1}{1 - e^{-\lambda}} - \frac{1}{\lambda} \right] \tag{3.222}$$

We are interested in this expression near $\lambda = 0$ which can be computed by a Taylor series expansion. We see that:

$$\left[\frac{1}{1 - e^{-\lambda}} - \frac{1}{\lambda} \right] \approx \frac{1}{\lambda \left[1 - \frac{1}{2}\lambda + \frac{1}{6}\lambda^2 + \mathcal{O}(\lambda^3) \right]} - \frac{1}{\lambda}$$

$$= \frac{1}{\lambda} \left[\left(\frac{1}{2}\lambda - \frac{1}{6}\lambda^2 \right) + \left(\frac{1}{2}\lambda - \frac{1}{6}\lambda^2 \right)^2 + \mathcal{O}(\lambda^3) \right]$$

$$= \frac{1}{2} + \frac{1}{12}\lambda + \mathcal{O}(\lambda^2) \tag{3.223}$$

It follows, quite remarkably, that:

$$\Delta E = -\frac{\pi}{2L} \left(\frac{1}{12} \right) + \mathcal{O}(\lambda) \to -\frac{\pi}{24L} \tag{3.224}$$

which is the same result we would have obtained by treating the sum over all positive integers as $(-1/12)$ as we mentioned before!

The above procedure of introducing a cut-off function $e^{-\lambda n}$ and taking the limit of $\lambda \to 0$ at the end of the calculation has one serious drawback. The $(-1/12)$ we obtained seems to have come from the Taylor series expansion of $e^{-\lambda}$ and hence it is not clear whether the same result will hold if we use some other cut-off function to regularize the sum and the integral. To settle this question, we need to evaluate expressions like

$$\Delta f \equiv \sum_{n=0}^{\infty} f(n) - \int_0^{\infty} dn \, f(n) \tag{3.225}$$

where $f(x)$ is a function which dies down rapidly for large values of the argument. It can be shown that (see Mathematical Supplement Sect. 3.8):

$$\Delta f = \sum_{n=0}^{\infty} f(n) - \int_0^{\infty} dn \, f(n) = - \sum_1^{\infty} \frac{B_k}{k!} \left(D^{k-1} f \right) \Big|_0 \tag{3.226}$$

where $D = d/dx$ and B_k (called *Bernoulli numbers*) are defined through the series expansion

$$\frac{t}{e^t - 1} = \sum_{k=0}^{\infty} \frac{t^k}{k!} B_k \tag{3.227}$$

Consider now an arbitrary cut-off function $F(\lambda x)$ instead of $\exp(-\lambda x)$ and let the modified function be $f(x) = x F(\lambda x)$ with the condition that $F(0) = 1$. Using Eq. (3.226) we can compute Δf with this cut-off function and take

[65] From a physical point of view, we could think of a factor like $e^{-\lambda n}$ as arising due to the finite conductivity of the metallic plates in the case of (1+3) electrodynamics which makes conductors less than perfect at sufficiently high frequencies.

the limit $\lambda \to 0$. Equation (3.226) requires us to compute the derivatives of $f(x)$ at the origin and take the limit $\lambda \to 0$. One can easily see that the first derivative evaluated at the origin is given by

$$\left[xF'(\lambda x)\lambda + F(\lambda x)\right]\Big|_{x=0} = 1 \qquad (3.228)$$

and the second derivative vanishes

$$\left[F'(\lambda x)\lambda + x\lambda^2 F''(\lambda x) + \lambda F'(x)\right]\Big|_{x=0} \to 0 \qquad (3.229)$$

when $\lambda \to 0$. All further derivatives will introduce extra factors of λ and hence will vanish. We therefore find that

$$\Delta f = -\frac{B_2}{2!}f'(0) = -\frac{1}{12} \qquad (3.230)$$

where we have used the fact that $B_2 = (1/6)$. This shows that, for a wide class of cut-off functions, we recover the same result. Knowing this fact, we could have even taken $f(n) = n$ in Eq. (3.226) and just retained the first derivative term (which is the only term that contributes) on the right hand side of Eq. (3.226).

Having done this warm-up exercise, let us turn our attention to the electromagnetic field in (1+3). We consider a region between two parallel conducting plates, each of area $L \times L$, separated by a distance a. We will assume that $L \gg a$ and will be interested in computing the force per unit area of the conducting plates by differentiating the corresponding expression for zero-point energy with respect to a. As in the previous case, we want to compute the zero-point energy in the presence of the plates and in their absence in the 3-dimensional volume $L^2 a$ and compute their difference.

The energy contained in this region in the absence of plates is given by the integral

$$\begin{aligned} E_0 &= 2\int \frac{L^2 d^2 k}{(2\pi)^2}\int \frac{a\,dk_3}{2\pi}\left[\frac{1}{2}\sqrt{k_1^2 + k_2^2 + k_3^2}\right] \\ &= \int \frac{L^2 d^2 k}{(2\pi)^2}\int_0^\infty dn\sqrt{k_1^2 + k_2^2 + \left(\frac{n\pi}{a}\right)^2} \end{aligned} \qquad (3.231)$$

where the overall factor 2 in front in the first line takes into account two polarizations and we have set $k_3 = (n\pi/a)$ with a continuum variable n to obtain the second line. Writing the transverse component of the wave vector as $k_\perp^2 \equiv k_1^2 + k_2^2 \equiv (\pi/a)^2 \mu$, we can re-write this expression as an integral over μ and n in the form

$$E_0 = \frac{\pi^2 L^2}{4a^3}\int_0^\infty d\mu \int_0^\infty dn\,[\mu + n^2]^{1/2} \qquad (3.232)$$

Both the integrals, of course, are divergent — as to be expected.

Let us next consider the situation in the presence of the conducting plates. This will require replacing the integral over n by a summation over n when $n \neq 0$. When $n = 0$, the corresponding result has to be multiplied by a factor $(1/2)$ because only one polarization state contributes. This comes about from the nature of the mode functions for the vector potential $\mathbf{A}$ in the presence of the plates. (See also Problem 4.) We will work in the gauge with $A^0 = 0$, $\nabla \cdot \mathbf{A} = 0$ and decompose the vector potential into parallel and normal components $\mathbf{A} = \mathbf{A}_\parallel + \mathbf{A}_\perp$ where $\mathbf{A}_\perp$ is along the z-axis while $\mathbf{A}_\parallel$ is parallel to the plates in the $x - y$ plane. If you impose the boundary condition, that the parallel component of the electric field and the normal component of the magnetic field should vanish on the plates, we get the conditions[66]

$$\frac{\partial \mathbf{A}_\parallel}{\partial t}\bigg|_{z=0} = \frac{\partial \mathbf{A}_\parallel}{\partial t}\bigg|_{z=a} = 0, \qquad B_z|_{z=0} = B_z|_{z=a} = 0 \qquad (3.233)$$

[66]Notation: The symbols $\parallel$ and $\perp$ do *not* mean curl-free and div-free parts in this context.

It is straightforward to determine the allowed modes in this case and you will find that there are two polarizations, each contributing the energy:

$$\omega_{k,n} = \sqrt{k_1^2 + k_2^2 + \left(\frac{n\pi}{a}\right)^2} \qquad (3.234)$$

when $n \neq 0$, and one polarization contributing when $n = 0$. Therefore, the corresponding expression in the presence of the plates is given by

$$\frac{E_1}{L^2} = \frac{\pi^2}{4a^3} \left[\sum_{n=1}^\infty \int_0^\infty d\mu \, (\mu + n^2)^{1/2} + \frac{1}{2} \int_0^\infty d\mu \, \mu^{1/2} \right] \qquad (3.235)$$

The difference between the energies per unit area is essentially determined by the combination

$$\frac{4a^3}{\pi^2} \frac{1}{L^2} [E_1 - E_0] = \sum_{n=1}^\infty f(n) + \frac{1}{2} f(0) - \int_0^\infty dn f(n) \qquad (3.236)$$

where

$$f(n) = \int_0^\infty d\mu \, \sqrt{\mu + n^2} = \int_{n^2}^\infty dv \, \sqrt{v} \qquad (3.237)$$

We will now use the result

$$\sum_{n=1}^\infty f(n) - \int_0^\infty dn f(n) + \frac{1}{2} f(0) = -\sum_{m=1}^\infty \frac{B_{2m}}{(2m)!} \left(D^{2m-1} f \right)\bigg|_0 \qquad (3.238)$$

which can be easily obtained from Eq. (3.226) (see Mathematical Supplement Sect. 3.8). We now have to regularize the expression in Eq. (3.236) by multiplying $f(n)$ by a cut-off function $F(\lambda n)$ and work out the derivatives at the origin and take the limit $\lambda \to 0$. We will, however, cheat at this stage and will work with $f(n)$ in Eq. (3.237) itself, because it leads to the same result. In this case, it is easy to see that

Exercise 3.21: Do this rigorously by introducing at cut-off function $F(\lambda x)$ — which will ensure convergence — and prove that you get the same result, independent of the choice of F.

$$f'(n)\big|_0 = -2n^2\big|_0 = 0; \quad f''(n)\big|_0 = -4n\big|_0 = 0; \quad f'''(n)\big|_0 = -4 \quad (3.239)$$

with all further derivatives vanishing. Hence we get the final result to be

$$\frac{\Delta E}{L^2} = \frac{\pi^2}{4a^3} \left[\frac{B_4}{6} \right] = \frac{\pi^2}{24a^3} \left(\frac{-1}{30} \right) = -\frac{\pi^2}{720a^3} \qquad (3.240)$$

where we have used the fact that $B_4 = -(1/30)$. The corresponding force of attraction (per unit area of the plates) is given by

$$f = -\frac{\partial \mathcal{E}}{\partial a} = -\frac{\pi^2}{240 \, a^4} \qquad (3.241)$$

[67]So does it mean that zero point energies are non-zero and normal ordering is a nonsensical procedure? While the above derivation provides the *most natural* explanation for Casimir effect, it is possible to obtain it by computing the direct Van der Waals like forces between the conductors. The fact that conductivity has a frequency dependence provides a natural high frequency cut-off. If you accept this alternative derivation (originally due to Liftshitz), then may be what you need to *really* understand is why zero point energies with dubious subtractions of infinities *also* give the same result. Nobody knows for sure.

It is this force (which works out to 10^{-8} N for $a = 1\mu$m, $L = 1$ cm) that has been observed in the lab.[67]

A more formal mathematical procedure for obtaining the Casimir effect in the case of the (1+1) scalar field or (1+3) electromagnetic field involves giving meaning to the zeta function

$$\zeta(s) = \sum_{n=1}^{\infty} \frac{1}{n^s} \qquad (3.242)$$

for negative values of s. We have already seen that, in the case of the scalar field in (1+1) dimension, we needed to evaluate the sum of all integers which is essentially $\zeta(-1)$. To see the corresponding result in the case of the electromagnetic field, consider the integral

$$\int_0^{\infty} d\mu \, (\mu + n^2)^{-\alpha} = \frac{1}{(\alpha - 1)} n^{-2(\alpha - 1)} \qquad (3.243)$$

which is well defined for sufficiently large α. We can therefore write

$$\int_0^{\infty} d\mu \, \mu^{-\alpha} = \lim_{\Lambda \to 0} \int_0^{\infty} d\mu \, (\mu + \Lambda^2)^{-\alpha} = \lim_{\Lambda \to 0} \left[\frac{1}{(\alpha - 1)} \Lambda^{-2(\alpha - 1)} \right] . \qquad (3.244)$$

Putting $\alpha = -1/2$ in this relation, it can be shown that:

$$\int_0^{\infty} d\mu \, \mu^{1/2} = \lim_{\Lambda \to 0} \left[\frac{1}{(-3/2)} \Lambda^3 \right] = 0 . \qquad (3.245)$$

[68]In the language of dimensional regularization, a pet trick in high energy physics, this means that several power law divergences can be "regularized" to vanish.

This means we need not worry[68] about the second integral within the square bracket in Eq. (3.235).

It follows that the quantity we needed to evaluate in Eq. (3.235) is given by

$$\sum_{n=1}^{\infty} \int_0^{\infty} d\mu \, \left(\mu + n^2 \right)^{-\alpha} = \frac{1}{(\alpha - 1)} \sum_{n=1}^{\infty} \frac{1}{n^{2(\alpha - 1)}} = \frac{1}{(\alpha - 1)} \zeta(2\alpha - 2) \qquad (3.246)$$

[69]In the previous discussions, we interpreted the *difference* between two divergent expressions as finite. Using dimensional regularization, we are giving meaning to the divergent E_0/L^2 in Eq. (3.235) directly without any subtraction.

in the limit of $\alpha \to -(1/2)$. That is, we need to give meaning to $\zeta(-3)$. It is actually possible to define $\zeta(s)$ for negative integral values of s by analytically continuing a suitable integral representation of $\zeta(s)$ in the complex plane to these values.[69] Such an analysis actually shows that $\zeta(1 - 2k) = -(B_{2k}/2k)$ using which we can obtain the same results as above.

Finally we mention that the Casimir effect has a clearer intuitive explanation in the Schrodinger picture. The ground state wave functional which we computed for a scalar field in Eq. (3.128) explicitly depended on using the running modes $\exp(i\mathbf{k} \cdot \mathbf{x})$ to go back and forth between the field configuration $\phi(\mathbf{x})$ and the oscillator variables $q_{\mathbf{k}}$. If you introduce metal plates into the vacuum, thereby modifying the boundary conditions, the vacuum functional will change. In other words, the ground state wave functional in the presence of the metal plates is actually quite different from that in the absence of the plates. (This should be obvious from the fact that Ψ should now vanish for any field configuration which does not obey the boundary conditions.) Since the ground states are different, it should be no surprise that the energies are different too.

Exercise 3.22: For a massless scalar field in the (1+1) dimension, find the vacuum functionals with and without the vanishing boundary conditions at $x = 0, L$.

3.6.4 Interaction of Matter and Radiation

The electromagnetic field, as we have mentioned before, is the first example in which one noticed the existence of quanta which, in turn, arose from the study of the interaction between matter and radiation. While this is probably not strictly a quantum field theoretic interaction — we will deal with a non-relativistic quantum mechanical system coupled to a quantized electromagnetic field — we will describe this briefly because of its historical and practical importance.

To do this, we have to study the coupling between (i) quantized electromagnetic fields and (ii) a charged particle (say, an electron in an atom) described by standard quantum mechanics. A charged particle in quantum mechanics will have its own dynamical variable $\mathbf{x}$ and momentum $\mathbf{p}$ obeying standard commutation rules. Depending on the nature of the system, such a charged particle can exist in different (basis) quantum states, each of which will be labeled by the eigenvalues of a complete set of commuting variables. (For example, the quantum state $|nlm\rangle$ of an electron in a hydrogen atom is usually labeled by three quantum numbers n, l and m.) For the sake of simplicity, let us assume that the quantum states of the charged particle are labeled by the energy eigenvalues $|E\rangle$; the formalism can be easily generalized when more labels are needed to specify the quantum state. The coupling between the electromagnetic field and the atomic system is described by the Hamiltonian

$$H_{\text{int}} = -\int d^3x \ \mathbf{J} \cdot \mathbf{A} \qquad (3.247)$$

where to the lowest order,[70] in the non relativistic limit, $\mathbf{J} = q\mathbf{v} \cong (q/m)\mathbf{p}$ $= (q/m)(-i\nabla)$. We are interested in the transitions caused between the quantum states of the electromagnetic field and matter due to this interaction.

Let the initial state of the system be $|E_i, \{n_{\mathbf{k}\alpha}\}\rangle$ where E_i represents the initial energy of the matter state and the set of integers $\{n_{\mathbf{k}\alpha}\}$ denote the quantum state of the electromagnetic field. We now turn on the interaction Hamiltonian H_{int}. Because of the coupling, the system can make a transition to a final state $|E_f, \{n'_{\mathbf{k}\alpha}\}\rangle$ where E_f represents the final energy of the matter state and a new set of integers $\{n'_{\mathbf{k}\alpha}\}$ denote the final quantum state of the electromagnetic field. To the lowest order in perturbation theory, the probability amplitude for this process is governed by the matrix element

$$Q \equiv \langle E_f, \{n'_{\mathbf{k}\alpha}\}|H_{\text{int}}|E_i, \{n_{\mathbf{k}\alpha}\}\rangle. \qquad (3.248)$$

To see the nature of this matrix element, let us substitute the expansion of the vector potential written in the form

$$\mathbf{A}(t, \mathbf{x}) = \sum_{\alpha=1,2} \int \frac{d^3\mathbf{k}}{(2\pi)^3} \left(a_{\mathbf{k}\alpha}\mathbf{A}_{\mathbf{k}\alpha} + a^\dagger_{\mathbf{k}\alpha}\mathbf{A}^*_{\mathbf{k}\alpha} \right);$$

$$\mathbf{A}_{\mathbf{k}\alpha} = \left(\frac{1}{2\omega_{\mathbf{k}}} \right)^{1/2} \boldsymbol{\epsilon}_{\mathbf{k}\alpha}e^{-ikx}. \qquad (3.249)$$

[70]The canonical momentum in the presence of $\mathbf{A}$ will have an extra piece $(q\mathbf{A}/c)$ which will lead to a term proportional to $q^2\mathbf{A}^2$ in $\mathbf{J} \cdot \mathbf{A}$. This quadratic term in $q\mathbf{A}$ is ignored in the lowest order.

into the interaction Hamiltonian; then H_{int} becomes the sum of two terms

$$
\begin{aligned}
H_{\text{int}} &= -\int d^3\mathbf{x}\, \mathbf{J} \cdot \mathbf{A} \tag{3.250} \\
&= -\int d^3\mathbf{x}\ \mathbf{J}. \sum_{\alpha=1,2} \int \frac{d^3\mathbf{k}}{(2\pi)^3} \left(a_{\mathbf{k}\alpha} \mathbf{A}_{\mathbf{k}\alpha} + a_{\mathbf{k}\alpha}^\dagger \mathbf{A}_{\mathbf{k}\alpha}^* \right) \equiv H_{ab} + H_{em}.
\end{aligned}
$$

Since the creation and annihilation operators can only change the energy eigenstate of the oscillator by one step, it is clear that the probability amplitude Q in Eq. (3.248) will be non zero only if the set of integers characterizing the initial and final states differ by unity for some oscillator labeled by $\mathbf{k}\alpha$. In other words, the lowest order transition amplitudes describe either the emission or the absorption of a *single* photon with a definite momentum and polarization. Since the creation operator $a_{\mathbf{k}\alpha}^\dagger$ changes the integer $n_{\mathbf{k}\alpha}$ to $(n_{\mathbf{k}\alpha} + 1)$, the term proportional to $a_{\mathbf{k}\alpha}^\dagger$ governs the emission; similarly, the term proportional to $a_{\mathbf{k}\alpha}$ governs the absorption of the photon. Let us work out the amplitude for emission in some detail.

The emission process, in which the quantum system makes the transition from $|E_i\rangle$ to $|E_f\rangle$ and the electromagnetic field goes from a state $|n_{\mathbf{k}\alpha}\rangle$ to $|n_{\mathbf{k}\alpha} + 1\rangle$, during the time interval $(0, T)$, is governed by the amplitude

$$
\begin{aligned}
\mathcal{A} &= \int_0^T dt\, \langle E_f | \langle n_{\mathbf{k}\alpha} + 1 | H_{\text{em}} | n_{\mathbf{k}\alpha} \rangle | E_i \rangle \\
&= -\int_0^T dt \int d^3\mathbf{x}\, \langle E_f | \mathbf{J} \cdot \mathbf{A}_{\mathbf{k}\alpha}^* | E_i \rangle \, (n_{\mathbf{k}\alpha} + 1)^{1/2} \tag{3.251}
\end{aligned}
$$

where we have used the fact that $\langle n_{\mathbf{k}\alpha} + 1 | \mathbf{A} | n_{\mathbf{k}\alpha} \rangle = \mathbf{A}_{\mathbf{k}\alpha}^* \, (n_{\mathbf{k}\alpha} + 1)^{1/2}$. Using the expansion for $\mathbf{A}_{\mathbf{k}\alpha}$ in Eq. (3.249) and the fact that the energy eigenstates have the time dependence $\exp(-iEt)$, the amplitude $\mathcal{A}$ can be written as

$$
\begin{aligned}
\mathcal{A} &= -\left(\frac{1}{2\omega_{\mathbf{k}}} \right)^{1/2} (n_{\mathbf{k}\alpha} + 1)^{1/2} \tag{3.252} \\
&\quad \times \int_0^T dt \int d^3\mathbf{x}\ \phi_f^*(\mathbf{x}) \left(\mathbf{J}.\boldsymbol{\epsilon}_{\mathbf{k}\alpha} e^{-i\mathbf{k}\cdot\mathbf{x}} \right) \phi_i(\mathbf{x})\, e^{-i(E_i - E_f - \omega)t}
\end{aligned}
$$

where $\phi_i(\mathbf{x}), \phi_f(\mathbf{x})$ denote the wave functions of the two states. Denoting the matrix element

$$
\int d^3\mathbf{x}\, \phi_f^*(\mathbf{x}) \left(\mathbf{J} e^{-i\mathbf{k}\cdot\mathbf{x}} \right) \phi_i(\mathbf{x}) = \frac{q}{m} \int d^3\mathbf{x}\, \phi_f^*(\mathbf{x})\, \mathbf{p}\, e^{-i\mathbf{k}\cdot\mathbf{x}} \phi_i(\mathbf{x}) \tag{3.253}
$$

(which is determined by the system emitting the photon) by the symbol $\mathbf{M}_{fi}$, the probability of transition $|\mathcal{A}|^2$ becomes

$$
|\mathcal{A}|^2 = P(T) = \left(\frac{1}{2\omega_{\mathbf{k}}} \right) |\boldsymbol{\epsilon}_{\mathbf{k}\alpha} \cdot \mathbf{M}_{fi}|^2 \, (n_{\mathbf{k}\alpha} + 1) \, |F(T)|^2 \tag{3.254}
$$

where

$$
|F(T)|^2 = \left| \int_0^T dt\, e^{-i(E_i - E_f - \omega)t} \right|^2 = \left[\frac{\sin(\mathcal{R}T/2)}{\mathcal{R}/2} \right]^2 \tag{3.255}
$$

with $\mathcal{R} \equiv (E_i - E_f - \omega)$. In the limit of $T \to \infty$, for any smooth function $S(\omega)$, we have the result

$$\int_0^\infty d\omega \, S(\omega) \frac{\sin^2[(\omega - \nu)T/2]}{[(\omega - \nu)/2]^2} \simeq 2T \, S(\nu) \int_{-\infty}^\infty \frac{\sin^2 \eta}{\eta^2} \, d\eta = 2\pi \, T \, S(\nu)$$

(3.256)

which shows that,

$$\lim_{T \to \infty} \frac{\sin^2[(\omega - \nu)T/2]}{[(\omega - \nu)/2]^2} \to 2\pi \, T \delta_D(\omega - \nu) \qquad (3.257)$$

in a formal sense. Hence Eq. (3.254) becomes, as $T \to \infty$,

$$P(T) \cong T \left(\frac{\pi}{\omega_{\mathbf{k}}} \right) |\boldsymbol{\epsilon}_{\mathbf{k}\alpha} \cdot M_{\text{fi}}|^2 (n_{\mathbf{k}\alpha} + 1) \, \delta_D(E_i - E_f - \omega). \qquad (3.258)$$

The corresponding *rate* of transition is $\mathcal{P} = P(T)/T$, which gives a finite *rate* for the emission of photons:[71]

$$\mathcal{P} \equiv \frac{dP}{dt} = \left(\frac{1}{2\omega_{\mathbf{k}}} \right) (n_{\mathbf{k}\alpha} + 1) |\boldsymbol{\epsilon}_{\mathbf{k}\alpha} \cdot \mathbf{M}_{fi}|^2 2\pi \delta_D(E_i - E_f - \omega). \qquad (3.259)$$

This expression gives the rate for emission of a photon with a specific wave vector $\mathbf{k}$ and polarization α. The delta function, $\delta_D(E_i - E_f - \omega)$ expresses conservation of energy and shows that the probability is non zero only if the energy difference between the states $E_i - E_f$ is equal to the energy of the emitted photon $\hbar\omega$.

Usually, we will be interested in the probability for emission of a photon in a frequency range $\omega, \omega + d\omega$ and in a direction defined by the solid angle element $d\Omega$. To obtain this quantity, we have to multiply the rate of transition by the density of states available for the photon in this range. The density of states (for unit volume) is given by

$$\frac{dN}{d\omega d\Omega} = \frac{d^3\mathbf{k}}{(2\pi)^3} \frac{1}{d\omega d\Omega} = \frac{1}{(2\pi)^3} \frac{k^2 dk d\Omega}{d\omega d\Omega} = \frac{k^2}{(2\pi)^3} = \frac{\omega_{\mathbf{k}}^2}{(2\pi)^3}. \qquad (3.260)$$

Hence

$$\begin{aligned}
\left[\frac{dP}{dt d\omega d\Omega} \right]_{\text{emi}} &= \frac{dP}{dt} \frac{dN}{d\omega d\Omega} = \frac{dP}{dt} \frac{\omega_{\mathbf{k}}^2}{(2\pi)^3} \\
&= \left(\frac{1}{2\omega} \right) (n_{k\alpha} + 1) |\boldsymbol{\epsilon}_{\mathbf{k}\alpha} \cdot \mathbf{M}_{fi}|^2 2\pi \delta_D(\omega_{\mathbf{k}} - \omega_{fi}) \frac{\omega_{\mathbf{k}}^2}{(2\pi)^3} \\
&\propto \omega_{\mathbf{k}} (n_{k\alpha} + 1) |\boldsymbol{\epsilon}_{\mathbf{k}\alpha} \cdot \mathbf{M}_{fi}|^2 \delta_D(\omega_{\mathbf{k}} - \omega_{fi}) \qquad (3.261)
\end{aligned}$$

with $\hbar\omega_{\text{fi}} \equiv E_i - E_f$.

The analysis for the absorption rate of photons is identical except that only the annihilation operator $a_{\mathbf{k}\alpha}$ contributes. Since $\langle n_{\mathbf{k}\alpha} - 1 | a_{\mathbf{k}\alpha} | n_{\mathbf{k}\alpha} \rangle = n_{\mathbf{k}\alpha}^{1/2}$, we get $n_{\mathbf{k}\alpha}$ rather than $(n_{\mathbf{k}\alpha} + 1)$ in the final result:

$$\left[\frac{dP}{d\Omega dt d\omega} \right]_{abs} \propto \omega_{\mathbf{k}} n_{k\alpha} |\boldsymbol{\epsilon}_{\mathbf{k}\alpha} \cdot \mathbf{M}_{fi}|^2 \delta_D(\omega_{\mathbf{k}} - \omega_{fi}). \qquad (3.262)$$

The probabilities for absorption and emission differ only in their dependence on $n_{\mathbf{k}\alpha}$. The probability for absorption scales in proportion to

---

[71] The integration over the infinite range of t implies, in practice, an integration over a range $(0, T)$ with $\omega T \gg 1$. If the energy levels have a characteristic width $\Delta\omega \ll \omega$, then the above analysis is valid for $\omega^{-1} \ll T \ll (\Delta\omega)^{-1}$.

$n_{\mathbf{k}\alpha}$. Clearly, if $n_{\mathbf{k}\alpha} = 0$, this probability vanishes; this is obvious since no photons can be absorbed if there were none in the initial state to begin with. But the probability for emission is proportional to $(n_{\mathbf{k}\alpha} + 1)$ and does not vanish even when $n_{\mathbf{k}\alpha} = 0$. Hence there is a non-zero probability for a system at an excited state to emit a photon and come down to a lower state *spontaneously*. If the initial state of the electromagnetic field has a certain number of photons already present, then the probability for emission is further enhanced. The emission of a photon by an excited system when no photons were originally present is called *spontaneous emission* and the emission of a photon in the presence of initial photons is called *stimulated emission*. Both these processes exist and contribute in electromagnetic transitions.[72]

Your familiarity with this result should not prevent you from appreciating it. Spontaneous emission is a conceptually non-trivial quantum field theoretic process and arises directly through the action of the "anti-photon" creating term in Eq. (3.172). The electron in the excited state in the atom, say, can be described perfectly well by the Schrödinger equation — until, of course, it pops down to the ground state emitting a photon. We originally had just one particle, the electron (in an external Coulomb field) to deal with, and the Schrödinger equation is adequate. But once it creates a photon, we have to deal with at least two particles of which one is massless and fully relativistic. So, this elementary process of emission of the (anti)photon by an excited atom has the key conceptual ingredient which we started out with, while explaining the need for quantum field theory. Without a quantized description of the electromagnetic field, we cannot account for an elementary process like the emission of a photon by an excited system in a consistent manner. This can also be seen from the fact that the initial state (of an electron in the excited atomic level) has no photons, and yet a photon appears in the final state. People loosely talk of this as "vacuum fluctuations of the electromagnetic field interacting with the electron"; but the way we have developed the arguments, it should be clear that this is a field theoretic effect[73] arising from the fact that negative energy solutions of the field equations for A_i are included in Eq. (3.172).

Finally, let us see how these results are related to the Planck spectrum of photons in a radiation cavity. If we consider quantized electromagnetic radiation in equilibrium with matter at temperature T, then in steady state we will expect the condition $N_{\mathrm{up}}\mathcal{R}_{\mathrm{em}} = N_{\mathrm{down}}\mathcal{R}_{\mathrm{ab}}$ to hold where N_{up} and N_{down} represent the two levels which we designate as up and down and $\mathcal{R}_{\mathrm{em}}$ and $\mathcal{R}_{\mathrm{ab}}$ are the rate of emission and absorption which we have computed above. In thermal equilibrium, the population of atoms in the two energy levels will satisfy the condition $N_{\mathrm{up}} = N_{\mathrm{down}}\exp(-\beta E) = N_{\mathrm{down}}\exp(-\beta\hbar\omega_{\mathbf{k}})$ where $E > 0$ is the difference in the energies of the two states and $\omega_{\mathbf{k}}$ is the frequency of the photon corresponding to this energy difference. On the other hand, we see from our expressions for $\mathcal{R}_{\mathrm{em}}$ and $\mathcal{R}_{\mathrm{ab}}$ that their ratio is given by $\mathcal{R}_{\mathrm{em}}/\mathcal{R}_{\mathrm{ab}} = [(n_{\mathbf{k}\alpha} + 1)/n_{\mathbf{k}\alpha}]$. The equilibrium condition now requires

$$e^{\beta E} = \frac{N_{\mathrm{down}}}{N_{\mathrm{up}}} = \frac{\mathcal{R}_{\mathrm{em}}}{\mathcal{R}_{\mathrm{ab}}} = \frac{n_{\mathbf{k}\alpha} + 1}{n_{\mathbf{k}\alpha}} \qquad (3.263)$$

which tells you the number density of photons in a cavity, say, if the radiation is in equilibrium with matter atoms in the cavity held at temperature

---

[72] An intuitive way of understanding stimulated emission is as follows. Consider an atom making a transition from the ground state $|G\rangle$ to an excited state $|E\rangle$ absorbing a single photon out of n photons present in the initial state, leaving behind a $(n-1)$ photon state. The fact that this absorption probability $\mathcal{P}\{|G;n\rangle \rightarrow |E;n-1\rangle\} \propto n \equiv Qn$ is proportional to n seems intuitively acceptable. Consider now the probability $\mathcal{P}'$ for the time reversed process $|E;n-1\rangle \rightarrow |G;n\rangle$. By principle of microscopic reversibility, we expect $\mathcal{P}' = \mathcal{P}$ giving $\mathcal{P}' \propto n \equiv Qn$. Calling $n-1 = m$, we get $\mathcal{P}'\{|E;m\rangle \rightarrow |G;m+1\rangle\} = Qn = Q(m+1)$. Clearly $\mathcal{P}'$ is non zero even for $m = 0$ with $\mathcal{P}'\{|E;0\rangle \rightarrow |G;1\rangle\} = Q$ which gives the probability for the spontaneous emission while Qm gives the probability for the stimulated emission. Thus, the fact that absorption probabilities are proportional to n while emission probabilities are proportional to $(n+1)$ originates from the principle of microscopic reversibility.

[73] Make sure you understand this point. The structure of the Klein-Gordon equation for a massless particle, $\Box\phi = 0$, is identical to Maxwell's equations $\Box A_i = 0$ in a particular gauge. The solutions for ϕ involve positive and negative energy modes and so do the solutions for A_i. Sometimes a lot of fuss is made over negative energy solutions, backward propagation in time, etc. for the $\Box\phi = 0$ equation while you have always been dealing quite comfortably with the $\Box A_i = 0$ equation, all your life. When you quantize the systems, they have the same conceptual structure.

T. We get the historically important Planck law for radiation in the cavity:

$$n_{\mathbf{k}\alpha} = \frac{1}{e^{\beta\hbar\omega_{\mathbf{k}}} - 1} \qquad (3.264)$$

The role played by the factors $n + 1$ and n in leading to this result is obvious. That, in turn, is due to the fact that electromagnetic field can be decomposed into a bunch of harmonic oscillators with the usual creation and annihilation operators.

3.7 Aside: Analytical Structure of the Propagator

This discussion essentially brings us to end of this Chapter 3, the main purpose of which was to interpret particles as excitations of an underlying quantum field. We illustrated these ideas using real and complex scalar fields and the electromagnetic field. But in all the cases we worked with an action which is quadratic in the field variables thereby describing free, noninteracting fields. We also found that in each of these cases one can obtain an expression for the propagator in terms of the vacuum expectation value of the time-ordered fields.

When we switch on interactions between the fields, the structure of propagators will change. As we shall see later, most of the physical processes we are interested in (for e.g., scattering of particles) can be interpreted in terms of the propagators of the interacting theory. Unfortunately, there is no simple way of computing these propagators when the interactions are switched on, and much of the effort in quantum theory is directed towards developing techniques for their computation. We shall take up many of these issues in the next two chapters. The purpose of this last, brief section is to introduce some exact results related to propagators which remain valid even in an interacting field theory.

It is possible to obtain some general results regarding the analytic structure of the propagators even in the exact theory, which will be useful later on while discussing QED. For the sake of simplicity, we shall discuss this aspect in the context of scalar field theory. The generalization to other fields is straightforward.

Let us write down the propagator $G(x, y) = \langle 0|T[\phi(x)\phi(y)]|0\rangle$ for $x^0 > y^0$ by inserting a complete set of states in between:

$$\langle 0|T[\phi(x)\phi(y)]|0\rangle = \langle 0|\phi(x)\phi(y)|0\rangle = \sum_n \langle 0|\phi(x)|n\rangle\langle n|\phi(y)|0\rangle \qquad (3.265)$$

where $|n\rangle$ is the eigenstate of the full Hamiltonian. Since the total momentum commutes with the Hamiltonian, we can label any such state with a momentum $\boldsymbol{p}$ and an energy E thereby defining[74] a quantity m_λ by the relation $E^2 - \boldsymbol{p}^2 \equiv m_\lambda^2$. The λ here merely labels such states since there could be several of them. We can therefore expand the complete set of states in terms of the momentum eigenstates by:

$$\sum_n |n\rangle\langle n| = |0\rangle\langle 0| + \sum_\lambda \int \frac{d^3\boldsymbol{p}}{(2\pi)^3} \frac{1}{2E_{\boldsymbol{p}}} |\boldsymbol{p}\rangle\langle\boldsymbol{p}| \qquad (3.266)$$

where $E_p^2(\lambda) = \boldsymbol{p}^2 + m_\lambda^2$. Plugging this into Eq. (3.265) and using the translation invariance and the scalar nature of the field, we can obtain

[74]Note that m_λ is not supposed to be the mass of any "fundamental" particle and could represent a bound state or a composite unbound state of many particles.

Exercise 3.23: Prove this result.

$$\langle 0|\phi(x)|\boldsymbol{p}\rangle_\lambda = \langle 0|\phi(0)|\boldsymbol{p}=0\rangle e^{-ip\cdot x} \qquad (3.267)$$

This, in turn, allows us to write the propagator for $x^0 > y^0$ in the form

$$\langle 0|\phi(x)\phi(y)|0\rangle = \sum_\lambda \int \frac{d^3\boldsymbol{p}}{(2\pi)^3} \frac{1}{2E_{\boldsymbol{p}}(\lambda)} e^{-ip(x-y)} \Big|\langle 0|\phi(0)|\boldsymbol{p}=0\rangle_\lambda\Big|^2 \quad (3.268)$$

Or, equivalently, as

$$\langle 0|\phi(x)\phi(y)|0\rangle = \sum_\lambda \int \frac{d^4p}{(2\pi)^4} \frac{ie^{-ip\cdot(x-y)}}{p^2 - m_\lambda^2 + i\epsilon} \Big|\langle 0|\phi(0)|\boldsymbol{p}=0\rangle_\lambda\Big|^2 \quad (3.269)$$

We have introduced a p^0 integral as usual which ensures that the correct pole will be picked up through the $i\epsilon$ prescription. Here, this is merely a mathematical trick and the propagator did not arise from any fundamental field theory. The advantage of this form is that it trivially works for the $x^0 < y^0$ case as well, allowing us to replace $\phi(x)\phi(y)$ by the time-ordered product $T[\phi(x)\phi(y)]$ in the left hand side. If we define the Fourier transform of the propagator in the usual manner, we find that the propagator in the Fourier space is given by

$$G(p) = \sum_\lambda \frac{i}{p^2 - m_\lambda^2 + i\epsilon} \Big|\langle 0|\phi(0)|\boldsymbol{p}=0\rangle_\lambda\Big|^2 = \int_0^\infty \frac{ds}{2\pi} \frac{\rho(s)i}{p^2 - s + i\epsilon} \quad (3.270)$$

where we have defined a function (called the *spectral density* of the theory) by

$$\rho(s) \equiv \sum_\lambda (2\pi)\delta(s - m_\lambda^2) \Big|\langle 0|\phi(0)|\boldsymbol{p}=0\rangle_\lambda\Big|^2 \qquad (3.271)$$

In Eq. (3.270), the exact propagator is re-expressed in an integral representation in terms of the spectral density of the theory $\rho(s)$. We see from the definition of $\rho(s)$ that it picks up contributions from the various states of the theory created by the action of the field on the vacuum. These states, depending on the mass threshold, will lead to non-analytic behaviour or branch-cuts in the propagators.[75] Conversely, if we can write the propagator $G(p)$ of the theory — at any order of approximation — in the form of Eq. (3.270), then we can read off the spectral density $\rho(s)$. The branch-cuts will now tell us about the properties of the multi-particle states in the theory. This is a useful strategy in some perturbative computations and we will have occasion to use this later on.

One curious application of this result is in determining the behaviour of two-point functions for large values of the momentum. This behaviour is important as regards divergences in perturbation theory which we will discuss in the next two chapters. These divergences arise because, while computing certain amplitudes using Feynman diagrams, we will integrate over the two-point functions in the momentum space. These integrals use the measure $d^4p \propto p^3 dp$, while the propagators decrease only as p^{-2} for large p, thereby leading to divergences. You might wonder whether these divergences arise, because our description is incorrect at sufficiently hight energies. Can we modify the theory in some simple manner to make the integrals convergent? This could very well be possible but not in any simple minded way. Let us illustrate this fact by a specific example.

The simplest hope is that we can make the divergences go away by modifying the theory at large scales to make the propagators die down

[75]Typically, they arise either through a square root or a logarithmic factor and invariably characterize the contribution of multi-particle states to the propagator in the theory.

faster than p^{-2}. As an example, consider a Lagrangian for a scalar field in which the kinetic energy term is modified to the form

$$L = -\frac{1}{2}\phi\left(\Box + c\frac{\Box^2}{\Lambda^2} + m^2\right)\phi + L_{\text{int}}(\phi) \qquad (3.272)$$

which would lead to a propagator in the momentum space which behaves as:

$$\Pi(p^2) = \frac{1}{p^2 - m^2 - c(p^4/\Lambda^2)} \qquad (3.273)$$

In this case, the propagator decays as p^{-4} for large p and one would think that the loop integrals would be better behaved. More generally, one could think of modifying $\Pi(p^2)$ such that it decreases sufficiently fast as $p^2 \to \infty$. Under Euclidean rotation, $p^2 \to -p_E^2$ and we would like $\Pi(-p_E^2)$ to go to zero faster than $(1/p_E^2)$.

We can actually show that this is not possible in a sensible[76] quantum theory. To do this, we go back to the spectral decomposition of the two-point function given by Eq. (3.270), which — as we emphasized — holds for an interacting field theory as well:

$$\langle 0|T[\phi(x)\phi(y)]|0\rangle = \int \frac{d^4p}{(2\pi)^4} e^{ip\cdot(x-y)} i\Pi(p^2) \qquad (3.274)$$

where

$$\Pi(p^2) \equiv \int_0^\infty dq^2 \frac{\rho(q^2)}{p^2 - q^2 + i\epsilon} \to \int_0^\infty dq^2 \frac{\rho(q^2)}{p_E^2 + q^2} = \Pi(-p_E^2) \qquad (3.275)$$

with the second result being valid in the Euclidean sector. From this, it follows that

$$\left|\Pi(-p_E^2)\right| = \left|\int_0^\infty dq^2 \frac{\rho(q^2)}{p_E^2 + q^2}\right| \geq \left|\int_0^{q_0^2} dq^2 \frac{\rho(q^2)}{p_E^2 + q_0^2}\right| \qquad (3.276)$$

for any q_0^2. If we now take the limit $p_E^2 \to \infty$, we will eventually have $p_E^2 > q_0^2$, leading to the result

$$\lim_{p_E^2 \to \infty} p_E^2 \left|\Pi(-p_E^2)\right| \geq \lim_{p_E^2 \to \infty} p_E^2 \left|\int_0^{q_0^2} dq^2 \frac{\rho(q^2)}{2p_E^2}\right| = \frac{A}{2} \qquad (3.277)$$

where

$$A = \int_0^{q_0^2} \rho(q^2) \, dq^2 \qquad (3.278)$$

defines some finite positive number. A propagator in Eq. (3.273) will violate this bound for any c and A at large enough p_E^2. Note that the positivity of $\rho(q^2)$ — which follows from its definition — is crucial for this definition. We therefore conclude that propagators cannot decrease[77] faster than $(1/p^2)$ for large p^2. Thus, simple modifications of the interactions, maintaining unitarity, Lorentz invariance and a local Lagrangian description will not help you to avoid the loop divergences.

[76]What constitute a 'sensible' theory in this context is, unfortunately, a matter of opinion, given our ignorance about physics at sufficiently high energies. Roughly speaking, one would like to have unitary, local, Lorentz invariant field theory.

[77]This is a rather powerful result obtained from very little investment!

3.8   Mathematical Supplement

3.8.1   Summation of Series

We will provide a proof of some of the relations used in Sect. 3.6.3 here.

Let $D = d/dx$ be the derivative operator using which we can write the Taylor series expansion for a function $f(x)$ as $f(x + n) = e^{nD} f(x)$ where n is an integer. Summing both sides over $n = 0$ to $n = N - 1$ we get

$$\sum_{n=0}^{N-1} f(x+n) = \sum_{n=0}^{N-1} e^{nD} f(x) = \frac{e^{ND} - 1}{e^D - 1} f(x)$$

$$= \frac{1}{(e^D - 1)} [f(x+N) - f(x)] \qquad (3.279)$$

On the right hand side, we introduce a factor $1 = D^{-1} D$ and use the Taylor series expansion of the function

$$\frac{t}{e^t - 1} = \sum_{k=0}^{\infty} \frac{t^k}{k!} B_k \qquad (3.280)$$

to obtain the result

$$D^{-1} \left[\frac{D}{(e^D - 1)} \right] [f(x+N) - f(x)] = D^{-1} \sum_{k=0}^{\infty} \frac{B_k}{k!} D^k [f(x+N) - f(x)] \qquad (3.281)$$

In the sum on the right hand side, we will separate out the $k = 0$ term from the others and simplify it as follows:

$$D^{-1}(f(x+N) - f(x)) = D^{-1} \left[\int_0^N f'(x+t) dt \right] = \int_0^N f(x+t)\, dt \qquad (3.282)$$

where, in obtaining the last result, we have used the fact that $D^{-1} f' = f$. Putting all these together, we get:

$$\sum_{n=0}^{N-1} f(x+n) - \int_0^N f(x+t) dt = \sum_1^{\infty} \frac{B_k}{k!} D^{k-1} [f(x+N) - f(x)] \qquad (3.283)$$

We will now take the limit of $N \to \infty$ assuming that the function and all its derivatives vanish for large arguments, thereby obtaining

$$\sum_{n=0}^{\infty} f(x+n) - \int_0^{\infty} f(x+t) dt = -\sum_1^{\infty} \frac{B_k}{k!} \left(D^{k-1} f(x) \right) \qquad (3.284)$$

If we now choose $x = 0$, we obtain Eq. (3.226) quoted in the text:

$$\Delta f = \sum_{n=0}^{\infty} f(n) - \int_0^{\infty} dn\, f(n) = -\sum_1^{\infty} \frac{B_k}{k!} D^{k-1} f \Big|_0 \qquad (3.285)$$

This result can be rewritten in a slightly different form which is often useful. To obtain this, consider the Taylor series expansion of the left hand side of Eq. (3.280) separating out the first two terms:

$$\frac{t}{e^t - 1} = \sum_{k=0}^{\infty} \frac{t^k}{k!} B_k = 1 - \frac{1}{2} t + \sum_{k=2}^{\infty} \frac{t^k}{k!} B_k \qquad (3.286)$$

Since the function

$$\frac{t}{e^t - 1} + \frac{t}{2} = \frac{t}{2}\left[\frac{e^t + 1}{e^t - 1}\right] \tag{3.287}$$

is even in t, the summation in the right hand side of Eq. (3.286) from $k = 2$ to ∞ has only even powers of t. In other words, all odd Bernoulli numbers other than B_1 vanish.[78] Therefore, one could equally well write the series expansion for Bernoulli numbers as

[78]This itself is an interesting result. In fact, when you have some time to spare, you should look up the properties of Bernoulli numbers. It is fun.

$$\frac{t}{e^t - 1} = 1 - \frac{t}{2} + \sum_{m=1}^{\infty} \frac{B_{2m}}{(2m)!} t^{2m} \tag{3.288}$$

which allows us to write Eq. (3.285) in the form

$$\sum_{n=0}^{\infty} f(n) - \int_0^{\infty} f(n)dn = -\sum_1^{\infty} \frac{B_k}{k!} \left(D^{k-1} f\right)\Big|_0 \tag{3.289}$$

$$= \frac{1}{2}f(0) - \sum_{m=1}^{\infty} \frac{B_{2m}}{(2m)!} \left(D^{2m-1} f\right)\Big|_0$$

Separating out the $n = 0$ term in the summation in the left hand side, this result can be expressed in another useful form:

$$\sum_{n=1}^{\infty} f(n) - \int_0^{\infty} dn f(n) + \frac{1}{2}f(0) = -\sum_{m=1}^{\infty} \frac{B_{2m}}{(2m)!} \left(D^{2m-1} f\right)\Big|_0 \tag{3.290}$$

This was used in the study of the Casimir effect in (1+3) dimensions for electrodynamics.

3.8.2 Analytic Continuation of the Zeta Function

We will next derive an integral representation for the zeta function $\zeta(s)$ which allows analytic continuation of this function for negative values of s. To do this, we begin with the integral representation for $\Gamma(s)$ and express it in the form

$$\Gamma(s) = \int_0^{\infty} d\bar{t}\, \bar{t}^{s-1} e^{-\bar{t}} = n^s \int_0^{\infty} dt\, t^{s-1} e^{-nt} \tag{3.291}$$

where we have set $\bar{t} = nt$ to arrive at the last expression. Summing the expression for $\Gamma(s)/n^s$ over all n, we get the result:

$$\Gamma(s) \sum_{n=1}^{\infty} \frac{1}{n^s} = \int_0^{\infty} dt \left(\frac{t^s}{e^t - 1} \frac{1}{t}\right) = \Gamma(s)\zeta(s) \tag{3.292}$$

Let us now consider the integral,

$$I(s) \equiv \int_C dz \frac{z^{s-1}}{e^z - 1} \tag{3.293}$$

in the complex plane, over a contour (see Fig. 3.6) consisting of the following paths: (i) Along the real line from $x = \infty$ to $x = \epsilon$ where ϵ is an infinitesimal quantity; (ii) On a circle of radius ϵ around the origin going from $\theta = 0$ to $\theta = 2\pi$; (iii) Along the real line from ϵ to ∞. Along the contour in (i) we get the contribution $-\Gamma(s)\zeta(s)$; it can be easily shown that the contribution

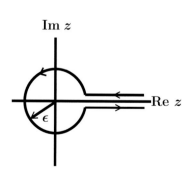

Figure 3.6: Contour to evaluate Eq. (3.293).

along the circle will only pick up the residue at the origin. Finally, along (iii) one obtains the contribution $e^{2\pi i s}\zeta(s)\Gamma(s)$. We therefore find that

$$
\begin{aligned}
\int_C dz \frac{z^{s-1}}{e^z - 1} &= \Gamma(s)\zeta(s)[e^{2\pi i s} - 1] = \Gamma(s)\zeta(s)e^{i\pi s}(2i\sin(\pi s)) \\
&= \zeta(s)e^{i\pi s}(2\pi i)\frac{1}{\Gamma(1-s)}
\end{aligned}
\tag{3.294}
$$

In arriving at the last expression, we have used the standard identity:

$$
\Gamma(s)\Gamma(1-s) = \frac{\pi}{\sin \pi s}
\tag{3.295}
$$

This allows us to express $\zeta(s)$ as a contour integral in the complex plane given by

$$
\zeta(s) = e^{-i\pi s}\Gamma(1-s)\frac{1}{2\pi i}\int_C dz \frac{z^{s-1}}{e^z - 1}
\tag{3.296}
$$

By studying the analytical properties of the right hand side, it is easy to show that this expression remains well defined for negative integral values of s. For example, when $s = -1$, the integrand in the contour integral can be expressed as $(1/z^3)[z/(e^z - 1)]$. The residue at the origin is therefore governed by the z^2 term in the power series expansion of $[z/(e^z-1)]$, giving a contribution proportional to B_2. Similarly, for $s = -3$ the integrand in the contour integral can be expressed as $(1/z^5)[z/(e^z - 1)]$. The residue at the origin is therefore governed by the z^4 term in the power series expansion of $[z/(e^z - 1)]$ giving a contribution proportional to B_4. In fact, it is easy to show, putting all the factors together, that:

Exercise 3.24: Prove Eq. (3.296).

$$
\zeta(1 - 2k) = \frac{B_{2k}}{2k} \qquad \text{for } k = 1, 2, ...
\tag{3.297}
$$

This gives $\zeta(-1) = -B_2/2 = -1/12$, $\zeta(-3) = -B_4/4 = 1/120$ etc. These results provide an alternate way of giving meaning to the divergent series which occur in the computation of the Casimir effect.

Chapter 4

Real Life I: Interactions

4.1 Interacting Fields

We have seen in the previous sections that a useful way of combining the principles of relativity and quantum theory is to introduce the concept of fields and describe particles as their quanta. So far, we have discussed the quantization of each of the fields separately and have also assumed that, when treated as individual fields, the Lagrangians are at most quadratic in the field variables. The study of such a real scalar field based on the action in Eq. (3.3) led to non-interacting, spinless, massive, relativistic particles obeying the dispersion relation $\omega_{\mathbf{p}}^2 = \mathbf{p}^2 + m^2$. Similarly, the study of the action in Eq. (3.41) led to spinless particles with mass m and charge q. This field will also be non-interacting unless we couple it to an electromagnetic vector potential A_j. Finally, we studied the free electromagnetic field based on Eq. (3.63) which led to massless spin-1 photons[1] which were also non-interacting unless coupled to charged matter.

The real world, however, consists of relativistic particles which are *interacting* with each other. If these particles arise as the quanta of free fields, their interactions have to be modeled by coupling the fields to one another. One prototype example could be a complex scalar field coupled to the electromagnetic field interacting through the Lagrangian:

$$L = (D_m \phi)^* D^m \phi - m^2 \phi^* \phi - \frac{1}{4} F^{ik} F_{ik} \tag{4.1}$$

which is essentially Eq. (3.58) with the electromagnetic Lagrangian in Eq. (3.63) added to it. When expanded this will read as:

$$
\begin{aligned}
L = & \left[\partial_m \phi \partial^m \phi^* - m^2 \phi^* \phi \right] - \left[\frac{1}{4} F^{ik} F_{ik} \right] \\
& + iq A_m (\phi \partial^m \phi^* - \phi^* \partial^m \phi) + q^2 |\phi|^2 A_m A^m
\end{aligned}
\tag{4.2}
$$

The first line of the above equation gives the sum of the two free Lagrangians for the complex scalar field and the electromagnetic field. The second line gives the coupling between the two fields with the strength of the coupling determined by the dimensionless constant q which is dimensionless in four dimensions.. This term makes the charged particles — which are the quanta of the complex scalar fields — interact through the electromagnetic field.

---

[1] We haven't actually discussed the spin of any of these fields rigorously and, for the moment, you can take it on faith that scalar fields are spin zero and the photon has spin 1. To do this properly will require the study of the Lorentz group which we will take up in Chapter 5.

It is also possible to have a field which interacts with itself. For example, a Lagrangian of the form

$$L = \frac{1}{2} \left[\partial_m \phi \partial^m \phi - m^2 \phi^2 \right] - \frac{\lambda}{4!} \phi^4 \qquad (4.3)$$

[2]This Lagrangian has the form $L = (1/2)(\partial\phi)^2 - V(\phi)$ with a potential $V(\phi) = (1/2)m^2\phi^2 + (\lambda/4!)\phi^4$. If you are studying classical or quantum *mechanics*, you would have treated the entire V as an external potential. But in quantum *field* theory we think of the $(1/2)m^2\phi^2$ part (that is, the quadratic term) as a part of the "free" field (as long as $m^2 > 0$) and the ϕ^4 as self-interaction. This is because the $m^2\phi^2$ leads to a linear term in the equations of motion and in the Fourier space, each mode evolves independently. The $m^2\phi^2$ term also contributes to the correct description of the quanta of the free field, with the m^2 part of the dispersion relation $\omega_{\mathbf{p}}^2 = \mathbf{p}^2 + m^2$ coming from this term.

describes a scalar field with a self-interaction[2] indicated by the last term proportional to ϕ^4. This term is again characterized by a dimensionless coupling constant λ (in four dimensions) and will allow, for example, the quanta of the scalar field to scatter off each other through this interaction.

The *real* quantum field theory involves studying such interactions and making useful predictions. Unfortunately, there are very few realistic interactions for which one can solve the Heisenberg operator equations exactly. So the art of making predictions in quantum field theory reduces to the art of finding sensible, controlled, approximation schemes to deal with such interactions.

These schemes can be broadly divided into those which are perturbative and those which are non-perturbative. The perturbative approach involves treating the interactions (characterized by the coupling constants q or λ in the above cases) as "small" and studying the physical processes order-by-order in the coupling constants. In some situations, one could use a somewhat different perturbation scheme but all of them, by definition, involve having a series expansion in a suitable small parameter. We will have a lot to say about this approach in the second half of this chapter and in the next chapter.

The non-perturbative approach, on the other hand, involves using some other approximation to capture certain aspects of the physical situation without relying on the smallness of the coupling constant. Obviously, this will be a context dependent scheme and is somewhat less systematic than the perturbative one. But — when it can be done — it provides important insights that could complement the perturbative approach. One possible way of obtaining non-perturbative results is by using a technique called the *effective action*. The first half of this chapter will be devoted to describing and illustrating this non-perturbative technique.

4.1.1   The Paradigm of the Effective Field Theory

To understand what is involved, let us introduce the Fourier transform $\phi(k)$ of $\phi(x)$ with each mode labeled by a given value of k. Roughly speaking, these wave vectors k^j describe the energy and momentum associated with the modes, with the larger values of k corresponding to higher energies. In the case of a quadratic interaction (corresponding to $V(\phi) \propto \phi^2$) the field equation will be linear in ϕ; then different modes do not talk to each other and we say that it is a *free* field. If $V(\phi)$ has terms higher than quadratic, then the modes for different k get coupled. For example if $V(\phi) \propto \phi^n$, then you can easily convince yourself that the interaction term in the Lagrangian will involve a product of n terms $\phi(k_1), \phi(k_2)...\phi(k_n)$ in the Fourier space. So, in an interacting field theory, physics at high energies is coupled to physics at low energies.

The fact that high energy interactions (e.g. collisions) between particles can lead to new particles now raises a fundamental issue of principle: Since, at any given time, we will have direct knowledge about interactions only below some energy scale (say, $E_{\max}$), accessible in the lab, how can we say anything about the low energy physics if the low energy modes are coupled to high energy modes? Don't we lose all predictability?

Actually we don't. The resolution of the above problem is closely related to the current paradigm in high energy physics called the *renormalization group* and the concept of effective field theory, which is in fact the proper language to describe quantum field theory.

To understand how this comes about, consider a more familiar situation. The hydrodynamics of a fluid can be described by a set of variables like density, velocity etc. which satisfy a certain set of equations.[3] These equations are non-linear, which again tells you that high wavenumber (momentum) modes of a fluid are coupled to low momentum modes of the fluid. Further, these high wavenumber modes probe small spatial scales and we know for a fact that, at sufficiently small scales, the hydrodynamical description is incorrect — because matter is made of discrete particles. This fact, however, has never prevented us from using the equations of hydrodynamics within its domain of validity, say, at scales sufficiently large compared to the intermolecular separation. The effect of all the small scale phenomena can be incorporated into a small set of parameters like the specific heat, the coefficients of viscosity etc. which, of course, need to be determined by experiments.

The situation is similar when we use fields to describe the interaction of relativistic particles, but with some interesting new complications. By and large, the low energy sector of the theory decouples from the high energy sector — in a well-defined manner — and can be described in terms of certain parameters in the low energy Lagrangian like the mass, charge etc., the values of which have to be determined by observations. More formally, let us assume that the description in a given field theory is valid only at energy scales below a particular value Λ. (This is analogous to saying that the hydrodynamic description breaks down at length scales comparable to the atomic size.) Let us divide the modes $\phi(k)$ into those below a particular energy $E < \Lambda$, (which we will call low energy modes, ϕ_{low}) and those with energies between E and Λ (which we will call high energy modes, ϕ_{high}) and ask what is the effect of ϕ_{high} on ϕ_{low}.

There is a formal method of "integrating out" ϕ_{high} from the path integral defining the theory and incorporating its effects by a modified effective interaction between ϕ_{low} modes. Among other things, this will have the effect of changing the values of the parameters in the original Lagrangian which were describing ϕ_{low}. The parameters like mass, charge etc. will now get modified to new values which will be functions of both E and Λ. In addition, this process will also introduce new terms into the Lagrangian which were not originally present.

Suppose we now lower the value of E. This will integrate out more and more high energy modes thereby causing the low energy parameters of the theory to "flow" or "run" as a function of E. The effect of high energy interactions in the low energy sector of the theory is essentially determined by the flow of these parameters. This flow has a natural description in terms of the physical meaning attributed to the parameters (like m, q, λ etc.) which occur in the Lagrangian.

As an example, let us consider the Lagrangian in Eq. (4.2) which has the parameter q occurring in it. In the earlier discussion, we identified this parameter as the charge of the quanta of the complex scalar field. But operationally, to determine this charge, we need to perform some experiments

[3]In fact, the density and velocity in fluid mechanics are also fields in the sense that they are functions of space and time governed by certain partial differential equations.

involving these quanta. One could, say, scatter these particles off each other electromagnetically, and by comparing the theoretically predicted scattering cross-section with the observationally determined one, we will be able to determine the strength of the coupling q. But what we measure in such a process is the value of q relevant for the scattering experiment, performed at some energy scale E, say, which includes *all* the effects of the electromagnetic interaction. There is no a priori reason to believe that this will be governed by the same parameter as introduced in the Lagrangian. One describes this process — viz., that the physical parameters can be different from the parameters introduced in the Lagrangian — by the term "renormalization" of the parameters in the Lagrangian. This is one of the features of an interacting theory and is not difficult to understand conceptually *by itself*.

The nature of the flow of each parameter decides its role in the theory. Consider, for example, a coupling constant in the theory which increases with E; then the theory will become more and more strongly coupled at high energies. This also means that if we fix the value of the coupling constant at some very large energy scale, then it will keep decreasing with energy and will become irrelevant at sufficiently low energy scales. Such coupling constants, even if they are generated during the process of integrating out the high energy modes, are not going to affect the low energy sector of the theory. We will have a description in terms of an effective field theory where the corrections due to such terms are suppressed (or vanish) at low energies compared to E.

On the other hand, if the coupling constant decreases with increasing energy, then the theory will be strongly coupled at low energies but will become "asymptotically free" at high energies.[4] It is also possible for the coupling constants to behave in a more complicated manner, like for example, first increase with energy and then decrease, or vice-versa.

The above description assumes that there is a cut-off scale Λ beyond which we will not trust the field theory description at all — just as one would not trust the hydrodynamic description at scales smaller than the atomic size. It is a philosophical issue whether we should adhere to such a point of view or not. If field theory is the ultimate truth[5] then one is justified in considering the limit $\Lambda \to \infty$. When you do that, something weird happens in quantum field theory: Many of the correction terms to the low energy Lagrangian will acquire divergent terms as their coefficients. When we compute the amplitude for a physically relevant process, we obtain a divergent result when there is no cut-off ($\Lambda \to \infty$). In a class of theories (called *perturbatively renormalizable theories*) these divergences can be accommodated by absorbing them in the parameters of the Lagrangian — that is, by renormalizing λ, m, q etc; no new term needs to be dealt with.

It should be stressed that the *conceptual* basis for renormalization has nothing to do with the divergences which arise either in perturbative or nonperturbative description.[6] For example, the effective mass of an electron in a crystal lattice is different from the mass of the electron outside the lattice, with both being finite; this could be thought of as a renormalization of the mass due to the effect of interactions with the lattice, which has nothing to do with any divergences. In this case, we can measure both the masses separately, but in quantum field theory, since the fields always have their interaction — which cannot be switched off, unlike the interaction of the electron with the lattice which is absent when we take the electron out

[4]In nature both examples are seen in model theories: quantum electrodynamics is a theory in which the coupling constant becomes stronger at higher energies while quantum chromodynamics (describing strong interaction of the quarks) is asymptotically free.

[5]No condensed matter physicist will set the spatial cut-off scale to zero and ask what happens. Some particle physicists can be enormously naive — or extremely optimistic depending on your point of view!

[6]The standard textbook approach will be to first introduce perturbation theory to study interacting fields. The amplitudes computed in the perturbation theory will be divergent and the concept of renormalization will be introduced to "remedy" this. We will do some non-perturbative examples first and discuss the renormalization and 'running' of coupling constants in the non-perturbative context to stress the conceptual separation between these effects and the existence of divergences in the perturbation theory.

— what we observe are the physical parameters that include all the effects of interaction.

To summarize (and emphasize!), the real reason for renormalization has to do with the following facts. We describe the theory by using a Lagrangian containing a certain number (say, N) of parameters λ_A^0 with $A = 1, 2, ..., N$ (like, e.g., the mass, charge, etc.) and a set of fields, neither of which, a priori, have any operational significance. The nature of the fields as well as the numerical values of the constants need to be determined in the laboratory through the measurement of physical processes. For example, in the study of the scattering of a ϕ-particle by another, one could relate the scattering cross-section to the parameters of the theory (like for e.g., the coupling constant λ). Measuring the scattering cross-section will then allow us to determine λ. But what we observe in the lab is the exact result of the full theory and not the result of the perturbative expansion truncated to some finite order! It is quite possible that part of the effect of the interaction is equivalent to modifying the original parameters λ_A^0 we introduced into the Lagrangian to the renormalized values λ_A^{ren}. (The interactions, of course, will have other effects as well, but *one* feature of the interaction *could be* the renormalization of the parameters.) When this is the case, no experiment will ever allow us to determine the numerical values of λ_A^0 we put into the Lagrangian and the physical theory is determined, operationally, in terms of the renormalized parameters λ_A^{ren}. This happens in all the interacting theories and is the basic reason behind renormalization.[7]

It is useful to consider a situation different from $\lambda\phi^4$ theory or QED to highlight another fact. Suppose you start a theory (with a set of N parameters λ_A^0 in its Lagrangian) and, when you compute an amplitude perturbatively, there are divergences in the theory which *cannot* be reabsorbed in any of these parameters. Sometimes the situation can be remedied by introducing another k parameters, say $\mu_B (B = 1, ...k)$ and when you study the perturbative structure of the new Lagrangian involving $(N + k)$ parameters (λ_A, μ_B), all the divergences can be reabsorbed into these parameters. We then might have a healthy theory using which we can make predictions. But for most Lagrangians you can write down, including some very sensible Lagrangians (like the one used in Einstein's theory of gravity), this process does not converge. You start with a certain number of parameters in the theory and the divergences which arise force you to add some more parameters; these lead to new interactions and when you rework the structure of the theory you find that fresh divergences are generated which will require still more parameters to be introduced etc., ad-infinitum. Such a theory will require an infinite number of parameters to be divergence-free, which is probably inappropriate for a *fundamental* theory.[8] Fortunately, a very wide class of theories in physics are renormalizable in the above sense. In particular, the $\lambda\phi^4$ theory and QED are classic examples of theories in which this program works.

After this preamble, we will study (i) the self-interacting scalar field and (ii) the coupled complex scalar-electromagnetic field system, using a nonperturbative approximation called the effective action method.

[7]Even if both λ_A^0 and λ_A^{ren} are finite quantities, we still need to introduce the concept of renormalization into the theory, acknowledge the fact that λ_A^0 are unobservable mathematical parameters and re-express physical phenomena in terms of λ_A^{ren}. It so happens that in $\lambda\phi^4$ theory, QED (and in many other theories) the parameters λ_A^0 and λ_A^{ren} differ by infinite values for reasons we do not quite understand. But the physical theory with finite λ_A^{ren} is quite well defined and agrees wonderfully with experiment. Thus, while renormalization has indeed helped us to take care of the infinities in these perturbative theories, the *raison d'être* for renormalization has nothing to do with divergences in perturbative expansion.

[8]It is important to appreciate that such theories could still be *very useful* and provide a perfectly sensible description of nature as effective theories valid below, say, certain energy scale. This is done by arranging the terms in the Lagrangian, arising from new parameters, in some sensible order so that their contributions are subdominant at low energies in a controlled manner. It is a moot point whether real nature might actually require an infinite number of parameters for its description if you want to study processes at arbitrarily high energies. Physicists hope it does not.

4.1.2   A First Look at the Effective Action

We will first introduce this concept in somewhat general terms and then illustrate it later on with specific examples. Consider a theory which describes the interaction between two systems having the dynamical variables Q and q. The action functional for the system, $A[Q, q]$ will then depend[9] on both the variables, possibly in a fairly complicated way. The full theory can be constructed from the exact propagator

$$K(Q_2, q_2; Q_1, q_1; t_2, t_1) = \int \mathcal{D}Q \int \mathcal{D}q \exp \frac{i}{\hbar} A[Q, q] \qquad (4.4)$$

which is quite often impossible to evaluate. The *effective action* method is an approximation scheme available for handling Eq. (4.4). This method is particularly appropriate if we are interested in a physical situation in which one of the variables, say, Q, behaves nearly classically while the other variable q is fully quantum mechanical. We then expect q to fluctuate rapidly compared to Q which is considered to vary more slowly in space and time. The idea is to investigate the effect of the quantum fluctuations of q on Q. In that case, we can approach[10] the problem in the following manner:

Let us suppose that we can do the path integral over q in Eq. (4.4) exactly for some given $Q(t)$. That is, we could evaluate the quantity

$$F[Q(t); q_2, q_1; t_2, t_1] \equiv \exp \frac{i}{\hbar} W = \int \mathcal{D}q \exp \frac{i}{\hbar} A[Q(t), q] \qquad (4.5)$$

treating $Q(t)$ as some specified function of time. If we could now do

$$K = \int \mathcal{D}Q \exp \frac{i}{\hbar} W[Q] \qquad (4.6)$$

exactly, we could have completely solved the problem. Since this is not possible, we will evaluate Eq. (4.6) by invoking the fact that Q is almost classical. This means that most of the contribution to Eq. (4.6) comes from the extremal paths satisfying the condition

$$\frac{\delta W}{\delta Q} = 0 \qquad (4.7)$$

Equation (4.7), of course, will contain some of the effects of the quantum fluctuations of q on Q, and is often called the 'semi-classical equation'.[11] The quantity W is called the 'effective action' in this context.

The original action for the field was expressible as a time integral over a Lagrangian in quantum mechanics (and as an integral over the four volume $d^4 x$ of a Lagrangian density in quantum theory) and, in this sense, is local. There is no guarantee that the effective action $W[q]$ can, in general, be expressed as an integral over a local Lagrangian. In fact, $W[q]$ will usually be nonlocal. But in several contexts in which we *can* compute W in some suitable approximation, it will indeed turn out to be local or can be approximated by a local quantity. In such cases it is also convenient to define an effective *Lagrangian* through the relation

$$W = \int L_{\text{eff}} dt \qquad (4.8)$$

[9]This notation is purely formal; the symbol Q, for example, could eventually describe a set of variables, like the components of a vector field. The detailed nature of these variables is not of importance at this stage; we will describe the concepts as though Q and q are quantum *mechanical* in nature and then generalize it to the fields.

[10]There is a formal way of introducing the idea of effective action, valid in more general contexts. We will describe it later on, at the end of Sect.4.5. Here we take a more intuitive approach.

[11]It is often quite easy to evaluate Eq. (4.6) in this approximation and thereby obtain an approximate solution to our problem. Usually we will be content with obtaining the solutions to Eq. (4.7), and will not even bother to calculate Eq. (4.6) in this approximation.

There is, however, one minor complication in the above formalism as it stands. The way we have defined our expressions, the quantities K and W depend on the boundary conditions (t_2, q_2, t_1, q_1). We would prefer to have an effective action which is completely independent of the q-degree of freedom. The most natural way of achieving this is to integrate out the effect of q for *all times* by considering the limit $t_2 \to +\infty, t_1 \to -\infty$ in our definition of the effective action. We will also assume, as is usual, that $Q(t)$ vanishes asymptotically. From our discussion in Sect. 1.2.3, we know that, in this limit, the path integral essentially represents the amplitude for the system to go from the ground state (of the q degree of freedom) in the infinite past to the ground state in the infinite future. Using the results of Sect. 1.2.3, we can write

$$F(q_2, q_1; +\infty, -\infty) \equiv \exp \frac{i}{\hbar} W[Q(t)] = N(q_2, q_1) \langle E_0, +\infty | -\infty, E_0 \rangle_{Q(t)}$$
(4.9)

where $\langle E_0, +\infty | -\infty, E_0 \rangle_{Q(t)}$ stands for the vacuum-to-vacuum amplitude in the presence of the external source $Q(t)$ and $N(q, q_2)$ is a normalization factor, independent of $Q(t)$. Taking logarithms, we get:

$$W[Q(t)] = -i\hbar \ln \langle E_0, +\infty | -\infty, E_0 \rangle_{Q(t)} + (\text{constant})$$
(4.10)

Since the constant term is independent of Q, it will not contribute in Eq. (4.7). Therefore, for the purposes of our calculation, we may take the effective action to be defined by the relation

$$W[Q(t)] \equiv -i\hbar \ln \langle E_0, +\infty | -\infty, E_0 \rangle_{Q(t)}$$
(4.11)

in which all reference to the quantum mode is eliminated. Note that $|E_0\rangle$ is the ground state of the original system; that is, the ground state with $Q = 0$.

This discussion also highlights another important feature of the effective action. We have seen in Sect. 2.1.2 that an external perturbation can cause transitions in a system from the ground state to the excited state. In other words, the probability for the system to be in the ground state in the infinite future (even though it started in the ground state in the infinite past) could be less than unity. This implies that our effective action W_0 need not be a real quantity! If we use this W directly in Eq. (4.7), we have no assurance that our solution Q will be real. The imaginary part of W contains information about the rate of transitions induced in the q-system by the presence of $Q(t)$; or — in the context of field theory — the rate of production of particles from the vacuum. The semi-classical equation is of very doubtful validity if these excitations drain away too much energy from the Q- mode. Thus we must confine ourselves to the situations in which

$$Im\, W \ll Re\, W$$
(4.12)

In that case, we can approximate the semi-classical equations to read[12]

$$\frac{\delta Re\, W}{\delta Q} = 0$$
(4.13)

The above discussion provides an alternative interpretation of the effective action which is very useful. Let us suppose that $Q(t)$ varies slowly enough for the adiabatic approximation to be valid. We then know — from

[12]In most practical situations, Eq. (4.12) will automatically arise because of another reason. Notice that the success of the entire scheme depends on our ability to evaluate the first path integral in Eq. (4.5). This task is far from easy, especially because we need this expression for an arbitrary $Q(t)$. Quite often, one evaluates this expression by assuming that the time variation of $Q(t)$ is slow compared to time scale over which the quantum variable q fluctuates. In such a case, the characteristic frequencies of the q-mode will be much higher than the frequency at which the Q-mode is evolving, and hence there will be very little transfer of energy from Q to q. The real part of W will dominate.

our discussion in Sect. 2.1.2 — that the 'vacuum to vacuum' amplitude is given by [see Eq. (2.32)]

$$\lim_{t_2 \to \infty} \lim_{t_1 \to -\infty} F(q_2, q_1; t_2, t_1) \propto \exp -\frac{i}{\hbar} \int_{-\infty}^{+\infty} E_0(Q) dt. \qquad (4.14)$$

This expression allows us to identify the effective Lagrangian as the ground state energy of the q-mode in the presence of Q:

$$L_{\text{eff}} = -E_0(Q) \qquad (4.15)$$

This result, which is valid when the time dependence of Q is ignored in the calculation of E_0, provides an alternative means of computation of the effective Lagrangian if the Q dependence of the ground state energy can be ascertained.

To summarize, the strategy in using this approach is as follows. We first integrate out the quantum degree of freedom q with asymptotic boundary conditions and obtain the $W[Q(t)]$ or $L_{\text{eff}}[Q(t)]$. The imaginary part of L_{eff} will contain information about particle production which could take place due to the interactions; the real part will give corrections to the original classical field equations for the Q degree of freedom.

Let us now proceed from quantum mechanics to field theory. In the field theoretic context, Q, q will become fields depending on space and time and in many contexts we are interested in, the relevant action can be expressed in the form, either exactly or as an approximation:

$$A[Q(x), q(x)] = \int d^4x \, L_0[Q(x)] - \frac{1}{2} \int d^4x \, q\hat{D}[Q]q \qquad (4.16)$$

[13] We have now set $\hbar = 1$

If we integrate out $q(x)$, we will end up getting an effective action[13] given by

$$Z[Q(x)] = \int \mathcal{D}q \, e^{iA} = \exp i \int d^4x \, (L_0 + L_{\text{eff}})$$

$$\Rightarrow \exp - \int d^4x \, (L_0 + L_{\text{eff}})_E \qquad (4.17)$$

where the last expression is valid in the Euclidean sector and

$$L_{\text{eff}} = -\frac{i}{2} \int_0^\infty \frac{ds}{s} \langle x|e^{-isD}|x\rangle \Rightarrow -\frac{1}{2} \int_0^\infty \frac{ds}{s} \langle x|e^{-sD_E}|x\rangle \qquad (4.18)$$

The first form of L_{eff} is in the Lorentzian space while the second form is valid in the Euclidean sector (see Eq. (2.66) and Eq. (2.77)). We have already seen that L_{eff} can be interpreted as (i) the vacuum energy of field oscillators (see Eq. (2.71) or (ii) the energy associated with particles in closed loops (see Eq. (1.107). This connects well with Eq. (4.14).

We can make some general comments about these expressions connecting up with our discussion in the last section. As we said before, L_{eff} can, in general, have a real and imaginary part. The imaginary part will indicate the probability for creation of particles while the real part will give corrections to $L_0(Q)$. Suppose that the real part of L_{eff} contained a term which has exactly the same form as a term in the original Lagrangian $L_0(Q)$, except for a different proportionality constant. For example, in the study of scalar field theory with $Q = \phi$, the $L_0(\phi)$ will contain a mass term

$-(1/2)m^2\phi^2$. Suppose when we consider the interaction of this field with itself or other fields and integrate out the other degrees of freedom, we end up getting an L_{eff} which has a term $-(\mu/2)m^2\phi^2$. Then the quantum corrected effective Lagrangian $L_0(\phi) + L_{\text{eff}}(\phi)$ will have a term which goes as $(1 + \mu)(1/2)m^2\phi^2$.

But since the quantum fluctuations can never be switched off, it doesn't make sense to think of the $(1/2)m^2\phi^2$ term and $(\mu/2)m^2\phi^2$ term as physically distinct entities because we will only be able to observe their total effect. In other words, the interaction modifies ("renormalizes") the parameter m^2 to a the value $m^2(1 + \mu)$. It is only this quantity which will be physically observable and the original m^2 we introduced in the "free" scalar field Lagrangian has no operational significance because "free" Lagrangians do not exist in nature. It follows that whenever L_{eff} contains terms which have the same structure as the terms in the original Lagrangian L_0, then these corrections will not be separately observable; instead they renormalize the original parameters in the Lagrangian L_0.

All these arguments make sense even if μ is a divergent, infinite quantity, as it will turn out to be. This divergence is disturbing and is possibly telling us that the theory needs a cut-off, say, Λ, but the fact that such a term will automatically get absorbed in the original parameters of the Lagrangian is a result which is independent of the divergence. In this case, we need to define the theory with a cut-off Λ beyond which we do not trust it. Then the L_{eff} can be interpreted as describing a low energy effective Lagrangian obtained after integrating out certain degrees of freedom, usually those with high energy or belonging to another field which we are not interested in. If we choose to integrate out modes with energies in the range $M < E < \Lambda$, then the correction μ will depend on both M and Λ. When we decrease M, integrating out more modes, the parameter μ and hence the physical mass will "run" with M. This is one way to approach the ideas of renormalization and running coupling constants through the effective action. We will see specific examples for such a behaviour in the coming sections.

4.2   Effective Action for Electrodynamics

As an illustration of the above ideas, let us consider a charged scalar field interacting with the electromagnetic field. In this case, we can take advantage of the fact that the electromagnetic field has a sensible classical limit in which it obeys the standard Maxwell equations. It makes sense to consider a nearly constant (very slowly varying) electromagnetic field background and ask how the coupling of this electromagnetic field with a charged scalar field affects the former. To answer this question, we need to compute the effective action for the electromagnetic field by integrating out the charged scalar field in the interaction term of the Lagrangian given by Eq. (3.53). That is, we need to compute

$$e^{iA_{\text{eff}}[F_{ik}]} = \int \mathcal{D}\phi\mathcal{D}\phi^* \; e^{i\int d^4x \; \phi[(i\partial - qA)^2 - m^2]\phi^*} \tag{4.19}$$

From Eq. (4.18), we see that we need to compute the matrix element $\langle x|e^{-isH}|x\rangle$ for the quantum mechanical Hamiltonian $H = -(i\partial - qA)^2 + m^2$. Once we have this matrix element, we can compute the effective Lagrangian L_{eff} using Eq. (4.18). This computation actually turns out to be easier than one would have imagined because one can use a series of tricks to evaluate it. We will now explain the procedure.

Since we are interested in the case in which the vector field A_j varies slowly in spacetime, we could ignore its second and higher derivatives and treat it as arising due to a constant F_{ik}; i.e., we study a background electromagnetic field with constant $\mathbf{E}, \mathbf{B}$. The gauge invariance implies that the L_{eff} which we compute can only depend on the quantities $(E^2 - B^2)$ and $\mathbf{E} \cdot \mathbf{B}$. We will define two constants a and b by the relation

$$a^2 - b^2 = E^2 - B^2; \qquad ab = \mathbf{E} \cdot \mathbf{B} \tag{4.20}$$

Exercise 4.1: Prove this claim.

so that $L_{\text{eff}} = L_{\text{eff}}(a, b)$. In the general case with arbitrary $\mathbf{E}$ and $\mathbf{B}$ for which a and b are not simultaneously zero, it is well known that, by choosing our Lorentz frame suitably, we can make E and B parallel, say along the y-axis. Such a field will have components $[\mathbf{E} = (0, E, 0); \mathbf{B} = (0, B, 0)]$ and can arise from the vector potential of the form $A_i = [-Ey, 0, 0, -Bx]$ in a specific gauge. In this particular case, $a^2 - b^2 = E^2 - B^2$ and $ab = \mathbf{E} \cdot \mathbf{B} = EB$. Therefore $E = a$ and $B = b$. Once we find the effective Lagrangian $L_{\text{eff}}(E, B)$ as a function of E and B in this particular case, the result for the most general case can be obtained by simply substituting $E = a$ and $B = b$.

For this choice of fields and vector potential, the H we need to deal with is given by

$$\begin{aligned} H &= -\left[(i\partial - qA)^2 - m^2\right] \\ &= -\left[(i\partial_t + qEy)^2 + \partial_x^2 + \partial_y^2 - (i\partial_z + qBx)^2 - m^2\right] \end{aligned} \tag{4.21}$$

While evaluating the matrix element $\langle x|e^{-isH}|x\rangle$, we can introduce a complete set of momentum eigenstates as far as the t and z coordinates are concerned, since the Hamiltonian is independent of these two coordinates. Taking

$$\langle p^0|t\rangle = e^{+ip^0 t}; \qquad \langle p^z|z\rangle = e^{-ip^z z} \tag{4.22}$$

we can replace $i\partial_t$ by $-p^0$ and $i\partial_z$ by p^z in the expression for H in Eq. (4.21) and integrate over p^0 and p^z. The coordinates x and y — which we will indicate together as $\mathbf{x}_\perp$ — cannot be handled by going to the momentum space since the Hamiltonian explicitly depends on these coordinates. So we will leave them as they are. The problem now reduces to evaluating

$$\langle x^i|e^{+is[(i\partial - qA)^2 - m^2]}|x^i\rangle = \int \frac{dp^0 dp^z}{(2\pi)^2}\, \langle \mathbf{x}_\perp|e^{-isH}|\mathbf{x}_\perp\rangle \tag{4.23}$$

where

$$H = -\left[(-p^0 + qEy)^2 + \partial_y^2 + \partial_x^2 - (p^z + qBx)^2 - m^2\right] \tag{4.24}$$

Introducing two new coordinates ℓ_1, ℓ_2 in place of x and y, by the definitions:

$$\ell_1 \equiv x + \frac{p^z}{qB}; \qquad \ell_2 \equiv y - \frac{p^0}{qE}, \tag{4.25}$$

H is reduced to the form

$$H = \left[-\partial_1^2 + q^2 B^2 \ell_1^2\right] + \left[-\partial_2^2 - q^2 E^2 \ell_2^2\right] + m^2 \tag{4.26}$$

This form is remarkably simple and should be familiar to you. It consists of two separate operators in square brackets (the m^2 term is just a constant added to the Hamiltonian) of which the first one is the Hamiltonian for a harmonic oscillator with mass $(1/2)$ and frequency $\omega = 2|qB|$. The second square bracket corresponds to an "inverted" harmonic oscillator (i,e., the one with the sign of ω^2 reversed so that $\omega \to i\omega$) with the same mass and frequency. The matrix element in Eq. (4.23) is now trivial and can be computed using the standard path integral kernel for a harmonic oscillator in the coincidence limit. For the ℓ_1 degree of freedom, this is given by the expression

Exercise 4.2: Revise your knowledge of the path integral for the harmonic oscillator and obtain this result.

$$K = \left(\frac{m\omega}{2\pi i \sin \omega t} \right)^{1/2} \exp \left[i \left(m\omega x^2 \right) \left(\tan \frac{\omega t}{2} \right) \right] \qquad (4.27)$$

evaluated with $m = (1/2)$, $\omega = 2|qB|$, $t = s$, $x = \ell_1$. The result for ℓ_2 is obtained by changing ω to $i\omega$ in this expression and is given by

$$K = \left(\frac{m\omega}{2\pi i \sinh \omega t} \right)^{1/2} \exp \left[-i \left(m\omega x^2 \right) \left(\tanh \frac{\omega t}{2} \right) \right] \qquad (4.28)$$

evaluated with $m = (1/2)$, $\omega = 2|qE|$, $t = s$, $x = \ell_2$. The integration over p^0 and p^z in Eq. (4.23) is equivalent to an integration over ℓ_1 and ℓ_2 with $dp^0 = -qEd\ell_2$, $dp^z = qBd\ell_1$. Doing these integrals and simplifying the expressions, we find that:

$$\langle x^i | e^{-isH} | x^i \rangle = \frac{1}{16\pi^2 i} \left(\frac{qB}{\sin qBs} \right) \left(\frac{qE}{\sinh qEs} \right) e^{-im^2 s} \qquad (4.29)$$

The effective Lagrangian is obtained from this kernel using the result

$$L_{\text{eff}} = 2 \times \left(-\frac{i}{2} \right) \int_0^\infty \frac{ds}{s} \langle x | e^{-iHs} | x \rangle \qquad (4.30)$$

where the extra factor of 2 takes care of the fact that the complex scalar field has twice the degrees of freedom as compared to a single field. This gives the final result to be

$$\begin{aligned} L_{\text{eff}} &= -\int_0^\infty \frac{ds}{(4\pi)^2} \frac{e^{-i(m^2 - i\epsilon)s}}{s^3} \left(\frac{|qE|s}{\sinh |qE|s} \right) \left(\frac{|qB|s}{\sin |qB|s} \right) \\ &= -\int_0^\infty \frac{ds}{(4\pi)^2} \frac{e^{-i(m^2 - i\epsilon)s}}{s^3} \left(\frac{|qa|s}{\sinh |qa|s} \right) \left(\frac{|qb|s}{\sin |qb|s} \right) \end{aligned} \qquad (4.31)$$

In arriving at the second expression, we have substituted $E = a$ and $B = b$ as explained earlier. It is understood that the combinations qa and qb (which arose from taking the positive square root of ω^2) should be treated as positive quantities irrespective of the individual signs of q, a, b. The integration over s is slightly off the real axis along $s(1 - i\epsilon)$ as described in Fig. 1.3. This requires the contour to be closed as shown in Fig. 4.1 on the quarter circle at infinity in the lower right quadrant. Equivalently, we can rotate the contour to go along the negative imaginary axis. This shows that the poles on the integrand along the *real axis* do not contribute to the integral because we are going below all these poles. On the other hand, any pole on the negative imaginary axis needs to be handled separately in this integral. We need to go around each of these poles in a small semi-circle in

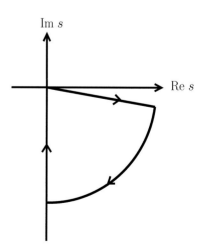

Figure 4.1: Integration contour for Eq. (4.31).

the complex plane which – as we will see — will lead to an imaginary part in L_{eff}.

From the integrand in Eq. (4.31), it is clear that the magnetic field leads to poles along the real axis due to the $\sin(qbs)$ factor, which are irrelevant; on the other hand, the electric field leads to poles along the negative imaginary axis due to the $\sinh(qas)$ factor, which, as we will see, will lead[14] to an imaginary contribution to L_{eff}. Our next task is to unravel the physics contained within L_{eff} which we will do in a step-by-step manner.

4.2.1 Schwinger Effect for the Charged Scalar Field

We will first extract the imaginary part of the effective action which represents the creation of particles from the vacuum. The simplest way to compute this is to rotate the contour from along the positive real axis to along the negative imaginary axis. Formally, this will lead to the expression:

$$L_{\text{eff}}(a,b) = \int_0^\infty \frac{ds}{(4\pi)^2} \frac{e^{-m^2 s}}{s^3} \left(\frac{qas}{\sin qas} \right) \left(\frac{qbs}{\sinh qbs} \right) \quad (4.32)$$

In this expression, the poles from the electric field lie along the path of integration, but, as described earlier, these poles are to be avoided by going in a semi-circle above them as illustrated in Fig. 4.2 for one particular pole. Each of the poles at $s = s_n = (n\pi/qa)$ with $n = 1, 2, \ldots$ are avoided by going around small semicircles of radius ϵ in the upper half plane. The n-th pole contributes to this semicircle the quantity

$$
\begin{aligned}
I_n &= \int_{\theta=\pi}^{\theta=0} \frac{(\epsilon e^{i\theta} id\theta)}{(4\pi)^2 s_n^2} e^{-m^2 s_n} \cdot \frac{qa}{\cos(n\pi)\epsilon e^{i\theta}} \left(\frac{qbs_n}{\sinh qbs_n} \right) \\
&= i(-1)^{n+1} \frac{(qa)^2}{16\pi^3} \left[\frac{1}{n^2} \exp\left(-\frac{m^2\pi}{qa} n \right) \right] \left(\frac{qbs_n}{\sinh qbs_n} \right) \quad (4.33)
\end{aligned}
$$

which is pure imaginary. So the total contribution to $\text{Im}\, L_{\text{eff}}$ is:

$$\text{Im}\, L_{\text{eff}} = \sum_{n=1}^\infty (-1)^{n+1} \cdot \frac{1}{2} \frac{(qa)^2}{(2\pi)^3} \frac{1}{n^2} \exp\left(-\frac{m^2\pi}{qa} n \right) \cdot \left(\frac{qbs_n}{\sinh qbs_n} \right) \quad (4.34)$$

It is now clear that $(\text{Im}\, L_{\text{eff}})$ arises because of non-zero a, i.e. (i) whenever there is an electric field in the direction of the magnetic field or (ii) if $\mathbf{E}$ is perpendicular to $\mathbf{B}$, but $E^2 > B^2$. (In this case, we can go to a frame in which the field is purely electric). If we set $B = 0$, we reduce the problem to that of a constant electric field, and in this case, the imaginary part is

$$\text{Im}\, L_{\text{eff}} = \sum_{n=1}^\infty \frac{1}{2} \frac{(qE)^2}{(2\pi)^3} \frac{(-1)^{n+1}}{n^2} \exp\left(-\frac{\pi m^2}{|qE|} n \right) \quad (4.35)$$

The interpretation of this result is similar to the one we provided in Sect. 2.1.2 in the case of an external source $J(x)$ creating particles from the vacuum (see Eq. (2.25)) but with some crucial differences. In that case, we considered a real scalar field $\phi(x)$ coupled to an externally specified c-number source $J(x)$ and found that $|\langle 0_+|0_-\rangle^J|^2$ can be related to number of quanta of the ϕ field produced per unit volume per unit time. As you can

[14]The L_{eff} can also be thought of as the vacuum energy of the field oscillators (as in Eq. (2.71)). It is possible to use this interpretation directly to obtain the effective Lagrangian for the electromagnetic field. We will describe this in Mathematical Supplement, Sect.4.9.1.

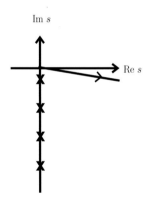

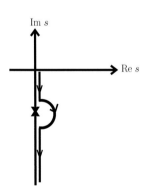

Figure 4.2: Integration contours for Eq. (4.33).

see, we are computing exactly the same physical quantity now. We have an externally specified electromagnetic field with a particular A_i which can act as a source for the charged quanta of a complex scalar field. If we think of the electric field, say, as being switched on adiabatically and switched off adiabatically, then the initial vacuum state will not remain a vacuum state at late times. The probability for the persistence of vacuum is again given by $|\langle 0_+|0_-\rangle^A|^2$ where the matrix element is now evaluated in an externally specified A_i. Writing L_{eff} as $(\mathcal{R} + i\mathcal{I})$ and noting that

$$Z[A_i] = \langle 0_+|0_-\rangle^A = \exp i \int d^4x \, (\mathcal{R} + i\mathcal{I}) , \qquad (4.36)$$

we get

$$|Z|^2 = |\langle 0_+|0_-\rangle^A|^2 = \exp -2 \int d^4x \, \mathcal{I} \qquad (4.37)$$

Thus, Im L_{eff} gives the number of *pairs* of charged scalar particles created by the electromagnetic field per unit time per unit volume.[15]

There are, however, two surprising features about this result which are worth noting.[16] First we find that a *static* electric field can produce particles from the vacuum, unlike the case of a time dependent $J(x)$ which produces particles in the example we saw in Sect. 2.1.2. Second — and more important — is the fact that the expression obtained in Eq. (4.35) is *non-analytic* in q. Conventional quantum electrodynamics (examples of which we will discuss in the next chapter) is based on a perturbative expansion in the coupling constant q as a series. The non-analyticity of the Schwinger effect shows that perturbation theory in q will never get you this result irrespective of your summing the series to an arbitrarily large number of terms. Thus the Schwinger effect is a genuinely non-perturbative result in quantum field theory.[17]

4.2.2 The Running of the Electromagnetic Coupling

Having disposed off the imaginary part of L_{eff}, let us consider its real part. The full action for the electromagnetic field — when we take into account its coupling to the charged scalar field — is given by the sum $L_0 + \text{Re } L_{\text{eff}}$ where $L_0 = (1/2)(E^2 - B^2)$ which will, for example, lead to corrections to the classical Maxwell equations. We certainly expect Re L_{eff} to be analytic[18] in the coupling constant q so that one could think of it as a small perturbative correction to L_0. The real part of Eq. (4.31) is given by

$$\text{Re } L_{\text{eff}} = -\int_0^\infty \frac{ds}{(4\pi)^2} \left(\frac{\cos m^2 s}{s^3} \right) \left(\frac{qas}{\sinh qas} \right) \left(\frac{qbs}{\sin qbs} \right) \qquad (4.38)$$

While one can work with this expression, the oscillation of the cosine function in the upper limit of the integral makes it a bit tricky; therefore, it is preferable to go back to the expression in Eq. (4.32) which we used previously:

$$L_{\text{eff}}(a,b) = \int_0^\infty \frac{ds}{(4\pi)^2} \frac{e^{-m^2 s}}{s^3} \left(\frac{qas}{\sin qas} \right) \left(\frac{qbs}{\sinh qbs} \right) \qquad (4.39)$$

We know how to handle the poles along the real axis arising from the sine function and leading to an imaginary part when the integration contour is deformed to a semi-circle. What remains is just a principal value of

[15]This result, viz. that a constant electric field can produce charged pairs from the vacuum, is known as the Schwinger effect and for once, it is named correctly; Schwinger did work out the clearest version of it.

[16]There are several ways of "understanding" this qualitatively, none of which are totally satisfactory; so it is best if you take it as a fact of life. You may legitimately wonder where the energy for the creation of particles is coming from. This is provided by the agency which is maintaining the constant electric field. Suppose you keep two charged capacitor plates maintaining a constant electric field in between. The pairs produced from the vacuum will move towards oppositely charged plates, shorting the field. The agency has to do work to prevent this from happening, which goes into the energy of the particles.

[17]That is why we began with this discussion before you lose your innocence in the perturbation expansions!

[18]If Re L_{eff} turns out to be non-analytic in q, we are in deep trouble and one cannot obtain the free electromagnetic field as the $q = 0$ limit. It is reassuring that this does not happen.

the integral which is real. With this understanding — which effectively ignores the poles along the real axis — we can continue to work with this expression. This is what we will do.

The first point to note about Re L_{eff} is that it is divergent near $s = 0$. There are two sources to this divergence, one trivial and the other conceptually important. The trivial source of divergence can be spotted by noticing that Re L_{eff} is divergent even when $\mathbf{E} = \mathbf{B} = 0$. This divergence is that of L_{eff} for a *free* complex scalar field, which essentially arises from the sum of the zero point energies of the field, which we studied in Sect. 2.2.1 (see Eq. (2.71)). We mentioned while discussing the expression in Eq. (2.73) that the integral representation for $\ln D$ should be interpreted as $\ln D/D_0$; this corresponds to subtracting out the L_{eff} in the absence of electromagnetic field. So, this divergence can be legitimately removed by simply subtracting out the value for $\mathbf{E} = \mathbf{B} = 0$. Thus, we modify Eq. (4.38) to

$$\text{Re } L_{\text{eff}} \equiv R = \int_0^\infty \frac{ds}{(4\pi)^2} \frac{e^{-m^2 s}}{s^3} \left[\frac{q^2 abs^2}{\sinh qbs \sin qas} - 1 \right] \qquad (4.40)$$

Since the subtracted term is a constant independent of $\mathbf{E}$ and $\mathbf{B}$, the equations of motion are unaffected.

The expression R is still logarithmically divergent near $s = 0$, since the quantity in the square brackets behaves as $[+(1/6)q^2 s^2(a^2 - b^2)]$ near $s = 0$. *But notice that this divergent term is proportional to $(a^2 - b^2) = \mathbf{E}^2 - \mathbf{B}^2$, which is the original — uncorrected — Lagrangian.* As we described right at the beginning, any term in L_{eff} which has the same structure as a term in the original Lagrangian cannot be observed separately in any physical process and merely renormalizes the parameters of the theory. To handle this, let us follow the same procedure we described earlier. We will write

$$L_{\text{total}} = L_0 + \text{Re } L_{\text{eff}} = (L_0 + L_c) + (\text{Re } L_{\text{eff}} - L_c) \equiv L_{\text{div}} + L_{\text{fin}} \qquad (4.41)$$

where $L_{\text{div}} \equiv L_0 + L_c$ is divergent and $L_{\text{fin}} \equiv \text{Re } L_{\text{eff}} - L_c = R - L_c$ is finite with L_c being the divergent part of R given by

$$L_c = \frac{1}{(4\pi)^2} \int_0^\infty \frac{ds}{s^3} e^{-m^2 s} \left[\frac{1}{6}(qs)^2 (a^2 - b^2) \right] \qquad (4.42)$$

$$= \frac{q^2}{6(4\pi)^2}(a^2 - b^2) \int_0^\infty \frac{ds}{s} e^{-m^2 s} \equiv \frac{Z}{2}(a^2 - b^2) = \frac{Z}{2}(\mathbf{E}^2 - \mathbf{B}^2)$$

where Z is given by[19]

$$Z = \frac{q^2}{48\pi^2} \int_0^\infty \frac{ds}{s} \, e^{-m^2 s} \qquad (4.43)$$

This integral, and hence Z, is divergent but that is not news to us. We can, as usual, combine L_c with our original Lagrangian L_0 and write the final result as $L_{\text{tot}} = L_{\text{div}} + L_{\text{fin}}$, where

$$L_{\text{div}} \equiv (L_0 + L_c) = \frac{1}{2}(1 + Z)(\mathbf{E}^2 - \mathbf{B}^2) \qquad (4.44)$$

and $L_{\text{fin}} \equiv (\text{Re } L_{\text{eff}} - L_c)$ is given by

$$L_{\text{fin}} = \frac{1}{(4\pi)^2} \int_0^\infty \frac{ds}{s^3} e^{-m^2 s} \left[\frac{q^2 s^2 ab}{\sin(qsa)\sinh(qsb)} - 1 - \frac{1}{6} q^2 s^2 (a^2 - b^2) \right] \qquad (4.45)$$

[19] The notation Z is rather conventional; in fact, when we study QED you will encounter Z_1, Z_2 and Z_3 all being divergent!

The quantity L_{fin} is perfectly well-defined and finite. [The leading term coming from the square bracket near $s = 0$ is proportional to s^4 and hence L_{fin} is finite near $s = 0$.] Further, it can be expanded in a Taylor series in q perturbatively, giving finite corrections to the classical electromagnetic Lagrangian. The leading order term has the form (with $\hbar$ and c re-introduced)

$$L_1 = \frac{q^4}{90} \frac{(\hbar/mc)^3}{mc^2} \left[\frac{7}{4} \left(E^2 - B^2 \right)^2 + (\mathbf{E} \cdot \mathbf{B})^2 \right] \qquad (4.46)$$

As regards L_{div}, we know that only the combination with a $(1 + Z)$ factor will be physically observable and not the individual term. To make this formal, we shall now redefine all our field strengths and charges by the rule

$$\mathbf{E}_{\text{phy}} = (1 + Z)^{1/2} \mathbf{E}; \quad \mathbf{B}_{\text{phy}} = (1 + Z)^{1/2} \mathbf{B}; \quad q_{\text{phy}} = (1 + Z)^{-1/2} q \quad (4.47)$$

This is, of course, same as scaling a and b by $(1+Z)^{1/2}$ leaving $(q_{\text{phy}} E_{\text{phy}}) = qE$ invariant. Since only the products qa, qb appear in L_{fin}, it can also be expressed in terms of $(q_{\text{phy}} E_{\text{phy}})$. Once we have introduced physical variables, L_{div} reduces to the standard electromagnetic Lagrangian in terms of physical variables, with corrections arising from L_{fin} also being expressed entirely in terms of physical variables.

All that remains is to study how the coupling constant runs in this case by evaluating the integral in Eq. (4.43) after regularizing it.[20] Here we shall use the simpler technique of calculating it with a cut-off. Since it is well defined for $s \to \infty$ and the divergence arises from the lower limit, we will first regularize it by introducing a cut-off $\Lambda \equiv (1/M^2)$ at the lower end.[21] An integration by parts then gives the result:

$$\int_{M^{-2}}^{\infty} \frac{ds}{s} e^{-m^2 s} = e^{-m^2/M^2} \ln M^2 + \int_{M^{-2}}^{\infty} m^2 \, ds \, e^{-m^2 s} \ln s \qquad (4.48)$$

We can now set $M^{-2} = 0$ in the second term and in e^{-m^2/M^2}. A simple calculation now gives:

$$\int_{M^{-2}}^{\infty} \frac{ds}{s} e^{-m^2 s} = \ln M^2 - \ln m^2 + \int_0^{\infty} du \, e^{-u} \ln u = \ln \frac{M^2}{m^2} - \gamma_E \quad (4.49)$$

where γ_E is Euler's constant defined in Eq. (2.82).

Since we are going to let $M \to \infty$, the finite term given by γ_E is not relevant and can be dropped. Also note that the original integral is invariant under $s \to m^2 s$ when there is no cut-off, with any constant scaling of s vanishing in the ds/s factor. This is also indicated by the fact that the integral reduces to logarithms, where any change of scale only adds a finite constant, which, according to our philosophy, can be dropped. That is, if we write

$$\ln(m^2/M^2) = \ln(\mu^2/M^2) + \ln(m^2/\mu^2) \qquad (4.50)$$

with some finite energy scale μ introduced by hand then — since we are dropping all finite terms compared to $\ln M$ — the factor $\ln(m^2/\mu^2)$ can be ignored. Thus the value of our integral is essentially $\ln(M^2/\mu^2)$ where μ is a finite energy scale which we have introduced into the problem and M^2 is a high energy cut-off, and we are supposed to let $M \to \infty$. Putting all these in, we find that

$$Z = \frac{q^2}{48\pi^2} \int_0^{\infty} \frac{ds}{s} e^{-m^2 s} = \frac{q^2}{48\pi^2} \ln \frac{M^2}{\mu^2} \qquad (4.51)$$

Exercise 4.3: Just for fun, do the Taylor series expansion and verify this expression. For a more challenging task, compute the corrections to Maxwell's equations due to this term.

[20]There are several ways to do this and later on, in the next chapter, we will do it using dimensional regularization. Right now, we will take a simple minded approach, of introducing a cut-off, which is adequate.

[21]Note that s and hence Λ have dimensions of $(\text{mass})^{-2}$. So M used for the cut-off represents a (formally infinite) energy scale.

Therefore, from Eq. (4.47) we find that the physical coupling constant is given by

$$q_{phys}^2(\mu) = q^2 \left(1 + \frac{q^2}{48\pi^2} \ln \frac{M^2}{\mu^2} \right)^{-1} \tag{4.52}$$

where, again, μ is a finite energy scale which we have introduced into the problem and M^2 is a high energy cut-off which is formally infinite. Further, the quantity q^2 in this expression is a parameter in the Lagrangian which has no operational significance.[22] To interpret this expression, we first take the reciprocal:

$$\frac{1}{q_{phys}^2(\mu)} = \frac{1}{q^2} + \frac{1}{48\pi^2} \ln \frac{M^2}{\mu^2} . \tag{4.53}$$

and think of the coupling constant q_{phys}^2 as the one observed using an experiment at energy scale μ. If we change this energy scale from μ to μ', the coupling constant will change by the relation

$$\frac{1}{q_{phys}^2(\mu')} = \frac{1}{q_{phys}^2(\mu)} + \frac{1}{48\pi^2} \ln \frac{\mu^2}{\mu'^2} \tag{4.54}$$

or, equivalently,

$$q_{phys}^2(\mu') = q_{phys}^2(\mu) \left(1 + \frac{q_{phys}^2(\mu)}{48\pi^2} \ln \frac{\mu^2}{\mu'^2} \right)^{-1} . \tag{4.55}$$

This expression, which is completely independent of the cut-off scale, relates the electromagnetic coupling constant at two different energy scales.[23] We can write this as:

$$\frac{1}{q_{phys}^2(\mu)} + \frac{1}{48\pi^2} \ln \mu^2 = \text{constant} \tag{4.56}$$

The variation of the coupling constant on the energy scale is usually characterized by something called the *beta function* of the theory, defined as

$$\beta(q) \equiv \mu \frac{\partial q}{\partial \mu} = \frac{q^3}{48\pi^2} \tag{4.57}$$

where we have omitted the subscript 'phys'. Any of these expressions will tell you that the electromagnetic coupling constant increases with the energy scale. That is, when you probe the charge distribution of a charged particle at higher and higher energies, its effective strength will increase.[24]

This is your first glimpse of handing the divergences in a theory by trading off unobservable parameters in a Lagrangian for operationally defined parameters. We have necessarily kept the discussion a bit cursory, because we will take up the renormalization of QED in much more detail, later on in the next chapter. Let us now move on to another example of the effective action.

4.3 Effective Action for the $\lambda\phi^4$ Theory

So far, we described the concept of effective action in terms of two sets of degrees of freedom as though they were distinct. There is, however, no real need for this and one can use a similar technique to study the effect of high

[22]Clearly, q_{phys}^2 can be finite when $M \to \infty$ only if this unobservable parameter in the Lagrangian is divergent. So it *is* nice that it is unobservable.

[23]You could not have done any of these if the divergent term in the effective Lagrangian was *not* proportional to the original electromagnetic Lagrangian. This is what allowed us to make the transformations in Eq. (4.47) in the first place. In a way, the rest of it is just a matter of detail.

[24]The β function in QED is a celebrated result and is usually given for spin-1/2 particles for which you get $q^3/12\pi^2$. (The spin-0 case differs by a factor 4.) As we shall see in Sect. 5.6.5, this result, for the spin-1/2 particle, can also be obtained by a similar analysis non-perturbatively.

energy modes in the low energy sector of the same field. As an example, consider a field Φ which has a non-trivial self interaction; for example, the Lagrangian could have the form $L = (1/2)(\partial\Phi)^2 - (1/2)m^2\Phi^2 - V(\Phi)$. In the absence of V this represents a free field with quanta of mass m. Let us now consider separating out the field Φ into two parts $\Phi = \phi_c + \phi_q$ where we think of ϕ_c as containing modes with low wavenumber and energy (in Fourier space) and ϕ_q as containing modes with high wave number and energy. We can now define an effective action $A_{\text{eff}}[\phi_c]$ for ϕ_c by integrating out ϕ_q. That is, the effective action for the low energy sector of the theory with the modes ϕ_c is now defined by the path integral over ϕ_q:

$$e^{iA_{\text{eff}}[\phi_c]} = \int' \mathcal{D}\phi_q \exp iA[\phi_c + \phi_q] \qquad (4.58)$$

Of course, an unconstrained path integral over ϕ_q in the right hand side of Eq. (4.58) is the same as an unconstrained path integral over $\phi = \phi_c + \phi_q$ (and hence will lead to a left hand side which is independent of ϕ_c), which is rather silly thing to do!. This is *not* what we want to do and an equation like Eq. (4.58) makes sense only if certain implicit restrictions are imposed on the path integral over ϕ_q. (This is indicated by a prime on the integral in Eq. (4.58).) The nature of the restrictions will depend on the context but here we assume that there is some sensible way of distinguishing between high energy and low energy modes. (The most common procedure is to assume ϕ_c as constant and expand $A[\phi_c + \phi_q]$ up to quadratic order in ϕ_q.)

In general, the path integral in Eq. (4.58) will lead to an $A_{\text{eff}}(\phi_c)$ with terms (a) which have the same structure as the ones we originally put in $A(\Phi)$ as well as (b) those which were not originally present in $A(\Phi)$. As an example of (a), suppose we get a a term proportional to ϕ^2 [say, $(\mu/2)m^2\phi^2$]. Then, the net effect will be to change the value of what we originally thought was the mass of the quanta, by $m^2 \to m^2(1+\mu)$. You will invariably find that the terms which are generated in $A_{\text{eff}}(\phi_c)$ come with divergent coefficients; for example, in the above case μ will depend on the cut-off scale and could diverge if the cut-off goes to infinity. As regards such divergent terms generated by the path integral in Eq. (4.58), which are of the same form as those that are already present in $A(\Phi)$, we are in good shape. All the divergences can now be absorbed by redefining the parameters in the original theory making them 'run'. Since only such modified parameters are physically relevant, these divergences need not bother us.

In addition to the divergent terms, the integration will also generate *finite* corrections to the original potential V, which, of course, are what we would like to study. These potentials will also have parameters which, as we described above, will run with the scale. This is broadly the situation in theories which are called renormalizable.

The path integral in Eq. (4.58) can also generate divergent terms which were not originally present in $A(\Phi)$. One can try adding these terms and redoing the computation to see whether self-consistency can be achieved at some stage. If that happens, we are back to the previous situation after a few iterations; it is just that we originally started with a potential which was missing some terms generated by quantum effects. Usually, however, this process does not converge. The path integral in Eq. (4.58) will produce divergent terms which are not in the action you had started with, irrespective of how hard you try. Such a theory is called non-renormalizable.

We will now illustrate the above concepts in a simple context of a self-interacting scalar field which is described classically by a Lagrangian of the form

$$L_0 = \frac{1}{2}\partial_a\Phi\partial^a\Phi - V(\Phi) \tag{4.59}$$

[25]If it has a $(1/2)m^2\phi^2$ part with real m, then that part should be conventionally clubbed as free field and only the rest of $V(\phi)$ will be usually called the "interaction". You will see soon that *in this example* it is simpler to work with the full $V(\Phi)$ rather than separate out the quadratic term.

where $V(\Phi)$ is a potential describing the interactions.[25] Writing $\Phi = \phi_c + \phi_q$, where ϕ_c is a constant, and expanding $A[\phi_c + \phi_q]$ up to quadratic order in ϕ_q, we get an expression like in Eq. (4.16), viz.[26]

$$A = \int d^4x L(\phi_c) - \frac{1}{2}\int d^4x \phi_q \hat{D}(\phi_c)\phi_q; \qquad \hat{D} = \Box + V''(\phi_c) \tag{4.60}$$

The computation of the effective action now requires evaluating the expression:

[26]A more formal procedure to define the effective action in this context is to think of ϕ_c as the expectation value $\langle\psi|\Phi|\psi\rangle$ and minimize the energy subject to this constraint. If you implement it with a Lagrange multiplier J added to the Hamiltonian H, then the minimum energy $E(J)$ will be related to ϕ_c by $(dE/dJ) = \phi_c$. Inverting this relation, we get $J(\phi_c)$ and doing a Legendre transform we get $\mathcal{E}(\phi_c) = E - J(dE/dJ)$. Then $\mathcal{E}(\phi_c)$ is essentially your effective potential. In this case we are integrating over quantum fluctuations keeping $\langle\psi|\Phi|\psi\rangle$ fixed. We will explore a related approach in Sect. 4.5.

$$
\begin{aligned}
Z[\phi_c] &= e^{iA_{\text{eff}}[\phi_c]} = \exp i\int d^4x\, L_{\text{eff}}(\phi_c) \\
&= \int \mathcal{D}\phi_q \exp -\frac{i}{2}\int d^4x\, \phi_q\left(\Box + \omega^2\right)\phi_q
\end{aligned} \tag{4.61}
$$

where $\omega^2 \equiv V''(\phi_c)$. But we have computed this in several forms before in Sect. 2.2.1. In Eq. (2.86) we have provided three different expressions for computing this L_{eff}. We will use the expression given by[27]

[27]Since we are assuming ϕ_c is a constant, $L_{\text{eff}} = V_{\text{eff}}(\phi_c)$, the effective potential in the Euclidean sector. The expression for $V_{\text{eff}}(\phi_c)$ will be the same in the Lorentzian sector as well, since analytic continuation in time does not change it.

$$L_{\text{eff}} = -\frac{1}{8\pi^2}\int_0^\infty \frac{d\lambda}{\lambda^3}e^{-\frac{1}{2}\omega^2\lambda} \equiv V_{\text{eff}}(\phi_c) \tag{4.62}$$

with $\omega^2 = V''(\phi_c)$. As we discussed in Sect. 2.2.1, this integral can be evaluated either with a cut-off or by dimensional regularization. Here we shall regularize the integral using a cut-off. That is, we will evaluate this integral with the lower limit set to some small value $\Lambda = 1/M^2$ and then consider the limit[28] of $M \to \infty$, $\Lambda \to 0$. The result of the integration is given by Eq. (2.83) which is reproduced here:

[28]Note that Λ has the dimensions of $(\text{mass})^{-2}$, so that $\Lambda = (1/M^2)$ where M corresponds to the high energy cut-off scale, as in the case of electrodynamics.

$$
\begin{aligned}
V_{\text{eff}} &= -\frac{1}{8\pi^2}\int_{M^{-2}}^\infty \frac{d\lambda}{\lambda^3}e^{-\frac{1}{2}\omega^2\lambda} \\
&= -\frac{1}{16\pi^2}\left[M^4 - \frac{\omega^2 M^2}{2} + \frac{1}{4}\omega^4\ln\frac{M^2}{\mu^2}\right] \\
&\quad + \frac{1}{64\pi^2}\omega^4\ln\frac{\omega^2}{2\mu^2} + \frac{1}{64\pi^2}\gamma_E\omega^4
\end{aligned} \tag{4.63}
$$

where γ_E is Euler's constant, defined in Eq. (2.82), and μ is an arbitrary energy scale we have introduced to separate out the divergent and finite terms.[29]

[29]Note that the V_{eff} is manifestly independent of μ because it cancels out in the two logarithms. This is precisely what we did to get Eq. (2.83). Of course, with proper interpretation, no observable effect can depend on μ^2.

We are supposed to take the limit of $M \to \infty$ in this expression. We see that the last two terms involving $\omega^4\ln\omega^2$ and ω^4 are finite but depend on the arbitrary scale μ we have introduced. The terms in the square bracket diverge as $M \to \infty$; there are quartic, quadratic and logarithmically divergent terms. Of these, the quartic term can be dropped because it is just an infinite constant. We can also drop the finite term $(1/2)\gamma_E\omega^4$ term since it only changes the value of M which is anyway arbitrary at this stage. But the next two terms, involving $M^2\omega^2$ and $-\omega^4\ln(M^2/\mu^2)$, cannot be dropped because they depend on ω^2 as well. The remaining expression is

$$V_{\text{eff}} = \frac{1}{32\pi^2}\left[V''M^2 + \frac{1}{2}(V'')^2\ln\frac{\mu^2}{M^2}\right] + \frac{1}{64\pi^2}(V'')^2\ln\left(\frac{V''}{2\mu^2}\right) \tag{4.64}$$

We see that when we integrate out the high energy modes from the theory, the resulting potential governing the dynamics of ϕ_c gets modified by the addition of the V_{eff} given by the above expression. We must think up some way of interpreting this. The last term $(1/2)(V'')^2 \ln(V''/\mu)$ in V_{eff} is finite and presumably can be interpreted if we can interpret the μ dependence. From the fact that μ^2 occurs in the combination $(-\ln(\mu/M)^2)$, we can think of this as arising from integrating out modes with energies in the range $\mu < E < M$. One striking feature of this expression is that even the finite term now describes a potential quite different from what we started out with. That is, the high energy modes which were integrated out have led to new interaction terms in the low energy sector. The real trouble, of course, arises from the divergent terms.

Let us examine the nature of the divergent part when the original potential was a n-th degree polynomial in ϕ with n coefficients $\lambda_1, \lambda_2 \cdots \lambda_n$:

$$V(\phi_c) = \lambda_1 \phi_c + \lambda_2 \phi_c^2 + \cdots + \lambda_n \phi_c^n = \sum_{k=1}^{n} \lambda_k \phi^k \qquad (4.65)$$

In the last equation, we have dropped the subscript c from ϕ_c for notational simplicity with the understanding that we will hereafter be dealing only with the low energy modes of the theory. Then V'' will be a polynomial of $(n-2)$-th degree and $(V'')^2$ will be a polynomial of $2(n-2)$-th degree. Two completely different situations arise depending on whether this is greater than n or not. When

$$2(n-2) \leq n \qquad \text{i.e.,} \qquad n \leq 4, \qquad (4.66)$$

all the terms in V_{eff} have the same degree or less compared to the terms in V. In that case, the divergent part of V_{eff} will again be a polynomial with degree n or less, but with coefficients which are divergent. This is precisely the idea of renormalization which we described earlier. When $n \leq 4$, the effect of integrating out high energy modes is to merely change the values of the coupling constants λ_j which exist in the original theory. But, as we explained before, the corrections due to high energy quantum fluctuations are always present in such a self-interacting field. Thus, what are actually observed experimentally are the renormalized coupling constants. So, when $n \leq 4$, we have a simple way of re-interpreting the theory and working out its physical consequences.

We shall do this in detail in a moment, but before that, let us consider the situation for $n > 4$. Then the effect of integrating out high energy modes is to introduce new terms in the low energy sector of the theory. For example, if the original potential had a ϕ^6 term, then $(V'')^2$ will have a ϕ^8 term. If we add a ϕ^8 term in the potential to tackle this, then that will generate a ϕ^{12} term, etc. By and large, the effect of high energy modes is to introduce in the low energy sector of the theory all possible terms which are not forbidden by any symmetry considerations.

The old fashioned view in quantum field theory was that such theories are just bad and should not be considered at all. This is, however, a bit drastic. Even though integrating out high energy modes leads to such terms, their effects are usually suppressed by a high energy scale below which these terms will not be important. In other words, one can still interpret the theory as a low energy effective theory with a domain of validity determined by a high energy scale. As an explicit example, consider

Exercise 4.4: Repeat the entire analysis for an arbitrary D and determine what is the value of n for which the above idea works as a function of D.

a term proportional to ϕ^6; since the Lagrangian should have the dimensions of $(\text{mass})^4$, this term will occur in the Lagrangian through a combination $V \propto \lambda(\phi^6/\mu^4)$ where μ^2 is a energy scale and λ dimensionless. The $V''^2 \propto \lambda^2(\phi^8/\mu^8)$ will occur in the two terms of V_{eff}. It is now natural to use the same μ which occurs in the Lagrangian to make the arguments of the logarithms in dimensionless. Then we see that these terms are suppressed by μ^4 factors. So it makes sense to treat this model as an effective theory, i.e as an approximation to a better theory, well below[30] the scale μ.

[30]As you can see, this is plain dimensional analysis but very useful and powerful! Whenever the Lagrangian contains a coupling constant with inverse dimensions of mass, the same conclusions apply.

Let us now see how the details work out when we take $n = 4$, which is the maximally allowed interaction which will not generate new terms. We will take (with an explicit subscript 0 added to the parameters of the original Lagrangian)

$$V(\phi) = \frac{1}{2}m_0^2\phi^2 + \frac{\lambda_0}{4!}\phi^4; \quad V'' = \left(m_0^2 + \frac{\lambda_0}{2}\phi^2\right) \quad (4.67)$$

Then, Eq. (4.64) becomes:

$$
\begin{aligned}
L_{\text{eff}} &= \frac{M^2}{32\pi^2}\left(m_0^2 + \frac{\lambda_0}{2}\phi^2\right) + \frac{1}{64\pi^2}\left(m_0^2 + \frac{\lambda_0}{2}\phi^2\right)^2 \ln\frac{\mu^2}{M^2} \\
&\quad + \frac{1}{64\pi^2}\left(m_0^2 + \frac{\lambda_0}{2}\phi^2\right)^2 \ln\frac{1}{2\mu^2}\left(m_0^2 + \frac{\lambda_0}{2}\phi^2\right) \quad (4.68)
\end{aligned}
$$

We will rearrange this expression, dropping infinite *constants* and absorbing divergent coefficients into $\ln M^2, M^2$ etc. This will allow us to write

$$
\begin{aligned}
8\pi^2 L_{\text{eff}} &= \frac{M^2\lambda_0}{8}\phi^2 + \left[\frac{\lambda_0^2}{32}\phi^4 + \frac{1}{8}\lambda_0 m_0^2\phi^2\right]\ln\frac{\mu^2}{M^2} \quad (4.69) \\
&\quad + \frac{1}{8}(m_0^2 + \frac{\lambda_0}{2}\phi^2)^2 \ln(m_0^2 + \frac{\lambda_0}{2}\phi^2)\frac{1}{2\mu^2} \\
&= \phi^2\left(\frac{M^2\lambda_0}{8} + \frac{1}{8}\lambda_0 m_0^2 \ln\frac{\mu^2}{M^2}\right) + \phi^4\left(\frac{\lambda_0^2}{32}\ln\frac{\mu^2}{M^2}\right) + 8\pi^2 V_{\text{finite}}(\phi; \mu^2)
\end{aligned}
$$

where the first two terms are divergent and the last one is finite when we send the cut-off to infinity ($M \to \infty$), with

$$V_{\text{finite}} = \frac{1}{64\pi^2}\left(m_0^2 + \frac{\lambda_0}{2}\phi^2\right)^2 \ln\frac{1}{2\mu^2}\left(m_0^2 + \frac{\lambda_0}{2}\phi^2\right). \quad (4.70)$$

[31]The magic of renormalizability, again. The divergent coefficients multiply the ϕ^2 and ϕ^4 terms which were originally present in the Lagrangian, allowing us to absorb them into the parameters of the theory. This is precisely what we saw in the case of electrodynamics as well.

Adding this correction term to the original $V(\phi)$, we get[31] the effective potential to be:

$$
\begin{aligned}
V &= \frac{1}{2}m_0^2\phi^2 + \frac{\lambda_0}{4!}\phi^4 + V_{\text{eff}} \\
&\equiv \frac{1}{2}m^2\phi^2 + \frac{\lambda}{4}\phi^4 + V_{\text{finite}}(\phi, \mu^2) \quad (4.71)
\end{aligned}
$$

where

$$
\begin{aligned}
m^2 &= m_0^2 + \frac{\lambda_0}{32\pi^2}\left(M^2 + m_0^2 \ln\frac{\mu^2}{M^2}\right) \\
\lambda &= \lambda_0 + \frac{3\lambda_0^2}{32\pi^2}\ln\frac{\mu^2}{M^2} \quad (4.72)
\end{aligned}
$$

and

$$V_{\text{finite}} = \frac{1}{64\pi^2}\left(m^2 + \frac{\lambda}{2}\phi^2\right)^2 \ln\frac{1}{2\mu^2}\left(m^2 + \frac{\lambda}{2}\phi^2\right). \tag{4.73}$$

in which we have replaced λ_0, m_0 by λ, m to the same order of accuracy as we are working with.[32]

Let us take stock of the situation. The net effect of integrating out high energy modes is to change the original potential to the form in Eq. (4.71) which differs in two crucial aspects from the original potential. First, there is a correction term in the form of V_{finite} which is non-polynomial (since it contains a log) that has been generated. This term is finite but depends on a parameter μ which is completely arbitrary and hence needs to be interpreted properly. We will come back to this issue. Second, all the divergent terms have been absorbed into the two parameters of the original Lagrangian, thereby changing m_0 and λ_0 to m and λ. In the expression for V_{finite}, we should actually be using m_0 and λ_0 but we will assume that, to the lowest order, we can replace them by m and λ.

The constants m^2 and λ are thought of as renormalized parameters and, as we said before, in this particular case ($n = 4$) we could reabsorb all the divergences by renormalizing the original parameters. The idea is that the observed mass and coupling constant determined by some scattering experiment, say, should be used for m and λ and the original parameters m_0 and λ_0 are unobservable.[33]

Let us now get back to the apparent μ dependence of V_{finite}. We know that this is an arbitrary scale which we introduced into the theory and V_{finite} really should not depend on μ; that is, if we change μ, then V_{finite} should not change in spite of appearances. The clue to this paradox lies in the fact that m^2 and λ in Eq. (4.72) also depend on μ. So, when we change μ, these parameters will also change, keeping V_{finite} invariant. In fact, this is apparent from the fact that μ was introduced into V_{eff} in two logarithmic terms such that it does not change the value of V_{eff}. All we have done is to rewrite this expression separating out the divergent and finite pieces, thereby making m^2, and λ functions of μ for a fixed M.

To see this more clearly, let us consider the expressions for λ for two different values of μ^2 given by $\mu^2 = \mu_1^2$ and $\mu^2 = \mu_2^2$. From Eq. (4.72), we find that

$$\lambda(\mu_1) - \lambda(\mu_2) = \frac{3\lambda^2(\mu_1)}{32\pi^2}\ln\frac{\mu_1^2}{\mu_2^2} \tag{4.74}$$

This expression is completely independent of the cut-off M and tells you how the strength of the coupling constant changes with the energy scale. If we use a particular value of μ in V_{finite} and the corresponding value of $\lambda(\mu)$ as the coupling constant, then the numerical value of V_{finite} will be independent of μ. Of course, we need to fix the value of λ at some energy scale through the experiment; once this is done, we know the value of the coupling constant at all other energy scales. The above result can also be stated in the form

$$\lambda(\mu) - \frac{3\lambda^2(\mu)}{32\pi^2}\ln\mu^2 = \text{constant} \tag{4.75}$$

or, more simply, as a differential equation

$$\mu\frac{\partial\lambda}{\partial\mu} = \frac{3\lambda^2}{16\pi^2} \equiv \beta(\lambda) \tag{4.76}$$

[32] Notice that the (divergent) correction to the bare mass m_0^2 is proportional to $\lambda_0 M^2$ where M is the high energy cut-off mass. The fact that one picks up a correction which is quadratic in M when one uses cut-off regularization is related to what is known as the hierarchy problem. If one considers the theory having a scalar field of mass m and a large energy scale M (which can act as a physical cut-off) then Eq. (4.72) tells us that the physical mass of the theory will be driven to M and we need to fine-tune the parameters of the theory to avoid this from happening. This is known as the hierarchy problem. We will see later that such quadratic divergences do not occur if we use dimensional regularization.

[33] The renormalization scheme we described works even if you change the sign of the $m^2\phi^2$ term. Then, $\phi = 0$ is not the minimum and we need to worry about spontaneous symmetry breaking — a key ingredient in the standard model of particle physics. What is nice is that the renormalizability continues to hold even in this context.

The logarithmic derivative in the left hand side of the coupling constant is the β function and the above result shows that $\beta > 0$, indicating that the coupling becomes stronger at higher energies.

Incidentally, we can play the same game with the mass parameter as well. From the first relation in Eq. (4.72), we find that

$$\mu \frac{\partial m^2}{\partial \mu} = \frac{\lambda m^2}{16\pi^2} \equiv \gamma(m) \tag{4.77}$$

which is sometimes called the gamma function of the flow. This tells you how m^2 runs with the scale μ to keep observable results independent of E. We will say more about these issues after developing the perturbation approach.

4.4   Perturbation Theory

In the last few sections, we discussed the behaviour of interacting fields non-perturbatively by using the concept of an effective action. This required us making some approximations in order to make the problem tractable but we did not have to assume that the coupling constant in the theory (λ in the $\lambda\phi^4$ interaction or q in electromagnetic coupling) is a small parameter. In fact, one of the results we obtained in the Schwinger effect is explicitly non-perturbative in the electromagnetic coupling constant.

A completely different way of tackling interacting fields is to deal with them perturbatively and calculate all physical observables in a series expansion in powers of the coupling constant. In this approach we imagine that there exists some kind of a free-field theory (containing, say, non-interacting photons described by a vector field $A_l(x)$ and charged, spin-zero, massive particles described by a complex scalar field $\phi(x)$) when the relevant coupling constant (in this case, the charge q of the scalar field) tends to zero. We then think of "switching on" the interaction by introducing a coupling between A_l and ϕ, the strength of which is controlled by q. Further assuming that the interaction can be treated as a perturbation of the free-field theory, one can compute the effect of the interaction order-by-order as a power series in the coupling constant. This is the essence of the perturbative approach to studying interacting field theory.

The key advantage of this approach is that the entire procedure can be made completely systematic in the form of a set of rules involving the *Feynman diagrams* of the theory. Once the rules have been written down from the structure of the theory, you can — in principle — compute any observable quantity in the perturbation series. When such computations are done, certain amplitudes become divergent and one again has to renormalize the parameters of the theory to obtain finite results. This, by itself, is not a serious issue because — as we have already emphasized — the parameters in the Lagrangian need to be interpreted in a scale dependent manner in any realistic theory. We have also seen that divergences arise even in the context of non-perturbative computations and hence perturbation theory cannot be blamed for these divergences. Moreover, the perturbative approach has proved to be enormously successful both in the case of quantum electrodynamics and in the study of electroweak interactions — which alone makes it a worthwhile object to study from the practical point of view.

One downside of the perturbative approach is that it is very difficult to rigorously demonstrate some of the constructs which form its basis. For example, it is not easy to define rigorously (though it *can* be done with a fair amount of mathematical formalism) the notion of a free asymptotic field and its properties or the notion of "switching on" the interaction. It is also not clear whether the perturbative series is convergent, asymptotic or divergent for a general interacting field theory. In addition, we obviously cannot handle any non-perturbative features of the theory (like the Schwinger effect or quark confinement) by such an approach.

On the whole, high energy physicists will claim — quite correctly — that the advantages of the perturbative approach outweigh the disadvantages. This is the point of view we will adopt and will now introduce the perturbative approach in the very simple context of a self-interacting $\lambda\phi^4$ theory.

4.4.1 Setting up the Perturbation Series

To understand how one goes about studying the $\lambda\phi^4$ theory in the perturbative approach and obtain the relevant Feynman diagrams etc., let us begin by briefly reviewing some of the results we obtained in Sect. 2.2.1 for the free-field theory and generalizing them suitably. Given an action functional $A(\phi)$ for the scalar field, we can obtain the functional Fourier transform $Z(J)$ of e^{iA} by the relation (see Eq. (2.54))

$$Z[J] = \int \mathcal{D}\phi \exp\left[iA[\phi] + i\int J(x)\phi(x)d^4x\right] \qquad (4.78)$$

If we compute the functional derivative of $Z(J)$ with respect to $J(x)$, we bring down one factor of $\phi(x)$ for each differentiation. For example, if we compute the functional derivative twice, we get:

$$
\begin{aligned}
\frac{1}{Z(0)}\left(\frac{\delta}{i\delta J(x_2)}\right)\left(\frac{\delta}{i\delta J(x_1)}\right)Z[J]\Big|_{J=0} &= \frac{1}{Z(0)}\int \mathcal{D}\phi\,\phi(x_2)\phi(x_1)\,e^{iA[\phi]} \\
&= \langle 0|T[\phi(x_2)\phi(x_1)]|0\rangle \\
&\equiv G(x_2,x_1) \equiv \langle x_2|x_1\rangle \qquad (4.79)
\end{aligned}
$$

where we have introduced the notation $G(x_1,x_2) \equiv \langle x_1|x_2\rangle$ for typographical simplicity.[34] The above equation was obtained in Sect. 2.2.1 for the case of free-fields, but it is obvious that the first equality is trivially valid for arbitrary $A[\phi]$. Further, since the time integral in the action goes from $t = -\infty$ to $t = +\infty$, we can easily demonstrate (by, for example, analytically continuing to the Euclidean sector) that the path integral expression reduces to the vacuum expectation value (VEV) of the time-ordered products even for an arbitrary action $A[\phi]$, so that the second equality in Eq. (4.79) also holds for any $A[\phi]$.

In a similar manner, one can obtain the functional average or the VEV of a product of n scalar fields $\phi(x_1)\phi(x_2)\ldots\phi(x_n) \equiv \phi_1\phi_2\ldots\phi_n$ [where we have simplified the notation by writing $\phi(x_j) = \phi_j$ etc.] by taking n functional

[34] For reasons described at the end of Sect. 1.4, do not attribute meanings to $|x_1\rangle$ and $|x_2\rangle$. This is just a notation.

derivatives with respect to $J(x_1), J(x_2), ...J(x_n)$. That is,

$$\frac{1}{Z(0)}\left(\frac{\delta}{i\delta J(x_1)}\right)\left(\frac{\delta}{i\delta J(x_2)}\right)...\left(\frac{\delta}{i\delta J(x_n)}\right)Z[J]\Big|_{J=0}$$
$$=\frac{1}{Z(0)}\int\mathcal{D}\phi\,\phi_1\phi_2...\phi_n\,\exp\left(iA[\phi]\right)=\langle 0|T[\phi_1\phi_2...\phi_n]|0\rangle$$
$$\equiv G(x_1,x_2,...x_n) \tag{4.80}$$

Recalling our interpretation in Sect. 2.2.1 of $\exp(iA)$ as being analogous to a probability distribution and $Z(J)$ as being analogous to the generating function for the probability distribution, we see that the above result gives the n-th moment of the random variable $\phi(x)$ in terms of the probability distribution as well as in terms of the derivatives of the generating function. It is also clear that if all the moments are given, one can reconstruct $Z(J)$ as a functional Taylor series by:

$$Z(J)=Z(0)\sum_{n=0}^{\infty}\frac{1}{n!}\int dx_1 dx_2...dx_n\,G(x_1,x_2,...x_n)\,J(x_1)J(x_2)...J(x_n) \tag{4.81}$$

Obviously, the set of all n-point functions $G(x_1,x_2,...x_n)$ contains the complete information about the theory and if we can calculate them, we can compute all other physical processes. Our first task is to set up a formalism by which these n-point functions can be computed in a systematic manner.[35]

Since we are interested in setting up a systematic procedure for computing various quantities, it will be useful to start from the free-field theory as a warm up. In this case, we know that $Z(J)$ can be explicitly computed and is given by:

$$Z(J)=Z(0)\exp-\frac{1}{2}\int d^4x_2\,d^4x_1 J(x_2)\langle x_2|x_1\rangle J(x_1) \tag{4.82}$$

The $n-$point functions can be determined by expanding the exponential in a power series and identifying the coefficients of $J(x_1)J(x_2)...J(x_n)$, remembering the fact that this product is completely symmetric. For example, the 4-point function, computed using Eq. (4.80), is given by:

$$G(x_1,x_2,x_3,x_4)=\langle x_1|x_2\rangle\langle x_3|x_4\rangle+\langle x_1|x_3\rangle\langle x_2|x_4\rangle+\langle x_1|x_4\rangle\langle x_2|x_3\rangle \tag{4.83}$$

and similarly for higher orders. We are essentially computing the moments of a Gaussian distribution for which all the information is contained in the two-point function. The odd moments will vanish and the even moments can be expressed as products of the two-point function, which is what Eq. (4.83) tells you.

With future applications in mind, we will associate a diagrammatic representation with these algebraic expressions. We will represent $G(x_1,x_2)$ by a line connecting the two events x_1 and x_2 as in Fig. 4.3. This term occurs in the series expansion for $Z(J)$ in the form $G(x_1,x_2)J(x_1)J(x_2)$ which can be interpreted as the source J creating a particle at one event and destroying it at another, with the amplitude $G(x_1,x_2)$ describing the propagation. This is, of course, completely consistent with our earlier discussion and interpretation of $\langle x_1|x_2\rangle$. Let us next consider $G(x_1,x_2,x_3,x_4)$ which can be represented by the figure in Fig. 4.4. This term occurs in the

[35]In the process we will also provide an intuitive picture of how the n-point functions are related to simple physical processes like the scattering of the quanta of the field, because of the interaction. Later on, in Sect. 4.6, we will provide a more rigorous proof of how the amplitude for scattering, etc. can be related to the n-point functions.

Figure 4.3: Diagrammatic representation of $G(x_1,x_2)$

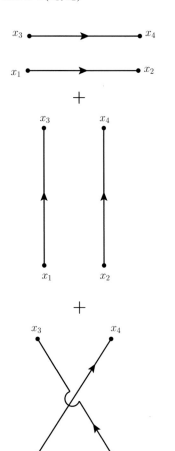

Figure 4.4: Diagrams contributing to $G(x_1,x_2,x_3,x_4)$

form $G(x_1, x_2, x_3, x_4)\ J(x_1)J(x_2)J(x_3)J(x_4)$ and one would like to interpret this as a pair of particles produced and destroyed by two sources and two sinks. The form of the expression in Eq. (4.83) clearly shows that the particles do not affect each other — as to be expected in the absence of any interactions — and this merely propagates from event to event in all possible combinations.

The Fig. 4.4 has three pieces and each of them have two *disconnected* components. It is obvious that when we look at $2n$-point functions we will get the same structure involving n particles, each propagating independently amongst the $2n$ events without any interactions. Any particular configuration will involve n lines of the kind in Fig. 4.4, each of which is disconnected from the other.

It is clear that if we try to draw all these diagrams with disconnected pieces and compute each term, we are duplicating the effort unnecessarily. This problem is easily solved by writing $Z(J)$ as $\exp[iW(J)]$ and working with $W(J)$. In the case of free field theory, $W(J)$ is a quadratic functional of J which leads to no higher than the two-point function. When we exponentiate $W(J)$ to get $Z(J)$, we automatically generate the correct products of disconnected diagrams, while $W(J)$ has only connected diagrams. That is, while

$$\langle 0|T[\phi(x_1)\cdots\phi(x_n)]|0\rangle = (-i\hbar)^n \frac{1}{Z[J]}\frac{\partial^n Z}{\partial J(x_1)\cdots\partial J(x_n)}\bigg|_{J=0} \qquad (4.84)$$

the $W(J)$, on the other hand, produces a similar quantity involving only those Feynman diagrams which are connected; i.e.,

$$(-i\hbar)^n \frac{\partial^n W[J]}{\partial J(x_1)\cdots\partial J(x_n)}\bigg|_{J=0} = -i\hbar\langle 0|T[\phi(x_1)\cdots\phi(x_n)]|0\rangle_{\text{connected}}$$

$$(4.85)$$

For example, when $n = 2$, we have the result

$$(-i\hbar)^2\frac{\partial^2 W}{\partial J_1 \partial J_2} = (-i\hbar)^3\frac{\partial}{\partial J_1}\left(\frac{1}{Z}\frac{\partial Z}{\partial J_2}\right) \qquad (4.86)$$

$$= (-i\hbar)^3\frac{1}{Z}\frac{\partial^2 Z}{\partial J_1 \partial J_2} - (-i\hbar)^3\left(\frac{1}{Z}\frac{\partial Z}{\partial J_1}\right)\left(\frac{1}{Z}\frac{\partial Z}{\partial J_2}\right)$$

$$= -i\hbar\left[\langle J|\phi(x_1)\phi(x_2)|J\rangle - \langle J|\phi(x_1)|J\rangle\langle J|\phi(x_2)|J\rangle\right]$$

where we have kept $J \neq 0$ to illustrate the point. The second term, which is subtracted out, involves all the disconnected pieces which we are not usually concerned with. In presence of the source, even the one-point function (which, of course, is connected) can be non-zero and is given by

$$\frac{\partial W[J]}{\partial J(x)} = -i\hbar\frac{1}{Z}\frac{\partial Z[J]}{\partial J(x)} = \langle J|\phi(x)|J\rangle \equiv \phi_J(x) \qquad (4.87)$$

where the last relation defines $\phi_J(x)$ as the expectation value of the field ϕ in a state describing a source $J(x)$.

This feature is quite generic in field theory[36] and hence one prefers to work with $W(J)$ rather than $Z(J)$. Alternatively, we can simply restrict ourselves to connected diagrams in our computation, which is what we will do eventually.

[36] It is not too difficult to prove; but we won't bother to do it since we will directly work with connected Feynman diagrams.

4.4.2   Feynman Rules for the $\lambda\phi^4$ Theory

Let us now consider the application of the above ideas to a self-interacting $\lambda\phi^4$ theory with the action

$$A[\phi] = \int d^4x \left[\frac{1}{2}(\partial_m\phi\partial^m\phi - m^2\phi^2) - \frac{\lambda}{4!}\phi^4 \right] \qquad (4.88)$$

and the corresponding generating function, in a condensed notation,

$$Z_\lambda(J) = \int D\phi \exp\left\{ i\int d^4x \left[\frac{1}{2}[(\partial\phi)^2 - m^2\phi^2] - \frac{\lambda}{4!}\phi^4 + J\phi \right] \right\} \quad (4.89)$$

Since we cannot evaluate this exactly when $\lambda \neq 0$, we will resort to a perturbation series in λ. If we expand the relevant part of the exponential as

$$\exp\left[-i\frac{\lambda}{4!}\int d^4x\phi^4 \right] = \sum_{n=0} \left(-i\frac{\lambda}{4!} \right)^n \qquad (4.90)$$
$$\times \int d^4x_1 d^4x_2...d^4x_n\phi^4(x_1)\phi^4(x_2)...\phi^4(x_n)$$

it is obvious that the $n-$th order term will insert the n factors of ϕ^4 (with integrals over each of them) inside the path integral. But this is exactly what we would have obtained if we had differentiated the remaining path integral (the one with $\lambda = 0$) n times with respect to $\delta^4/\delta J^4$. Therefore we get:

$$Z_\lambda[J] = \sum_{n=0} \left(-i\frac{\lambda}{4!} \right)^n \int d^4x_1 d^4x_2...d^4x_n \frac{\delta^4}{\delta J(x_1)^4} \frac{\delta^4}{\delta J(x_2)^4} .... \frac{\delta^4}{\delta J(x_n)^4} Z_0[J]$$
$$(4.91)$$

The generating function $Z_0[J]$ for $\lambda = 0$ is of course known and is given by Eq. (4.82). Therefore, we can write, in a compact notation:

$$Z_\lambda[J] = Z_0[0] \exp\left[-i\frac{\lambda}{4!}\int d^4x \frac{\delta^4}{\delta J(x)^4} \right]$$
$$\times \exp -\frac{1}{2}\int d^4x_2\, d^4x_1 J(x_2)\langle x_2|x_1\rangle J(x_1) \qquad (4.92)$$

The exponential operator is defined through the power series in λ given explicitly in Eq. (4.91). The expression in Eq. (4.92) formally solves our problem. Once we expand the exponential operator in a power series in λ, we can compute the nth order contribution (involving λ^n) by carrying out the functional differentiation in $Z_0(J)$. This is a purely algorithmic procedure which will provide the perturbative expansion for $Z(J)$. This, in turn, leads to a functional series expansion for $Z(J)$ in J as shown in Eq. (4.81), the coefficients of which will give the n-point functions.

Before we discuss the nature of this expansion, we will also describe an alternative way of obtaining the same result. If one expands Eq. (4.89) as a power series in $J(x)$ directly, we will get

$$
\begin{aligned}
Z_\lambda[J] &= Z_0[0] \sum_{n=0}^{\infty} \frac{1}{n!} \int dx_1 dx_2 \cdots dx_n J(x_1) \cdots J(x_n) G^{(n)}(x_1, \cdots, x_n) \\
&= \sum_{n=0}^{\infty} \frac{1}{n!} \int dx_1 dx_2 \cdots dx_n \ J(x_1) \cdots J(x_n) \int D\phi \ \phi(x_1) \cdots \phi(x_n) \\
&\quad \times \exp\left\{ i \int d^4x \left[\frac{1}{2} \left[(\partial\phi)^2 - m^2\phi^2 \right] - \frac{\lambda}{4!}\phi^4 \right] \right\} \qquad (4.93)
\end{aligned}
$$

which allows us to read off the different n-point functions as standard path integral averages like

$$
\begin{aligned}
G(x_1, x_2) &\equiv \frac{1}{Z_0(0)} \int D\phi \ \phi(x_1)\phi(x_2) \\
&\quad \times \exp\left\{ i \int d^4x \left[\frac{1}{2} \left[(\partial\phi)^2 - m^2\phi^2 \right] - \frac{\lambda}{4!}\phi^4 \right] \right\} \\
&= \langle 0|T[\phi(x_1)\phi(x_2)]|0\rangle \qquad (4.94)
\end{aligned}
$$

and

$$
\begin{aligned}
G(x_1, x_2, x_3, x_4) &\equiv \frac{1}{Z_0(0)} \int D\phi \ \phi(x_1)\phi(x_2)\phi(x_3)\phi(x_4) \\
&\quad \times \exp\left\{ i \int d^4x \left[\frac{1}{2} \left[(\partial\phi)^2 - m^2\phi^2 \right] - \frac{\lambda}{4!}\phi^4 \right] \right\} \\
&= \langle 0|T[\phi(x_1)\phi(x_2)\phi(x_3)\phi(x_4)]|0\rangle \qquad (4.95)
\end{aligned}
$$

etc. These have the same structure as the 2-point and 4-point functions in Eq. (4.79) and Eq. (4.80) but are now constructed for the interacting theory with $\lambda \neq 0$. For the same reason, we cannot evaluate these expressions exactly and have to resort to a perturbative expansion in λ in Eq. (4.94) and Eq. (4.95).

The two approaches (repeated functional differentiation with respect to $J^4(x)$ of the non-interacting generating function, or evaluation of the path integral average in Eq. (4.80) in a perturbative series in λ) are mathematically identical; they also involve substantially the same amount of effort in calculation. The functional series expansion of $Z(J)$ in J, however, gives a clearer physical picture of particles being created and destroyed by the source $J(x)$. This allows us to interpret the term $G(x_1, x_2)J(x_1)J(x_2)$, for example, as the creation and destruction of a particle at x_1 and x_2 with $G(x_1, x_2)$ giving the propagation amplitude in the interaction theory. Similarly, the term $G(x_1, x_2, x_3, x_4)J(x_1)J(x_2)J(x_3)J(x_4)$ can be interpreted as the creation of a pair of particles at x_1 and x_2, say, with their subsequent destruction at x_3 and x_4. In the free-field theory, we know that the pair of particles will propagate in between, ignoring each other. But in the interacting field theory, this process will also involve scattering of the particles due to the interaction term.

To gain some insight into the explicit computation of these amplitudes, let us try our hand in computing $G(x_1, x_2, x_3, x_4)$ to the lowest order at λ. This is given by the path integral average:

$$
\begin{aligned}
G(x_1, x_2, x_3, x_4) &= \frac{1}{Z_0(0)} \left(-\frac{i\lambda}{4!} \right) \int d^4x \int D\phi \, \phi(x_1)\phi(x_2)\phi(x_3)\phi(x_4)\phi(x)^4 \\
&\quad \times \exp\left\{ i \int d^4x \left[\frac{1}{2} \left[(\partial\phi)^2 - m^2\phi^2 \right] \right] \right\} \qquad (4.96)
\end{aligned}
$$

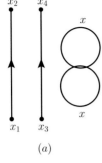

(a)

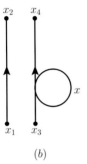

(b)

(c)

Figure 4.5: Three types of diagrams involved in the computation of Eq. (4.96)

This is a path integral average of eight factors of ϕ with four factors evaluated at the same event x and the other four factors evaluated at (x_1, x_2, x_3, x_4). Because we are doing perturbation theory, the path integral is over the free particle action which is just a quadratic expression; in other words, we just have to evaluate the 8-point function of the *free-field* theory. Since the only non-trivial n-point function in the free-field theory is the 2-point function $\langle x_1 | x_2 \rangle$, we know that each term in the 8-point function will be the product of four 2-point functions taken in different combinations. Let us sort out the kind of terms which will emerge in such an expression.

If we combine x_1 with any of x_2, x_3 or x_4, we will get a factor like $\langle x_1 | x_2 \rangle$, say. (This process is usually called 'contracting' $\phi(x_1)$ with $\phi(x_2)$.) The remaining x_3 and x_4 can either be combined with each other or combined with the xs in $\phi^4(x)$. In either case, we see that there is a factor $\langle x_1 | x_2 \rangle$ represented by a single line disconnected from the rest (see Fig. 4.5(a), (b)). The Fig. 4.5(a) represents the particles propagating without any interaction from x_1 to x_2 and x_3 to x_4 with the four factors of $\phi^4(x)$, contracted pairwise amongst themselves, leading to the closed figure-of-eight loop shown in the figure. Similarly, Fig. 4.5(b) shows one particle propagating from x_1 to x_2 while we combine x_3 with an x in one of the $\phi(x)$ of the $\phi^4(x)$ term, x_4 with an x in another $\phi(x)$ of the $\phi^4(x)$ term and combine the remaining two $\phi(x)$ together to get the closed loop corresponding to $\langle x | x \rangle$. The processes in Fig. 4.5(a), (b) clearly belong to the disconnected type with (at least) one particle propagating freely without interaction. What we are interested in, of course, is the situation in which the two particles *do* interact, which can happen only if we combine each of the x_1, x_2, x_3 and x_4 with one each of the $\phi(x)$ in the $\phi^4(x)$ term. This will lead to the amplitude we are after, given by

$$G(x_1, x_2, x_3, x_4) \propto (-i\lambda) \int d^4x \, \langle x_1 | x \rangle \, \langle x_2 | x \rangle \, \langle x_3 | x \rangle \, \langle x_4 | x \rangle \qquad (4.97)$$

This is represented by the diagram in Fig. 4.5(c). This gives the lowest order amplitude for scattering of the scalar particles by one another.

Just to understand the structure of these terms and their diagrammatic representation better, let us see how the *same* result will be obtained if we calculate the process using the technique of functional differentiation with respect to $J^4(x)$. The linear term in λ arises from the expansion of the operator

$$\exp\left[-i\frac{\lambda}{4!}\int d^4x \, \frac{\delta^4}{\delta J^4(x)}\right] = 1 - i\frac{\lambda}{4!}\int d^4x \, \frac{\delta^4}{\delta J^4(x)} \qquad (4.98)$$

which should act on the free-field generating function $Z_0(J)$ again expanded in a functional Taylor series in J. We want to isolate the term which will be linear in λ and will have four factors of J surviving after the functional differentiation. Since the linear term in λ in Eq. (4.98) involves four differentiations with respect to J, the only term which can contribute is the term involving eight factors of J in the expansion of $Z_0(J)$. Given the fact that $Z_0(J)$ is an exponential of a quadratic, we are looking for the fourth term in the expansion of the exponential which will contain eight factors of J. This will lead to the computation of the term

$$\mathcal{A} = -i\frac{\lambda}{4!}\int d^4x \, \frac{\delta^4}{\delta J^4(x)} \frac{1}{4!}\left[-\frac{1}{2}\int dx_1 dx_2 \langle x_1 | x_2 \rangle J(x_1) J(x_2)\right]^4 \qquad (4.99)$$

Expanding it out, we see that this term has the structure

$$\mathcal{A} \propto -i\lambda \int d^4x \, \frac{\delta^4}{\delta J^4(x)} \int dx_1....dx_8 \, J_1 J_2 J_3 J_4 J_5 J_6 J_7 J_8$$
$$\times \langle x_1|x_5\rangle\langle x_2|x_6\rangle\langle x_3|x_7\rangle\langle x_4|x_8\rangle \qquad (4.100)$$

in obvious notation. The structure of different terms arising from the functional differentiation depends on how the different J_ks are paired with the differentiating J_x. When the four J_x factors hit J_5, J_6, J_7 and J_8, we get the connected diagram Fig. 4.5(c) which we want. This has the algebraic structure:

$$\mathcal{A}_c \propto -i\lambda \int d^4x \int dx_1....dx_4 \, J_1 J_2 J_3 J_4 \langle x_1|x\rangle\langle x_2|x\rangle\langle x_3|x\rangle\langle x_4|x\rangle \quad (4.101)$$

This result is easy to understand. We think of two particles being created at x_1 and x_2 and propagating to the same event x with the amplitude $\langle x_1|x\rangle \, \langle x_2|x\rangle$ (recall that $\langle a|b\rangle$ is symmetric in a and b), scattering with an amplitude $(-i\lambda)$ and then propagating from x to x_3, x_4 where they get destroyed by the external source again. For the purpose of calculating the 4-point function $G(x_1, x_2, x_3, x_4)$, we can imagine replacing the source J_k by a Dirac-delta function at the event.

On the other hand, the functional differentiation will also contain a term, say, in which the four factors of J_x in $(\delta^4/\delta J_x^4)$ hit J_3, J_4, J_7, J_8. This will lead to a term of the form

$$\mathcal{A}_a \propto -i\lambda \int dx_1 dx_2 dx_5 dx_6 \, J_1 J_2 J_5 J_6 \langle x_1|x_5\rangle\langle x_2|x_6\rangle \int dx \langle x|x\rangle\langle x|x\rangle$$
$$(4.102)$$

which corresponds to Fig. 4.5(a). We interpret this as two particles created by the sources at x_1 and x_2, propagating freely to x_5 and x_6 respectively. Simultaneously, at some other event x, there arises a vacuum fluctuation in terms of two closed loops with amplitude $-i\lambda\langle x|x\rangle \, \langle x|x\rangle$. Clearly, this is a disconnected process. One can similarly see that if two of the J_xs hit x_1 and x_5 respectively and others are paired off distinctively, we will pick up just one factor of $\langle x|x\rangle$ and one freely propagating particle, corresponding to Fig. 4.5(b) which is yet another disconnected process.

You should now be able to understand how the correspondence between the Feynman diagrams and the functional differentiation works in order to produce the same algebraic expressions. The expansion of $Z_0(J)$ in J involves a bunch of Js at different events in spacetime connected by $\langle x_i|x_j\rangle$ factors which we denote by a line connecting x_i and x_j. For example, the 6-th order term in the expansion will involve 6 factors of $\langle x_i|x_j\rangle$ connecting 12 events of which Fig. 4.6(a) shows a representative set. We have to operate on these terms with the operator on the left hand side of Eq. (4.98) which, when expanded out, will have a whole series of terms each involving products of $(\delta^4/\delta J_x^4)$. This operation essentially combines four ends of the lines to a single event on which the differentiation acts. For example, the term of order λ^2 in our operator, involving

$$\mathcal{O} = \lambda^2 \int d^4x \int d^4x' \left[\frac{\delta}{\delta J(x)}\right]^4 \left[\frac{\delta}{\delta J(x')}\right]^4 \qquad (4.103)$$

can (i) pick up four ends x_2, x_5, x_4, x_7 of Fig. 4.6(a) and will connect them up at an event x and (ii) pick up x_6, x_9, x_{11} and x_8 of Fig. 4.6(a) and

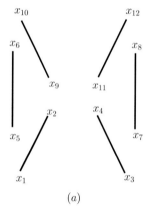

(a)

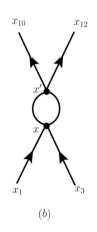

(b)

Figure 4.6: Diagrams relevant to the computation of Eq. (4.102).

combine them at an event x'. This leads to the diagram in Fig. 4.6(b). The locations x and x' are then integrated out to give the final amplitude.

In all the above cases we have been essentially concentrating on representative terms and diagrams in each category. But given the fact that many of the products involved are symmetric in their arguments, one needs to keep count of a symmetry factor for a term of the diagram. As an illustration, consider the correction to the 2-point function to order λ which requires us to compute

$$\mathcal{B} = -\frac{i\lambda}{4!} \int d^4z \, \langle \phi_x \phi_y \phi_z^4 \rangle \tag{4.104}$$

where we have further simplified the notation by using $\langle .... \rangle$ to denote the path integral average. We can contract the six factors of ϕ pair-wise in $^6C_2 = 15$ different ways and each one of them will give a contribution in terms of products of three amplitudes $\langle x_i | x_j \rangle$. However, a little thought shows that only two of these 15 are really different. If we contract ϕ_x with ϕ_y, there are three ways of contracting the remaining four ϕs with each other and all of these will give identical contribution. This way of combining the factors leads to the disconnected diagram shown in Fig. 4.7(a). In writing the corresponding algebraic expression, we have to multiply it by a factor 3. The other possibility is to combine ϕ_x with one of the ϕ_z^4 (which can be done in 4 ways), ϕ_y with another one of ϕ_z^4 which can be done in 3 ways, with the remaining two ϕ_z being combined with each other (which is unique). This gives rise to the diagram in Fig. 4.7(b). The corresponding algebraic expression will, therefore, be

$$\mathcal{B} = 3 \times \left(-\frac{i\lambda}{4!} \right) \langle x|y \rangle \int d^4z \, \langle z|z \rangle \langle z|z \rangle + 12 \times \left(-\frac{i\lambda}{4!} \right) \int d^4z \, \langle x|z \rangle \langle y|z \rangle \langle z|z \rangle \tag{4.105}$$

From these examples, you should be able to figure out the rules governing the Feynman diagrams by which every term in the perturbation series can be computed. After one draws the relevant diagrams, for a given process, we see that we have to do the following: (a) Associate the amplitude $\langle x|y \rangle$ with a line joining the events x and y. (b) Associate a factor $-i\lambda$ with every vertex at which four lines meet. Further, integrate over the spacetime event associated with the vertex. (c) Introduce the correct symmetry factor for the diagram.

4.4.3   Feynman Rules in the Momentum Space

While the above rules certainly work, it is often easier to work in the momentum space by introducing the Fourier transforms of all the expressions, especially the n-point functions. We will then label each line by a four-momentum vector k (instead of the events at the end points) and associate the propagation amplitude in momentum space $i(k^2 - m^2 + i\epsilon)^{-1}$ with that line. It is obvious that, because of translational invariance, the Fourier transform $G(k_1, k_2, k_3, k_4)$, say, of $G(x_1, x_2, x_3, x_4)$ will contain a Dirac-delta function on the sum of the momenta k_i. It is convenient to omit this in the overall factor and also assume the momentum conservation at each vertex. With these modifications, the Feynman rules in momentum space will be the following:

(a) Draw the diagram in the momentum space associating a momentum label with each of the lines. That is, we label each line with a momentum

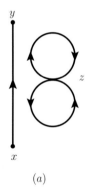

(a)

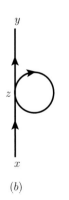

(b)

Figure 4.7: Diagrams relevant to the computation in Eq. (4.104).

k and associate with it the factor $i(k^2 - m^2 + i\epsilon)^{-1}$.

(b) Assume that momentum is conserved at each vertex. This is most conveniently done by having in-going and out-going momentum vectors at each vertex (denoted by suitable arrows) and ensuring that the net sum of in-going momenta is equal to the sum of the out-going momenta.

(c) Momenta associated with internal lines are integrated over with the measure $d^4k/(2\pi)^4$.

(d) Associate the correct symmetry factor with the diagram.

With these rules, the momentum space version of Fig. 4.5(c) will become Fig. 4.8. The corresponding algebraic expression will correspond to the amplitude

$$\mathcal{A} = -i\lambda(2\pi)^4\delta^{(4)}(k_1 + k_2 - k_3 - k_4)\prod_{a=1}^{4}\left(\frac{i}{k_a^2 - m^2 + i\epsilon}\right) \qquad (4.106)$$

It is obvious that nothing is lost in omitting the propagator factors corresponding to the external lines and dropping the overall momentum conservation factor. This is usually done while writing down the algebraic expressions corresponding to the Feynman diagram. In the case of Fig. 4.8, we will now get the amplitude to be remarkably simple:

$$\mathcal{A} = -i\lambda \qquad (4.107)$$

To understand how this transition from the real space to momentum space Feynman diagram works, let us do one more example — which we will anyway need later, to study the running coupling constant — in some detail. Consider the evaluation of the 4-point function $G(x_1, x_2, x_3, x_4)$ to order λ^2 in the coupling constant. A typical diagram, among the set of diagrams which will contribute, is shown in Fig. 4.10. This arises from all possible pairings of the product of twelve ϕs in the expression $\langle\phi(x_1)\phi(x_2)\phi(x_3)\phi(x_4)\phi(x)^4\phi(y)^4\rangle$. Let us first determine the overall symmetry factor for this diagram. One can combine $\phi(x_1)$ with one of the $\phi(x)$ and $\phi(x_2)$ with another $\phi(x)$ in 4×3 possible ways; $\phi(x_3)$ and $\phi(x_4)$ can similarly be combined with two of the $\phi(y)$-s in another 4×3 possible ways; the remaining two of the $\phi(x)$ can combine with the remaining two $\phi(y)$ in two possible ways; finally, we can do the whole thing in two different ways by exchanging $x \leftrightarrow y$, which gives another factor 2. Thus we get two factors of 4! which nicely cancel with the $(4!)^2$ which arises in the coefficient $(1/2)(-i\lambda/4!)^2$ (when we expand the exponential to quadratic order in λ) leaving just $(1/2)(-i\lambda)^2$. To this order, our real space Feynman rule will translate the figure Fig. 4.9 to the algebraic expression

$$G(x_1, x_2, x_3, x_4) \quad = \quad \frac{1}{2}(-i\lambda)^2\int d^4y \qquad (4.108)$$

$$\times \int d^4x \,\langle x_1|x\rangle\,\langle x_2|x\rangle\,\langle x|y\rangle\,\langle x|y\rangle\,\langle y|x_3\rangle\,\langle y|x_4\rangle$$

When we go to Fourier space, we will work with the Fourier transform of $G(x_1, x_2, x_3, x_4)$ with respect to its arguments. It is convenient to associate p_1, p_2 with x_1, x_2 and $-k_1, -k_2$ with x_3 and x_4, so that we can think of in-going and out-going momenta with proper signs. We will denote the Fourier transform of $\langle x|y\rangle$ by $G(k) = i(k^2 - m^2 + i\epsilon)^{-1}$. The Fourier

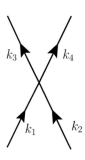

Figure 4.8: Momentum space representation of the vertex.

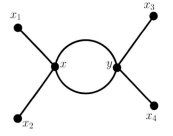

Figure 4.9: Diagram in real space corresponding to Eq. (4.108).

transform will, therefore, be given by

$$\frac{1}{2}(-i\lambda)^2 \int d^4y \int d^4x \int d^4x_1 \int d^4x_2 \int d^4x_3 \int d^4x_4$$

$$\times \langle x_1|x\rangle \langle x_2|x\rangle \langle x|y\rangle \langle x|y\rangle \langle y|x_3\rangle \langle y|x_4\rangle$$

$$\times \exp\left(i[p_1x_1 + p_2x_2 - k_1x_3 - k_2x_4]\right)$$

$$= \frac{1}{2}(-i\lambda)^2 G(p_1)\,G(p_2)\,G(k_1)\,G(k_2) \int d^4x\, d^4y \langle x|y\rangle^2$$

$$\times \exp\left(i[(p_1 + p_2)x - (k_1 + k_2)y]\right)$$

$$= \frac{1}{2}(-i\lambda)^2 G(p_1)\,G(p_2)\,G(k_1)\,G(k_2)(2\pi)^4\delta(p_1 + p_2 - k_1 - k_2)$$

$$\times \int \frac{d^4k}{(2\pi)^4}G(k)G(k_1 + k_2 - k) \qquad (4.109)$$

You will get similar expressions if we join x_1 and x_3 to x and x_2 and x_4 to y or if you join x_1 and x_4 to x and x_2 and x_3 to y. These three situations appear in momentum space as the three diagrams in Fig. 4.10. It is now clear that we again obtain four factors of G for the external legs of the diagram, one integration over the internal line momentum, an overall Dirac delta function expressing momentum conservation, individual momentum conservation at each vertex and an overall symmetry factor. This is precisely what we would have got if we had worked directly with the diagrams in Fig. 4.10. You should be able to convince yourself that similar features arise for arbitrarily complicated diagrams when we translate them from real space to momentum space.

4.5 Effective Action and the Perturbation Expansion

Incidentally, the definition of $Z(J)$ and $W(J)$ — which we introduced in order to facilitate the perturbation series — is also useful in another context. We saw earlier that one way of obtaining non-perturbative results is by using the concept of effective action which we defined in Sect. 4.1.1 from a fairly physical point of view. There is a more formal way of defining the effective action which you will find in several textbooks based on $Z(J)$ and $W(J)$. This formal definition will be useful, for e.g., to relate the approach based on the effective action with that based on perturbation theory etc.

To do this, we begin from the partition function $Z(J)$ defined by the standard relation

$$Z[J] = e^{iW[J]} = \int \mathcal{D}\phi \exp\left\{i\left[A[\phi] + \int d^4x\, J\phi\right]\right\} \qquad (4.110)$$

The Green's functions of the theory can be obtained either from the functional derivatives of Z or from the functional derivatives of W with respect to J. Usually we set $J = 0$ after the differentiations so as to obtain the time ordered expectation values in the *vacuum* state. But if we keep $J \neq 0$, we obtain the relevant expectation values in a state $|J\rangle$ describing an external source $J(x)$.

We are now in a position to define the effective action $A_{\text{eff}}[\phi]$ as a Legendre transform. To do this, we invert the equation Eq. (4.87) viz.,

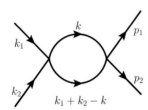

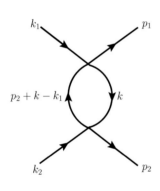

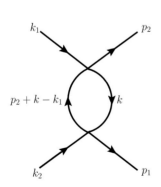

Figure 4.10: Three momentum space diagrams related to the same process.

$(\partial W[J]/\partial J(x)) \equiv \phi_J(x)$, to express J as a function of ϕ_J and omit the subscript J on ϕ_J to obtain the function $J = J(\phi)$. We then define the effective action as the Legendre transform of W by

$$A_{\text{eff}}[\phi] \equiv W[J(\phi)] - \int d^4x \, J[\phi(x)]\phi(x) \tag{4.111}$$

In this relation, $J(\phi)$ is treated as a functional of ϕ so that the left hand side is a functional of ϕ, as it should. The standard relations which occur in any such Legendre transform, viz.:

$$\frac{\delta A_{\text{eff}}[\phi]}{\delta \phi(x)}\bigg|_{\phi=\phi_J} = -J(x); \qquad \frac{\delta W[J]}{\delta J(x)} = \phi_J(x) \tag{4.112}$$

hold in this case as well and

$$W[J] = A_{\text{eff}}[\phi(J)] + \int d^4x \, J(x)\phi[J(x)] \tag{4.113}$$

gives the inverse Legendre transform. You can easily verify that the $A_{\text{eff}}(\phi)$ defined by the above rule is the same as the effective action we have defined earlier for quadratic actions.

One can now understand the properties of A_{eff} better through the above approach. We recall that the functional Taylor series expansion of $W(J)$ gives all the connected end point functions $G_c^{(n)}$ via the relation:

$$W[J] = -i \sum_{n=0}^{\infty} \frac{i^n}{n!} \int d^4x_1 \cdots d^4x_n J(x_1) \cdots J(x_n) G_c^{(n)}(x_1, \cdots, x_n) \tag{4.114}$$

Taking a cue from this, we can also expand $A_{\text{eff}}(\phi)$ in a functional Taylor series with the coefficient $\Gamma^{(n)}$ defined through the relation[37]

$$A_{\text{eff}}[\phi] = \sum_{n=1}^{\infty} \frac{i^n}{n!} \int d^4x_1 \cdots \int d^4x_n \Gamma^{(n)}(x_1, \cdots x_n)\phi(x_1) \cdots \phi(x_n) \tag{4.115}$$

Since we require $\delta A_{\text{eff}}/\delta\phi = 0$ when $J = 0$ with $\phi_c = 0$, the above expansion implies that $\Gamma^{(1)} = 0$. To understand what this expansion means, let us concentrate on $\Gamma^{(2)}(x_1, x_2)$. We differentiate the second relation in Eq. (4.112) with respect to $\phi(y)$. This gives

$$\delta^4(x-y) = \frac{\delta}{\delta\phi(y)}\left(\frac{\delta W[J]}{\delta J(x)}\right) = \int d^4z \, \frac{\delta J(z)}{\delta\phi(y)} \frac{\delta^2 W[J]}{\delta J(z)\delta J(x)} \tag{4.116}$$

But, from the first relation in Eq. (4.112), we have

$$-\frac{\delta J(z)}{\delta\phi(y)} = \frac{\delta^2 A_{\text{eff}}[\phi]}{\delta\phi(z)\delta\phi(y)} \tag{4.117}$$

Using this in Eq. (4.116) and evaluating it with $\phi = J = 0$, we obtain a simple relation between $\Gamma^{(2)}$ and $G_c^{(2)}$, given by:

$$-\delta^4(x-y) = \int d^4z \, \Gamma^{(2)}(y, z) \, G_c^{(2)}(z, x) \quad \text{or} \quad \Gamma^{(2)} = -\left\{G_c^{(2)}\right\}^{-1} \tag{4.118}$$

In other words, the quadratic part of the effective action gives the inverse of the connected two-point function. For example, we will show in Sec 4.7.1

[37] For simplicity, we have assumed that $\langle 0|\phi|0\rangle$ is zero; otherwise we need to make a shift $\phi \to \phi - \phi_c$.

that the sum over a set of loop diagrams in Fig. 4.12 will give $G_c^{(2)} = (p^2 - m^2 - \Sigma(p))^{-1}$ (see Eq. (4.145)). Then the relation in Eq. (4.118) tells us that $\Gamma^{(2)}$ is then given by $\Gamma^{(2)}(p) \propto (p^2 - m^2 - \Sigma(p))$ in the momentum space. In other words, the effective action succinctly summarizes the information contained in the sum of all the diagrams in Fig. 4.12. This should be contrasted with a perturbation expansion which will lead[38] to the individual terms in the expansion in Eq. (4.145).

An alternate way of obtaining a similar result is as follows. If we substitute the expression in Eq. (4.113) for $\exp[iW(J)]$ in Eq. (4.110) and use the first relation in Eq. (4.112) in the right hand side, we get the result

$$\exp i \left(A_{\text{eff}} + \int J\phi \, d^4x \right) = \int D\phi \, \exp i \left(A[\phi] - \int dx \, \phi(x) \frac{\delta A_{\text{eff}}[\phi]}{\delta \phi(x)} \right) \tag{4.119}$$

which can be written as

$$\exp i A_{\text{eff}} = \int D\phi \, \exp i \left(A[\phi] - \int dx (\phi - \psi) \frac{\delta A_{\text{eff}}[\psi]}{\delta \psi(x)} \right) \tag{4.120}$$

If we now shift ϕ to $\psi + \phi$, we can rewrite this result as

$$\exp i A_{\text{eff}} = \int D\phi \, \exp i \left(A[\psi + \phi] - \int dx \, \phi(x) \frac{\delta A_{\text{eff}}[\psi]}{\delta \psi(x)} \right) \tag{4.121}$$

At first sight, you might think that this relation is not of much value since both sides contain A_{eff}. However, the second term in the right hand side is a higher order correction and hence we can iteratively calculate the result as an expansion in $\hbar$.

For example, at one loop order we obtain an extra factor from the second term, which is given by

$$\exp i A_{\text{eff}}^{(1)}[\phi] \propto \int D\phi \, \exp \left\{ \frac{i}{2} \int d^4x \, \phi(x) \int d^4y \, \phi(y) \frac{\delta^2 A_0[\phi]}{\delta\phi(x)\delta\phi(y)} \right\} \tag{4.122}$$

where A_0 is the tree-level action. In a theory with

$$A_0[\phi] = \int dx \left(\frac{1}{2}\partial_m\phi\partial^m\phi - \frac{1}{2}m^2\phi^2 - V(\phi) \right) \tag{4.123}$$

we get the correction to be

$$\exp i A_{\text{eff}}^{(1)} \propto \int D\phi \, \exp \left\{ \frac{i}{2} \int d^4x \, \phi \left(\Box + m^2 + V''(\phi) \right) \phi \right\} \tag{4.124}$$

This leads to an effective action of the form

$$A_{\text{eff}}^{(1)} = A_0[\phi] + \frac{i}{2} \ln \text{Det} \left[\Box + m^2 + V''(\phi) \right] \tag{4.125}$$

This is indeed the expression we have used before to compute the effective action.

4.6   Aside: LSZ Reduction Formulas

The results obtained so far can be made much more rigorous in many different ways, one of which is called the LSZ (short for Lehmann-Symanzik-Zimmermann) formalism.[39]   While we do not want to be unnecessarily formal, we will describe very briefly the ideas involved in this formalism in order to clarify certain aspects of our previous discussion.

For the sake of definiteness, let us consider a physical process in which two particles with momenta p_1 and p_2 scatter with each other and produce a bunch of particles with momenta $p_3, p_4, ...p_n$. We introduce an S-matrix element — which is the matrix element of an operator S — denoted by $\langle p_3, p_4, ...p_n | S | p_1, p_2 \rangle$ which gives the amplitude for this process. We have seen that such amplitudes can be computed in terms of Feynman diagrams which, in turn, are short hand notations for algebraic expressions involving vacuum expectation values of the time-ordered product of field operators like the N point function:

$$G(x_1, x_2, \ldots x_N) = \langle 0 | T[\phi(x_1)\phi(x_2)\ldots\phi(x_N)] | 0 \rangle \qquad (4.126)$$

The procedure based on the Feynman diagrams suggests that the S-matrix element $\langle p_3, p_4, ...p_n | S | p_1, p_2 \rangle$ is directly related to the vacuum expectation values of the time ordered products like $G(x_1, x_2, \ldots x_N)$ defined by Eq. (4.126). The LSZ reduction formula provides the formal relation between these two. For the process which we were discussing, one can show that this relation is given by

$$\langle p_3 \cdots p_n | S | p_1 p_2 \rangle = \left[i \int d^4 x_1 e^{-i p_1 x_1} (\Box_1 + m^2) \right] \cdots \qquad (4.127)$$

$$\times \left[i \int d^4 x_n e^{-i p_n x_n} (\Box_n + m^2) \right] \langle 0 | T \{ \phi(x_1)\phi(x_2)\phi(x_3) \cdots \phi(x_n) \} | 0 \rangle$$

We will now describe briefly how such a relation comes about.

In general, the field will be described by an expansion of the form

$$\phi(x) = \phi(t, \boldsymbol{x}) = \int \frac{d^3 \boldsymbol{p}}{(2\pi)^3} \frac{1}{\sqrt{2\omega_{\boldsymbol{p}}}} \left[a_{\boldsymbol{p}}(t) e^{-i\boldsymbol{p}\cdot\boldsymbol{x}} + a_{\boldsymbol{p}}^\dagger(t) e^{i\boldsymbol{p}\cdot\boldsymbol{x}} \right] \qquad (4.128)$$

which is an operator in Heisenberg picture. In a free field theory, $a_{\boldsymbol{p}}(t)$ will evolve as $e^{-i\omega_p t}$; but in an interacting theory, the evolution of these operators will be non-trivial. Let us assume that all the interactions happen during the time interval $-T < t < T$. At very early times ($t \to -\infty$) and at very late times ($t \to +\infty$), the system is described by a free field theory which satisfies the condition

$$\langle 0 | \phi(t = \pm\infty, \boldsymbol{x}) | p \rangle \propto e^{i\boldsymbol{p}\cdot\boldsymbol{x}} \qquad (4.129)$$

Further, since all interactions take place only during the interval $-T < t < T$, the creation and annihilation operators evolve like in the free-field theory at $t \to \pm\infty$. So, asymptotically, we can construct the initial and final states by acting on the vacuum with suitable creation operators. More specifically, we have,

$$|p_1, p_2\rangle \equiv |i\rangle = \sqrt{2\omega_1}\sqrt{2\omega_2} \, a_{\boldsymbol{p}_1}^\dagger(-\infty) a_{\boldsymbol{p}_2}^\dagger(-\infty) | 0 \rangle \qquad (4.130)$$

$$|p_3, p_4 \cdots p_n\rangle \equiv |f\rangle = \sqrt{2\omega_3} \cdots \sqrt{2\omega_n} \, a_{\boldsymbol{p}_3}^\dagger(\infty) \cdots a_{\boldsymbol{p}_n}^\dagger(\infty) | 0 \rangle \qquad (4.131)$$

Therefore, the S-matrix element is just the amplitude given by

$$\langle f | S | i \rangle = \sqrt{2^n \omega_1 \cdots \omega_n} \, \langle 0 | a_{\boldsymbol{p}_3}(\infty) \cdots a_{\boldsymbol{p}_n}(\infty) a_{\boldsymbol{p}_1}^\dagger(-\infty) a_{\boldsymbol{p}_2}^\dagger(-\infty) | 0 \rangle \qquad (4.132)$$

We first re-write this expression in a slightly different way which will prove to be convenient. To begin with, we insert a time-ordering operator and express it in the form

$$\langle f|S|i\rangle = \sqrt{2^n \omega_1 \cdots \omega_n}\, \langle 0|T\left\{a_{\boldsymbol{p}_3}(\infty)\cdots a_{\boldsymbol{p}_n}(\infty)a^\dagger_{\boldsymbol{p}_1}(-\infty)a^\dagger_{\boldsymbol{p}_2}(-\infty)\right\}|0\rangle \tag{4.133}$$

We next notice that this expression can be written in an equivalent form as

$$\begin{aligned}
\langle f|S|i\rangle \;=\;& \sqrt{2^n \omega_1 \cdots \omega_n} \\
& \times \langle 0|T\left\{[a_{\boldsymbol{p}_3}(\infty) - a_{\boldsymbol{p}_3}(-\infty)]\cdots[a_{\boldsymbol{p}_n}(\infty) - a_{\boldsymbol{p}_n}(-\infty)]\right. \\
& \left. \times\, [a^\dagger_{\boldsymbol{p}_1}(-\infty) - a^\dagger_{\boldsymbol{p}_1}(\infty)][a^\dagger_{\boldsymbol{p}_2}(-\infty) - a^\dagger_{\boldsymbol{p}_2}(\infty)]\right\}|0\rangle
\end{aligned} \tag{4.134}$$

This trick works because the time-ordering moves all the unwanted $a^\dagger(\infty)$ operators associated with the initial state to the left where they annihilate on $\langle f|$ while all the unwanted $a(-\infty)$ operators are pushed to the right where they annihilate on $|i\rangle$. So, in order to determine $\langle f|S|i\rangle$ we only need to find a suitable expression for $[a_{\boldsymbol{p}}(\infty) - a_{\boldsymbol{p}}(-\infty)]$ (and its Hermitian conjugate) for each of the creation and annihilation operators. Our aim is to manipulate this expression and obtain Eq. (4.127).

The key to this result is the purely algebraic relation

$$i\int d^4x\, e^{ipx}(\Box + m^2)\phi(x) = \sqrt{2\omega_{\boldsymbol{p}}}\,[a_{\boldsymbol{p}}(\infty) - a_{\boldsymbol{p}}(-\infty)] \tag{4.135}$$

in which the four-momentum is on-shell. To obtain this relation, we will assume that all fields die off at spatial infinity, allowing a suitable integration by parts. Then, we have the results

$$i\int d^4x\, e^{ipx}(\Box + m^2)\phi(x) = i\int d^4x\, e^{ipx}(\partial_t^2 + \omega_{\boldsymbol{p}}^2)\phi(x) \tag{4.136}$$

and

$$\begin{aligned}
\partial_t[e^{ipx}(i\partial_t + \omega_{\boldsymbol{p}})\phi(x)] \;=\;& [i\omega_{\boldsymbol{p}}e^{ipx}(i\partial_t + \omega_{\boldsymbol{p}}) + e^{ipx}(i\partial_t^2 + \omega_{\boldsymbol{p}}\partial_t)]\phi(x) \\
=\;& ie^{ipx}(\partial_t^2 + \omega_{\boldsymbol{p}}^2)\phi(x)
\end{aligned} \tag{4.137}$$

Exercise 4.5: Prove these.

which can be proved by simple algebra. Combining these, we get

$$\begin{aligned}
i\int d^4x\, e^{ipx}(\Box + m^2)\phi(x) \;=\;& \int d^4x\, \partial_t[e^{ipx}(i\partial_t + \omega_{\boldsymbol{p}})\phi(x)] \tag{4.138} \\
=\;& \int dt\, \partial_t\left[e^{i\omega_{\boldsymbol{p}}t}\int d^3\boldsymbol{x}\, e^{-i\boldsymbol{p}\cdot\boldsymbol{x}}(i\partial_t + \omega_{\boldsymbol{p}})\phi(x)\right]
\end{aligned}$$

Since the integrand is a total time derivative and the integration is over the entire range $-\infty < t < \infty$, the result is only going to depend on the behaviour of the integrand in the asymptotic limits. By construction, $a_{\boldsymbol{p}}(t)$ and $a^\dagger_{\boldsymbol{p}}(t)$ are time-independent asymptotically. If we now use this fact and the expansion in Eq. (4.128), we can easily show that

$$\int d^3\boldsymbol{x}\, e^{-i\boldsymbol{p}\cdot\boldsymbol{x}}(i\partial_t + \omega_{\boldsymbol{p}})\phi(x) = \sqrt{2\omega_{\boldsymbol{p}}}\, a_{\boldsymbol{p}}(t)\, e^{-i\omega_{\boldsymbol{p}}t} \tag{4.139}$$

which allows us to write

$$\begin{aligned}
i\int d^4x\, e^{ipx}(\Box + m^2)\phi(x) \;=\;& \int dt\, \partial_t\left[\left(e^{i\omega_{\boldsymbol{p}}t}\right)\left(\sqrt{2\omega_{\boldsymbol{p}}}\, a_{\boldsymbol{p}}(t)\, e^{-i\omega_{\boldsymbol{p}}t}\right)\right] \\
=\;& \sqrt{2\omega_{\boldsymbol{p}}}\,[a_{\boldsymbol{p}}(\infty) - a_{\boldsymbol{p}}(-\infty)] \tag{4.140}
\end{aligned}$$

Taking the Hermitian conjugate, we also have the result

$$-i \int d^4 x \, e^{-ipx} (\Box + m^2) \phi(x) = \sqrt{2\omega_{\boldsymbol{p}}} \left[a_{\boldsymbol{p}}^\dagger(\infty) - a_{\boldsymbol{p}}^\dagger(-\infty) \right] \qquad (4.141)$$

That is all we need. Substituting Eq. (4.141) into Eq. (4.134), we immediately obtain the relation in Eq. (4.127).

What does this relation mean? The time-ordered correlation function contains a large amount of information about the dynamics of the field, much of which is irrelevant for the amplitude we are interested in. The operator $\Box + m^2$, which becomes $-p^2 + m^2$ in Fourier space, actually vanishes in the asymptotic states where the field becomes free. Therefore, these factors in Eq. (4.127) will remove all the terms in the time-ordered product *except* those containing the factor $(p^2 - m^2)^{-1}$. In other words, we are essentially picking out the residue from each pole in the correlation function. Thus, the S-matrix projects out the one-particle asymptotic states from the time-ordered product of the fields. The LSZ reduction formula encodes a careful cancellation between the zeros resulting from $\Box + m^2$ acting on the fields and the $(\Box + m^2)^{-1}$ arising from the one-particle states. This formal relation is what is pictorially represented through the Feynman diagrams.

4.7 Handling the Divergences in the Perturbation Theory

When we studied the self-interacting scalar field from a non-perturbative perspective — using the effective action — in Sect. 4.3, we found that the effective theory can be expressed in terms of certain physical parameters $m_{\text{phy}}, \lambda_{\text{phy}}$ which are different from the parameters m_0, λ_0 which appear in the original Lagrangian.[40] We also found that for $m_{\text{phy}}, \lambda_{\text{phy}}$ to remain finite, the bare parameters m_0, λ_0 have to be divergent. In particular, the strength of the interaction in the theory, measured by the coupling constant λ_{phys}, needs to be defined operationally at some suitable energy scale. Once this is done, it gets determined at all other scales through the equation Eq. (4.75). The question arises as to how we are led to these results when we approach the problem perturbatively.

The answer to this question relies on the concept of perturbative renormalization which is an important technical advance in the study of quantum field theory. The main application of this technique — leading to definite predictions which have been observationally verified — occurs in QED which we will study in the next chapter. At that time, we will also describe the conceptual aspects of the perturbative renormalization in somewhat greater detail. The purpose of this section is to briefly introduce these ideas in the context of $\lambda\phi^4$ theory as a preparation for the latter discussion in QED.[41]

The really nontrivial aspects of any quantum field theory, in particular $\lambda\phi^4$ theory or QED, arise only when we go beyond the lowest order perturbation theory (called the tree-level). This is because, as long as there are no internal loops, the translation of a Feynman diagram into an algebraic expression does not involve any integration. But when we consider a diagram involving one or more internal loops, we have to integrate over the momenta corresponding to these loops. The propagators corresponding to the internal loops will typically contribute a $(1/p^2)$ or $(1/p^4)$ factor

[40] We have added the subscripts 'phy' and 0 to these parameters for later notational convenience.

[41] The technical details of renormalization depend crucially on the kind of theory you are interested in. From this point of view, $\lambda\phi^4$ theory is somewhat too simple to capture all the nuances of this technique. This is one reason for keeping this discussion rather brief in this section.

to the integral at large momenta and the integration measure will go as $p^3 dp$. This could lead to the contributions from these diagrams, to diverge at large p. In fact, this disaster *does* occur in $\lambda\phi^4$ theory and QED, and the diagrams involving even a single loop produce divergent results. It is necessary to understand and reinterpret the perturbation theory when this occurs.

The situation described above is the usual motivation given in textbooks for the procedure called renormalization; it is thought of as a practical necessity needed to "save" the perturbative calculation technique. From a conceptual point of view, this reasoning is at best misleading and at worst incorrect. As we have stressed several times before, the process of renormalization, a priori, has very little to do with the existence or removal of divergences in perturbative field theory. We have also seen examples of these divergences in nonperturbative calculations based on the concept of effective action.[42]

It is now clear that, to implement the program of perturbative renormalization, we need to undertake the following steps. (a) Identify processes or Feynman diagrams which will lead to a divergent contribution to a probability amplitude. (b) Regularize this divergence in some sensible manner; this is usually done by introducing some extra parameter ϵ, say, such that the contribution is finite and well defined when $\epsilon \neq 0$ and diverges in the limit of $\epsilon \to 0$. (c) See whether all the divergent terms can be made to disappear by changing the original parameters λ_A^0 in the Lagrangian of the theory to the physically observable set λ_A^{ren}. This is best done keeping ϵ finite, so that you are dealing with regularized finite expressions. (d) Re-express the theory and, in particular, the non-divergent parts of the amplitude in terms of the physical parameters λ_A^{ren}. This will form the basis for comparing the theory with observations.

In any given theory, say e.g., $\lambda\phi^4$ theory or QED, we will need to carry out this program to all orders in perturbation theory in order to convince ourselves that the theory is well defined.[43] This is indeed possible, but we shall concentrate on illustrating this procedure at the lowest order, which essentially involves integration over a single internal momentum variable. This will introduce the necessary regularization procedure and the conceptual details which are involved.

4.7.1   One Loop Divergences in the $\lambda\phi^4$ Theory

After this preamble, let us consider the Feynman diagrams in the $\lambda\phi^4$ theory which contain a single loop. There are essentially two such diagrams of which the second one comes in three avatars. The first one shown in Fig. 4.11 while the second one — which has three related forms — is shown in Fig. 4.10. The analytic expression corresponding to Fig. 4.11 is given by the integral

$$-i\Sigma(p^2) = -i\Sigma(0) = -i\frac{\lambda}{2}\int \frac{d^4\ell}{(2\pi)^4}\frac{i}{\ell^2 - m_0^2 + i\epsilon} \qquad (4.142)$$

On the other hand, the contribution from the sum of the three diagrams in Fig. 4.10 is given by

$$i\mathcal{M} = -i\lambda + \mathcal{A}(p_1 + p_2) + \mathcal{A}(k_1 - p_1) + \mathcal{A}(k_1 - p_2) \qquad (4.143)$$

[42]The effective action is closely related to the expansion in loops. One may therefore try to argue that, ultimately, the source of all divergences is in the loops. But the fact that the effective action can also lead to results which are non-analytic in the coupling constant — like the Schwinger effect — suggests that it is better not to try and interpret *everything* in quantum field theory in terms of the perturbative expansion.

[43]When we study the $\lambda\phi^4$ theory at one loop order — which we will do in the next section — we will find that the mass m and the coupling constant λ gets renormalized and this is adequate to handle all the *one loop* divergences. But when we proceed to the next order, viz. $\mathcal{O}(\lambda^2)$, we will encounter a new type of divergence which can be handled through the renormalization of the field, usually called the wave function renormalization. This is described in Problem 9 but we will discuss a similar renormalization in detail when we study QED.

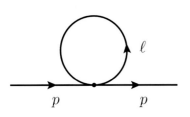

Figure 4.11: One loop correction to the propagator.

where

$$\mathcal{A}(p) \equiv \frac{(-i\lambda)^2}{2} \int \frac{d^4k}{(2\pi)^4} \frac{i}{k^2 - m_0^2 + i\epsilon} \frac{i}{(p-k)^2 - m_0^2 + i\epsilon} \qquad (4.144)$$

It is obvious that both the integrals are divergent; we also note that $\Sigma(p^2)$ is actually independent of p and is just a divergent constant.[44] The contribution from Fig. 4.11 corrects the propagator from its expression at the tree-level while the contribution from Fig. 4.10 changes the value of the coupling constant.

Let us begin with the role played by the diagram in Fig. 4.11. To understand it in proper context, it is convenient to introduce the concept of one-particle, irreducible (1PI) diagrams. The 1PI means that you start with a single line and end with a single line and draw in between all possible diagrammatic structures which cannot be cut apart by just cutting a single line. Using this concept, one can draw a series of 1PI diagrams which will correct the propagator in a geometric series as shown in Fig. 4.12. Translating this geometric series in diagrams into equivalent algebraic expressions, we find that the Feynman propagator gets corrected as follows:

$$
\begin{aligned}
iG(p) &= iG_0(p) + iG_0(p)\left[-i\Sigma(p)\right]iG_0(p) + \cdots \\
&= \frac{i}{p^2 - m_0^2 + i\epsilon} + \frac{i}{p^2 - m_0^2 + i\epsilon}\left[-i\Sigma(p)\right]\frac{i}{p^2 - m_0^2 + i\epsilon} + \cdots \\
&= \frac{i}{p^2 - m_0^2 + i\epsilon}\left[\frac{1}{1 + i\Sigma(p^2)\frac{i}{p^2 - m_0^2 + i\epsilon}}\right] \\
&= \frac{i}{p^2 - m_0^2 - \Sigma(p^2) + i\epsilon} \qquad (4.145)
\end{aligned}
$$

Since $\Sigma(p^2) = \Sigma(0)$ is a constant, it is obvious that the net effect of the diagram in Fig. 4.11 — when used in the geometric series in Fig. 4.12 — is to change the bare mass as $m_0^2 \to m_0^2 + \Sigma(0)$.

Our next job is to evaluate $\Sigma(p)$ by some regularization procedure. As usual, one could do this either by introducing a cut-off or by dimensional regularization. This time we shall use the procedure of dimensional regularization in order to illustrate an important difference with respect to the cut-off regularization which we used earlier when we studied the effective potential.

To do this, it is convenient at this stage to switch to the n-dimensional Euclidean momentum space. As we described earlier, dimensional regularization involves[45] working in n-dimensions and analytically continuing the results to all values of n and then taking the limit of $n \to 4$. By and large, this procedure has the advantage that it maintains the symmetries of the theory.

Such a dimensional continuation of the integral in Eq. (4.142) will require us to essentially evaluate the integral given by

$$I \equiv \lambda\mu^{4-n}\left[\int \frac{d^n\ell}{(2\pi)^n}\frac{1}{\ell^2 + m_0^2}\right] \qquad (4.146)$$

The pre-factor μ^{4-n}, where μ has the dimensions of energy, is introduced in order to keep λ dimensionless[46] even in n-dimensions, just as it was in $n = 4$. It is also convenient to define a parameter ϵ by $2\epsilon = 4 - n$. The

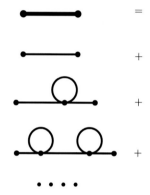

Figure 4.12: Diagrammatic representation of the geometric progression evaluated in Eq. (4.145).

[45]We will say more about this procedure when we discuss QED in the next chapter.

[46]Since we want $d^n x(\partial\phi)^2$ to be dimensionless, ϕ must have the dimensions $\mu^{1-\epsilon}$ where μ is an energy scale. Further, $\lambda\phi^4$ must have the dimensions of $(\partial\phi)^2$. This will require λ to have the dimensions of $\mu^{2\epsilon} = \mu^{4-n}$ in n-dimensions. It is this factor which we scale out in order to keep λ dimensionless.

integral in Eq. (4.146) is most easily evaluated by writing the integrand $(1/D)$ as

$$\frac{1}{D} = \int_0^\infty ds \exp(-sD), \qquad (4.147)$$

Exercise 4.6: Fill in the algebraic details in Eq. (4.148) to Eq. (4.152).

carrying out the Gaussian integrations over ℓ and then performing the integration over the parameter s. Analytically continuing back to Lorentzian spacetime, we find that our result is given by

$$-i\Sigma(0) = \frac{-i\lambda m^2}{32\pi^2} \left(\frac{4\pi\mu^2}{m^2}\right)^\epsilon \Gamma(-1+\epsilon) \qquad (4.148)$$

This result is finite for finite ϵ, but of course we require the limit of $\epsilon \to 0$ which will allow us to isolate the divergent terms in a clean fashion. We first note that

$$\left(\frac{4\pi\mu^2}{m^2}\right)^\epsilon = \exp\left[\epsilon \ln\left(\frac{4\pi\mu^2}{m^2}\right)\right] \simeq 1 + \epsilon \ln\left(\frac{4\pi\mu^2}{m^2}\right) \qquad (4.149)$$

while the Gamma function has the expansion near $\epsilon = 0$ given by

$$\Gamma(-1+\epsilon) = -\left[\frac{1}{\epsilon} + 1 - \gamma_E + \mathcal{O}(\epsilon)\right] \qquad (4.150)$$

Putting these in, we find the final result which we are after, viz.,

$$-i\Sigma(0) = \frac{i\lambda m^2}{32\pi^2} \left[\frac{1}{\epsilon} + 1 - \gamma_E + \ln\left(\frac{4\pi\mu^2}{m^2}\right) + \mathcal{O}(\epsilon)\right] \qquad (4.151)$$

This result displays the structure of the divergence (which arises from the $(1/\epsilon)$ term) and a dependence on the arbitrary mass scale μ through the logarithmic term. We can take care of the divergence by replacing m_0^2 by $m^2 = m_0^2 + \delta m^2$ which can be written as

$$m_0^2 = m^2 - \delta m^2 = m^2 \left[1 + \frac{\lambda}{32\pi^2}\frac{1}{\epsilon}\right] \qquad (4.152)$$

[47] Do not confuse this mass scale μ with the cut-off scale Λ which was used to regularize divergences earlier. In dimensional regularization, the ultraviolet regulator is actually provided by the parameter $2\epsilon = 4 - n$ which measures the deviation from 4-dimensions. The $\epsilon \to 0$ limit is equivalent to an ultraviolet regulator scale Λ going to infinity. The μ is not a regulator, but, as we shall see, allows us to define a renormalization point (usually a finite mass scale of the order of the other mass scales in the theory) used to define renormalized quantities.

We, of course, do not want the physical results to depend on the arbitrary scale μ; this would require m and λ to vary with μ in a particular way.[47] We will comment on this fact after discussing the renormalization of λ.

You might have noticed that the relation between m_0 and m is now different from what we found in Eq. (4.72). In Eq. (4.72) we had a logarithmic divergence proportional to λm^2 and a quadratic divergence proportional to λ. But in Eq. (4.152) we only have a single divergence proportional to λm^2 and the quadratic divergence has disappeared. As we mentioned earlier, this is a general feature of dimensional regularization. To understand this better, consider the n-dimensional Euclidean momentum integral of the form

$$I_k(n, m^2) = \int \frac{d^n p_E}{(2\pi)^n} \frac{1}{(p_E^2 + m^2)^k} \qquad (4.153)$$

This integral is most easily evaluated by using the following integral representation for the integrand:

$$\frac{1}{a^k} = \frac{1}{\Gamma(k)} \int_0^\infty dt \; t^{k-1} e^{-at} \qquad (a > 0) \qquad (4.154)$$

and performing the Gaussian integrals over the momentum. This will give the result

$$I_k(n, m^2) = \frac{1}{(4\pi)^{2+\frac{n-4}{2}}} \frac{\Gamma(k - 2 - \frac{n-4}{2})}{\Gamma(k)(m^2)^{k-2-\frac{n-4}{2}}} \qquad (4.155)$$

We can now take the limit of $n \to 4$ and obtain

$$I_k(n, m^2) \to \frac{(m^2)^{2-k}}{16\pi^2} \frac{2}{n-4} + \text{finite part} \qquad (4.156)$$

We see that the nature of the divergence is *independent* of k and always has a $1/(n-4)$ behaviour.

On the other hand, if we compute the same integral using a momentum space cut-off at some large value M, then we get the result which clearly depends on k:

$$\int_{|p_E|<M} \frac{d^4 p_E}{(2\pi)^4} \frac{1}{(p_E^2 + m^2)^k} \propto \begin{Bmatrix} \frac{m^2}{8\pi^2}\left[\frac{M^2}{m^2} - \log\left(\frac{M^2}{m^2}\right)\right] & k = 1 \\ \\ \frac{1}{8\pi^2}\left[\log\left(\frac{M^2}{m^2}\right) - \frac{1}{2}\right] & k = 2 \\ \\ \frac{m^{4-2k}}{8\pi^2(k-1)(k-2)} & k > 2 \end{Bmatrix}$$
$$(4.157)$$

We see that the $k = 1$ and $k = 2$ cases (which prop up repeatedly in quantum field theory) have different structures of the divergence as compared to what we found with dimensional regularization.[48]

In our case, we are interested in the $k = 1$ integral. If we do it using a cut-off — as we did when we computed the effective potential — we pick up two kinds of divergences, one quadratic and one logarithmic, as seen in Eq. (4.157). On the other hand, if we had used the dimensional regularization, we would not have picked up the quadratic divergence as can be seen from Eq. (4.156). In some sense, this issue is irrelevant, because the divergent terms are being absorbed into the parameters of the theory, and no physical effects, including the running of the parameters, will depend on whether we use dimensional regularization or cut-off regularization. It is nevertheless important to appreciate that dimensional regularization makes quadratic divergences disappear.[49]

4.7.2 Running Coupling Constant in the Perturbative Approach

Let us next consider the contributions from the three diagrams in Fig. 4.9 which is given by the expression in Eq. (4.143) and Eq. (4.144), reproduced here for convenience:

$$i\mathcal{M} = -i\lambda + \mathcal{A}(p_1 + p_2) + \mathcal{A}(k_1 - p_1) + \mathcal{A}(k_1 - p_2) \qquad (4.158)$$

where we have defined with

$$\mathcal{A}(p) \equiv \frac{(-i\lambda)^2}{2} \int \frac{d^4 k}{(2\pi)^4} \frac{i}{k^2 - m^2 + i\epsilon} \frac{i}{(p-k)^2 - m^2 + i\epsilon} \qquad (4.159)$$

Physically this represents the amplitude for a $\phi\phi \to \phi\phi$ scattering[50] correct to $\mathcal{O}(\lambda^2)$. To the lowest order, we just have $(-i\lambda)$ and to the next order we have the contribution from the three diagrams in Fig. 4.9.

Exercise 4.7: Obtain the expressions in Eq. (4.157).

[48]This is just another fact of life. You cannot do anything about it except be aware that the pattern of divergence can depend on the regularization procedure you use and make sure no observable results are affected.

[49]Notice that, in the case of the integral $I_1(n, m^2)$, we have a $\Gamma((2-n)/2)$ making its appearance. This has, besides the pole at $n = 4$, another one at $n = 2$. The original quadratic divergence transforms itself into the existence of two such poles in the resulting integral.

[50]The arguments of $\mathcal{A}$ in Eq. (4.158) occur frequently in problems involving the scattering of two particles of masses m_1, m_2 from the initial momenta (p_1, p_2) to the final momenta (p_3, p_4). For such a process, one defines the *Mandelstam variables* $s \equiv (p_1 + p_2)^2$, $t \equiv (p_1 - p_3)^2$, $u \equiv (p_1 - p_4)^2$. You can easily verify that $\sqrt{s}$ is the center of mass energy for the scattering and $s + t + u$ is equal to the sum of the squares of the masses of the particles. They play an important role in the study of the kinematics of the scattering.

The most interesting (or disturbing, depending on your point of view) feature of the expression in Eq. (4.158) is that the integral defining $\mathcal{A}$ is logarithmically divergent. The way to handle this divergence, again, is as we described before. The parameters in the Lagrangian — and in particular the coupling constant λ — have no intrinsic physical significance and have to be defined through some operational procedure like the $\phi\phi \to \phi\phi$ scattering, etc. When such an experiment is performed, one would work at some energy[51] scale E. With such a scattering experiment, one can measure the amplitude $\mathcal{M}$ which will define for us the effective coupling constant $\lambda_{\text{eff}}(E)$ defined through the relation $i\mathcal{M} = -i\lambda_{\text{eff}}(E)$. This effective coupling constant could, of course, depend on the energy scale of the scattering experiment which is indicated by an explicit functional dependence in the expression. When such an experiment is performed, one always gets finite answers for $\mathcal{M}$ and the experimental result takes into account the interaction to all orders in the perturbation. So, we need to do is to re-express all the answers in terms of the experimentally determined λ_{eff} (at some scale). If the divergences disappear when the scattering amplitudes are re-expressed in terms of physically relevant parameters, then we have a good theory which can be used to make useful predictions. We will now see how this comes about.[52]

Our first task is to regularize the divergent integral in Eq. (4.159) and isolate the divergences. To do this, we will again use dimensional regularization which requires us to work in n dimensions after analytically continuing to the Euclidean sector. Assuming $p^2 < m^2$ and rotating to the Euclidean plane in n-dimensions, we need to essentially evaluate the integral:

$$F(p^2) = -\int \frac{d^n k}{(2\pi)^n} \frac{1}{k^2 + m^2} \frac{1}{(k-p)^2 + m^2} \qquad (4.160)$$

This naive procedure of analytic continuation does work in this case; but it is worth noting the following subtlety (to which we will come back later). Since we want to analytically continue k^0 to imaginary values, let us examine the analytic structure of the integrand in Eq. (4.159), made of the product of two propagators. The first factor has the poles at $\omega_{\mathbf{k}} - i\epsilon$ and $-\omega_{\mathbf{k}} + i\epsilon$. These two poles are the usual ones in the second and fourth quadrant and do not prevent us from performing the Wick rotation in $k^0 e^{i\theta}$ from $\theta = 0$ to $\theta = \pi/2$. The poles of the second factor are at $(p_0 + \sqrt{(\mathbf{k} - \mathbf{p})^2 + m^2} - i\epsilon)$ and $(p_0 - \sqrt{(\mathbf{k} - \mathbf{p})^2 + m^2} + i\epsilon)$. It is easy to see that, if $0 < p_0 < m$, then these poles are also in the second and fourth quadrant and hence do not create any obstacles for the Wick rotation. Therefore, we can first determine the value of the integral for $p^2 < m^2$ and then analytically continue to the whole energy plane. This will require a transition across a Riemann sheet (because of square roots) which can be handled — as we shall see in the Mathematical Supplement 4.9.2 – by evaluating the imaginary part of the integral explicitly if required. We will now proceed with above integral in Eq. (4.160).

It is convenient at this stage to introduce a parametric integration to combine the denominators of the two individual terms.[53] Using again the trick in Eq. (4.147), we obtain

$$F(p^2) = -\int_0^\infty ds \int_0^\infty dt\, e^{-(s+t)m^2} \qquad (4.161)$$
$$\times \int \frac{d^n k}{(2\pi)^n} \exp\left[-\left((s+t)k^2 - 2kpt + p^2 t\right)\right]$$

---

[51] For example, if we scatter two particles at very low momentum, we can say that the energy scale is set by the rest mass m of the particles, or if we scatter two particles at energies far larger than the rest mass energy, we could take the center of mass energy $\sqrt{s}$ as setting the scale for this scattering, etc.

[52] We have, of course, come across the same issue in Sect. 4.3 and have explained the philosophy behind tackling this problem. It is not surprising that the same issue arises in the perturbative approach as well. In fact, you will see that we will get exactly the same result for the "running" of λ, which is gratifying given the cavalier attitude we have taken towards mathematical rigour.

[53] This is usually called the Feynman parameterization in textbooks, though it has appeared in an earlier work of Julian Schwinger in the exponential form. The discussion about priorities in a conversation between Schwinger and Harold (Hypothetically Alert Reader Of Limitless Dedication; a character, created by Schwinger, who sets records straight) in page 338 of the book *Sources, Fields and Particles*, Vol. I is amusing to listen to:

Harold: Is it not true, however, that the usual intent of that device, to replace space-time integrations by invariant parametric integrals, was earlier exploited by you in a related exponential version, and that the elementary identity combining two denominators, in fact, appears quite explicitly in a paper of yours, published in the same issue that contains Feynman's contribution?

Schwinger: Yes.

in which one can immediately perform the Gaussian integrals. This requires completing the squares by writing

$$(s+t)k^2 - 2kpt + p^2t = (s+t)\left(k - \frac{pt}{s+t}\right)^2 + \frac{st}{s+t}p^2 \tag{4.162}$$

and shifting the integration variable to

$$\bar{k} = k - \frac{pt}{s+t} \tag{4.163}$$

We then get the result

$$F(p^2) = -\frac{1}{(4\pi)^{n/2}} \int_0^\infty ds \tag{4.164}$$

$$\times \int_0^\infty dt \frac{1}{(s+t)^{n/2}} \exp\left\{-\left[(s+t)m^2 + \left(\frac{st}{s+t}\right)p^2\right]\right\}$$

To proceed further, it is useful to do a change of variable which can be done formally as follows. We first insert a factor of unity

$$1 = \int_0^\infty dq\, \delta(q - s - t) \tag{4.165}$$

and introduce two variables α and β in place of s and t by the definitions $s = \alpha q$, $t = \beta q$. This leads to

$$F(p^2) = -\frac{1}{(4\pi)^{n/2}} \int_0^\infty dq \int_0^\infty q\, d\alpha \int_0^\infty q\, d\beta \frac{1}{q^{n/2}} \delta(q(1 - \alpha - \beta))$$
$$\times \exp\left[-q(m^2 + \alpha\beta p^2)\right] \tag{4.166}$$

Using $\delta(qx) = \delta(x)/q$, we can do the q integration, obtaining

$$\int_0^\infty dq\, q^{1-n/2} \exp\left[-q(m^2 + \alpha\beta p^2)\right] = \Gamma\left(2 - \frac{n}{2}\right)\left[m^2 + \alpha\beta p^2\right]^{\frac{n}{2}-2} \tag{4.167}$$

Therefore,

$$F(p^2) = -\frac{\mu^{n-4}}{(4\pi)^2}\Gamma\left(2 - \frac{n}{2}\right)\int_0^1 d\alpha \left[\frac{m^2 + \alpha(1-\alpha)p^2}{4\pi\mu^2}\right]^{\frac{n}{2}-2} \tag{4.168}$$

In the last expression we have introduced an arbitrary mass scale μ by multiplying and dividing by[54] the factor μ^{n-4}. This makes the integral dimensionless which will be convenient later on.

> [54]That ensures the result is independent of μ; right? Now watch the fun.

We now need to take the $n \to 4$ limit of this expression when we get a divergence from the Gamma function which has a simple pole at this point. In this limit, we have the relation

$$\Gamma\left(2 - \frac{n}{2}\right) = \frac{\Gamma\left(3 - \frac{n}{2}\right)}{2 - \frac{n}{2}} \to \frac{1}{2 - \frac{n}{2}} + \Gamma'(1) \equiv \frac{1}{\epsilon} - \gamma_E \tag{4.169}$$

with $\epsilon = 2 - (n/2)$ and γ_E being Euler's constant. Using this in Eq. (4.168) and pulling out a μ^{n-4} factor to provide the dimensions, we can express our result as

$$F(p^2) = \mu^{n-4}\left[-\frac{1}{(4\pi)^2}\frac{1}{\epsilon} + F_{\text{fin}}(p^2)\right] \tag{4.170}$$

where the finite part is:

$$F_{\text{fin}}(p^2) = -\frac{1}{(4\pi)^2} \int_0^1 d\alpha \left\{ \Gamma(\epsilon) \left[\frac{4\pi\mu^2}{m^2 + \alpha(1-\alpha)p^2} \right]^\epsilon - \frac{1}{\epsilon} \right\} \quad (4.171)$$

These expressions clearly isolate the divergent and finite parts of the function. In fact, using $x^\epsilon = e^{\epsilon \ln x} \approx 1 + \epsilon \ln x$ for small ϵ and the expansion in Eq. (4.169), we can easily show that the finite part is given by the integral representation:

$$F_{\text{fin}}(p^2) = \frac{1}{(4\pi)^2} \left\{ \gamma_E - \int_0^1 d\alpha \ln \left[\frac{4\pi\mu^2}{m^2 + \alpha(1-\alpha)p^2} \right] \right\} \quad (4.172)$$

in the Euclidean space. Analytically continuing back to the Lorentzian space, p^2 will become $-p^2$ and we will get

$$F_{\text{fin}}(p^2) = \frac{1}{(4\pi)^2} \left\{ \gamma_E - \int_0^1 d\alpha \ln \left[\frac{4\pi\mu^2}{m^2 - \alpha(1-\alpha)p^2} \right] \right\} \quad (4.173)$$

We can evaluate the integral explicitly, but before we do that, let us once again demonstrate how the running of the coupling constant arises from this expression. The total scattering amplitude in Eq. (4.158) now reads as

$$-\mathcal{M} = \lambda + \frac{1}{2}\lambda^2 \left[-\frac{3}{(16\pi^2)}\frac{1}{\epsilon} + F_{\text{fin}}(s) + F_{\text{fin}}(t) + F_{\text{fin}}(u) \right] + \mathcal{O}(\lambda^3) \quad (4.174)$$

with $s = (p_1 + p_2)^2, t = (p_1 - p_3)^2, u = (p_1 - p_4)^2$. We now define a physical coupling constant λ_{phy} by the relation

$$\lambda = \lambda_{\text{phy}}\mu^{2\epsilon}\left(1 + \frac{3}{2}\frac{\lambda_{\text{phy}}}{16\pi^2}\frac{1}{\epsilon}\right) + \mathcal{O}(\lambda_{\text{phy}}^3) \to \lambda_{\text{phy}}\left(1 + \frac{3}{2}\frac{\lambda_{\text{phy}}}{16\pi^2}\frac{1}{\epsilon}\right) + \mathcal{O}(\lambda_{\text{phy}}^3)$$
$$(4.175)$$

with λ_{phy} being finite, so that the scattering amplitude becomes:[55]

$$-\mathcal{M} = \lambda_{\text{phy}} + \frac{1}{2}\lambda_{\text{phy}}^2 \left[F_{\text{fin}}(s) + F_{\text{fin}}(t) + F_{\text{fin}}(u) \right] + \mathcal{O}(\lambda_{\text{phy}}^3) \quad (4.176)$$

This expression is now perfectly finite and we have taken care of the divergence.

As usual, the λ_{phy} depends on the scale μ which is lurking inside F_{fin}. Obviously, we cannot have the amplitude $\mathcal{M}$ to depend on this arbitrary scale μ. This is taken care of by arranging the μ dependence of λ_{phy} to be such that $\mathcal{M}$ becomes independent of μ. This leads to the condition:

$$0 = -\frac{d\mathcal{M}}{d\mu} = \frac{d\lambda_{\text{phys}}}{d\mu} + \lambda_{\text{phys}}\frac{d\lambda_{\text{phys}}}{d\mu}\left[F_{\text{fin}}(s) + F_{\text{fin}}(t) + F_{\text{fin}}(u)\right] \quad (4.177)$$

$$+ \frac{1}{2}\lambda_{\text{phy}}^2\frac{d}{d\mu}\left[F_{\text{fin}}(s) + F_{\text{fin}}(t) + F_{\text{fin}}(u)\right] + \mathcal{O}(\lambda_{\text{phy}}^3)$$

Since $(dF_{\text{fin}}/d\mu) = -(1/16\pi^2)(2/\mu)$, this condition reduces to

$$\mu\frac{d\lambda_{\text{phys}}}{d\mu} = \frac{3}{16\pi^2}\lambda_{\text{phys}}^2 + \mathcal{O}(\lambda_{\text{phy}}^3) \equiv \beta\lambda_{\text{phys}}^2 + \mathcal{O}(\lambda_{\text{phy}}^3) \quad (4.178)$$

with $\beta \equiv (3/16\pi^2) > 0$. This is exactly the same "running" of the coupling constant obtained earlier Eq. (4.76). If we formally integrate this relation, we get[56]

[55]Algebra alert: For most purposes, we can set $\mu^{2\epsilon}$ in Eq. (4.175) to unity when $\epsilon \to 0$, which is what we have done here. But occasionally, the fact that $\mu^{2\epsilon} \approx 1 + \epsilon \ln \mu$ can be relevant to regularize certain results.

[56]If you expand the denominator of Eq. (4.179) in a Taylor series, you can easily see that this reduces to Eq. (4.74). Writing the result as in Eq. (4.179) involves pretending that Eq. (4.178) is exact. We shall comment on the domain of validity of expressions like Eq. (4.179) in a moment.

$$\lambda_{\mathrm{phys}}(\mu) = \frac{\lambda_{\mathrm{phys}}(\mu_0)}{1 - \dfrac{3}{16\pi^2}\lambda_{\mathrm{phys}}(\mu_0)\ln\dfrac{\mu}{\mu_0}} \equiv \frac{\lambda_{\mathrm{phys}}(\mu_0)}{1 - \beta\lambda_{\mathrm{phys}}(\mu_0)\ln\dfrac{\mu}{\mu_0}} \qquad (4.179)$$

If you expand the denominator of Eq. (4.179), it will lead to an expansion in powers of logarithms, given by:

$$\lambda_{\mathrm{phy}}(\mu) = \lambda_0 + \frac{3\lambda_0^2}{(4\pi)^2}\ln\frac{\mu}{\mu_0} + \frac{9\lambda_0^3}{(4\pi)^4}\ln^2\frac{\mu}{\mu_0} + \cdots \qquad (4.180)$$

where $\lambda_0 \equiv \lambda_{\mathrm{phy}}(\mu_0)$. The first term in this expansion is the coupling constant λ_0 at the scale μ_0 which we take as fixed. The second term with a single logarithm represents the momentum dependence arising from the one loop correction to the 4-point function. In fact, it is easy to see that the higher order terms provide an expansion in terms of the number of loops. An n-loop diagram will have $n+1$ vertices, giving rise to a factor λ^{n+1} and the n momentum integrals will give a factor $(4\pi)^{-2n}$. The momentum dependence of the type $\ln^n(\mu/\mu_0)$ arises from regions of integrals where all the loop momenta have similar scales.

Since the β function was obtained in a perturbation series, you may wonder whether it is legitimate to integrate the resulting equation Eq. (4.178) and obtain an expression like the one in Eq. (4.179) which contains arbitrary powers of the coupling constant. To see what is involved, let us write the the series expansion of Eq. (4.179) formally as:

$$\lambda_{\mathrm{phy}}(\mu) = \lambda_0\left[1 + \sum_{n=1}^{\infty} c_n(\mu)\lambda_0^n\right]; \quad c_n(\mu) = \left(\beta\ln\frac{\mu}{\mu_0}\right)^n \qquad (4.181)$$

In the case of the $\lambda\phi^4$ theory, $\beta = (3/16\pi^2) > 0$, but let us consider a more general situation for a moment. The series in Eq. (4.181) (as well as the result in Eq. (4.179)) can indeed be interpreted as arising due to summing up a subset of Feynman diagrams in a geometric progression. To check the limits of its validity, we need to know what the higher loop corrections do to this expansion. The effect of higher loop corrections to the β functions, it turns out, is to modify $c_n(\mu)$ to the form

$$c_n(\mu) = \left[\beta\ln\frac{\mu}{\mu_0} + \mathcal{O}(\ln\ln(\mu/\mu_0))\right]^n \qquad (4.182)$$

which, in turn, is equivalent to adding additional terms proportional to $\ln\ln\mu$ to the denominator of Eq. (4.179). This means the term with $(\ln\mu)^n$ is unaffected by higher order corrections to β, but at each order, c_n will also pick up terms of the order $(\ln\mu)^{n-1}$ and smaller. These are the terms which are missed when we use the one loop β function.

It follows that the re-summation of a special class of diagrams leading to something like Eq. (4.179) is useful when $\ln(\mu/\mu_0) \gg 1$; in this case, we are actually picking the dominant term at high energies at each order of n. But this, in turn, means that the procedure is really useful only when $\beta < 0$. This is required because we simultaneously demand $\ln(\mu/\mu_0) \gg 1$ (to validate the re-summation) and $\lambda_{\mathrm{phys}} \ll 1$ (to validate the perturbation theory).[57] This suggests that, by using the expression in Eq. (4.179), one can improve significantly the accuracy of the perturbative expansion at the scale μ, differing significantly from the scale μ_0 at which the original coupling constant λ_0 was defined. In fact, even at the one loop level we

[57] In both $\lambda\phi^4$ theory — and in QED, as we will see later — we have $\beta > 0$. But we do have $\beta < 0$, in asymptotically free theories like QCD, which makes the coupling constant smaller at higher energies. In such theories, it is computationally easier to isolate the divergent parts — containing the poles in $(d-4)$ — than to calculate the finite parts. The coefficient of the divergent part can be used to integrate the β function equation and thus obtain the running of the coupling constant in the leading log approximation. This turns out to be quite useful computationally.

notice that the effective coupling constant is not really λ_0 but $\lambda_0 \ln(\mu/\mu_0)$. So even if λ_0 was small at the scale μ_0, the factor $\lambda_0 \ln(\mu/\mu_0)$ can become large when μ is not comparable to μ_0. The denominator in Eq. (4.179) tells us that λ_{phy} can be actually small even if $\lambda_0 \ln(\mu/\mu_0)$ is large (if $\beta < 0$), thereby providing us with a better expansion parameter.

Let us now close the discussion on the running of m, which we came across at the end of the last section. Once we determine the running of λ, we can determine the running of m, from Eq. (4.152). Using the condition that m_0 cannot depend on the arbitrary scale λ, we get:

$$\mu \frac{\partial m_0^2}{\partial \mu} = 0 = \mu \frac{\partial m^2}{\partial \mu} \left[1 + \frac{\lambda}{32\pi^2 \epsilon} \right] + m^2 \mu \frac{\partial \lambda}{\partial \mu} \frac{1}{32\pi^2} \frac{1}{\epsilon} \qquad (4.183)$$

which can be solved to give:

$$\mu \frac{\partial m^2}{\partial \mu} = -m^2 \mu \frac{\partial \lambda}{\partial \mu} \frac{1}{32\pi^2} \frac{1}{\epsilon} \left[1 - \frac{\lambda}{32\pi^2} \frac{1}{\epsilon} \right] \qquad (4.184)$$

In this expression, we need to careful about evaluating $\mu(\partial \lambda/\partial \mu)$ because of the $(1/\epsilon)$ term. Using the first expression in Eq. (4.175), we have $\mu(\partial \lambda/\partial \mu) = -2\epsilon\lambda \left[1 - (3\lambda/32\pi^2\epsilon) \right]$, which leads to the result

$$
\begin{aligned}
\mu \frac{\partial m^2}{\partial \mu} &= -m^2(-2\epsilon\lambda) \left[1 - \frac{3\lambda}{32\pi^2} \frac{1}{\epsilon} \right] \frac{1}{32\pi^2} \frac{1}{\epsilon} \left[1 - \frac{\lambda}{32\pi^2} \frac{1}{\epsilon} \right] \\
&\Rightarrow \quad m^2 \left[\frac{\lambda}{16\pi^2} + \mathcal{O}(\lambda^2) \right]
\end{aligned} \qquad (4.185)
$$

where the last relation holds when $\epsilon \to 0$. This is precisely what we obtained earlier in Eq. (4.77) from the effective action approach. With this scaling, all physical results will be independent of the parameter μ.

All that remains is to evaluate the finite part, $F_{\text{fin}}(p^2)$ in Eq. (4.173). If we do not worry about the analytic structure, Riemann sheets etc., this can be done by using the result

$$\int_0^1 d\alpha \ln \left[1 + \frac{4}{b} \alpha(1 - \alpha) \right] = -2 + \sqrt{1+b} \ln \left(\frac{\sqrt{1+b}+1}{\sqrt{1+b}-1} \right) \qquad (4.186)$$

to obtain the final answer as:

$$F_{\text{fin}}(s) = \frac{1}{(4\pi)^2} \left\{ \ln \left(\frac{m^2 e^{\gamma_E}}{4\pi\mu^2} \right) + \sqrt{1 - \frac{4m^2}{s}} \ln \left[\frac{\sqrt{s - 4m^2} + \sqrt{s}}{\sqrt{s - 4m^2} - \sqrt{s}} \right] - 2 \right\} \qquad (4.187)$$

The original integral in Eq. (4.186) is valid only when $b > 0$, but we have quietly continued the result to all values of p^2. While the final result is correct, this misses some interesting physics contained in the branch cuts of the scattering amplitude, which is discussed in the Mathematical Supplement, Sect. 4.9.2.

4.8 Renormalized Perturbation Theory for the $\lambda\phi^4$ Model

In the previous sections, we studied the renormalization of the $\lambda\phi^4$ model in the perturbative approach. We started with the bare Lagrangian of the form

$$\mathcal{L} = \frac{1}{2}(\partial_m \phi_0)^2 - \frac{1}{2}m_0^2 \phi_0^2 - \frac{\lambda_0}{4!}\phi_0^4 \tag{4.188}$$

where we have now introduced the subscript 0 to make it explicit that this Lagrangian is written in terms of bare quantities. You have already seen that the bare parameters m_0 and λ_0 get renormalized to m_R and λ_R (now denoted with subscript R for emphasis) order by order in the perturbation theory. We haven't seen any evidence for the renormalization of the field ϕ itself, but it turns out that this will be required when we study two-loop diagrams (see Problem 9). Just to be formally correct, we have also put a subscript 0 to the field ϕ. We can now set up a perturbation expansion in λ_0 which will lead to the results discussed earlier.

The purpose of this brief section is to redo the analysis somewhat differently — using renormalized quantities and counter-terms — which has significant formal advantage. As you will see, a similar procedure works in the case of QED as well and learning it in the context of $\lambda\phi^4$ theory is somewhat simpler. In this spirit, we shall not work through the algebra again, but will merely borrow the results from the previous discussions, concentrating on the conceptual features.

We will begin by rescaling the field by $\phi_0 = Z^{1/2}\phi_R$ and introducing the three parameters $(\delta_Z, \delta_m, \delta_\lambda)$ by the definitions

$$Z = 1 + \delta_Z, \qquad m_0^2 Z = m_R^2 + \delta_m, \qquad Z^2 \lambda_0 = \lambda_R + \delta_\lambda \tag{4.189}$$

This reduces the Lagrangian to the form

$$\mathcal{L} = \frac{1}{2}(\partial_m \phi_R)^2 - \frac{1}{2}m_R^2 \phi_R^2 - \frac{\lambda_R}{4!}\phi_R^4 + \frac{1}{2}\delta_Z(\partial_m \phi_R)^2 - \frac{1}{2}\delta_m \phi_R^2 - \frac{\delta_\lambda}{4!}\phi_R^4 + \rho \tag{4.190}$$

where we have added a constant ρ to cancel out any other infinite *constants* we might encounter. This Lagrangian splits into two parts. We would expect the first three terms to remain finite at all orders of perturbation theory once we choose the constants $(\delta_Z, \delta_m, \delta_\lambda)$ and ρ appropriately. What is more, we will now do the perturbation theory with the renormalized coupling constant λ_R, which makes a lot of physical sense.[58] We already know the Feynman diagrams corresponding to the first three terms of the Lagrangian in Eq. (4.190). The three counter-terms involving $(\delta_Z, \delta_m, \delta_\lambda)$ will generate new Feynman diagrams of very similar structure, as indicated in Fig. 4.13.

With this preamble, let us again look at the four-point and two-point functions of the theory. In the four-point function, we have the standard diagrams which led to the results in Eq. (4.158) and Eq. (4.159). We will now have to add to it the first diagram in Fig. 4.13, contributing a term $-i\delta\lambda$. With a slight change in notation, the final amplitude in Eq. (4.158) can be written as

$$i\mathcal{M} = -i\lambda_R + (-i\lambda_R)^2 \left[i\mathcal{B}(s) + i\mathcal{B}(t) + i\mathcal{B}(u) \right] - i\delta_\lambda \tag{4.191}$$

where

$$\mathcal{B}(p^2) = \frac{i}{2} \int \frac{d^4 k}{(2\pi)^4} \frac{i}{k^2 - m_R^2 + i\epsilon} \frac{i}{(p_1 + p_2 + k)^2 - m_R^2 + i\epsilon} \tag{4.192}$$

We know that the integral for $\mathcal{B}(p^2)$ is infinite. As before, we shall handle it using dimensional regularization which will lead to (see Eq. (4.173) and

[58]Note, however, that at this stage we have not identified the renormalized parameters with the physical mass and the physical coupling constants, though they will be closely related.

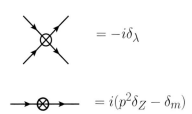

Figure 4.13: The diagrams generated by the counter-terms and their algebraic equivalents.

Eq. (4.170))

$$\mathcal{B}(p^2) = -\frac{1}{32\pi^2} \int_0^1 dx \left(\frac{2}{\epsilon} + \log\left[\frac{4\pi e^{-\gamma_E}\mu^2}{m_R^2 - x(1-x)p^2} \right] \right) \tag{4.193}$$

Substituting back, and doing a little bit of cleaning up, we get the final result to be

$$iM = -i\lambda_R + \frac{i\lambda_R^2}{32\pi^2} \int_0^1 dx \left[\frac{6}{\epsilon} + \right. \tag{4.194}$$
$$\left. \log\left(\frac{(4\pi e^{-\gamma_E}\mu^2)^3}{[m_R^2 - x(1-x)s][m_R^2 - x(1-x)t][m_R^2 - x(1-x)u]} \right) \right] - i\delta_\lambda$$

The integral in this expression is divergent, but we can cancel the divergence by choosing δ_λ appropriately. Since one could have subtracted any finite quantity along with the divergent result, the finite part — after eliminating the divergence — has some ambiguity. This ambiguity is usually resolved by choosing what is known as a subtraction scheme which we will now describe.

There are essentially three subtraction schemes which are used extensively in the literature. The first one, called minimal subtraction (MS) consists of cancelling out only the divergent part. This is done by choosing

$$\delta_\lambda = \frac{\lambda_R^2}{32\pi^2} \int_0^1 dx \frac{6}{\epsilon} = \frac{3\lambda_R^2}{16\pi^2\epsilon} \tag{4.195}$$

thereby leading to a well defined final result given by

$$iM = -i\lambda_R + \frac{i\lambda_R^2}{32\pi^2} \tag{4.196}$$
$$\times \int_0^1 dx \log\left(\frac{(4\pi e^{-\gamma_E}\mu^2)^3}{[m_R^2 - x(1-x)s][m_R^2 - x(1-x)t][m_R^2 - x(1-x)u]} \right)$$

The result, of course, depends on the arbitrary scale μ, but we know how to handle this by letting λ_R run with the scale.

The second subtraction scheme is a slight variant on MS usually called $\overline{\text{MS}}$ in which one takes out the 4π and γ_E. This involves the choice

$$\delta_\lambda = \frac{\lambda_R^2}{32\pi^2} \int_0^1 dx \left[\frac{6}{\epsilon} + 3\log(4\pi e^{\gamma_E}) \right] = \frac{\lambda_R^2}{32\pi^2} \left[\frac{6}{\epsilon} + 3\log(4\pi e^{\gamma_E}) \right] \tag{4.197}$$

leading to the final expression

$$iM = -i\lambda_R + \frac{i\lambda_R^2}{32\pi^2} \tag{4.198}$$
$$\times \int_0^1 dx \log\left(\frac{(\mu^2)^3}{[m_R^2 - x(1-x)s][m_R^2 - x(1-x)t][m_R^2 - x(1-x)u]} \right)$$

The third possibility is to choose δ_λ such that λ_R is actually a physically relevant observable quantity. To do this, we choose what is called a renormalization point at which the coupling constant λ_R is actually measured in some experiment. One choice, for example, could be $s_0 = 4m_P^2$, $t_0 = u_0 = 0$ where m_P is called the pole mass.[59] We then demand that, at the renor-

---

[59]This is not a priori equal to m_R; we will relate to m_R in a moment.

malization point, $\mathcal{M}$ should be equal to the observed value λ_R. This leads to the condition

$$-i\lambda_R = -i\lambda + \frac{i\lambda_R^2}{32\pi^2} \int_0^1 dx \left[\frac{6}{\epsilon} + \log\left(\frac{(4\pi e^{-\gamma_E}\mu^2)^3}{m_R^4 [m_R^2 - 4x(1-x)m_P^2]} \right) \right] - i\delta_\lambda \tag{4.199}$$

which can be solved for δ_λ. Substituting back, the result we get now is given by

$$\begin{aligned} i\mathcal{M} &= -i\lambda_R + \frac{i\lambda_R^2}{32\pi^2} \\ &\times \int_0^1 dx \log\left(\frac{m_R^4 [m_R^2 - 4x(1-x)m_P^2]}{[m_R^2 - x(1-x)s][m_R^2 - x(1-x)t][m_R^2 - x(1-x)u]} \right) \end{aligned} \tag{4.200}$$

Exercise 4.8: Prove Eqs. (4.196) to (4.200).

Notice that, in this procedure — in contrast to the MS and $\overline{\text{MS}}$ schemes — the final result is independent of the arbitrary scale μ. The result in Eq. (4.200) has a very direct physical meaning. In this expression, λ_R is *defined to* be the coupling constant measured in an experiment at the renormalization point characterized by the scale m_P. Once your experimentalist friend tells you what m_P and λ_R are, you can predict the value of $\mathcal{M}$ without any further arbitrariness.

Let us next consider what happens to the two-point function in this approach. Here, we will stick with the first two diagrams in Fig. 4.12 and add to them the diagram arising from the counter-term (given by the second one in Fig. 4.13). Adding up the relevant contributions and using the standard dimensional regularization, we can compute the resulting amplitude to be

$$\begin{aligned} i\mathcal{M} &= \frac{i}{p^2 - m_R^2 + i\epsilon} + \left(\frac{im_R^2\lambda_R}{32\pi^2} \left[\frac{2}{\epsilon} + 1 + \log\left(\frac{4\pi e^{-\gamma_E}\mu^2}{m_R^2} \right) \right] \right. \\ &\left. + i(p^2\delta_Z - \delta_m) \right) \left(\frac{i}{p^2 - m_R^2 + i\epsilon} \right)^2 \end{aligned} \tag{4.201}$$

We now have two parameters, δ_Z and δ_m, to play around with and remove the divergences. As in the previous case, we need to choose a subtraction scheme and the three schemes described earlier will again lead to three different final expressions. First, in the MS scheme, we cancel out only the divergent part which can be achieved by the choice

$$\delta_Z = 0, \quad \delta_m = \frac{m_R^2\lambda_R}{16\pi^2\epsilon} \tag{4.202}$$

leading to the result

$$i\mathcal{M} = \frac{i}{p^2 - m_R^2 + i\epsilon} + \frac{im_R^2\lambda_R}{32\pi^2} \left[1 + \log\left(\frac{4\pi e^{-\gamma_E}\mu^2}{m_R^2} \right) \right] \left(\frac{i}{p^2 - m_R^2 + i\epsilon} \right)^2 \tag{4.203}$$

The μ dependence of the result remains and has to be compensated by the running of m_R as we have seen earlier. In the $\overline{\text{MS}}$ scheme, we take out the 4π and γ_E as well, by choosing:

$$\delta_Z = 0, \quad \delta_m = \frac{m_R^2\lambda_R}{32\pi^2} \left[\frac{2}{\epsilon} + \log\left(4\pi e^{-\gamma_E} \right) \right] \tag{4.204}$$

which leads to the final result

$$i\mathcal{M} = \frac{i}{p^2 - m_R^2 + i\epsilon} + \frac{im_R^2\lambda_R}{32\pi^2}\left[1 + \log\left(\frac{\mu^2}{m_R^2}\right)\right]\left(\frac{i}{p^2 - m_R^2 + i\epsilon}\right)^2 \tag{4.205}$$

The third procedure will now involve interpreting m_R as the physical mass of the particle, which requires the propagator to have the structure

$$i\mathcal{M} = \frac{i}{p^2 - m_R^2 + i\epsilon} + (\text{terms that are regular at } p^2 = m_R^2) \tag{4.206}$$

Exercise 4.9: Prove Eqs. (4.202) to (4.208).

That is, we demand the final propagator to have a simple pole at m_R with a residue i. These conditions require the choice

$$\delta_Z = 0, \quad \delta_m = \frac{m_R^2\lambda_R}{32\pi^2}\left[\frac{2}{\epsilon} + 1 + \log\left(\frac{4\pi e^{-\gamma_E}\mu^2}{m_R^2}\right)\right] \tag{4.207}$$

thereby leading to the simple result where the pole mass is indeed m_R:

$$i\mathcal{M} = \frac{i}{p^2 - m_R^2 + i\epsilon} \tag{4.208}$$

Notice that we could do everything consistently keeping $\delta_Z = 0$. In other words, we do not have to perform any renormalization of the field strength ϕ to this order. This is not true at higher orders (see Problem 9), but with just the three counter-terms $(\delta_Z, \delta_m, \delta_\lambda)$ one can handle the divergences at all orders of perturbation. This is why you call the $\lambda\phi^4$ theory as renormalizable.

4.9   Mathematical Supplement

4.9.1 L_{eff} from the Vacuum Energy Density

We have seen earlier that L_{eff} can be thought of as the vacuum energy of the field oscillators (see Eq. (2.71)). It is possible to use this interpretation directly to obtain the effective Lagrangian for the electromagnetic field along the following lines.

First, note that, in the case of a pure magnetic field, the H in Eq. (4.21) reduces to the harmonic oscillator Hamiltonian. Using the standard eigenvalues for harmonic oscillators, we can sum over the zero point energies to obtain L_{eff}. Differentiating this expression twice with respect to m^2, summing up the resulting series and integrating again, one can identify the finite part which will lead to the result in Eq. (4.31) with $E = 0$. The expression for the pure electric field can be obtained by noting that L_{eff} must be invariant under the transformation $E \to iB$, $B \to -iE$. Getting the complete expression requires a little more more work: It involves noting that the eigenvalues of H in Eq. (4.21) depend on E only through the combination $\tau = (qE)^{-1}(m^2 + qB(2n + 1))$ and have the form $E_0 = \sum_{n=0}^{\infty}(2qB)G(\tau)$. This expression can be summed up using Laplace transform techniques, thereby leading to the result in Eq. (4.31).

It is easy to obtain L_{eff} for a pure magnetic field. Let us suppose that the background field satisfies the conditions $\mathbf{E} \cdot \mathbf{B} = 0$ and $\mathbf{B^2} - \mathbf{E^2} > 0$. In such a case, the field can be expressed as purely magnetic in some

Lorentz frame. Let $\mathbf{B} = (0, B, 0)$; we choose the gauge such that $A^i = (0, 0, 0, -Bx)$. The Klein-Gordon equation

$$[(i\partial_\mu - qA_\mu)^2 - m^2]\phi = 0 \tag{4.209}$$

can now be separated by taking

$$\phi(t, \mathbf{x}) = f(x) \exp i(k_y y + k_z z - \omega t). \tag{4.210}$$

where $f(x)$ satisfies the equation

$$\frac{d^2 f}{dx^2} + [\omega^2 - (qBx - k_z)^2] f = (m^2 + k_y^2) f \tag{4.211}$$

This can be rewritten as

$$-\frac{d^2 f}{d\xi^2} + q^2 B^2 \xi^2 f = \epsilon f \tag{4.212}$$

where

$$\xi = x - \frac{k_z}{qB}; \qquad \epsilon = \omega^2 - m^2 - k_y^2 \tag{4.213}$$

Equation Eq. (4.212) is that of a harmonic oscillator with mass $(1/2)$ and frequency $2(qB)$. So, if $f(x)$ has to be bounded for large x, the energy ϵ must be quantized:

$$\epsilon_n = 2(qB)\left(n + \frac{1}{2}\right) = \omega^2 - (m^2 + k_y^2) \tag{4.214}$$

Therefore, the allowed set of frequencies is

$$\omega_n = \left[m^2 + k_y^2 + 2qB\left(n + \frac{1}{2}\right)\right]^{1/2} \tag{4.215}$$

The ground state energy per mode is $2(\omega_n/2) = \omega_n$ because the complex scalar field has twice as many degrees of freedom as a real scalar field. The total ground state energy is given by the sum over all modes k_y and n. The weightage factor for the discrete sum over n in a magnetic field is obtained by the correspondence:

$$\frac{dk_x}{2\pi} \frac{dk_y}{2\pi} \rightarrow \sum_n \left(\frac{qB}{2\pi}\right) \frac{dk_y}{2\pi} \tag{4.216}$$

Hence, the ground state energy is

$$E_0 = \sum_{n=0}^{\infty} \left(\frac{qB}{2\pi}\right) \int_{-\infty}^{+\infty} \frac{dk_y}{(2\pi)} \left[(k_y^2 + m^2) + 2qB(n + \frac{1}{2})\right]^{1/2} = -L_{\text{eff}} \tag{4.217}$$

This expression, as usual, is divergent. To separate out a finite part we will proceed as follows: Consider the quantity

$$I \equiv -\left(\frac{2\pi}{qB}\right) \frac{\partial^2 E_0}{\partial(m^2)^2} = \left(\frac{2\pi}{qB}\right) \frac{\partial^2 L_{\text{eff}}}{\partial(m^2)^2}. \tag{4.218}$$

which can be evaluated in the following manner:

$$
\begin{aligned}
I &= \frac{1}{4}\sum_{n=0}^{\infty}\int_{-\infty}^{+\infty}\frac{1}{2\pi}\frac{dk_y}{\left[k_y^2+m^2+2qB(n+\frac{1}{2})\right]^{3/2}} \qquad (4.219) \\[2mm]
&= \frac{1}{4\pi}\sum_{n=0}^{\infty}\frac{1}{[m^2+2qB(n+\frac{1}{2})]} = -\frac{1}{4\pi}\sum_{n=0}^{\infty}\int_0^{\infty}d\eta\, e^{-\eta(m^2+2qB(n+\frac{1}{2}))} \\[2mm]
&= \frac{1}{4\pi}\int_0^{\infty}d\eta\, e^{-\eta m^2}e^{-qB\eta}\frac{1}{1-e^{-2qB\eta}} \\[2mm]
&= \frac{1}{4\pi}\int_0^{\infty}d\eta\frac{e^{-\eta m^2}}{e^{qB\eta}-e^{-qB\eta}} = \frac{1}{8\pi}\int_0^{\infty}d\eta\frac{e^{-\eta m^2}}{\sinh qB\eta} = \left(\frac{2\pi}{qB}\right)\frac{\partial^2 L_{\text{eff}}}{\partial(m^2)^2}
\end{aligned}
$$

The L_{eff} can be determined by integrating the expression twice with respect to m^2. We get[60]

$$
L_{\text{eff}} = \frac{qB}{(4\pi)^2}\int_0^{\infty}\frac{d\eta}{\eta^2}\frac{e^{-\eta m^2}}{\sinh qB\eta} = \int_0^{\infty}\frac{d\eta}{(4\pi)^2}\cdot\frac{e^{-\eta m^2}}{\eta^3}\cdot\frac{qB\eta}{\sinh qB\eta} \qquad (4.220)
$$

If the L_{eff} has to be Lorentz and gauge invariant then it can only depend on the quantities (E^2-B^2) and $\mathbf{E}\cdot\mathbf{B}$. We will again define two constants a and b by the relation

$$
a^2-b^2 = E^2-B^2; \qquad ab = \mathbf{E}\cdot\mathbf{B} \qquad (4.221)
$$

Then $L_{\text{eff}} = L_{\text{eff}}(a,b)$. In the case of the pure magnetic field we are considering, $a=0$ and $b=B$. Therefore, the L_{eff} can be written in a manifestly invariant way as:

$$
L_{\text{eff}} = \int_0^{\infty}\frac{d\eta}{(4\pi)^2}\cdot\frac{e^{-\eta m^2}}{\eta^3}\cdot\frac{qb\eta}{\sinh qb\eta} \qquad (4.222)
$$

Because this form is Lorentz invariant, it must be valid in any frame in which $E^2-B^2 < 0$ and $\mathbf{E}\cdot\mathbf{B}=0$. In all such cases,

$$
L_{\text{eff}} = \int_0^{\infty}\frac{d\eta}{(4\pi)^2}\frac{e^{-\eta m^2}}{\eta^3}\frac{q\eta\sqrt{B^2-E^2}}{\sinh q\eta\sqrt{B^2-E^2}}. \qquad (4.223)
$$

The L_{eff} for a pure electric field can be determined from this expression if we analytically continue the expression even for $B^2 < E^2$. We will find, for $B=0$,

$$
L_{\text{eff}} = \int_0^{\infty}\frac{d\eta}{(4\pi)^2}\frac{e^{-\eta m^2}}{\eta^3}\frac{q\eta E}{\sin q\eta E} \qquad (4.224)
$$

The same result can be obtained by noticing that a and b are invariant under the transformation $E\to iB, B\to -iE$. Therefore, $L_{\text{eff}}(a,b)$ must also be invariant under these transformations: $L_{\text{eff}}(E,B) = L_{\text{eff}}(iB,-iE)$. This allows us to get Eq. (4.224) from Eq. (4.220).

We will now consider the general case with arbitrary $\mathbf{E}$ and $\mathbf{B}$ for which a and b are not simultaneously zero. It is well known that by choosing our Lorentz frame suitably, we can make E and B parallel, say along the y-axis. We will describe this field $[\mathbf{E}=(0,E,0);\mathbf{B}=(0,B,0)]$ in the gauge $A_i = [-Ey,0,0,-Bx]$. The Klein-Gordon equation becomes

$$
[(i\partial_\mu - qA_\mu)^2 - m^2]\phi \qquad (4.225)
$$

$$
= \left[\left(i\frac{\partial}{\partial t}+qEy\right)^2 + \frac{\partial^2}{\partial x^2} + \frac{\partial^2}{\partial y^2} - \left(i\frac{\partial}{\partial z}+qBx\right)^2 - m^2\right]\phi = 0
$$

Separating the variables by assuming

$$\phi(t, \mathbf{x}) = f(x, y) \exp -i(\omega t - k_z z) \qquad (4.226)$$

we get

$$\left[\left(\frac{\partial^2}{\partial x^2} + \frac{\partial^2}{\partial y^2} \right) + (\omega + qEy)^2 - (k_z - qBx)^2 \right] f = m^2 f \qquad (4.227)$$

which separates out into x and y modes. Writing

$$f(x, y) = g(x)Q(y) \qquad (4.228)$$

where $g(x)$ satisfies the harmonic oscillator equation

$$\frac{d^2 g}{dx^2} - (k_z - qBx)^2 g = -2qB \left(n + \frac{1}{2} \right) g \qquad (4.229)$$

we get

$$\frac{d^2 Q}{dy^2} + (\omega + qEy)^2 Q = \left[m^2 + 2qB \left(n + \frac{1}{2} \right) \right] Q \qquad (4.230)$$

Changing to the dimensionless variable

$$\eta = \sqrt{qE} y + \frac{\omega}{\sqrt{qE}} \qquad (4.231)$$

we obtain

$$\frac{d^2 Q}{d\eta^2} + \eta^2 Q = \frac{1}{qE} \left(m^2 + 2qB(n + \frac{1}{2}) \right) Q. \qquad (4.232)$$

To proceed further, we use a trick. The above expression shows that the only dimensionless combination which occurs in the presence of the electric field is $\tau = (qE)^{-1}(m^2 + qB(2n + 1))$. Thus, purely from dimensional considerations, we expect the ground state energy to have the form

$$E_0 = \sum_{n=0}^{\infty} (2qB)G(\tau) \qquad (4.233)$$

where G is a function to be determined. Introducing the Laplace transform F of G, by the relation

$$G(\tau) = \int_0^{\infty} F(k)e^{-k\tau} dk \qquad (4.234)$$

we can write

$$L_{\text{eff}} = (2qB) \sum_{n=0}^{\infty} \int_0^{\infty} dk F(k) \exp \left[-\frac{k}{qE}(m^2 + qB(2n + 1)) \right] \qquad (4.235)$$

Summing the geometric series, we obtain

$$\begin{aligned} L_{\text{eff}} &= 2(qB)(qE) \int_0^{\infty} ds F(qEs)e^{-sm^2} e^{-qBs} \cdot \frac{1}{1 - e^{-2qBs}} \\ &= 2(qB)(qE) \int_0^{\infty} ds \frac{F(qEs)e^{-sm^2}}{e^{qBs} - e^{-qBs}} \\ &= (qB)(qE) \int_0^{\infty} ds \frac{F(qEs)}{\sinh qBs} e^{-m^2 s} \qquad (4.236) \end{aligned}$$

We now determine F by using the fact that L_{eff} must be invariant under the transformation $E \to iB, B \to -iE$. This means that

$$L_{\text{eff}} = (qB)(qE) \int_0^\infty ds\, e^{-m^2 s} \frac{F(iqBs)}{-\sinh(iqEs)} \tag{4.237}$$

Comparing the two expressions and using the uniqueness of the Laplace transform with respect to m^2, we get

$$\frac{F(qEs)}{\sinh qBs} = -\frac{F(iqBs)}{\sinh(iqEs)} \tag{4.238}$$

Or, equivalently,

$$F(qEs)\sin qEs = F(iqBs)\sin(iqBs) \tag{4.239}$$

Since each side depends only on either E or B alone, each side must be a constant independent of E and B. Therefore

$$F(qEs)\sin qEs = F(iqBs)\sin(iqBs) = \text{constant} = A(s) \tag{4.240}$$

giving

$$L_{\text{eff}} = (qB)(qE) \int_0^\infty ds \frac{e^{-m^2 s} A(s)}{\sin qEs \sinh qBs} \tag{4.241}$$

The $A(s)$ can be determined by comparing this expression with, say, Eq. (4.220) in the limit of $E \to 0$. We have

$$\begin{aligned}
L_{\text{eff}}(E=0, B) &= qB \int_0^\infty \frac{ds}{s} e^{-m^2 s} \cdot \frac{A(s)}{\sinh qBs} \\
&= qB \int_0^\infty \frac{ds}{(4\pi)^2 s^2} e^{-m^2 s} \frac{1}{\sinh qBs}
\end{aligned} \tag{4.242}$$

implying

$$A(s) = \frac{1}{(4\pi)^2 s} \tag{4.243}$$

Thus we arrive at the final answer

$$L_{\text{eff}} = \int_0^\infty \frac{ds}{(4\pi)^2} \frac{e^{-m^2 s}}{s^3} \left(\frac{qEs}{\sin qEs}\right) \left(\frac{qBs}{\sinh qBs}\right) \tag{4.244}$$

In the situation we are considering, $\mathbf{E}$ and $\mathbf{B}$ are parallel making $a^2 - b^2 = E^2 - B^2$ and $ab = \mathbf{E} \cdot \mathbf{B} = EB$. Therefore $E = a$ and $B = b$. Thus, our result can be written in a manifestly invariant form as

$$L_{\text{eff}}(a, b) = \int_0^\infty \frac{ds}{(4\pi)^2} \frac{e^{-m^2 s}}{s^3} \left(\frac{qas}{\sin qas}\right) \left(\frac{qbs}{\sinh qbs}\right) \tag{4.245}$$

This result will be now valid in any gauge or frames with a and b determined in terms of $(\mathbf{E^2} - \mathbf{B^2})$ and $(\mathbf{E} \cdot \mathbf{B})$.

4.9.2 Analytical Structure of the Scattering Amplitude

In the text, we computed the $\phi\phi \to \phi\phi$ scattering amplitude to the one loop order, ie., to $\mathcal{O}(\lambda^2)$. There is a curious connection between the imaginary part of any scattering amplitude evaluated at one loop order and the real part of the scattering amplitude at the tree-level. Since tree-level amplitudes do not have divergences, the amplitude we computed at one loop level can have a divergence only in the real part. We will first prove this connection which arises from the unitarity of the S-matrix and then briefly describe how it relates to the analytical structure of the amplitude we have computed.

Let us introduce the S-matrix somewhat formally and consider its structure. Given the fact that the states evolve according to the law $|\Psi; t\rangle = e^{-iHt}|\Psi; 0\rangle$ we can identify the S-matrix operator for a finite time t as $S = \exp(-itH)$. Since the Hamiltonian is Hermitian with $H^\dagger = H$, we have the unitarity condition for the S-matrix given by $S^\dagger S = 1$. It is conventional to write $S = 1 + iT$ where the T matrix will be related to the amplitude of transition between an initial and final states $\mathcal{M}(i \to f)$ by an overall momentum conservation factor:

$$\langle f|T|i\rangle = (2\pi)^4\, \delta^4(p_i - p_f)\mathcal{M}(i \to f) \qquad (4.246)$$

(This $\mathcal{M}$ and hence the T is what you would compute using, say, the Feynman diagrams). While S is unitary, T satisfies a more complicated constraint, viz.,

$$i(T^\dagger - T) = T^\dagger T \qquad (4.247)$$

Using Eq. (4.247) and Eq. (4.246) we immediately obtain:

$$\begin{aligned} \langle f|i(T^\dagger - T)|i\rangle &= i\langle i|T|f\rangle^* - i\langle f|T|i\rangle \qquad (4.248)\\ &= i(2\pi)^4\, \delta^4(p_i - p_f)\left(\mathcal{M}^*(f \to i) - \mathcal{M}(i \to f)\right) \end{aligned}$$

Introducing a complete set of states, we can express the right hand side of Eq. (4.247) as:

$$\begin{aligned} \langle f|T^\dagger T|i\rangle &= \sum_\psi \int d\Gamma_\psi \langle f|T^\dagger|\psi\rangle\langle\psi|T|i\rangle \\ &= \sum_\psi (2\pi)^4\, \delta^4(p_f - p_\psi)(2\pi)^4\, \delta^4(p_i - p_\psi) \\ &\qquad \times \int d\Gamma_\psi \mathcal{M}(i \to \psi)\mathcal{M}^*(f \to \psi) \qquad (4.249) \end{aligned}$$

Therefore, we conclude that the unitarity of the S-matrix implies the relation

$$\mathcal{M}(i \to f) - \mathcal{M}^*(f \to i) \qquad (4.250)$$
$$= i\sum_\psi \int d\Gamma_\psi (2\pi)^4\, \delta^4(p_i - p_\psi)\mathcal{M}(i \to \psi)\mathcal{M}^*(f \to \psi)$$

which goes under the name generalized optical theorem because of historical reasons. Notice that the left hand side is linear in the amplitudes $\mathcal{M}$ while the right hand side is quadratic in the amplitudes. This relation must hold order-by-order in any perturbation theory in some coupling constant, say, λ. This implies that, at $\mathcal{O}(\lambda^2)$, for example, the left hand side must be a loop amplitude if it has to match a tree-level amplitude on the right hand side.[61]

We will now compute the imaginary part of the scattering amplitude and determine its analytical structure. In the process, we will also be

[61] Therefore, the imaginary parts of loop amplitudes at a given order are related to products of tree-level amplitudes and a theory without loops will be inconsistent within this formalism.

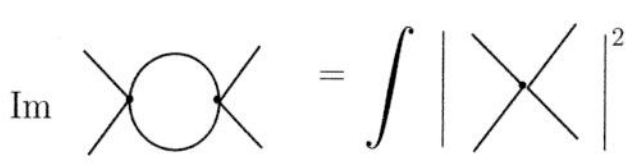

Figure 4.14: The result in diagrammatic form.

able to verify the above relation in this simple case — which, in terms of Feynman diagrams, is given in Fig. 4.14. Let us now get back to the explicit evaluation of $F_{\text{fin}}(p^2)$ in Eq. (4.173). We see that the integrand in Eq. (4.173) involves the logarithm of the factor $m^2 - \alpha(1-\alpha)p^2$. We write this expression as

$$m^2 - \alpha(1-\alpha)p^2 = m^2 + \alpha(1-\alpha)\mathbf{p}^2 - \alpha(1-\alpha)p_0^2 \qquad (4.251)$$

The sum of the first two terms is greater than or equal to m^2 for $0 \le \alpha \le 1$ while the second term is less than or equal to p_0^2. Therefore, the whole expression is greater than or equal to $(m^2 - p_0^2)$, making the argument of the logarithm positive in the original range $0 < p_0 < m$ which we considered and also, obviously, for $p^2 < 0$. The logarithm is well defined in this range and our aim now is to analytically continue the expression for all values of p_0. To do this when the argument of the logarithm becomes negative, we have to use the proper Riemann sheet to define the function in the entire complex plane. We will now show how this can be done.

We continue working in the Euclidean space (and have again dropped the subscript E in p_E^2) and do an integration by parts to obtain

$$\int_0^1 d\alpha \, \ln\left[\frac{m^2 + \alpha(1-\alpha)p^2}{4\pi\mu^2}\right] = \ln\left(\frac{m^2}{4\pi\mu^2}\right) - p^2 \int_0^1 d\alpha \, \frac{\alpha(1-2\alpha)}{m^2 + \alpha(1-\alpha)p^2} \qquad (4.252)$$

Introducing a Dirac delta function into the integrand by

$$1 = \int_0^\infty ds' \, \delta\left(s' - \frac{m^2}{\alpha(1-\alpha)}\right) \qquad (4.253)$$

we can write the integral in the form of a dispersion relation:

$$-\int_0^1 d\alpha \, \frac{\alpha(1-2\alpha)}{m^2 + \alpha(1-\alpha)p^2} = \int_0^\infty ds' \, \frac{\rho(s')}{s' - s} \qquad (4.254)$$

where

$$\begin{aligned}
\rho(s') &= -\int_0^1 d\alpha \, \frac{1-2\alpha}{1-\alpha} \, \delta\left(s' - \frac{m^2}{\alpha(1-\alpha)}\right) \\
&= -\frac{1}{s'} \int_0^1 d\alpha \, \alpha(1-2\alpha) \, \delta\left(\alpha(1-\alpha) - \frac{m^2}{s'}\right) \qquad (4.255)
\end{aligned}$$

and we have set $p^2 = -s$. Factorizing the quadratic form in the delta function, it is easy to see that $\rho(s')$ is given by

$$\rho(s') = \begin{cases} 0 & (s' < 4m^2), \\[2ex] \dfrac{1}{s'}\sqrt{1 - \dfrac{4m^2}{s'}} & (s' \ge 4m^2) \end{cases} \qquad (4.256)$$

Since s' plays a role similar to m^2, the prescription $m^2 \to m^2 - i\epsilon$ is the same as an $i\epsilon$ prescription on s. Using this, we can express our result in the form

$$F_{\text{fin}}(s) = \frac{1}{(4\pi)^2} \ln\left(\frac{m^2 e^{\gamma_E}}{4\pi\mu^2}\right) + \frac{s}{(4\pi)^2} \int_{4m^2}^\infty ds' \, \frac{\rho(s')}{s - s' + i\epsilon} \qquad (4.257)$$

We now see, using the result

$$\text{Im} \ \frac{1}{x - i\epsilon} = \pi \, \delta(x) \tag{4.258}$$

for $s > 4m^2$, that the imaginary part of the amplitude is given by

$$\text{Im} \ F_{\text{fin}}(s) = -\frac{s}{16\pi} \rho(s) = -\frac{1}{16\pi} \sqrt{1 - \frac{4m^2}{s}} \tag{4.259}$$

It is easy to show, by explicit computation, that this expression can also be written in the form

Exercise 4.10: Do this.

$$\text{Im} \ F(p) = -\frac{1}{2} \int \frac{d^3\mathbf{k}}{(2\pi)^3 2E_{\mathbf{k}}} \frac{d^3\mathbf{q}}{(2\pi)^3 2E_{\mathbf{q}}} (2\pi)^4 \delta(p - k - q) \equiv -\frac{1}{2} \Gamma_{(2)}(p) \tag{4.260}$$

where $\Gamma_{(2)}(p)$ is the volume of the two-body phase space. From this relation, it is easy to verify that our general theorem relating the imaginary part of the loop amplitude to the real part of tree-level amplitude does hold.

Further, because of the analytic structure of the scattering amplitude, F_{fin} is uniquely determined by: (i) its discontinuity across the cut, viz., $2i \, \text{Im} F_{\text{fin}}$, (ii) its asymptotic behaviour and (iii) its value at some point. The asymptotic behaviour is given by

$$\lim_{|s|\to\infty} \ s \int_{4m}^{\infty} ds' \frac{\rho(s')}{s - s'} = \lim_{|s|\to\infty} \ s \int_{\text{const.}}^{\infty} ds' \left[\frac{1}{s'} - \frac{1}{s' - s} \right] = \ln s + \text{const.} \tag{4.261}$$

while the value at $s = 0$ is given by

$$F_{\text{fin}}(0) = \frac{1}{(4\pi)^2} \ln \left(\frac{m^2 e^{\gamma_E}}{4\pi\mu^2} \right) \tag{4.262}$$

A function which has all these properties is indeed given by:

$$F_{\text{fin}}(s) = \frac{1}{(4\pi)^2} \left\{ \ln \left(\frac{m^2 e^{\gamma_E}}{4\pi\mu^2} \right) + \sqrt{1 - \frac{4m^2}{s}} \ln \left[\frac{\sqrt{s - 4m^2} + \sqrt{s}}{\sqrt{s - 4m^2} - \sqrt{s}} \right] - 2 \right\} \tag{4.263}$$

which is, therefore, the required expression for the finite part of the scattering amplitude.

Chapter 5

Real Life II: Fermions and QED

5.1 Understanding the Electron

The formalism of quantum field theory we have developed in the previous chapters allows us to think of (what we usually call) particles as quantum excitations of (what we usually call) fields. The clearest example is that of a photon which arises as an excitation of the electromagnetic field. We have also discussed how we can describe the interaction of the particles, say, e.g., their scattering, starting from a field theoretical description. We did all these using a scalar field and the electromagnetic field as prototypes.

It turns out that we run into new issues when we try to describe *fermionic* particles — the simplest and the most important one being the electron — using this formalism. We need to cope with several new features and extend our formalism to cover these. The purpose of this — rather long — last chapter is to orient you towards understanding these issues and illustrate the (suitably extended) formalism in the case of quantum electrodynamics, which describes the quantum interaction of electrons with photons.

The first issue in dealing with the electron, which hits you even in NRQM, is that it is a spin-half particle. A spinless non-relativistic particle can be studied using the Schrodinger equation $i\dot{\psi} = H\psi$ where $|\psi(t, \boldsymbol{x})|^2$ will give you the probability to find the particle around the event $(t, \boldsymbol{x})$. But an electron could be found around this event with spin up (along, say, the $z-$axis) or spin down or, more generally, in a linearly superposed spin state. So, at the least, we need to describe it with a two component wave function $(\psi_+(t, \boldsymbol{x}), \psi_-(t, \boldsymbol{x}))$ corresponding to the spin up and spin down states.

If you think of (ψ_+, ψ_-) as a column vector, then the Hamiltonian H in the Schrodinger equation $i\dot{\psi} = H\psi$ probably should be some kind of a 2×2 matrix. What is more, the doublet (ψ_+, ψ_-) must have some specific transformation properties when we rotate the coordinate systems, changing the direction of the $z-$axis. (Such a non-trivial transformation of the wave function was not required in the study of spinless particles in NRQM.)[1]

What we have said so far, in some sense, is just the kinematic aspect of the spin. Spin also has a dynamical aspect in the sense that it carries

---

[1] You might have learnt all these in a QM course and might have even known that the description of spin is quite closely related to the 3-dimensional rotation group $SO(3)$ or — more precisely — to $SU(2)$. If not, don't worry. We will do all these and more as we go along.

with it a magnetic moment, which couples to any external magnetic field. Observations tell us that the magnetic moment associated with the spin angular momentum $s = \hbar/2$ is given by $\mu = e\hbar/2m = (e/2m)(2s)$. On the other hand, the orbital angular momentum L of a charged particle (like an electron) produces a magnetic moment[2] given by $(e/2m)L$. So the spin angular momentum couples with twice as much strength as the orbital angular momentum. Why the electron spin should behave like an angular momentum, and why it should produce a coupling with twice as much strength are questions for which NRQM has no clear answer; they just need to be taken as phenomenological inputs while constructing the Hamiltonian for the electron in an electromagnetic field. As we shall see, making things consistent with relativity helps in this matter to a great extent.

The second dynamical aspect related to electron (and its spin) is a lot more mysterious. To bring this sharply in focus, let us consider a system with two electrons (like e.g., the Helium atom). Suppose you manage to solve for the energy eigenstates of the (non-relativistic) Hamiltonian for this system, obtaining a set of wave functions $\psi_E(\boldsymbol{x}_1, \boldsymbol{x}_2)$. (We have suppressed the spin indices for the sake of simplicity.) Your experimental friend will tell you that not all these solutions are realized in nature. Only those solutions in which the wave function is antisymmetric with respect to the exchange $\boldsymbol{x}_1 \Leftrightarrow \boldsymbol{x}_2$ occur in nature. This is a characteristic property of all fermions (including the electron) which should also be introduced as an additional principle[3] (the Pauli exclusion principle) in NRQM.

You could easily see that Pauli exclusion principle is going to throw a spanner into the works when we try to think of an electron as a one-particle excitation of a field, if we proceed along the lines we did for, say, the photon. In that approach, we start with a vacuum state $|0\rangle$, and act on it by a creation operator $a_\alpha^\dagger$ to produce a one-particle state $a_\alpha^\dagger|0\rangle \propto |1_\alpha\rangle$, act on $|0\rangle$ twice to produce a two-particle state $a_\alpha^\dagger a_\beta^\dagger|0\rangle \propto |1_\alpha, 1_\beta\rangle$ etc. Here, α, β etc. specify the properties of the excitations in that state like the spin, momentum etc. Since $a_\alpha^\dagger$ commutes with $a_\beta^\dagger$, these states are necessarily *symmetric* with respect to the interchange of particles; i.e., $|1_\alpha, 1_\beta\rangle = |1_\beta, 1_\alpha\rangle$. In particular, one can take the limit of $\alpha \to \beta$ leading to a state with two particles with the same properties (spin, momentum ...), by using $(a_\alpha^\dagger)^2|0\rangle \propto |2_\alpha\rangle$.

This procedure simply won't do if we have to create electrons starting from the vacuum state. To begin with we need $|1_\alpha, 1_\beta\rangle = -|1_\beta, 1_\alpha\rangle$ to ensure antisymmetry. If we denote the creation operator for electron by $b_\alpha^\dagger$, then we must have $b_\alpha^\dagger b_\beta^\dagger = -b_\beta^\dagger b_\alpha^\dagger$. So the creation operator for the fermions must *anti-commute* rather than commute:

$$\{b_\alpha^\dagger, b_\beta^\dagger\} = 0 \tag{5.1}$$

In particular, this ensures that $(b_\alpha^\dagger)^2|0\rangle = 0$ so that you cannot create two electrons in the same state α thereby violating the Pauli exclusion principle.

A couple of operators anticommuting — rather than commuting — may not disturb you too much, given the sort of strange things you have already seen in QFT. But the situation is a bit more serious than that. Recall that we build the field by multiplying the creation and annihilation operators by functions like $\exp(\pm ipx)$. This form of the mode function is dictated by translational invariance and the fact that we are using the momentum basis; you need this mode function to describe particles with momentum

[3]This behaviour of the fermions can be understood from fairly sophisticated arguments of quantum field theory. Occasionally someone comes up with a fraudulent "proof" of this result within NRQM, which never stands up to close scrutiny. No amount of your staring at the relevant non-relativistic Hamiltonian will allow you to discover the Pauli exclusion principle. It is also not like the relativistic corrections to a non-relativistic Hamiltonian which can be understood in some Taylor series in v/c. The "strength" of the Pauli exclusion principle cannot be measured in terms of some small parameter in the problem which takes you from QFT to QM continuously! Here is an $\mathcal{O}(1)$ effect which manifests itself in the approximate theory and is actually a relic of the more exact theory. *This is a peculiar situation with no other parallel in theoretical physics.* When a result can be stated quite simply, but requires very elaborate machinery for its proof, it is usually because we have not understood the physics at an appropriately deep level.

$\boldsymbol{k}$ and energy $\omega_k \equiv (\boldsymbol{k}^2 + m^2)^{1/2}$. So, for the field $\psi(x)$ describing the electron, you would expect an expansion which goes something like[4]

$$\psi(x) = \sum_\alpha \int \frac{d^3\boldsymbol{k}}{(2\pi)^3 \sqrt{2\omega_k}} \left(b_{\alpha\boldsymbol{k}} u_{\alpha\boldsymbol{k}} \; e^{-ikx} + d^\dagger_{\alpha\boldsymbol{k}} v_{\alpha\boldsymbol{k}} \; e^{ikx} \right) \tag{5.2}$$

Here we have separated out the spin variable α from the momentum variable $\boldsymbol{k}$ and introduced a creation operator $b^\dagger_{\alpha\boldsymbol{k}}$ capable of creating an electron (with spin orientation α and momentum $\boldsymbol{k}$) and an operator $d^\dagger_{\alpha\boldsymbol{k}}$ for creating its antiparticle, viz., the positron. The c-number coefficients $u_{\alpha\boldsymbol{k}}$ and $v_{\alpha\boldsymbol{k}}$ take care of the normalization and the correct Lorentz transformation properties in the presence of spin. (This is just kinematics — but important kinematics — which we will describe in detail later on.) The key point to note is that $\psi(x)$ will inherit the *anti-commuting* properties of $b^\dagger_{\alpha\boldsymbol{k}}$ and $d_{\alpha\boldsymbol{k}}$. In fact, we would expect something like $\psi^2(x) = 0$ to hold for the field describing an electron, which is rather strange.[5]

We will now take up these issues and develop a field-theoretic formalism for electrons. For this purpose, we will start by asking fairly innocuous questions.

5.2 Non-relativistic Square Roots

5.2.1 Square Root of $p_\alpha p^\alpha$ and the Pauli Matrices

We have seen that the two key new features related to electrons involve (i) the relation between its spin and magnetic moment and (ii) some kind of inherent anticommutativity of the operators. There is a rather curious way in which these ideas can be introduced even in NRQM which will help us to understand the relativistic aspects better. With this motivation, we will start by asking a question which, at first sight, has nothing to do with the electron (or even with physics!).

Consider a 3-vector, say, the momentum $\boldsymbol{p}$ of a particle with components p^α in a Cartesian coordinate system. The magnitude of this vector $|\boldsymbol{p}| \equiv (-\eta_{\alpha\beta} p^\alpha p^\beta)^{1/2}$ scales[6] linearly with its components in the sense that, if you rescale all the components by a factor λ, the magnitude $|\boldsymbol{p}|$ changes by the same factor λ. This prompts us to ask the question: Can we write the magnitude as a linear combination of the components in the form $|\boldsymbol{p}| \equiv (-\eta_{\alpha\beta} p^\alpha p^\beta)^{1/2} = A_\alpha p^\alpha$ with some triplet A_α that is independent of the vector in question? For this result to hold, we must have [7]

$$-\eta_{\alpha\beta} p^\alpha p^\beta = A_\alpha A_\beta p^\alpha p^\beta = \frac{1}{2} \{A_\alpha, A_\beta\} p^\alpha p^\beta \tag{5.3}$$

which requires the triplet A_α to satisfy the *anti-commutation* rule

$$\{A_\alpha, A_\beta\} = -2\eta_{\alpha\beta} = 2\delta_{\alpha\beta} \tag{5.4}$$

Obviously, the A_αs cannot be ordinary numbers if they have to behave in this way. The simplest mathematical structures you would have come across, which will not commute under multiplication, are matrices. So we might ask whether there exist three matrices A_α which satisfy these relations. Indeed there are, as you probably know from your NRQM course,

[4]This is exactly in analogy with what we did for the complex scalar field in Sect. 3.5, Eq. (3.148) and for the photon in Sect. 3.6.1 , Eq. (3.172). For the complex scalar field, we did not have to worry about the spin while for the photon we did not have to worry about a distinct antiparticle. For the electron, we need to introduce both the complications.

[5]Unlike the electromagnetic field which *definitely* has a classical limit or a scalar field for which we can cook up a classical limit if push comes to shove, there is no way you are going to have a *classical* wave field the quanta of which will be the electrons — or any fermions for that matter. Fermions, in that sense, are rather strange beasts of which the world is made of.

[6]With our signature, $(-\eta_{\alpha\beta})$ is just a fancy way of writing a *unit* 3×3 matrix.

[7]In this discussion, we will not distinguish between the subscripts and superscripts of Greek indices corresponding to spatial coordinates when no confusion is likely to arise.

and these are just the standard Pauli matrices; that is, $A_\alpha = \sigma_\alpha$ where, for definiteness, we take them to be:

$$\sigma_x = \begin{pmatrix} 0 & 1 \\ 1 & 0 \end{pmatrix}, \quad \sigma_y = \begin{pmatrix} 0 & -i \\ i & 0 \end{pmatrix}, \quad \sigma_z = \begin{pmatrix} 1 & 0 \\ 0 & -1 \end{pmatrix} \quad (5.5)$$

Before proceeding further, let us pause and recall some properties of the Pauli matrices which will be useful in our discussion.

(1) They satisfy the following identity: (Again, as far as Pauli matrices are concerned, we will not worry about the placement of the index as superscript or subscript; $\sigma_\alpha = \sigma^\alpha$ etc., by definition.)

$$\sigma_\alpha \sigma_\beta = \delta_{\alpha\beta} + i\epsilon_{\alpha\beta\gamma}\sigma^\gamma \quad (5.6)$$

From which it follows that[8]:

$$[\sigma_\alpha, \sigma_\beta] = 2i\epsilon_{\alpha\beta\gamma}\sigma^\gamma; \quad \{\sigma_\alpha, \sigma_\beta\} = 2\delta_{\alpha\beta} \quad (5.7)$$

(2) For any pair of vectors $\boldsymbol{a}, \boldsymbol{b}$ we have:

$$(\boldsymbol{\sigma} \cdot \boldsymbol{a})(\boldsymbol{\sigma} \cdot \boldsymbol{b}) = \boldsymbol{a} \cdot \boldsymbol{b} + i\boldsymbol{\sigma} \cdot (\boldsymbol{a} \times \boldsymbol{b}) \quad (5.8)$$

(3) Any *arbitrary* function of $Q \equiv a + \boldsymbol{b} \cdot \boldsymbol{\sigma}$ can be reduced to a linear function of $\boldsymbol{\sigma}$! That is, for any arbitrary[9] function $f(Q)$, we have

$$f(Q) = A + \boldsymbol{B} \cdot \boldsymbol{\sigma} \quad (5.9)$$

with

$$A = \frac{1}{2}[f(a+b) + f(a-b)], \quad \boldsymbol{B} = \frac{\boldsymbol{b}}{2b}[f(a+b) - f(a-b)] \quad (5.10)$$

Getting back to our discussion of square roots, we see that the square root of $|\boldsymbol{p}|^2$ can be thought of as the 2×2 *matrix*, $p \equiv \sigma_\alpha p^\alpha$, in the sense that $(\sigma_\alpha p^\alpha)(\sigma_\beta p^\beta) = |\boldsymbol{p}|^2$. Since the left hand side is a 2×2 matrix, we need to interpret such relations as matrix identities, introducing the unit 2×2 matrix wherever required.

5.2.2 Spin Magnetic Moment from the Pauli Equation

This might appear to be an amusing curiosity, but it will soon turn out to be more than that. We know that the Hamiltonian for a free electron is usually taken to be $H = \boldsymbol{p}^2/2m = (-i\partial_\alpha)^2/2m$. But with our enthusiasm for taking novel square roots, one could also write this free-particle Hamiltonian and the resulting Schrodinger equation[10] in the form:

$$H = \frac{(-i\boldsymbol{\sigma} \cdot \boldsymbol{\nabla})(-i\boldsymbol{\sigma} \cdot \boldsymbol{\nabla})}{2m}; \quad i\dot{\psi} = \frac{(-i\sigma^\alpha \partial_\alpha)(-i\sigma^\beta \partial_\beta)}{2m}\psi \quad (5.11)$$

Because of the identity Eq. (5.8) and the trivial fact $\boldsymbol{p} \times \boldsymbol{p} = 0$, it does not matter whether you write $\boldsymbol{p}^2$ or $(\boldsymbol{p} \cdot \boldsymbol{\sigma})(\boldsymbol{p} \cdot \boldsymbol{\sigma})$ in the Hamiltonian for a *free* electron. But suppose we use this idea to describe — not a free electron — but an electron in a magnetic field $\boldsymbol{B} = \boldsymbol{\nabla} \times \boldsymbol{A}$. We know that the gauge invariance requires the corresponding Hamiltonian to be obtained by the

[8]Mnemonic: You know that $\boldsymbol{J} = \boldsymbol{\sigma}/2$ satisfies the standard angular momentum commutation rules $[J_\alpha, J_\beta] = i\epsilon_{\alpha\beta\gamma}J_\gamma$ which explains the factor 2.

[9]You would have seen a special case of this expressed as $\exp -[(i\theta/2)(\boldsymbol{\sigma} \cdot \boldsymbol{n})] = \cos(\theta/2) - i(\boldsymbol{\sigma} \cdot \boldsymbol{n})\sin(\theta/2)$ in NRQM; but it is more general.

[10]Or rather, the Pauli equation, after Pauli who came up with this.

replacement $\boldsymbol{p} \to \boldsymbol{p} - e\boldsymbol{A}$. So we now have to describe an electron using the Hamiltonian

$$
\begin{aligned}
H_{em} &= \frac{1}{2m}\boldsymbol{\sigma}\cdot(\boldsymbol{p}-e\boldsymbol{A})\,\boldsymbol{\sigma}\cdot(\boldsymbol{p}-e\boldsymbol{A}) \\
&= \frac{1}{2m}(\boldsymbol{p}-e\boldsymbol{A})^2 + \frac{i}{2m}\boldsymbol{\sigma}\cdot[(\boldsymbol{p}-e\boldsymbol{A})\times(\boldsymbol{p}-e\boldsymbol{A})] \\
&= \frac{1}{2m}(\boldsymbol{p}-e\boldsymbol{A})^2 - \frac{e\hbar}{2m}\boldsymbol{\sigma}\cdot\boldsymbol{B}
\end{aligned}
\tag{5.12}
$$

where we have used the identity

$$
(\boldsymbol{p}\times\boldsymbol{A}) + (\boldsymbol{A}\times\boldsymbol{p}) = -i\hbar(\boldsymbol{\nabla}\times\boldsymbol{A})
\tag{5.13}
$$

The first term in the Hamiltonian in Eq. (5.12) is exactly what you would have obtained if you had used $H = \boldsymbol{p}^2/2m$ as the Hamiltonian and made the replacement $\boldsymbol{p} \to \boldsymbol{p} - e\boldsymbol{A}$; that would have been the correct description of a *spinless* charged particle in a magnetic field. For a constant magnetic field $\boldsymbol{B}$ with $\boldsymbol{A} = (1/2)\boldsymbol{B}\times\boldsymbol{x}$, this term will expand to give

$$
\begin{aligned}
\frac{1}{2m}(\boldsymbol{p}-e\boldsymbol{A})^2 &= \frac{1}{2m}\left(\boldsymbol{p}-\frac{e}{2}\boldsymbol{B}\times\boldsymbol{x}\right)^2 \approx \frac{p^2}{2m} - \frac{e}{2m}(\boldsymbol{B}\times\boldsymbol{x})\cdot\boldsymbol{p} \\
&\simeq \frac{p^2}{2m} - \frac{e}{2m}\boldsymbol{B}\cdot\boldsymbol{L} + \mathcal{O}(B^2)
\end{aligned}
\tag{5.14}
$$

Exercise 5.1: Prove the following claim. One needs to be careful in general, while computing $(\boldsymbol{p}-e\boldsymbol{A})^2$, because $\boldsymbol{A}(\boldsymbol{x})$ and $\boldsymbol{p}$ won't commute. But when we use the gauge $\nabla\cdot\boldsymbol{A}=0$, as we have done, $\boldsymbol{A}\cdot\boldsymbol{p}=\boldsymbol{p}\cdot\boldsymbol{A}$ and we can be careless.

with $\boldsymbol{L} = \boldsymbol{x}\times\boldsymbol{p}$ being the orbital angular momentum. One can see that, to the linear order in the magnetic field, the orbital angular momentum contributes a magnetic moment $\boldsymbol{m} = (e/2m)\boldsymbol{L}$.

But we have a surprise in store when we look at the second term in Eq. (5.12). This tells you that the spin $\boldsymbol{s} \equiv (\hbar/2)\boldsymbol{\sigma}$ of the electron contributes a magnetic moment $\boldsymbol{m} = (e/m)\boldsymbol{s}$, making the coupling of the spin to the magnetic field is twice as strong as that of orbital angular momentum. When we take into account both the orbital and spin angular momentum, the relevant coupling term in the Hamiltonian will be given by[11]

$$
H_{\text{coupling}} = -\frac{e}{2m}(\boldsymbol{L}+2\boldsymbol{s})\cdot\boldsymbol{B}
\tag{5.15}
$$

This is precisely what is seen in spectroscopic observations. So our trick of taking the square root by $\sqrt{\boldsymbol{p}^2}=\boldsymbol{p}\cdot\boldsymbol{\sigma}$ and constructing the Hamiltonian from this, produces the correct coupling of the electron spin to the magnetic field observed in the lab!

[11] The ratio between the magnetic moment and the angular momentum is usually written as $eg/2m$ and we see that $g=1$ for the orbital angular momentum while $g=2$ for the electron spin. (Actually it is $g=2+(\alpha/\pi)$ but that part of the story comes later.) In normal units, the ratio is $(e\hbar/2mc)g$.

5.2.3 How does ψ Transform?

So, clearly, the way you take the square root of $\boldsymbol{p}^2$ changes physics and it is worth looking more closely[12] at the modification in Eq. (5.11). Our modification of the Schrodinger equation will require interpreting ψ as a two-component column vector ψ_A (called a *spinor*) with $A=1,2$. The first worry you should have with such an equation is what happens when we rotate the coordinate system by the standard linear transformation $x^\alpha \to x'^\alpha$, where

[12] More precisely, we should be looking at Eq. (5.12) but this distinction is unimportant for what we are going to do.

$$
x'^\alpha = M_\gamma{}^\alpha x^\gamma; \quad M_\gamma{}^\alpha \equiv \frac{\partial x'^\alpha}{\partial x^\gamma}; \quad M_\gamma{}^\alpha M_\alpha{}^\beta = \delta_\gamma^\beta
\tag{5.16}
$$

which leaves $x_\alpha x^\alpha$ invariant. For a general rotation, the form of the matrix $M_\gamma{}^\alpha$ is a bit complicated, but — to study the invariance properties — we can get away by just looking at infinitesimal rotations. Under such a rotation, by an infinitesimal angle θ around an axis represented by a unit vector $\boldsymbol{n}$, the coordinates change according to $\boldsymbol{x}' = \boldsymbol{x} + (\theta)\boldsymbol{n} \times \boldsymbol{x} \equiv \boldsymbol{x} + \boldsymbol{\theta} \times \boldsymbol{x}$ where $\boldsymbol{\theta} \equiv \theta\boldsymbol{n}$. This can be written in component form as:

$$x'^\alpha = x^\alpha + [\boldsymbol{\theta} \times \boldsymbol{x}]^\alpha = x^\alpha + \epsilon^\alpha{}_{\beta\gamma}\theta^\beta x^\gamma \equiv x^\gamma[\delta^\alpha_\gamma + \omega_\gamma{}^\alpha] \equiv M_\gamma{}^\alpha x^\gamma \quad (5.17)$$

where we have defined

$$\omega_{\gamma\alpha} \equiv \epsilon_{\gamma\alpha\beta}\theta^\beta; \quad \theta^\mu = \frac{1}{2}\epsilon^{\mu\alpha\beta}\omega_{\alpha\beta} \qquad (5.18)$$

The antisymmetry of $\omega_{\alpha\beta}$ is required for (and guarantees) $|\boldsymbol{x}|^2$ to remain invariant under such a rotation.

Before proceeding further, we should confess to the crime we are committing by talking about "rotations about an axis"[13] . To every rotation described by three parameters (say, $\theta_\alpha \equiv \theta n_\alpha$ with $\boldsymbol{n}^2 = 1$) we can also associate a 3-dimensional antisymmetric tensor, $\omega_{\alpha\gamma}$, defined in Eq. (5.18) with three independent components, such that $\omega_{12} = \theta n_3$ etc. While θn_3 makes you think of "rotation about the z-axis", the ω_{12} makes you think of "rotation in the xy-plane", which is more precise and generalizes easily to $D = 4$.

For this approach to give sensible results, we would like quantities like $\sigma^\alpha \partial_\alpha \psi$ to transform like ψ does; that is, the σ^α should behave "like a vector" under rotation. The operator ∂_α does change under the rotation to $\partial'_\beta = (\partial x^\alpha/\partial x'^\beta)\partial_\alpha$ but the σ^α, being a mere set of numbers, does *not* change. Further we do expect the components of ψ to undergo a linear transformation[14] of the form $\psi' = R\psi$ where R is a 2×2 matrix. (Physically, ψ_1 should give you the amplitude for a spin-up state along, say, the original $z-$ axis which has now been rotated. So ψ better change when we rotate coordinates.) The question is whether we can associate a 2×2 matrix R with each rotation such that $\sigma^\alpha \partial_\alpha \psi \to R[\sigma^\alpha \partial_\alpha \psi]$ under that rotation.

We can do this if we can find a matrix R such that, for a general rotation specified by a rotation matrix $M_\gamma{}^\alpha$ in Eq. (5.16), the following relation holds[15]:

$$R^{-1}\sigma^\alpha R = M_\gamma{}^\alpha \sigma^\gamma \qquad (5.19)$$

On multiplying both sides of this equation by ∂'_α and using $M_\gamma{}^\alpha\partial'_\alpha = (\partial x'^\alpha/\partial x^\gamma)\partial'_\alpha = \partial_\gamma$ we immediately get

$$R^{-1}[\sigma^\alpha \partial'_\alpha]R = \sigma^\gamma \partial_\gamma; \quad \sigma^\alpha \partial'_\alpha = R[\sigma^\gamma \partial_\gamma]R^{-1} \qquad (5.20)$$

Multiplying the second relation by $\psi' = R\psi$, side by side, we get

$$\sigma^\alpha \partial'_\alpha \psi' = R[\sigma^\gamma \partial_\gamma \psi] \qquad (5.21)$$

which is what we wanted. In fact, all the powers of this operator, viz. $[\sigma^\gamma \partial_\gamma]^n$, — especially the square, which occurs in H — transform in the same manner. So the Pauli equation will indeed retain its form under rotations.

All that we have to do now is to find an R such that Eq. (5.19) holds. This is again easy to do by considering infinitesimal rotations (and 'exponentiating' the result to take care of finite rotations). If we write $R = 1 - iA$

---

[13] Hopefully you know that it is best to think of rotations as "rotations in a plane". When $D = 3$, thinking of a rotation in the xy plane is equivalent to thinking of it as a rotation about the z-axis; but if you are in a $D = 4$ Euclidean space (x, y, z, w), you can still talk about rotations in the xy-plane but now you don't know whether it is a rotation about the z-axis or w-axis! All these happen for the same reason as why you can define a cross product $\boldsymbol{A} \times \boldsymbol{B}$ giving another vector in $D = 3$ but not in other dimensions; it is related to the curious fact that $3 - 2 = 1$.

[14] Sanity check: If you start with a linear transformation for the components as $\psi'_1 = a\psi_1 + b\psi_2; \psi'_2 = c\psi_1 + d\psi_2$ (with 4 complex or 8 real parameters) you will find that the bilinear form $\psi^1\phi^2 - \psi^2\phi^1$ transforms to $(ad - bc)$ times itself. This requires it to represent a spin-zero state and hence we must have $(ad - bc) = 1$. Further, the demand that the $|\psi_1|^2 + |\psi_2|^2$ should be a scalar, will require $a = d^*; b = -c^*$. These 1+4 conditions on the 8 parameters, show that the linear transformation has 3 free parameters, exactly what a rotation has.

[15] The right hand side is what you will write as σ'^α if the σ^α-s transformed like the components of a vector. The left hand side tells you that this is almost true, except for a similarity transformation by a matrix R.

(where A is a first order infinitesimal matrix that needs to be determined), we have, for any matrix Q, the result $R^{-1}QR = Q + i[A, Q]$. So the left hand side of Eq. (5.19) is $R^{-1}\sigma^\alpha R = \sigma^\alpha + i[A, \sigma^\alpha]$. The right hand side of Eq. (5.19), for an infinitesimal rotation, is $(\sigma^\alpha + \epsilon^\alpha{}_{\beta\gamma}\theta^\beta\sigma^\gamma)$. (This is same as Eq. (5.17) with x replaced by σ.) So the A has to satisfy the condition $[\sigma_\alpha, A] = i\epsilon_{\alpha\beta\gamma}\theta^\beta\sigma^\gamma$. From the first relation in Eq. (5.7), we immediately see that the choice $A = (1/2)\sigma_\alpha\theta^\alpha = (1/2)\boldsymbol{\theta} \cdot \boldsymbol{\sigma}$ satisfies this relation! So we have succeeded in finding an R such that Eq. (5.19) holds:

$$R = 1 - (i\theta/2)(\boldsymbol{\sigma} \cdot \boldsymbol{n}) \tag{5.22}$$

The result for rotation by a finite angle is given by the usual exponentiation:

$$R(\theta) = \exp -\frac{i\theta}{2}(\boldsymbol{\sigma} \cdot \boldsymbol{n}) = \cos\left(\frac{\theta}{2}\right) - i(\boldsymbol{\sigma} \cdot \boldsymbol{n})\sin\left(\frac{\theta}{2}\right) \tag{5.23}$$

In the process we have discovered how the two components of the spinor ψ_A transform under this rotation: $\psi' = R\psi$ (interpreted as a matrix equation). Since $\psi' = R\psi$ looks like $x' = Mx$, you might think spinors are 'like vectors'; well, it is true, but, only to a limited extent. If you rotate about any axis by 2π, vectors — and most of the stuff in the lab — do not change but spinors flip sign, thanks to the $\theta/2$ in R! Clearly, something peculiar is going on and we will come back to this issue later while discussing the Lorentz group.[16]

This transformation law $\psi'^A = R^A{}_B\psi^B$ for the two-component spinors ψ^A could in fact be taken as the definition of a 2-spinor. This will *demand* the complex conjugate to transform as $\psi^* \to R^*\psi^*$ but the probability density $|\psi_1|^2 + |\psi_2|^2 = \psi^\dagger\psi$ (summed over the spin states) should be a scalar under rotations. This is assured because $\boldsymbol{\sigma}^\dagger = \boldsymbol{\sigma}$ and $R^\dagger = R^{-1}$ making $(\psi^\dagger\psi)' = (\psi^\dagger R^\dagger)(R\psi) = (\psi^\dagger R^{-1})(R\psi) = (\psi^\dagger\psi)$.

As we said before it is a crime to think of "rotations about an axis" but that issue is easily remedied. We have

$$\sigma_\alpha\theta^\alpha = \sigma_\alpha\left(\frac{1}{2}\epsilon^{\alpha\beta\gamma}\omega_{\beta\gamma}\right) = \frac{1}{2}\left(\epsilon^{\alpha\beta\gamma}\sigma_\alpha\right)\omega_{\beta\gamma} = -\frac{i}{4}[\sigma^\beta, \sigma^\gamma]\omega_{\beta\gamma} \tag{5.24}$$

which allows us to write R in Eq. (5.23) as

$$R = \exp\left[-\frac{i}{2}\omega_{\alpha\beta}S^{\alpha\beta}\right]; \quad S^{\alpha\beta} \equiv -\frac{i}{4}[\sigma^\alpha, \sigma^\beta] \tag{5.25}$$

Instead of thinking of σ_α being associated with rotations "about the x^α-axis", we now think of $S^{\beta\gamma}$ as being associated with rotations "in the $\beta\gamma$ plane" which is a much better way to think of rotations.[17] After this preamble, let us move along and see what relativity does to all these.

5.3 Relativistic Square Roots

5.3.1 Square Root of $p_a p^a$ and the Dirac Matrices

The reason we spent so much of time with the "curiosity" $\boldsymbol{p}^2 = (\boldsymbol{p} \cdot \boldsymbol{\sigma})(\boldsymbol{p} \cdot \boldsymbol{\sigma})$ is because it generalizes rather trivially to the relativistic context. You just go from $D = 3$ to $D = 4$, the rotation group to the Lorentz group and Pauli matrices to Dirac matrices! Let us see how.

[16]There is a simple connection between the factor -1 arising from a 2π rotation and the antisymmetry of the wave functions describing a pair of electrons. To see this, consider two electrons located at x and $-x$ with spins along the z-axis. Such a state is represented by the product of two spinors $\Psi_{12} \equiv \psi_{\text{up}}(x)\psi_{\text{up}}(-x)$. If we now perform a rotation around the z-axis by π, we will end up interchanging the two particles without affecting their spins. Such a rotation will introduce a factor $R = -i$ for each spinor so that $\Psi_{12} \to -\Psi_{21}$. Therefore, the wave function picks up a factor -1 when the particles are interchanged.

[17]A precise characterization of what the phrase "being associated with" means will require some group theory. We will do it in Sect. 5.4.

In the relativistic case, we would have loved to write down some kind of "Schrodinger equation" of the form $i\dot\phi = H\phi$ with H given by $\sqrt{\boldsymbol{p}^2 + m^2}$ (see Sect. 1.4.6). Taking a cue from the previous analysis, one would like to express this square root as a linear function of $\boldsymbol{p}$ and m in the form $(\alpha m + \boldsymbol{\beta}\cdot\boldsymbol{p})$. This will lead to what we want but there is a more elegant way of doing it, maintaining formal relativistic invariance, which is as follows.

We know that the content of the wave equation describing the relativistic free particles is given by $(p^2 - m^2)\psi(x) = 0$ with $p_a \equiv i\partial_a$. Our aim now is to think of the square root $\sqrt{p^2}$, in the relativistic case, as being given by $\gamma^i p_i$ where the four entities γ^i need to be determined.[18] Then we can write the relation $\sqrt{p^2} = \sqrt{m^2}$ as $\gamma^a p_a = m$; so our wave equation can now take the form:

$$(\gamma^a p_a - m)\psi \equiv (i\gamma^a\partial_a - m)\psi = 0 \qquad (5.26)$$

[18]Notation alert: We treat the index a of γ^a like a four-vector index so that we can define $\gamma_a \equiv \eta_{ab}\gamma^b$ etc. Purely algebraically, we will then have $\gamma^a p_a = \gamma_b p^b$ etc., which is convenient.

This is quite straightforward and we play the same game which led to Eq. (5.3) but now in 4-dimensions. We need

$$\eta^{ij} p_i p_j = (\gamma^i p_i)(\gamma^j p_j) = (\gamma^i\gamma^j)p_i p_j = \frac{1}{2}\left\{\gamma^i, \gamma^j\right\} p_i p_j \qquad (5.27)$$

As in the case of Eq. (5.4), we now need to discover a set of four *anticommuting* matrices which satisfy the condition

$$\left\{\gamma^i, \gamma^j\right\} = 2\eta^{ij} \qquad (5.28)$$

Exercise 5.2: Do it yourself!

[19]Very imaginatively called gamma matrices.

Though we will not bother to prove it, you cannot satisfy these relations with 2×2 or 3×3 matrices. The smallest dimension in which the relativistic square root works is in terms of 4×4 matrices. One useful set of matrices[19] which satisfy Eq. (5.28) is given by

$$\gamma^0 = \begin{pmatrix} 0 & 1 \\ 1 & 0 \end{pmatrix}, \qquad \gamma^\alpha = \begin{pmatrix} 0 & \sigma^\alpha \\ -\sigma^\alpha & 0 \end{pmatrix} \qquad (5.29)$$

(This particular choice of matrices which obey Eq. (5.28) is called the Weyl representation). This can be written more compactly as

$$\gamma^m = \begin{pmatrix} 0 & \sigma^m \\ \bar\sigma^m & 0 \end{pmatrix} \qquad (5.30)$$

where we have upgraded the Pauli matrices to 4-dimensions with the notation:

$$\sigma^m = (1, \sigma^\alpha), \qquad \bar\sigma^m = (1, -\sigma^\alpha) \qquad (5.31)$$

Here, σ^α are the standard Pauli matrices and 1 stands for the unit matrix. As you might have guessed, the form of γ-matrices is far from unique and which set you choose to use depends on what you want to do. Another set (called the Dirac representation) which we will occasionally use is given by:

$$\gamma^0 = \begin{pmatrix} 1 & 0 \\ 0 & -1 \end{pmatrix}, \qquad \gamma^\alpha = \begin{pmatrix} 0 & \sigma^\alpha \\ -\sigma^\alpha & 0 \end{pmatrix} \qquad (5.32)$$

It is convenient, for future purposes, to define the *commutators* of the gamma matrices by the relation:

$$\sigma^{mn} \equiv \frac{i}{2}[\gamma^m, \gamma^n] \qquad (5.33)$$

In the Weyl representation, the explicit form of these commutators are given by

$$\sigma^{0\alpha} = -i \begin{pmatrix} \sigma^\alpha & 0 \\ 0 & -\sigma^\alpha \end{pmatrix}, \qquad \sigma^{\alpha\beta} = \epsilon^{\alpha\beta\kappa} \begin{pmatrix} \sigma^\kappa & 0 \\ 0 & \sigma^\kappa \end{pmatrix} \qquad (5.34)$$

So, in this representation, $\sigma^{\alpha\beta}$ is just block diagonal matrix built from Pauli matrices; for e.g., $\sigma^{12} = \text{dia}(\sigma^3, \sigma^3)$.

So we have successfully taken a relativistic square root and (given a set of γ-s), we can explicitly write down the equation Eq. (5.26). We will now enquire what an equation like Eq. (5.26) could possibly mean and how does it behaves if we perform a rotation (which we needed to worry about even in $D = 3$) or a Lorentz boost (which is a $D = 4$ feature!).

5.3.2   How does ψ Transform?

To begin with, the ψ in Eq. (5.26) is a 4-component[20] object ψ_α with $\alpha = 1, 2, 3, 4$ which is acted upon by the 4×4 matrix from the left. Since we are already familiar with a 2-component spinor describing the electron from the previous section, let us write the column vector ψ in terms of a pair of quantities ψ_L and ψ_R, each having two components, and investigate this pair.[21] That is, we write

$$\psi = \begin{pmatrix} \psi_L \\ \psi_R \end{pmatrix} \qquad (5.35)$$

If we expand out the 4×4 matrix equation Eq. (5.26) — which we will hereafter call the Dirac equation, giving credit where credit is due — into a pair of 2×2 matrix equations, we will get:

$$(i\partial_t + i\boldsymbol{\sigma} \cdot \boldsymbol{\nabla})\psi_R = m\psi_L; \quad (i\partial_t - i\boldsymbol{\sigma} \cdot \boldsymbol{\nabla})\psi_L = m\psi_R \qquad (5.36)$$

This shows that we are indeed dealing with a *pair* of 2-component objects coupled through the mass term; when $m = 0$ they separate into a pair of uncoupled equations. Our next task is to figure out what happens to this equation under rotations and Lorentz boosts.

Rotations are easy. If we rewrite Eq. (5.36) in the form:

$$i(\boldsymbol{\sigma} \cdot \boldsymbol{\nabla})\psi_R = m\psi_L - i\partial_t\psi_R; \quad -i(\boldsymbol{\sigma} \cdot \boldsymbol{\nabla})\psi_L = m\psi_R - i\partial_t\psi_L \qquad (5.37)$$

the only non-trivial term affected by rotations is $\boldsymbol{\sigma} \cdot \boldsymbol{\nabla} = \sigma^\alpha \partial_\alpha$. But we already know from the previous section how this term changes under rotations! We see that the Eq. (5.37) will be form-invariant under rotations if ψ_R and ψ_L transform as the standard 2-spinors we introduced in the previous section. That is, under pure rotations,

$$\psi_L \to R\psi_L; \quad \psi_R \to R\psi_R \qquad (5.38)$$

with (see Eq. (5.25)):

$$R = \exp\left\{(-i\boldsymbol{\theta}) \cdot \frac{\boldsymbol{\sigma}}{2}\right\} = \exp[-\frac{i}{2}\omega_{\alpha\beta}S^{\alpha\beta}] \qquad (5.39)$$

where $S^{\alpha\beta} \equiv -(i/4)[\sigma^\alpha, \sigma^\beta]$. This solves part of the mystery: Our ψ is actually made of a pair of objects each of which transforms as an ordinary 2-spinor as far as spatial rotations go.[22]

[20]These, of course, are not spacetime indices.

[21]The subscripts L and R will not make sense to you at this stage but will become clearer later on; right now, just think of them as labels.

[22]Once again, there is a peculiar sign flip on the 2π rotation which we need to get back to later on.

Of course, we now have to ensure invariance under the full Lorentz transformations — not just spatial rotations. This is again very easy because the idea is essentially the same as the one we encountered in the last section. Recall that the rotational invariance worked because, for a rotation described by $x^\alpha \to x'^\alpha \equiv M_\gamma{}^\alpha x^\gamma$, we could find a matrix R such that

$$R^{-1}\sigma^\alpha R = M_\gamma{}^\alpha \sigma^\gamma \qquad (5.40)$$

holds. This immediately implies that $R(\boldsymbol{\sigma}\cdot\boldsymbol{\nabla})R^{-1} = (\boldsymbol{\sigma}\cdot\boldsymbol{\nabla}')$ and we could keep the equation invariant by postulating $\psi \to R\psi$.

Now we have to consider the more general class of transformations $x^a \to x'^a \equiv L^a{}_b x^b$ representing both rotations and Lorentz transformations, which keep $x_a x^a$ invariant.[23] The infinitesimal version is given by $x'^a \equiv [\delta^a_b + \omega^a{}_b]x^b$ where the six parameters of the antisymmetric tensor $\omega_{ab} = -\omega_{ba}$ now describe 'rotations' in the six planes (xy, yz, zx, tx, ty, tz); i.e., three rotations and 3 boosts. Everything will work out fine, if we can find a 4×4 "rotation matrix" $\mathcal{R}$ such that

$$\mathcal{R}^{-1}\gamma^m \mathcal{R} = L^m{}_n \gamma^n = \frac{\partial x'^m}{\partial x^n}\gamma^n \qquad (5.41)$$

[Compare with Eq. (5.19).] As before, multiplying by ∂'_m will lead[24] to the results:

$$\mathcal{R}^{-1}[\gamma^a \partial_a']\mathcal{R} = \gamma^c \partial_c; \quad \gamma^a \partial_a' = \mathcal{R}[\gamma^c \partial_c]\mathcal{R}^{-1} \qquad (5.42)$$

If we now postulate the ψ transforms under the Lorentz transformation as

$$\psi' = \mathcal{R}\psi \qquad (5.43)$$

we can again multiply equations Eq. (5.42) and Eq. (5.43) side by side, to obtain

$$\gamma^a \partial_a' \psi' = \mathcal{R}[\gamma^c \partial_c \psi] \qquad (5.44)$$

This will keep the Dirac equation invariant under a Lorentz transformation. So, if we find a matrix $\mathcal{R}$ that satisfies Eq. (5.41), we are through; as a bonus we find how ψ transforms under Lorentz transformations. (And since these include rotations as a special case, we can check out the previous results.)

Can we find such an $\mathcal{R}$? Very easy. You can again write $\mathcal{R} = 1 - iA$ and play the same game we did to arrive at Eq. (5.23). But since the R for 3-D rotations is given by Eq. (5.23) with 3 rotational parameters in $\omega_{\alpha\beta}$ and $S^{\alpha\beta}$ is made from the commutators of the Pauli matrices, it is a good guess[25] to try:

$$\mathcal{R} \equiv \exp\left(-\frac{i}{2}\omega_{mn}S^{mn}\right); \quad S^{mn} \equiv \frac{i}{4}[\gamma^m, \gamma^n] \qquad (5.45)$$

You can now verify that, with the ansatz in Eq. (5.45), the condition in Eq. (5.41) holds, which in turn implies Eq. (5.42). So, the Dirac equation will be Lorentz invariant provided the ψ transforms as:

$$\psi \to \mathcal{R}\psi = \exp\left(-\frac{i}{2}\omega_{mn}S^{mn}\right)\psi \qquad (5.46)$$

What does this transformation mean? We already know that under pure rotations, the ψ_L and ψ_R of which ψ is made of, transform in identical manner, just like non-relativistic two-spinors. To see what happens

[23] We will continue to call this transformation 'Lorentz transformation', even though it includes rotations as a subset; once we do this, we will be able to reproduce — and check — the previous results for rotations as a special case.

[24] Make sure you appreciate the smooth transition $3 \to 4$. Here we have merely retyped Eq. (5.20) changing Pauli matrices to Dirac matrices, $R \to \mathcal{R}$ and Greek indices to Latin indices!

[25] This is again essentially Eq. (5.25) retyped with $R \to \mathcal{R}$, Pauli matrices to Dirac matrices and Greek $\to$ Latin, except for a sign flip in $S^{\alpha\beta}$ compared to S^{mn}. This is just to take care of the fact that $[\gamma^\alpha, \gamma^\beta]$ for spatial indices is dia $(-[\sigma^\alpha, \sigma^\beta], -[\sigma^\alpha, \sigma^\beta])$ with a minus sign, because of the way we have defined the γ-matrices. So the minus sign in the definition of S^{mn} in Eq. (5.45) vis-a-vis the definition of $S^{\alpha\beta}$ in Eq. (5.25) ensures that when we consider pure rotations, we will get back the previous results.

Exercise 5.3: Prove the claim. (Hint: First prove $[\gamma^l, S^{mn}] = (i/2)[\gamma^l, \gamma^m\gamma^n] = i[\eta^{lm}\gamma^n - \eta^{ln}\gamma^m]$; then, if you use the infinitesimal version of Lorentz transformation, it is straight forward.)

under Lorentz transformations, let us consider a pure boost with $\omega_{\alpha\beta} = 0, \omega_{0\alpha} \equiv \eta_\alpha$ so that we can write $(i/2)\omega_{mn}S^{mn} = i\omega_{0\alpha}S^{0\alpha}$ and use the fact that (in the Weyl representation) $S^{0\alpha} = -(i/2)dia(\sigma^\alpha, -\sigma^\alpha)$ to obtain $(i/2)\omega_{mn}S^{mn} = (1/2)dia(\eta \cdot \sigma, -\eta \cdot \sigma)$. Substituting this into the transformation law Eq. (5.46) and using the block-diagonal nature of the matrices, we find that ψ_L and ψ_R transform *differently* under a Lorentz boost:

$$\psi_L \to \mathcal{R}_L \psi_L \equiv \exp\left(-\frac{1}{2}\eta \cdot \sigma\right)\psi_L; \quad \psi_R \to \mathcal{R}_R \psi_R \equiv \exp\left(\frac{1}{2}\eta \cdot \sigma\right)\psi_R \tag{5.47}$$

Thus, under pure rotations, ψ_L and ψ_R transform identically[26] but they transform differently under Lorentz boost.

For practical computations, it is convenient to write $\eta = r n$ where r is the rapidity of the Lorentz boost given by $(v/c) = \tanh(r)$. In this case, the relevant transformation matrices become:

$$\mathcal{R}_L = \exp\left(-\frac{r}{2}n \cdot \sigma\right), \qquad \mathcal{R}_R = \exp\left(+\frac{r}{2}n \cdot \sigma\right) \tag{5.48}$$

In fact, using Eq. (5.9), we can explicitly evaluate these transformation matrices to get

$$\mathcal{R}_L^2 = \exp(-r n \cdot \sigma) = \cosh(r) - \sinh(r)(n \cdot \sigma) = \gamma - \beta\gamma(n \cdot \sigma) \tag{5.49}$$

and similarly for $\mathcal{R}_R^2$. This leads to the alternative expressions for the transformation matrices:

$$\mathcal{R}_L = \sqrt{\gamma - \beta\gamma n \cdot \sigma}, \qquad \mathcal{R}_R = \sqrt{\gamma + \beta\gamma n \cdot \sigma} \tag{5.50}$$

We can now demystify the subscripts L and R by considering how rotations and Lorentz boosts are affected by the parity transformation, $P \equiv (t \to t, x \to -x)$. Under parity, rotations are unaffected while Lorentz boosts flip sign. So under the parity transformation, ψ_L and ψ_R are interchanged. That is,

$$P : \psi_R(t, x) \to \psi_L(t, -x); \quad P : \psi_L(t, x) \to \psi_R(t, -x) \tag{5.51}$$

This can be written more concisely in terms of the 4-spinor ψ, using the γ^0 matrix, as

$$P : \psi(t, x) \to \gamma^0 \psi(t, -x) \tag{5.52}$$

From this, it is trivial to verify that if $\psi(t, x)$ satisfies the Dirac equation, the parity transformed spinor defined as $\gamma^0 \psi(t, -x)$ also satisfies[27] the Dirac equation:

$$(i\gamma^0\partial_t + i\gamma^\alpha\partial_\alpha - m)\gamma^0\psi(t, -x) = \gamma^0(i\gamma^0\partial_t - i\gamma^\alpha\partial_\alpha - m)\psi(t, -x) = 0 \tag{5.53}$$

Putting together the results in Eq. (5.39) and Eq. (5.47), we can write down how ψ_L, ψ_R transform under the most general Lorentz transformations, specified by ω_{mn} made of $\omega_{\alpha\beta} \equiv \epsilon_{\alpha\beta\gamma}\theta^\gamma, \omega_{0\alpha} \equiv \eta_\alpha$. This is given, in full glory, by:

$$\psi_L \to \exp\left\{(-i\theta - \eta) \cdot \frac{\sigma}{2}\right\}\psi_L \tag{5.54}$$

and

$$\psi_R \to \exp\left\{(-i\theta + \eta) \cdot \frac{\sigma}{2}\right\}\psi_R \tag{5.55}$$

Interesting, but rather strange. We will have more to say about this in Sect. 5.4.

[26]Since NRQM only cares about rotations as a symmetry, we need not distinguish between ψ_L and ψ_R in the context of NRQM. In (relativistic) literature, the index of ψ_L is denoted by $\dot{\alpha}$ and one often uses the terminology of 'dotted' and 'undotted' spinors rather than that of L and R spinors.

[27]Algebra alert: The extra minus sign arising from γ^0 moving through γ^α is nicely compensated by the derivative acting on $-x$ instead of x.

5.3.3 Spin Magnetic Moment from the Dirac Equation

Let us next check whether the Dirac equation gives the correct magnetic moment for the electron, which we obtained from the Pauli equation in the last section. As in the non-relativistic case, this fancy square root of p^2 really comes into its own only when we introduce the electromagnetic coupling. In the relativistic case, this will change Eq. (5.26) to the form[28]

$$(i\gamma^m D_m - m)\psi = 0; \quad D_m \equiv \partial_m - ieA_m \qquad (5.56)$$

To see what this implies for the electromagnetic coupling, multiply this equation by $(i\gamma^a D_a + m)$ and use the results

$$\gamma^m \gamma^n D_m D_n = \frac{1}{2}\left(\{\gamma^m, \gamma^n\} + [\gamma^m, \gamma^n]\right) D_m D_n = D_m D^m - i\sigma^{mn} D_m D_n$$

$$i\sigma^{mn} D_m D_n = \frac{i}{2}\sigma^{mn}[D_m, D_n] = \frac{e}{2}\sigma^{mn} F_{mn} \qquad (5.57)$$

Simple algebra now gives

$$\left(D_m D^m - \frac{e}{2}\sigma^{mn} F_{mn} + m^2\right)\psi = 0 \qquad (5.58)$$

The first term $D^2 \equiv D_m D^m$ is exactly what we would have got for the spinless particle and, as before, we do not expect anything new to emerge from this term. Expanding it out, we can easily see that the spatial part will give, as before, the contribution

$$(D_\alpha)^2 = \boldsymbol{\nabla}^2 - e\boldsymbol{B} \cdot (\boldsymbol{x} \times \boldsymbol{p}) + \mathcal{O}(A_\alpha^2) \qquad (5.59)$$

The orbital angular momentum will contribute to the Hamiltonian a magnetic moment $\boldsymbol{m} = (e/2m)\boldsymbol{L}$.

As before, we find that there is indeed an extra term $(e/2)\sigma^{mn} F_{mn}$ which arises due to our fancy square-rooting. To see its effect, let us again consider a magnetic field along the $z-$axis, in which case the $(e/2)\sigma^{mn} F_{mn}$ will contribute a term

$$\left(\frac{e}{2}\right)\sigma^3(F_{12} - F_{21}) = \left(\frac{e}{2}\right)2\sigma^3 B = 2e\boldsymbol{B} \cdot \boldsymbol{s}; \qquad \boldsymbol{s} = \boldsymbol{\sigma}/2 \qquad (5.60)$$

Clearly, we get an extra factor of 2 in the contribution of the spin to the magnetic moment; so this result is not an artifact of the non-relativistic discussion in the previous section.

To summarize, we have discovered a new class of objects ψ built from a *pair* of 2-spinors which have nice Lorentz transformation properties. This puts them in the same "respectability class" as scalars, four vectors, tensors, etc. using which one can construct relativistically invariant field theories. At this stage, you will find it rather mysterious that a new class of objects like ψ_α even exists and — since they do — whether there are more such objects to be discovered. The answer to this question involves classifying the representations of the Lorentz group and introducing the fields which can "carry" these representations. This is something which we will turn our attention to in the next section. We will approach the entire problem from a slightly more formal angle which will tie up the loose ends and clarify what is really going on.

[28] The argument for this replacement is exactly the same as that in the case of complex scalar field we studied earlier, in Sect. 3.1.5. Under the transformation $\psi \to \exp[-iq\alpha(x)]\psi$, $A_m \to A_m + \partial_m\alpha$, we see that $D_m\psi \to \{\exp[-iq\alpha(x)]\}D_m\psi$, maintaining gauge invariance.

5.4 Lorentz Group and Fields

5.4.1 Matrix Representation of a Transformation Group

Consider the set of all linear transformations, $x^a \rightarrow \bar{x}^a \equiv L^a{}_b x^b$, of the Cartesian 4-dimensional coordinates which preserves the interval $s^2 = \eta_{ab} x^a x^b = t^2 - |\boldsymbol{x}|^2$. This condition

$$\eta_{mn} \bar{x}^m \bar{x}^n = \eta_{mn}(L^m{}_r x^r)(L^n{}_s x^s) = \eta_{rs} x^r x^s \qquad (5.61)$$

will hold for all x^a if the matrix L satisfies the condition

$$\eta_{rs} = \eta_{mn} L^m{}_r L^n{}_s; \quad \eta = L^T \eta L \qquad (5.62)$$

Taking the determinant, we have $[\det L]^2 = 1$ and we will take[29] $\det L = 1$ which corresponds to, what is called, *proper* Lorentz transformations. Further, if you expand out the zero-zero component of Eq. (5.62), it is easy to show that $(L^0{}_0)^2 \geq 1$ which divides the proper Lorentz transformations into two disconnected components with $L^0{}_0 \geq 1$ ("orthochronous") and the one with $L^0{}_0 \leq 1$ ("non-orthochronous"). We will stick with the orthochronous, proper, Lorentz transformations.[30] This is what you would have usually thought of as the Lorentz transformations from a special relativity course.

A subset of these transformations will be pure rotations in the three dimensional space with $\bar{x}^0 = x^0, \bar{x}^\alpha = L^\alpha{}_\beta x^\beta$; another subset will be pure Lorentz boosts along some direction. The successive application of two rotations will lead to another rotation so they form a group by themselves; but the successive application of two Lorentz boosts along two different directions will, in general, lead to a rotation plus a Lorentz boost.[31] This fact, along with the existence of the identity transformation and the inverse transformation, endows the above set of linear transformations with a group structure, which we will call the Lorentz group. The composition law for two group elements $L_1 \circ L_2$ will be associated with the process of making the transformation with L_2 followed by L_1. As we said before, the group structure is determined by identifying the element of the group L_3 which corresponds to $L_1 \circ L_2$ for all pairs of elements.

This description is rather abstract for the purpose of physics! By and large, you can get away with what is known as the *matrix representation* of the group rather than with the abstract group itself. We can provide a matrix representation to any group (and, in particular, to the Lorentz group which we are interested in) by associating a matrix with each element of the group such that, the group composition law is mapped to the multiplication of the corresponding matrices. That is, the group element $g_1 \circ g_2$ will be associated with a matrix that is obtained by multiplying the matrices associated with g_1 and g_2.

In the case of the Lorentz group, a set of $k \times k$ matrices $D(L)$ will provide a $k-$dimensional representation of the Lorentz group if $D(L_1)D(L_2) = D(L_1 \circ L_2)$ for any two elements of the Lorentz group L_1, L_2 where L_1, L_2 etc. could correspond to either Lorentz boosts or rotations. We can also introduce a set of k quantities Ψ_A with $A = 1, 2, ...k$, forming a column vector, on which these matrices can act such that, under the action of an element L_1 of Lorentz group (which could be either a Lorentz transformation or a rotation), the Ψ_As undergo a linear transformation of the form $\Psi'_A = D^B_A(L_1)\Psi_B$ where $D^B_A(L_1)$ is a $k \times k$ matrix representing the element L_1.

[29] Transformations with $\det L = -1$ can always be written as a product of those with $\det L = +1$ and a discrete transformation that reverses the sign of an odd number of coordinates.

[30] The non-orthochronous transformations can again be obtained from the orthochronous ones with a suitable inversion of coordinates.

[31] So Lorentz boosts *alone* do not form a group. You may have learnt this from a course in special relativity; if not, just work it out for yourself. We will obtain this result in a more sophisticated language later on.

The simplest example of such a set of $k-$tuples Ψ_A occurs when $k = 4$. Here we identify Ψ_A with just the components of any four vector v^a and the matrices D_A^B with the Lorentz transformation matrices $L^b{}_a$. This, of course, should work since it is this transformation which was used in the first place, to define the group.[32] It is also possible to have a trivial "matrix" representation with $k = 1$ where we map all the group elements to identity.

In general, however, $D_A^B(L_1)$ can be a $k \times k$ matrix where k need not be 4 (or 1). In such a case, we have effectively generalized the idea of Lorentz transformation from 4-component objects to k-component objects Ψ_A which have nice transformation properties under Lorentz transformation. In the case of $k = 1$, the one-component object can be thought of as a real scalar field $\phi(x)$ which transforms under the Lorentz transformation trivially as $\bar{\phi}(\bar{x}) = \phi(x)$. In the case of $k = 4$, it could be a 4-vector field $A^j(x)$ which transforms as $\bar{A}^j(\bar{x}) = L^j{}_k A^k(x)$. In both these cases, we could say that the fields "carry" the representation of the Lorentz group.

So you see that, if we have a $k-$dimensional representation of the Lorentz group, we can use it to study a $k-$component field which will have nice properties and could lead to a relativistically invariant theory. Hence, the task of identifying all possible kinds of fields which can exist in nature, reduces to finding all the different $k \times k$ matrices which can form a representation of the Lorentz group. We will now address this task.

5.4.2   Generators and their Algebra

The key trick one uses to accomplish this is the following. You really do not have to find all the $k \times k$ matrices which satisfy our criterion. We only have to look at matrices which are infinitesimally close to the identity matrix and understand their structure. This is because any finite group transformation we are interested in can be obtained by "exponentiating" an infinitesimal transformation close to the identity element. If we write the matrix corresponding to an element close to the identity as $D \equiv 1 + \epsilon^A G_A$ where ϵ^A are a set of infinitesimal scalar parameters, then a finite transformation can be obtained by repeating the action of this element N times and taking the limit $N \to \infty, \epsilon^A \to 0$ with $N\epsilon^A \equiv \theta^A$ remaining finite. We then get

$$\lim_{\substack{N \to \infty \\ \epsilon \to 0}} \left(1 + \epsilon^A G_A\right)^N = \exp(N\epsilon^A G_A) = \exp(\theta^A G_A) \qquad (5.63)$$

which represents some generic element of the group.[33]

[33]The set of elements G_A which represent elements close to the identity element are called the generators of the group.

[34]This will miss out some global aspects of the group but they can be usually figured out e.g., by looking at the range of ϵ^A. The generators define what is known as the Lie algebra of the Lie group and for most purposes we will just concentrate on the Lie algebra of the Lorentz group.

So our task now reduces to finding the matrix representation of the *generators* G_A of the group[34] rather than the matrix representation for all the elements of the group. What is the property the matrices representing the generators of the group need to satisfy? To find this out, consider two matrices $U_1 = 1 + i\epsilon_1 G_1$ and $U_2 = 1 + i\epsilon_2 G_2$ representing two elements of the group close to the identity (with the factor i introduced for future convenience). We then know that the element $U_3 = 1 + i\epsilon_3 G_3$ defined by the operation $U_3 = U_1^{-1} U_2 U_1$ is also another element of the group close to the identity. An elementary calculation shows that $G_3 = G_2 + i\epsilon_1 [G_2, G_1]$ which, in turn, implies that the commutator $[G_2, G_1]$ of two generators is also another generator. For a group with N generators, the element close to identity can be expressed in the form $U = 1 + i\epsilon^A G_A$, and the above analysis tells us that the commutator between any pair of generators must be a linear combination of other generators:

$$[G_A, G_B] = i f_{AB}{}^C G_C \qquad (5.64)$$

where the set of numbers $f_{AB}{}^C$ are called the structure constants of the group. These structure constants determine the entire part of the group which is continuously connected to the identity since any finite element in that domain can be obtained by exponentiating a suitable element close to the identity.

So, all we need to do is to: (i) Determine the structure constants for the relevant group and (ii) discover and classify all the $k \times k$ matrices which obey the commutation rule given by Eq. (5.64) with the structure constants appropriate for the Lorentz group. We will now turn to these tasks.

5.4.3   Generators of the Lorentz Group

Let us start with a Lorentz transformation which is close to the identity transformation represented by the transformation matrix $L^m{}_n = \delta^m_n + \omega^m{}_n$. The condition in Eq. (5.62) now requires $\omega_{mn} = -\omega_{nm}$. Such an antisymmetric 4×4 matrix has 6 independent elements, so we discover that the Lorentz group has 6 parameters.[35] This also means that there will be six generators J_A (with $A = 1, 2, ..., 6$) for the Lorentz group and a generic group element close to the identity can be written in the form $1 + \theta^A J_A$ where the six parameters θ^A corresponds to the angles of rotation in the six planes. We can get any other group element by exponentiating this element close to the identity.

Since we have chosen to represent the 6 infinitesimal parameters by the second rank antisymmetric tensor ω_{mn} (which has six independent components) rather than by a six component object θ^A, it is also convenient to write the generators J^A in terms of another second rank antisymmetric tensor as J^{mn} with $J^{mn} = -J^{nm}$. We are interested in the structure constants of the Lorentz group, for which, we need to work out the commutators of the generators $[J^{mn}, J^{rs}]$. Since the structure constants are independent of the representation, we can work this out by looking at the defining representations of the Lorentz group (the $k = 4$ one), for which we know the explicit matrix form of the Lorentz transformation. In this representation, J^{mn} is represented by a 4×4 matrix with components $(J^{mn})^r{}_s$, the explicit form of which is given by[36]

$$(J^{mn})^r{}_s = i \left(\eta^{mr} \delta^n_s - \eta^{nr} \delta^m_s \right) \qquad (5.65)$$

This is a 4-dimensional representation under which all the components of the four-vector gets mixed up, and hence this is an irreducible representation.

You can now work out the form of $[J^{mn}, J^{rs}]$ using the explicit form of matrices in Eq. (5.65) and obtain:

$$[J^{mn}, J^{rs}] = i \left(\eta^{nr} J^{ms} - \eta^{mr} J^{ns} - \eta^{ns} J^{mr} + \eta^{ms} J^{nr} \right) \qquad (5.66)$$

This is the formal result which completely determines the structure of the Lorentz group. All we need to do is to find $k \times k$ matrices for the six generators in J^{mn} such that these matrices satisfy the same commutation rule as given above. That will also allow us to determine and classify all possible matrix representations of the Lorentz group. Obviously, there

[35]This, of course, makes sense. Since one can think of Lorentz boost along, say, x-axis as a rotation (albeit with a complex angle) in the tx plane, the 6 parameters correspond to rotations in the 6 planes xy, yz, zx, tx, ty, tz. We see the virtue of 'rotations in planes' compared to 'rotations about an axis'.

[36]This is easily verified by noticing that, under an infinitesimal transformation, any four vector changes as $\delta V^m = \omega^m{}_n V^n$ which can be written as $\delta V^r = -(i/2)\omega_{mn}(J^{mn})^r{}_s V^s$ Substituting the form of the matrix in Eq. (5.65) reproduces the correct result.

Exercise 5.4: Work this out. It builds character, if not anything else.

should be a cleverer way of doing this — and there is. Since it also helps you to connect up the structure of Lorentz group with what you know about angular momentum in QM, we will follow that route.

The trick is to think of a representation in the space of (scalar) functions and come up with a set of differential operators (which act on functions) and could represent J^{mn}. Since a pure rotation in the $\alpha\beta$ plane will have as generator the standard angular momentum operator [37], $J_{\alpha\beta} = i(x_\alpha \partial_\beta - x_\beta \partial_\alpha)$ it is a safe guess that the Lorentz group generators are just:

$$J_{mn} = i(x_m \partial_n - x_n \partial_m) \qquad (5.67)$$

This can be obtained more formally by, say, considering the change $\delta_0\phi \equiv \bar{\phi}(x) - \phi(x)$ of a scalar field $\phi(x)$ at a given x^a under an infinitesimal Lorentz transformation $\bar{x}^m = x^m + \delta x^m = \bar{x}^m + \omega^m{}_n x^n$. From $\bar{\phi}(\bar{x}) = \bar{\phi}(x + \delta x) = \phi(x)$, we find[38] that $\bar{\phi}(x) = \phi(x - \delta x) = \phi(x) - \delta x^m \partial_m \phi$. This leads to the result

$$\begin{aligned} \delta_0\phi &= \phi'(x) - \phi(x) = -\delta x^m \partial_m \phi = -\omega^{mn} x_n \partial_m \phi \\ &= -\frac{1}{2}\omega^{mn}(x_n \partial_m - x_m \partial_n)\phi \equiv -\frac{i}{2}\omega^{mn} J_{mn}\phi \end{aligned} \qquad (5.68)$$

In arriving at the third equality, we have used the explicit form for $\delta x^m = \omega^m{}_n x^n$ and in getting the fourth equality we have used the antisymmetry of ω^{mn}. The last equality identifies the generators of the Lorentz transformations as guessed in Eq. (5.67).

It is now easy to figure out the matrix representations of the Lorentz group. In Eq. (5.67) we get back the standard angular momentum operators of quantum mechanics for $J_{\alpha\beta}$ while $J_{0\alpha}$ leads to a Lorentz boost in the α-direction. Introducing the six parameters (θ^μ, η^μ) in place of $\omega_{\alpha\beta}$ by the standard trick $\omega_{\alpha\beta} \equiv \epsilon_{\alpha\beta\mu}\theta^\mu, \omega^{\alpha 0} \equiv \eta^\alpha$ and defining six new generators $\boldsymbol{J}, \boldsymbol{K}$ by:

$$J_\alpha = -\frac{1}{2}\epsilon_{\alpha\beta\gamma}J^{\beta\gamma}, \qquad K^\alpha = J^{\alpha 0} \qquad (5.69)$$

we can express[39] the relevant combination $(1/2)\omega_{mn}J^{mn}$ as:

$$\frac{1}{2}\omega_{mn}J^{mn} = \frac{1}{2}\omega_{\alpha\beta}J^{\alpha\beta} + \frac{1}{2}\omega_{\alpha 0}J^{\alpha 0} \times 2 = \boldsymbol{\theta} \cdot \boldsymbol{J} - \boldsymbol{\eta} \cdot \boldsymbol{K} \qquad (5.70)$$

which actually has the structure of $\theta^A G_A$ with the six generators and six parameters explicitly spelt out. A generic element L of the Lorentz group will be given by exponentiating the element close to the identity, and can be expressed in the form:

$$L = \exp[(-i/2)\omega_{mn}J^{mn}] = \exp[-i\boldsymbol{\theta} \cdot \boldsymbol{J} + i\boldsymbol{\eta} \cdot \boldsymbol{K}] \qquad (5.71)$$

Clearly, J^α are the generators for the spatial rotations; the K^α are the generators of the Lorentz boosts.

Using the representation in Eq. (5.67) and the definitions in Eq. (5.69) we find that J_α and K_α obey the following commutation rules.

$$[J_\alpha, J_\beta] = i\epsilon_{\alpha\beta\gamma}J^\gamma; \quad [J_\alpha, K_\beta] = i\epsilon_{\alpha\beta\gamma}K^\gamma; \quad [K_\alpha, K_\beta] = -i\epsilon_{\alpha\beta\gamma}J^\gamma \qquad (5.72)$$

These relations have a simple interpretation. The first one is the standard commutation rule for angular momentum operators. Since these commutators of $\boldsymbol{J}$ close amongst themselves, it is obvious that the spatial rotations alone form a sub-group of the Lorentz group. The second one is equivalent to saying that K_α behaves like a 3-vector under rotations. The crucial relation is the third one which shows that the commutator of two boosts is a rotation (with an important minus sign); so Lorentz boosts alone do not form a group.

To provide an explicit matrix representations for the Lorentz group, we only have to provide a matrix representation for the infinitesimal generators of the Lorentz group in Eq. (5.72). That is, we have to find all matrices which satisfy these commutation relations.

5.4.4 Representations of the Lorentz Group

To do this, we introduce the linear combination $a_\alpha = (1/2)(J_\alpha + iK_\alpha)$ and $b_\alpha = (1/2)(J_\alpha - iK_\alpha)$. This allows the commutation relation in Eq. (5.72) to be separated into two sets

$$[a_\alpha, a_\beta] = i\epsilon_{\alpha\beta\mu}a^\mu; \qquad [b_\alpha, b_\beta] = i\epsilon_{\alpha\beta\mu}b^\mu; \qquad [a_\mu, b_\nu] = 0 \qquad (5.73)$$

These are the familiar commutation rules for a *pair* of independent angular momentum matrices in quantum mechanics. We therefore conclude that each irreducible representation of the Lorentz group is specified by two numbers n, m, each of which can be an integer or half-integer with the dimensionality $(2n+1)$ and $(2m+1)$. So the representations can be characterized in increasing dimensionality as $(0,0), (1/2,0), (0,1/2), (1,0), (0,1), \ldots$ etc.

We have already stumbled across two of these representations, one corresponding to the trivial representation by a 1×1 matrix ('scalar') and the other corresponding to the defining representation with the generators represented by 4×4 matrices ('four-vector') given explicitly in Eq. (5.65). The really interesting case, for our purpose, arises when we study the representations corresponding to $(1/2,0)$ and $(0,1/2)$. These representations have dimension 2 and hence must be represented by 2×2 matrices. From the commutation relations in Eq. (5.73), it is clear that we can take these matrices to be two copies of the standard Pauli matrices. They will act on two component objects ψ_α with $\alpha = 1, 2$ treated as a column vector. We will denote these spinors by the L, R notation (introduced in the last section), in anticipation of the fact that they will turn out to be identical to the beasts we came across earlier. That is, we use the notation $(\psi_L)_\alpha$ with $\alpha = 1, 2$ for a spinor in the representation $(1/2, 0)$ and by $(\psi_R)_\alpha$ with $\alpha - 1, 2$ for a spinor in the representation $(0, 1/2)$. We could call ψ_L as the left-handed Weyl spinor and ψ_R as the right-handed Weyl spinor.[40] To make the connection with the last section, we have to explicitly determine how these spinors transform, which requires determining the explicit form of the generators J, K while acting on the Weyl spinors. In the case of the $(1/2, 0)$ representation, b is represented by $\sigma/2$ while $a = 0$. This leads to the result[41]

$$J = a + b = \frac{\sigma}{2}; \qquad K = -i(a - b) = i\frac{\sigma}{2} \qquad (5.74)$$

We can now write down the explicit transformation of the left-handed Weyl spinor using Eq. (5.71) which leads to

$$\psi_L \to L_{(1/2)L}\psi_L = \exp\left\{(-i\boldsymbol{\theta} - \boldsymbol{\eta}) \cdot \frac{\sigma}{2}\right\}\psi_L \qquad (5.75)$$

[40] As we shall see, they transform differently under the action of the Lorentz group, a fact we already know from the discussion in the last section. These are also called 'dotted' and 'undotted' spinors in the literature.

[41] Note that, in this representation the generator K is *not* Hermitian, which is consistent with a theorem that noncompact groups have no non-trivial unitary representations of finite dimensions. This is, in turn, related to the fact that a and b are *complex* combinations of J and K. This has some important group theoretical consequences which, however, we will not get into.

In the case of the $(0, 1/2)$ representation, we have $\boldsymbol{J} = \boldsymbol{\sigma}/2$ and $\boldsymbol{K} = -i\boldsymbol{\sigma}/2$ and the corresponding transformation law is

$$\psi_R \to L_{(1/2)R}\psi_R = \exp\left\{(-i\boldsymbol{\theta} + \boldsymbol{\eta}) \cdot \frac{\boldsymbol{\sigma}}{2}\right\}\psi_R \qquad (5.76)$$

This is precisely what we found earlier for ψ_L and ψ_R in the last section working out separately for rotations (see Eq. (5.38)) and boosts (see Eq. (5.47)). The infinitesimal forms of these relations are given by

$$\delta\psi_R = \frac{1}{2}[(-i\boldsymbol{\theta} + \boldsymbol{\eta}) \cdot \boldsymbol{\sigma}]\psi_R; \qquad \delta\psi_L = \frac{1}{2}[(-i\boldsymbol{\theta} - \boldsymbol{\eta}) \cdot \boldsymbol{\sigma}]\psi_L \qquad (5.77)$$

The corresponding results for the adjoint spinor are

$$\delta\psi_R^\dagger = \frac{1}{2}(i\boldsymbol{\theta} + \boldsymbol{\eta}) \cdot (\psi_R^\dagger\boldsymbol{\sigma}); \qquad \delta\psi_L^\dagger = \frac{1}{2}(i\boldsymbol{\theta} - \boldsymbol{\eta}) \cdot (\psi_L^\dagger\boldsymbol{\sigma}) \qquad (5.78)$$

Before concluding this section, let us settle this business of something becoming the negative of itself under a 2π rotation. To be precise, this can never happen under the action of the Lorentz group. Almost by definition, the rotation group, which is a subgroup of Lorentz group, map 2π rotations to identity elements. So, if we are really obtaining the representations of the Lorentz group, we cannot have the funny business of something becoming the negative of itself under a 2π rotation. Actually, by exponentiating the elements of the Lie algebra, we have in fact generated a different group called SL(2,C) which is known as the universal covering group of the Lorentz group. To be precise, spinors transform under the representations of SL(2,C) rather than under the representations of the Lorentz group SO(1,3).

The reason this is not a problem in QFT is because one has the freedom in the choice of the phase in quantum theory so that two amplitudes differing by a sign lead to the same probability. Mathematically, this means that we need not have an exact representation of the Lorentz group but can be content with what is known as the *projective representations* of the Lorentz group. If we have two matrices, $M(g1)$ and $M(g2)$ corresponding to two group elements $g1$ and $g2$, then the usual requirement of a representation is that $M(g1)M(g2) = M[g1 \circ g2]$. But if we allow the projective representations, we can also have $M(g1)M(g2) = e^{i\phi}M[g1 \circ g2]$ where ϕ could, in general, depend on $g1$ and $g2$. It turns out that the projective representations of SO(1,3) are the same as the representations of the SL(2,C) and that is how we get spinors in our midst. The flip of sign $\psi \to -\psi$ under the $\theta = 2\pi$ rotation is closely related to the fact that the 3-D rotation group is not simply connected.

While all this is frightfully interesting, it does not make any difference to what we will be doing. So we will continue to use phrases like "spinor representations of Lorentz group" etc. without distinguishing between projective representations and honest-to-God representations.

5.4.5 Why do we use the Dirac Spinor?

As a bonus, we also now know how the 4-spinor ψ built from a pair of two spinors (ψ_L, ψ_R) transforms under Lorentz transformations. From the discussion in the previous section, this has to be in accordance with the rule in Eq. (5.46) reproduced below:

$$\psi \to L_{1/2}\psi = \exp\left(-\frac{i}{2}\omega_{mn}S^{mn}\right)\psi; \qquad S^{mn} \equiv \frac{i}{4}[\gamma^m, \gamma^n] \qquad (5.79)$$

Formally, one says that the Dirac spinor (which is another name for this 4-spinor) transforms under the *reducible* representation $(1/2, 0) \oplus (0, 1/2)$ of the Lorentz group. For this to work, it is necessary that S^{mn} defined in Eq. (5.79) must satisfy the commutation rules for the generators of the Lorentz group in Eq. (5.66); i.e., if we set $J^{mn} = S^{mn}$ in Eq. (5.66), it should be identically satisfied. You can verify that this is indeed true. This shows that S^{mn} can also provide a representation for the Lorentz group algebra, albeit a reducible one. We saw earlier [see Eq. (5.65)] that the transformation of the four-vectors provides an *irreducible* 4-dimensional representation of the Lorentz group. What we have here is another, different 4-dimensional representation of the Lorentz group but a reducible one.

Exercise 5.7: Verify this.

But how come the Dirac equation uses the 4-spinor ψ (belonging to a *reducible* representation) rather than either ψ_L or ψ_R which belong to the *irreducible* representations? The reason has to do with parity. We saw earlier that, under the parity transformation, ψ_L and ψ_R are interchanged[42] but ψ goes to $\gamma^0 \psi$ keeping the Dirac equation invariant. (See Eq. (5.53).) In other words, if we use ψ_L and ψ_R separately, we do not get a basis which is true to parity transformations. While parity is indeed violated by the weak interactions, the physics at scales below 100 GeV or so, should conserve parity. So the field we choose should provide a representation for both Lorentz transformation and parity transformation. This is the fundamental reason why we work with ψ rather than with ψ_L or ψ_R individually.[43]

In the Weyl representation we have used, $L_{1/2}$ came up as a block diagonal matrix and one could separate out the Dirac spinor into L and R components quite trivially. But we know that one could have equally well used some other representation related to the Weyl representation by:

$$\gamma^m \to U\gamma^m U^{-1}; \qquad \psi \to U\psi \qquad (5.80)$$

in which case it is not obvious how one could define the left and right handed components in terms of ψ. Is there an invariant way of effecting this separation?

Indeed there is and this brings us to yet another gamma matrix[44] defined as

$$\gamma^5 \equiv -i\gamma^0\gamma^1\gamma^2\gamma^3 \qquad (5.81)$$

You can easily verify that $\{\gamma^5, \gamma^m\} = 0$, $(\gamma^5)^2 = +1$. Further, γ^5 commutes with S^{mn} so that it behaves as a scalar under rotations and boosts. Since $(\gamma^5)^2 = 1$, we can now define two Lorentz invariant projection operators

$$P_L \equiv \frac{1}{2}(1 - \gamma^5); \quad P_R \equiv \frac{1}{2}(1 + \gamma^5) \qquad (5.82)$$

which obey the necessary properties $P_L^2 = P_L$, $P_R^2 = P_R$, $P_L P_R = 0$ of projection operators. These operators P_L and P_R allow us to project out ψ_L and ψ_R from ψ by the relations

$$\psi_L = \frac{1}{2}(1 - \gamma^5)\psi; \quad \psi_R = \frac{1}{2}(1 + \gamma^5)\psi \qquad (5.83)$$

One can trivially check that, in the Weyl representation, γ^5 is given by

$$\gamma^5 = \begin{pmatrix} 1 & 0 \\ 0 & -1 \end{pmatrix} \qquad (5.84)$$

[42] More generally, under $(t, \boldsymbol{x}) \to (t, -\boldsymbol{x})$, we have $\boldsymbol{K} \to -\boldsymbol{K}$ and $\boldsymbol{J} \to \boldsymbol{J}$. This leads to the interchange $\boldsymbol{a} \leftrightarrow \boldsymbol{b}$. Therefore, under the parity transformation, an object in the (n, m) representation is transformed to an object in the (m, n) representation.

[43] In fact, in the standard model of the electroweak interaction, which we will not discuss, the left and right handed components of spin-1/2 particles enter the theory in a different manner in order to capture the fact that weak interactions do not conserve parity.

[44] With this notation we do not have a γ^4 and you jump from γ^3 to γ^5. This is related to calling time, the zero-th dimension rather than the fourth dimension! If time was the fourth dimension, we would have γ^a with $a = 1, 2, 3, 4$, and the next in line would have been γ^5!

and hence these relations are obvious. Since everything is done in an invariant manner, this projection will work in any basis.

We thus discover that the four component Dirac spinor — we (and Dirac) stumbled on by taking a fancy square root — actually arises from a much deeper structure. Just to appreciate this fact, let us retrace the steps which we followed in this section to arrive at the Dirac spinor.

(1) We note that the set of all Lorentz boosts plus rotations form a group with six parameters that specify the six rotational "angles" in the six planes of the 4-dimensional space.

(2) We identify the six generators $J^{mn} = -J^{nm}$ of this group and determine their commutator algebra — which is given by Eq. (5.66).

(3) We look for $k \times k$ matrices which satisfy the same commutator algebra as these generators to determine all the matrix representations of the Lorentz group. In addition to the trivial representation ($k = 1$), we already know a representation by 4×4 matrices which is used in the transformation law for the normal four-vectors.

(4) We separate out the Lorentz algebra into two decoupled copies of angular momentum algebra given by Eq. (5.73). This allows us to classify all the representations of the Lorentz group by (n, m) where n and m are either integers or half-integers.

(5) This takes us to the spinorial representations of the Lorentz group $(1/2, 0)$ and $(0, 1/2)$. The corresponding 2-spinors ψ_L and ψ_R transform identically under rotations but differently under Lorentz boosts. These are the basic building blocks from which all other representations can be constructed.

(5) One can obtain a *reducible* representation of Lorentz group by building a 4-spinor out of the pair (ψ_L, ψ_R). This is the Dirac spinor which appears in the Dirac equation.

5.4.6   Spin of the Field

Finally, let us clarify how the notion of spin arises from the Lorentz transformation property for any field. As an example, let us consider the local variation of the left-handed Weyl spinor under the transformation $x \to x' = x + \delta x$, $\psi_L(x) \to \psi'_L(x') = \mathcal{R}\psi_L(x)$, which is given by

$$
\begin{aligned}
\delta_0 \psi_L &\equiv \psi'_L(x) - \psi_L(x) = \psi'_L(x' - \delta x) - \psi_L(x) \\
&= \psi'_L(x') - \delta x^r \partial_r \psi_L(x) - \psi_L(x) \\
&= (\mathcal{R} - 1)\psi_L(x) - \delta x^r \partial_r \psi_L(x)
\end{aligned}
\tag{5.85}
$$

We now see that the variation has two parts. One comes from the variation of the coordinates δx^r, which is nonzero even for a *spinless* scalar field, as we have seen before. This will lead to the standard orbital angular momentum contribution given by

$$
-\delta x^r \partial_r \psi_L = -\frac{i}{2}\omega_{mn} J^{mn}\psi_L; \qquad J^{mn} \equiv i(x^m \partial^n - x^n \partial^m)
\tag{5.86}
$$

On the other hand, in the term involving $(\mathcal{R} - 1)$, the Taylor expansion of the Lorentz transformation group element $\mathcal{R} = \exp[-(i/2)\omega_{mn}S^{mn}]$ will give an *additional* contribution. So, the net variation is given by $\delta_0 \psi_L = -(i/2)\omega_{mn}\mathcal{J}^{mn}\psi_L$ where $\mathcal{J}^{mn} = J^{mn} + S^{mn}$.

This result is completely general and describes how we get the correct spin for the field from its Lorentz transformation property. The orbital part J^{mn} comes from kinematics and has the same structure for all fields (including the spinless scalar) while the extra contribution to the angular momentum — arising from the spin — depends on the specific representation of the Lorentz group which the field represents. In the special case of the left-handed Weyl spinor, we know the explicit form of S^{mn} from Eq. (5.75). This tells us that

$$S^\alpha = \frac{1}{2}\epsilon^{\alpha\beta\gamma}S^{\beta\gamma} = \frac{\sigma^\alpha}{2}; \qquad S^{\alpha 0} = i\frac{\sigma^\alpha}{2} \qquad (5.87)$$

The left-handed Weyl spinors correspond to $\boldsymbol{S} = \boldsymbol{\sigma}/2$. In fact, the right-handed Weyl spinor will also lead to the same result except for a flip of sign in $S^{\alpha 0}$.

5.4.7 The Poincare Group

The quantum fields which we study (like the scalar, spinor, vector etc.) are directly related to the representations of the Lorentz group in the manner we have outlined above. But to complete the picture, we also need to demand the theory to be invariant under spacetime translations in addition to the Lorentz transformations.

A generic element of the spacetime translation group has the structure $\exp\{-ia_m P^m\}$ where a_m defines the translation $x^m \to x^m + a^m$ and P^m are the standard generators of the translation. Lorentz transformations, along with spacetime translations, form a larger group usually called the *Poincare group*. Let us briefly study some properties of this group.

The first thing we need to do is work out all the new commutators involving P^m. Since translations commute, we have the trivial relation $[P^m, P^n] = 0$. To find the commutation rule between P^m and J^{rs}, we can use the following trick. Since energy is a scalar under spatial rotations, we must have $[J^\alpha, P^0] = 0$. Similarly, since P^α is a 3-vector under rotations, we have $[J^\alpha, P^\beta] = i\epsilon^{\alpha\beta\gamma}P^\gamma$. What we need is a 4-dimensional, Lorentz covariant commutation rule which incorporates these two results. This rule is unique and is given by

Exercise 5.8: Prove this.

$$[P^m, J^{rs}] = i(\eta^{mr}P^s - \eta^{ms}P^r) \qquad (5.88)$$

Along with the standard commutation rules of the Lorentz group, this completely specifies the structure of the Poincare group. In terms of the $1+3$ split, these commutation rules read (in terms of $J^\alpha, K^\alpha, P^0 \equiv H$ and P^α):

$$\left[J^\alpha, J^\beta\right] = i\epsilon^{\alpha\beta\gamma}J^\gamma, \quad \left[J^\alpha, K^\beta\right] = i\epsilon^{\alpha\beta\gamma}K^\gamma, \quad \left[J^\alpha, P^\beta\right] = i\epsilon^{\alpha\beta\gamma}P^\gamma$$
$$(5.89)$$
$$\left[K^\alpha, K^\beta\right] = -i\epsilon^{\alpha\beta\gamma}J^\gamma, \quad \left[P^\alpha, P^\beta\right] = 0, \quad \left[K^\alpha, P^\beta\right] = iH\delta^{\alpha\beta} \qquad (5.90)$$
$$[J^\alpha, H] = 0, \quad [P^\alpha, H] = 0, \quad [K^\alpha, H] = iP^\alpha \qquad (5.91)$$

The physical meaning of these commutation rules is obvious. Equation (5.89) tells us that J^α generates spatial rotations under which K^α and P^α transform as vectors. Equation (5.91) tells us that J^α and P^α commute with the Hamiltonian (which is the time translation generator) and hence are conserved quantities. On the other hand, K^α is not conserved which is why the eigenvalues of K^α do not assume much significance.

At the classical level, relativistic invariance is maintained if we use fields which transform sensibly under the Poincare group. At the quantum level, these fields become operators and they act on states in a Hilbert space. In particular, the one-particle state (which emerges from the vacuum through the action of a field) should exhibit proper behaviour under the action of the Poincare group in order to maintain relativistic invariance at the quantum level. So we need to understand how the Poincare group is represented in the one-particle states of the Hilbert space.

One particle states, which carry such a representation, will be labelled by the Casimir operators — which are the operators that commute with all the generators— of the Poincare group. One Casimir operator is obviously $P^2 \equiv P_m P^m$ which clearly commutes with all other generators. If we label[45] the one-particle state by $|\boldsymbol{p}, s\rangle$ where $\boldsymbol{p}$ is the momentum of the particle and s denotes all other variables, then P^2 has the value m^2 in this one-particle state. It follows that one of the labels which characterize the one particle state is the mass m of the particle, which makes perfect sense.

There is another Casimir operator for the Poincare group, given by $W^2 = W_i W^i$ where

$$W^m = -\frac{1}{2}\epsilon^{mnrs} J_{nr} P_s \qquad (5.92)$$

is called the *Pauli-Lubanski* vector. So, we can use its eigenvalue to further characterize the one-particle states. Let us first consider the situation when $m \neq 0$. We can then compute this vector in the rest frame of the particle, getting $W^0 = 0$, and:

$$W^\alpha = \frac{m}{2}\epsilon^{0\alpha\beta\gamma} J^{\beta\gamma} = \frac{m}{2}\epsilon^{\alpha\beta\gamma} J^{\beta\gamma} = mJ^\alpha \qquad (5.93)$$

Clearly, the eigenvalues of W^2 are proportional to those of J^2 which we know are given by $j(j+1)$ where j is an integer or half integer. Identifying this with the spin of a one-particle state with mass m and spin j, we have the result

$$-W_m W^m |\boldsymbol{p}, j\rangle = m^2 j(j+1)|\boldsymbol{p}, j\rangle, \qquad (m \neq 0) \qquad (5.94)$$

So, the massive one particle states are labelled by[46] the mass m and the spin j. As a bonus we see that a massive particle of spin j will have $2j+1$ degrees of freedom.

The massless case, in contrast, turns out to be more subtle. We can now bring the momentum to the form $P^m = (\omega, 0, 0, \omega)$. Let us ask what further subgroup of Poincare group will leave this vector unchanged. (Such a subgroup is called the Little Group.) An obvious candidate is the group of rotations in the (x, y) plane corresponding to SO(2) and this obvious choice will turn out to be what we will use. But there is actually another — less evident[47] — Lorentz transformation which leaves $P^m = (\omega, 0, 0, \omega)$ invariant. The generator for this transformation is given by

$$\Lambda = e^{-i(\alpha A + \beta B + \theta C)} \qquad (5.95)$$

where α, β and θ are parameters and A, B, C are matrices with the following elements

$$C^m{}_n = (J^3)^m{}_n; \quad A^m{}_n = (K^1 + J^2)^m{}_n; \quad B^m{}_n = (K^2 - J^1)^m{}_n \qquad (5.96)$$

These operators J^3, A, B obey the commutation rules:

$$[J^3, A] = +iB, \quad [J^3, B] = -iA, \quad [A, B] = 0 \qquad (5.97)$$

Exercise 5.9: Prove that W^2 commutes with all generators. (Hint: This is fairly easy. W^2, being a Lorentz invariant scalar, clearly commutes with J^{mn}. Further, from its explicit form and the antisymmetry of ϵ^{mnrs}, it is easy to see that it commutes with P^m as well.)

[46]There is an intuitive way of understanding this result. When $m \neq 0$, suitable Lorentz transformations can reduce P^m to the form $P^m = (m, 0, 0, 0)$. Having made this choice, we still have the freedom of performing spatial rotations. That is, the space of one-particle states with momentum fixed as $P^m = (m, 0, 0, 0)$ can still act as a basis for the representation of spatial rotations. Since we also want to include spinor representations, the relevant group is SU(2). This brings in the second spin label $j = 0, 1/2, 1, \ldots$

[47]The most straightforward way of obtaining this result is to consider an infinitesimal Lorentz transformation with parameters ω^{mn} and impose the condition $\omega^{mn} P_n = 0$ with $P^m = (\omega, 0, 0, \omega)$. Solving these equations will determine the structure of the group element generating the transformation. Identifying with the result in Eq. (5.96) is then a matter of guess work.

This is formally the same algebra as that generated by the operators $p^x, p^y, L^z = xp^y - yp^x$ which describe translations and rotations of the Euclidean plane. Also note that the matrices $A^m{}_n$ and $B^m{}_n$ (with one upper and one lower index) are non-Hermitian; this is as it should be, since they provide a finite dimensional representation of the non-compact Lorentz generators. For our purpose, what is important is the expression for W^2 which is now given by

$$-W_m W^m = \omega^2 (A^2 + B^2) \qquad (5.98)$$

In principle, we should now consider one-particle states $|\boldsymbol{p}; a, b\rangle$ with labels corresponding to eigenvalues (a, b) of A and B as well. It is then fairly easy to show that, unless $a = b = 0$, one can find a *continuous* set of eigenvalues for labeling this state with an internal degree of freedom θ (see Problem 11). While it is a bit strange that nothing in nature seems to correspond to this structure, the usual procedure is to choose the one-particle states with eigenvalues $a = b = 0$. Therefore, from Eq. (5.98), we conclude that the massless one-particle states have[48] $W^2 = 0$.

Once we have eliminated this possibility, what remains as the Little Group are the rotations in the $x - y$ plane, viz., SO(2). The irreducible representations of SO(2) are one-dimensional and the generator is the angular momentum J^3. So the massless representations can be labelled by the eigenvalue h of J^3 which represents the angular momentum in the direction of propagation of the massless particle. (This is called *helicity*.) Taking a cue from standard angular momentum lore, one would have thought that $h = 0, \pm 1/2, \pm 1$ etc. Again, this result turns out to be true but its proof requires using the fact that the universal covering group of the Lorentz group is SL(2,C) and that is a double covering.[49]

One — rather non-obvious — conclusion we arrive at is that massless particles are characterized by a single value h of the helicity. The helicity $h = \boldsymbol{P} \cdot \boldsymbol{J}$ is a pseudoscalar. So if the interactions conserve parity, to each state of helicity h there must exist another state of helicity $-h$ and the question arises whether both represent the same particle. Since electromagnetic interactions conserve parity, it seems sensible to represent the photon through a representation of Poincare group *and* parity, corresponding to the two states $h = \pm 1$. Then the massless photon will have two degrees of freedom. In contrast, for massless neutrinos which participate only in the weak interactions (we always ignore gravity, because we don't understand it!) which do not conserve parity, we can associate $h = -1/2$ with a neutrino and $h = +1/2$ with an antineutrino.

So, guided by physical considerations plus the group structure, we associate with different particles, the relevant states in the Hilbert space such that they transform under unitary representations of the Poincare group. As we said before, all such representations should necessarily be infinite dimensional; a single particle state $|\psi\rangle$ will be an eigenstate of P_a with $P_a|\psi\rangle = p_a|\psi\rangle$ for the momentum eigenvalues p_a with $p^0 > 0$ and $p^2 \geq 0$. The $|\psi\rangle$ will transform as $|\psi\rangle \to \exp(i\theta_{ab} S^{ab})|\psi\rangle$ where the boost and rotation are specified by the parameters θ_{ab} and S^{ab} are the generators of Lorentz group relevant to the particle.[50] This Hilbert space is complete with respect to these states in the sense that

$$1 = \sum_\psi \int d\Gamma_\psi |\psi\rangle\langle\psi|; \qquad d\Gamma_\psi \equiv \prod_{j \in \psi} \frac{d^3 p_j}{(2\pi)^3} \frac{1}{2E_j} \qquad (5.99)$$

[48] This agrees well with the limit $m \to 0$ in Eq. (5.94). One could have taken the easy way out and argued based on this idea. However, it is intriguing that a continuous one-parameter representation is indeed allowed by the Little Group structure for massless particles. Nobody really knows what to make of it, so we just ignore it.

[49] The standard proof (for SU(2)) that j_z is quantized is purely algebraic. One considers the matrix elements of $j_x + ij_y$ and using the angular momentum commutation relations, shows that a contradiction will ensue unless j_z is quantized. Unfortunately, for the Little Group of massless particles, we do not have any j_x, j_y and we are treating the generator j_z of SO(2) as a single entity. This is why the usual proof fails.

[50] Multi-particle states behave in a similar way with the transformations induced by the representations corresponding to each of the particles.

where $d\Gamma_\psi$ is the Lorentz invariant phase space element (except for an overall delta function) for the total momentum.

5.5   Dirac Equation and the Dirac Spinor

5.5.1   The Adjoint Spinor and the Dirac Lagrangian

Once we think of $\psi(x)$ as a field with definite Lorentz transformation properties, it is natural to ask whether one can construct a Lorentz invariant action for this field, varying which we can obtain the Dirac equation — rather than by taking dubious square-roots — just as we did for the scalar and the electromagnetic fields. Let us now turn to this task.

This would require constructing scalars, out of the Dirac spinor, which could be used to build a Lagrangian. Normally, given a column vector ψ, one would define an adjoint row vector $\psi^\dagger \equiv (\psi^*)^T$ and hope that $\psi^\dagger\psi$ will be a scalar. This will not work in the case of the Dirac spinor. Under a Lorentz transformation, we have $\psi \to L_{1/2}\psi$, $\psi^\dagger \to \psi^\dagger L_{1/2}^\dagger$; but $\psi^\dagger\psi$ does not[51] transform as a Lorentz scalar because $L_{1/2}^\dagger L_{1/2} \neq 1$.

[51] This is related to the fact mentioned earlier, viz. that the representation of the Lorentz algebra is not unitary.

However, this situation can be easily remedied as follows. Since in the Weyl representation of the gamma matrices we have $(\gamma^0)^\dagger = \gamma^0$ and $(\gamma^\alpha)^\dagger = -\gamma^\alpha$, it follows that $\gamma^0\gamma^m\gamma^0 = (\gamma^m)^\dagger$, and hence

$$(S^{mn})^\dagger = -(i/4)\left[(\gamma^n)^\dagger,(\gamma^m)^\dagger\right] = (i/4)[(\gamma^m)^\dagger,(\gamma^n)^\dagger] \tag{5.100}$$

leading to the result

$$\gamma^0(S^{mn})^\dagger\gamma^0 = \frac{i}{4}[\gamma^0(\gamma^m)^\dagger\gamma^0,\gamma^0(\gamma^n)^\dagger\gamma^0] = \frac{i}{4}[\gamma^m,\gamma^n] = S^{mn} \tag{5.101}$$

Therefore,

$$\begin{aligned}(\gamma^0 L_{1/2}\gamma^0)^\dagger &= \gamma^0\exp(i\theta_{mn}S^{mn})^\dagger\gamma^0 = \exp(-i\theta_{mn}\gamma^0 S^{mn\dagger}\gamma^0)\\ &= \exp(-i\theta_{mn}S^{mn}) = L_{1/2}^{-1}\end{aligned} \tag{5.102}$$

Exercise 5.10: Verify that $\bar\psi\gamma_m\psi$ transforms as a vector.

giving us

$$\psi^\dagger\gamma^0\psi \to (\psi^\dagger L_{1/2}^\dagger)\gamma^0(L_{1/2}\psi) = (\psi^\dagger\gamma^0 L_{1/2}^{-1}L_{1/2}\psi) = \psi^\dagger\gamma^0\psi \tag{5.103}$$

Therefore, if we define a *Dirac adjoint spinor* $\bar\psi$ by the relation

$$\bar\psi(x) = \psi^\dagger(x)\gamma^0 \tag{5.104}$$

it follows that $\bar\psi\psi$ transforms as a scalar under the Lorentz transformations. In a similar manner, we can show that $\bar\psi\gamma^m\psi$ transforms as a four-vector under the Lorentz transformations.

Given these two results, we can construct the Lorentz invariant action for the Dirac field to be

$$\mathcal{A} = \int d^4x\,\bar\psi(x)\,(i\gamma^m\partial_m - m)\psi(x) \equiv \int d^4x L_D \tag{5.105}$$

Since ψ is complex, we could vary ψ and $\bar\psi$ independently in this action; varying $\bar\psi$ leads to the Dirac equation, while varying ψ leads to the equation for the adjoint spinor which, of course, does not tell you anything more.[52]

[52] To be precise, we only want the equation of motion to be Lorentz invariant rather than the action being a scalar. (The so called $\Gamma\Gamma$ action in GR is not generally covariant but its variation gives the covariant Einstein's equations.) If you had constructed an action A' by using $\psi^\dagger$ rather than $\bar\psi$ in Eq. (5.105), you can still obtain the Dirac equation for ψ by varying $\psi^\dagger$. But if you vary ψ you will end up getting a wrong equation which is not consistent with the Dirac equation. (Incidentally, this is a good counter example to the folklore myth that equations obtained from a variational principle will be automatically consistent.) The real problem with A', as far as variational principle goes, is that it is not real while A is. While we can be cavalier about the action being a Lorentz scalar, we do not want it to become complex.

(If you want a more symmetric treatment of ψ and $\bar{\psi}$, you can write down another action by integrating Eq. (5.105) by parts and allowing ∂_m to act on $\bar{\psi}$. If you then add the two and divide by 2, you end up getting a more symmetric form of the action.)

It is also instructive to rewrite the Dirac Lagrangian in terms of the two spinors with $\psi = (\psi_L, \psi_R)$. A simple calculation gives

$$L_D = i\psi_L^\dagger \bar{\sigma}^m \partial_m \psi_L + i\psi_R^\dagger \sigma^m \partial_m \psi_R - m(\psi_L^\dagger \psi_R + \psi_R^\dagger \psi_L) \qquad (5.106)$$

where σ^m and $\bar{\sigma}^m$ are defined in Eq. (5.31). You could easily verify that $\psi_L^\dagger \bar{\sigma}^m \psi_L$ and $\psi_R^\dagger \sigma^m \psi_R$ transforms as a four-vector under the Lorentz transformations and the pair $\psi_L^\dagger \psi_R$ and $\psi_R^\dagger \psi_L$ are invariant. For example, using Eq. (5.77) and Eq. (5.78), it is easy to see that $\psi_R^\dagger \psi_R$ is not Lorentz invariant because

$$\delta(\psi_R^\dagger \psi_R) = \frac{1}{2}\psi_R^\dagger[(-i\boldsymbol{\theta}+\boldsymbol{\eta})\cdot\boldsymbol{\sigma}]\psi_R + \frac{1}{2}(i\boldsymbol{\theta}+\boldsymbol{\eta})\cdot(\psi_R^\dagger\boldsymbol{\sigma})\psi_R = \boldsymbol{\eta}\cdot(\psi_R^\dagger\boldsymbol{\sigma}\psi_R) \neq 0 \tag{5.107}$$

On the other hand, $\psi_L^\dagger \psi_R$ is indeed Lorentz invariant:

$$\delta(\psi_L^\dagger \psi_R) = \psi_L^\dagger \frac{1}{2}[(-i\boldsymbol{\theta}+\boldsymbol{\eta})\cdot\boldsymbol{\sigma}]\psi_R + \frac{1}{2}(i\boldsymbol{\theta}-\boldsymbol{\eta})\cdot(\psi_L^\dagger\boldsymbol{\sigma}\psi_R) = 0 \quad (5.108)$$

From these, we can form two real combinations $\psi_L^\dagger \psi_R + \psi_R^\dagger \psi_L$ and $i(\psi_L^\dagger \psi_R - \psi_R^\dagger \psi_L)$. But since parity interchanges ψ_L and ψ_R, only the first combination is a scalar (while the second one is a pseudoscalar). This explains the structure of the terms which occur in the Dirac Lagrangian in Eq. (5.106) in terms of the two-spinors.

Varying ψ_L^* and ψ_R^* in Eq. (5.106) we again get the Dirac equation in terms of ψ_R and ψ_L obtained earlier in Eq. (5.36):

$$\bar{\sigma}^m i\partial_m \psi_L = m\psi_R, \qquad \sigma^m i\partial_m \psi_R = m\psi_L \qquad (5.109)$$

Acting by $\sigma^m i\partial_m$ on both sides of the first equation, and using the second and the identity $\{\sigma^m, \bar{\sigma}^n\} = 2\eta^{mn}$, we immediately obtain

$$(\Box + m^2)\psi_L = 0 \qquad (5.110)$$

and a similar result for ψ_R. Thus each component of the Dirac equation satisfies the Klein-Gordon equation with mass m.

5.5.2 Charge Conjugation

Another important transformation of the Dirac spinor corresponds to an operation called *charge conjugation*. To see this in action, let us start with the Dirac equation in the presence of an electromagnetic field

$$[i\gamma^m(\partial_m - ieA_m) - m]\psi = 0 \qquad (5.111)$$

obtained by the usual substitution $\partial_m \to D_m \equiv \partial_m - ieA_m$. Taking the complex conjugate of this equation will give

$$[-i\gamma^{m*}(\partial_m + ieA_m) - m]\psi^* = 0 \qquad (5.112)$$

But since $-\gamma^{m*}$ obeys the same algebra as γ^m, it must be possible to find a matrix M such that $-\gamma^{m*} = M^{-1}\gamma^m M$; if we now further define

Exercise 5.11: Prove that the global symmetry of the Dirac Lagrangian under $\psi \to e^{i\alpha}\psi$, leads to a conserved current proportional to the $\bar{\psi}\gamma^m\psi$ we met earlier.

$\psi_c \equiv M\psi^*$, then it follows that ψ_c satisfies the same Dirac equation with the sign of the charge reversed:

$$[i\gamma^m(\partial_m + ieA_m) - m]\psi_c = 0 \qquad (5.113)$$

It is easy to verify that $\gamma^2\gamma^{m*}\gamma^2 = \gamma^m$. Therefore, we can define the charge conjugated[53] spinor ψ_c as:

$$\psi_c \equiv \gamma^2\psi^* \qquad (5.114)$$

Roughly speaking, if ψ describes the electron, ψ_c will describe the positron.

[53]The only imaginary gamma matrix is γ^2 in both Dirac and Weyl representations.

5.5.3   Plane Wave solutions to the Dirac Equation

We saw earlier (see Eq. (5.110)) that each component of the Dirac equation satisfies the Klein-Gordon equation with mass m. So a good basis for the solutions of the Dirac equation is provided by the plane wave modes $\exp(\pm ipx)$. Let us consider the solutions of the form $u(p,s)e^{-ipx}$ and $v(p,s)e^{ipx}$ for ψ where

$$(\gamma^a p_a - m)u(p,s) = 0; \quad (\gamma^a p_a + m)v(p,s) = 0 \qquad (5.115)$$

The variable $s = \pm 1$ takes care of the two possibilities: spin-up and spin-down. The two spinors u and v behave like ψ under the Lorentz transformation and, in particular, $\bar{u}u$ and $\bar{v}v$ are Lorentz scalars. In the rest frame with $p^i = (m, 0)$, we have $\gamma^j p_j - m = m(\gamma^0 - 1)$. Hence, the spinor solution to the Dirac equation in the rest frame $u(\boldsymbol{p} = 0, s)$ will satisfy $(\gamma^0 - 1)u = 0$. All we now need to do is to solve this in the rest frame and normalize the solutions in a Lorentz invariant manner; then we will have a useful set of plane wave solutions for the Dirac equation. The expressions for $u(p,s)$ and $v(p,s)$ in an arbitrary frame can be obtained by using a Lorentz boost since we know how the Dirac spinor changes under a Lorentz boost. There are many ways to do this and we will illustrate two useful choices.

In the Weyl basis, $(\gamma^0 - 1)u = 0$ implies that u is made of a pair of identical two-spinors (ζ, ζ). (We have suppressed the spin dependence $\zeta(s)$ in the rest frame to keep the notation simple.) Similarly, v is made of a pair of spinors $(\eta, -\eta)$. For example, we can choose this set to be

$$u_\uparrow = \begin{pmatrix} 1 \\ 0 \\ 1 \\ 0 \end{pmatrix} \quad \text{and} \quad u_\downarrow = \begin{pmatrix} 0 \\ 1 \\ 0 \\ 1 \end{pmatrix}; \quad v_\uparrow = \begin{pmatrix} -1 \\ 0 \\ 1 \\ 0 \end{pmatrix} \quad \text{and} \quad v_\downarrow = \begin{pmatrix} 0 \\ 1 \\ 0 \\ -1 \end{pmatrix} \qquad (5.116)$$

One convention for normalizing this spinor is by $u^\dagger u = 2E_{\boldsymbol{p}}$ which implies $\zeta^\dagger\zeta = m$ in the rest frame. We will rescale the spinor and take it to have the form $\zeta = \sqrt{m}\,\xi$ in the rest frame. Under a Lorentz boost, these spinors transform as

$$u(p,s) = \mathcal{R}u(m,s) = \begin{pmatrix} \mathcal{R}_L & 0 \\ 0 & \mathcal{R}_R \end{pmatrix}\begin{pmatrix} \sqrt{m}\,\xi \\ \sqrt{m}\,\xi \end{pmatrix} = \begin{pmatrix} \sqrt{m}\,\mathcal{R}_L\xi \\ \sqrt{m}\,\mathcal{R}_R\xi \end{pmatrix} \qquad (5.117)$$

Using the result in Eq. (5.50) and using $\gamma = (E/m), \gamma\beta\boldsymbol{n} = \boldsymbol{p}/m$, we find that the boosted spinors have the form

$$u(p,s) = \begin{pmatrix} \sqrt{E - \boldsymbol{p}\cdot\boldsymbol{\sigma}}\,\xi \\ \sqrt{E + \boldsymbol{p}\cdot\boldsymbol{\sigma}}\,\xi \end{pmatrix}; \quad v(p,s) = \begin{pmatrix} \sqrt{E - \boldsymbol{p}\cdot\boldsymbol{\sigma}}\,\eta \\ -\sqrt{E + \boldsymbol{p}\cdot\boldsymbol{\sigma}}\,\eta \end{pmatrix} \qquad (5.118)$$

Another choice is to work in the Dirac representation and use the normalization $\bar{u}u = 1$. If we choose the Dirac representation of the γ-matrices, then the four independent spinors we need: $u(p, +1), v(p, +1)$ and $v(p, -1), u(p, -1)$ can be taken to be the column vectors

$$u = \begin{pmatrix} 1 \\ 0 \\ 0 \\ 0 \end{pmatrix} \text{ and } \begin{pmatrix} 0 \\ 1 \\ 0 \\ 0 \end{pmatrix}; \quad v = \begin{pmatrix} 0 \\ 0 \\ 1 \\ 0 \end{pmatrix} \text{ and } \begin{pmatrix} 0 \\ 0 \\ 0 \\ 1 \end{pmatrix} \tag{5.119}$$

This choice corresponds to the implicit normalization conditions of the form

$$\bar{u}(p, s)u(p, s) = 1, \quad \bar{v}(p, s)v(p, s) = -1, \quad \bar{u}v = 0, \quad \bar{v}u = 0 \tag{5.120}$$

Obviously, Lorentz invariance plus basis independence tells us that these relations should hold in general.[54] If you now do the Lorentz boost and work through the algebra you will get the solution in an arbitrary frame to be:

$$u(p, s) = \frac{\gamma^a p_a + m}{\sqrt{2m(E_{\mathbf{p}} + m)}}u(0, s); \quad v(p, s) = \frac{-\gamma^a p_a + m}{\sqrt{2m(E_{\mathbf{p}} + m)}}v(0, s) \tag{5.121}$$

It is often useful to express these normalization conditions in a somewhat different form. To do this, note that in the rest frame, we have the relation

$$\sum_s u_a(p, s)\bar{u}_b(p, s) = \frac{1}{2}\left(\gamma^0 + 1\right)_{ab}$$

$$\sum_s v_a(p, s)\bar{v}_b(p, s) = \frac{1}{2}\left(\gamma^0 - 1\right)_{ab} \tag{5.122}$$

To express this in a more general form, we have to rewrite the right hand side as a manifestly Lorentz invariant expression. It is possible to do this rather easily if you know the result! The trick is to write[55]

$$\sum_s u_a(p, s)\bar{u}_b(p, s) = \left(\frac{\gamma^a p_a + m}{2m}\right)_{ab}$$

$$\sum_s v_a(p, s)\bar{v}_b(p, s) = \left(\frac{\gamma^a p_a - m}{2m}\right)_{ab} \tag{5.123}$$

and verify that it holds in the rest frame. If you did not know the result, you could reason as follows: There is a theorem which states that any 4×4 matrix can be expressed as a linear combination of a set of 16 matrices[56]

$$\{1, \gamma^m, \sigma^{mn}, \gamma^m\gamma^5, \gamma^5\} \tag{5.124}$$

So the left hand side of Eq. (5.123), being a 4×4 matrix, should be expressible as a linear combination of these matrices; all we need to do is to reason out which of these can occur and with what coefficients. You cannot have γ^5 or $\gamma^m\gamma^5$ because of parity; you cannot have σ^{mn} because of Lorentz invariance when you realize that you only have a single Lorentz vector p^a available. Therefore, the right hand side must be a linear combination of $\gamma^a p_a$ and m. You can fix the relative coefficient by operating from the left by $\gamma^a p_a$. The overall normalization needs to be fixed by taking $\alpha = \beta$ and summing over α.

[54] Notice that $\bar{u}u = 1$ while $\bar{v}v = -1$; we do not have a choice in this matter. Also note that, under the charge conjugation, the individual spinor components change as $(u_\uparrow)^c = v_\downarrow, (u_\downarrow)^c = v_\uparrow$, and $(v_\uparrow)^c = u_\downarrow, (v_\downarrow)^c = u_\uparrow$. In other words, charge conjugation takes particles into antiparticles and flips the spin.

Exercise 5.12: Obtain Eq. (5.121).

[55] If we use the alternative normalisation of $\bar{u}u = 2m$ in the rest frame, we need to redefine u and v by scaling up a factor $(\sqrt{2m})$ so that you don't have the factor $2m$ on the right hand side of Eq. (5.123). This is particularly useful if you want to study a massless fermion.

[56] A simple way to understand this result is as follows. We know that $\gamma^m\gamma^n = \pm\gamma^n\gamma^m$ where the sign is plus if $m = n$ and is minus if $m \neq n$. This allows us to reorder any product of gamma matrices as $\pm\gamma^0\ldots\gamma^0\gamma^1\ldots\gamma^1\gamma^2\ldots\gamma^2\gamma^3\ldots\gamma^3$. In each of the sets, because γ^a squares to ±1, we can simplify this product to a form $\pm(\gamma^0\text{or }1)\times(\gamma^1\text{or }1)\times(\gamma^2\text{or }1)\times(\gamma^3\text{or }1)$. So the net result of multiplying any number of gamma matrices will be — upto a sign or $\pm i$ factor — one of the following 16 matrices: 1 or γ^m; or $\gamma^m\gamma^n$ with $m \neq n$ which can be written $(1/2)\gamma^{[m}\gamma^{n]}$; or a product like $\gamma^l\gamma^m\gamma^n$ with different indices which, in turn, can be expressed in terms of $(1/6)\gamma^{[l}\gamma^m\gamma^{n]} = i\epsilon^{lmnr}\gamma_r\gamma^5$; or $\gamma^0\gamma^1\gamma^2\gamma^3 = -i\gamma^5$. These are clearly 16 linearly independent 4×4 matrices. Given the fact that any 4×4 matrix has only 16 independent components, it follows that any matrix can be expanded in terms of these.

Another combination which usually comes up in the calculations is $\bar{u}(p')\gamma_m u(p)$. There is an identity (called the *Gordon identity*) which allows us to express terms of the kind $\bar{u}(p')\gamma_m u(p)$ in a different form involving $p + p'$ and $q \equiv p - p'$. This identity is quite straightforward to prove along the following lines:[57]

$$
\begin{aligned}
\bar{u}(p')\gamma_m u(p) &= \frac{1}{2m}\bar{u}(p')(\gamma^n p'_n \gamma_m + \gamma_m \gamma^n p_n)u(p) \\
&= \frac{1}{2m}\bar{u}(p')(p'^n \gamma_n \gamma_m + \gamma_m \gamma_n p^n)u(p) \\
&= \frac{1}{2m}\bar{u}(p')(p'^n[\eta_{mn} + i\sigma_{mn}] + [\eta_{mn} - i\sigma_{mn}]p^n)u(p) \\
&= \frac{1}{2m}\bar{u}(p')[(p'_m + p_m) + i\sigma_{mn}q^n]u(p) \qquad (5.125)
\end{aligned}
$$

The existence of the σ^{mn} term tells you that, when coupled to an electromagnetic potential in the Fourier space through $\bar{u}(p')\gamma_m u(p)A^m(q)$ where $q = p' - p$ is the relevant momentum transfer, we are led to a coupling involving the term

$$
\sigma^{mn}q_n A_m = \frac{1}{2}\sigma^{mn}(q_n A_m - q_m A_n) = \frac{1}{2}\sigma^{mn}F_{nm} \qquad (5.126)
$$

in the interaction. This is the real source of a magnetic moment coupling arising in the Dirac equation. We will need these relations later on while studying QED.

5.6 Quantizing the Dirac Field

5.6.1 Quantization with Anticommutation Rules

Now that we have found the mode functions which satisfy the Dirac equation, it is straightforward to elevate the Dirac field as an operator and attempt its quantization. The mode expansion in plane waves will be

$$
\psi(x) = \int \frac{d^3\boldsymbol{p}}{(2\pi)^{3/2}(E_{\boldsymbol{p}}/m)^{1/2}} \sum_s \left[b(\boldsymbol{p},s)u(\boldsymbol{p},s)e^{-ipx} + d^\dagger(\boldsymbol{p},s)v(\boldsymbol{p},s)e^{ipx}\right]
$$

$$(5.127)$$

As usual, the px in the exponentials are evaluated on-shell with $p^0 = E_p = (\boldsymbol{p}^2 + m^2)^{1/2}$. In the right hand side, we have integrated over the momentum $\boldsymbol{p}$ and summed over the spin s. The overall normalization is slightly different from what we are accustomed to (e.g., in the case of a scalar field) and has a $m^{1/2}$ factor; but this is just a matter of choice in our normalisation of the mode functions. Obviously one could rescale u and v to change this factor. The b and $d^\dagger$ are distinct, just as in the case of the complex scalar field we encountered in Sect. 3.5 (see Eq. (3.148)), because ψ is complex. Here b annihilates the electrons and $d^\dagger$ creates the positrons; both operations have the effect of reducing the amount of charge by $e = -|e|$.

The key new feature in the quantization, of course, has to do with the replacement of the commutation relations between the creation and the annihilation operators by *anti-commutation* relations. That is, for the creation and annihilation operators for the electron, we will impose the conditions

$$
\{b(\boldsymbol{p},s), b^\dagger(\boldsymbol{p}',s')\} = \delta^{(3)}(\boldsymbol{p} - \boldsymbol{p}')\delta_{ss'}
$$

$$
\{b(\boldsymbol{p},s), b(\boldsymbol{p}',s')\} = 0; \qquad \{b^\dagger(\boldsymbol{p},s), b^\dagger(\boldsymbol{p}',s')\} = 0 \quad (5.128)
$$

There is a corresponding set of relations for the creation and annihilation operators for the positrons involving d and $d^\dagger$.

The rest of the discussion proceeds in the standard manner. One starts with a vacuum state annihilated by b and d and constructs the one particle state by acting on it with $b^\dagger$ or $d^\dagger$. You cannot go any further because $(b^\dagger)^2 = (d^\dagger)^2 = 0$ showing that you cannot put two electrons (or two positrons) in the same state. Since we have done much of this in the case of bosonic particles, we will not pause and describe them in detail, except to highlight the key new features.

An obvious question to ask is what happens to the *field* commutation rules, when we impose Eq. (5.128). You would have guessed that they will also obey anticommutation rules rather than commutation rules. This is in fact easy to verify for, say, the anticommutator $\{\psi(x), \bar{\psi}(x')\}$ at equal times $t = t'$. The calculation is simplified by using translational invariance and setting $t = t' = 0$ as well as $\boldsymbol{x}' = 0$. This will give, with straightforward algebra using the anticommutation rule for the creation and annihilation operators, the result

$$\{\psi(\boldsymbol{x}, 0), \bar{\psi}(0)\} = \int \frac{d^3\boldsymbol{p}}{(2\pi)^3(E_{\boldsymbol{p}}/m)} \tag{5.129}$$
$$\times \sum_s \left[u(\boldsymbol{p}, s)\bar{u}(\boldsymbol{p}, s)e^{-i\boldsymbol{p}\cdot\boldsymbol{x}} + v(\boldsymbol{p}, s)\bar{v}(\boldsymbol{p}, s)e^{i\boldsymbol{p}\cdot\boldsymbol{x}} \right]$$

We now use the result in Eq. (5.123) to obtain[58]

$$\{\psi(\boldsymbol{x}, 0), \bar{\psi}(0)\} = \int \frac{d^3\boldsymbol{p}}{(2\pi)^3(2E_{\boldsymbol{p}})} \left[(\gamma^a p_a + m)e^{-i\boldsymbol{p}\cdot\boldsymbol{x}} + (\gamma^a p_a - m)e^{i\boldsymbol{p}\cdot\boldsymbol{x}} \right]$$
$$= \int \frac{d^3\boldsymbol{p}}{(2\pi)^3(2E_{\boldsymbol{p}})} \, 2p^0\gamma^0 e^{-i\boldsymbol{p}\cdot\boldsymbol{x}} = \gamma^0\delta^{(3)}(\boldsymbol{x}) \tag{5.130}$$

In other words,[59]

$$\{\psi_a(\boldsymbol{x}, t), i\psi_b^\dagger(\boldsymbol{0}, t)\} = i\delta^{(3)}(\boldsymbol{x})\,\delta_{ab} \tag{5.131}$$

In the same way, one can also show that anticommutators $\{\psi, \psi\} = 0$ and $\{\psi^\dagger, \psi^\dagger\} = 0$. One could have also postulated these anticommutation rules for the field operators and obtained the corresponding results for the creation and annihilation operators.

This anticommutation rule which we use for fermions has several nontrivial implications all round. As a first illustration of the peculiarities involved, let us compute the Hamiltonian for the field in terms of the creation and annihilation operators. We begin with the standard definition of the Hamiltonian density given by

$$\mathcal{H} = \pi \frac{\partial\psi}{\partial t} - L = \bar{\psi}(i\gamma^\alpha\partial_\alpha + m)\psi \tag{5.132}$$

Note that the last expression has only the spatial derivatives which can be converted into a time derivative using the Dirac equation. This allows us to express the Hamiltonian in the form:

$$H = \int d^3x\, \mathcal{H} = \int d^3x\, \bar{\psi}i\gamma^0\frac{\partial\psi}{\partial t} \tag{5.133}$$

Plugging in the mode expansion for ψ and carrying out the spatial integration, it is easy to reduce[60] this to the form:

[58] We have also used the fact that the integral vanishes when the integrand is an odd function of $\boldsymbol{p}$.

[59] From the structure of the Dirac Lagrangian, $L = \bar{\psi}(i\gamma^m\partial_m - m)\psi$, we find that the canonical momentum is indeed given by $\pi_\psi = [\partial L/\partial(\partial_0\psi)] = i\bar{\psi}\gamma^0 = i\psi^\dagger$. So the choice of variables for imposing the commutation rules makes sense; but, of course, it is the anticommutator rather than the commutator which we use.

[60] Make sure you understand the manner in which various terms come about. Very schematically, $\bar{\psi}$ contributes $(b^\dagger + d)$ while ψ contributes $(b - d^\dagger)$ where the minus sign arises from $\partial_t\psi$. So the product goes like $(b^\dagger + d)(b - d^\dagger) \sim b^\dagger b - dd^\dagger$ because the condition $\bar{v}u = 0$ kills the cross terms.

$$H = \int \frac{d^3\boldsymbol{p}}{(2\pi)^3} \sum_s E_{\boldsymbol{p}} \left[b^\dagger(\boldsymbol{p},s)b(\boldsymbol{p},s) - d(\boldsymbol{p},s)d^\dagger(\boldsymbol{p},s) \right] \tag{5.134}$$

When we re-order the second term with $d^\dagger$ to the left of d, we have to use the anticommutation rule and write

$$-d(\boldsymbol{p},s)d^\dagger(\boldsymbol{p},s) = d^\dagger(\boldsymbol{p},s)d(\boldsymbol{p},s) - \delta^{(3)}(\boldsymbol{0}) \tag{5.135}$$

The last term is the Dirac delta function in *momentum* space, evaluated for zero momentum separation; that is:

$$\delta^{(3)}(\boldsymbol{0}) = \lim_{\boldsymbol{p} \to 0} \int d^3\boldsymbol{x} \exp[i\boldsymbol{p} \cdot \boldsymbol{x}] = \int d^3\boldsymbol{x} \tag{5.136}$$

which is just the (infinite) normalisation volume. Substituting Eq. (5.135) into Eq. (5.134), we get the final expression for the Hamiltonian to be

$$H = \int \frac{d^3\boldsymbol{p}}{(2\pi)^3} \sum_s E_{\boldsymbol{p}} \left[b^\dagger(\boldsymbol{p},s)b(\boldsymbol{p},s) + d^\dagger(\boldsymbol{p},s)d(\boldsymbol{p},s) \right]$$
$$- \int d^3\boldsymbol{x} \int \frac{d^3\boldsymbol{p}}{(2\pi)^3} \sum_s E_{\boldsymbol{p}} \tag{5.137}$$

The first two terms are the contributions from electrons and positrons of energy $E_{\boldsymbol{p}}$ corresponding to the momentum $\mathbf{p}$ and spin s. This makes complete sense. The last term has the form:

$$E_0 = - \int d^3\boldsymbol{x} \int \frac{d^3\boldsymbol{p}}{(2\pi)^3} \sum_s 2 \left(\frac{1}{2} E_{\boldsymbol{p}} \right) \tag{5.138}$$

This is our old friend, the zero-point energy, with $(1/2)E_{\boldsymbol{p}}$ being contributed by an electron and a positron separately, for each spin state. But the curious fact is that it comes with an *overall minus sign* unlike the bosonic degrees of freedom which contribute $+(1/2)E_{\boldsymbol{p}}$ per mode. This is yet another peculiarity of the fermionic degree of freedom.[61]

In our approach, we postulated the Pauli exclusion principle for fermions and argued that this requires the creation and annihilation operators to anticommute rather than commute. This, in turn, leads to the anticommutation rules for the field with all the resulting peculiarities. A somewhat more formal approach, usually taken in textbooks, is to start from the anticommutation rules for the field and obtain the anticommutation rules for the creation and annihilation operators as a result. In such an approach, one would first show that something will go wrong if we use the usual commutation rules rather than anticommutation rules for the field operators. What goes wrong is closely related to the expression for the energy we have obtained, which — once we throw away the zero-point energy — is a positive definite quantity. If you quantize the Dirac field using the *commutators*, you will find that the term involving $d^\dagger d$ in Eq. (5.137) will come with a negative sign and the energy will be unbounded from below. This is the usual motivation given in textbooks for using anticommutators rather than commutators for quantizing the Dirac field.[62] It is possible to provide a rigorous proof that one should quantize fermions with anticommutators and bosons with commutators by using very formal mechanisms of quantum field theory; but we will not discuss it.[63]

[61] So, if nature has the same number of bosonic and fermionic degrees of freedom, then the total zero-point energy will vanish — which would be rather pleasant. This is what happens in theories with a hypothetical symmetry, called *supersymmetry*, which the particle physicists are fond of. So far, we have seen no sign of it, having discovered only roughly half the particles predicted by the theory; but many live in hope.

Exercise 5.13: Try this out.

[62] You may object saying that even the expression in Eq. (5.137) is infinitely negative because of the zero-point energy. But remember that you are not supposed to raise issues about zero-point energy in a QFT course; or rather, you make it go away by introducing the normal ordering prescription we had discussed earlier.

[63] The fact that such a simple, universal rule — viz., that fermions obey the Pauli exclusion principle — requires the formidable machinery of QFT for its proof, is probably because we do not understand fermions at a sufficiently deep level. But most physicists would disagree with this view point.

5.6.2 Electron Propagator

Having quantized the Dirac field, we can now determine its propagator. A cheap, but useful, trick to obtain the propagator is the following. We know that when any field Ψ obeys an equation of the kind $\hat{K}(\partial_a)\Psi(x) = 0$, the propagator in momentum space is given by $(1/K(p_a))$ and the real space propagator, fixed by translational invariance, is just:

$$G(x - y) = \int \frac{d^4p}{(2\pi)^4} \frac{e^{-ip\cdot(x-y)}}{K(p)} \tag{5.139}$$

For example, the Klein-Gordon field has $K(p) = p^2 - m^2 + i\epsilon$ and the corresponding propagator is precisely what we find[64] using the prescription in the above equation. (See e.g., Eq. (1.96).) For the Dirac field, $\hat{K}(\partial_a) = i\gamma^a p_a - m$ and hence $K(p) = \gamma^a p_a - m + i\epsilon$. Hence, the propagator in momentum space (which is conventionally written with an extra i-factor) is given by:

$$iS(p) \equiv \frac{i}{\gamma^a p_a - m + i\epsilon} = i\frac{(\gamma^a p_a + m)}{p^2 - m^2 + i\epsilon} \tag{5.140}$$

where the second expression is obtained by multiplying both the numerator and the denominator by $(\gamma^a p_a + m)$. This propagator, like everything else in sight, is a 4×4 matrix. The corresponding real space propagator is given by

$$iS(x) = \int \frac{d^4p}{(2\pi)^4} e^{-ip\cdot x} \frac{i}{\gamma^a p_a - m + i\epsilon} = \int \frac{d^4p}{(2\pi)^4} e^{-ip\cdot x} \frac{i(\gamma^a p_a + m)}{p^2 - m^2 + i\epsilon} \tag{5.141}$$

In fact, one can write this even more simply as

$$S(x) = (i\gamma^a \partial_a + m) \int \frac{d^4p}{(2\pi)^4} \frac{e^{-ip\cdot x}}{p^2 - m^2 + i\epsilon} = (i\gamma^a \partial_a + m)G(x) \tag{5.142}$$

where $G(x)$ is the scalar field propagator.

If we want to do this more formally (without the cheap trick), we have to define the propagator in terms of the time ordered product $T[\psi(x)\bar{\psi}(0)]$ of the Dirac operators, in a Lorentz invariant manner. The definition of this operator was quite straightforward for bosonic fields, leading to a relativistically invariant expression. But in the case of fermionic fields we again have to be a little careful in its definition.

To see what is involved, let us take the propagator in Eq. (5.141), obtained above through our trick, and try to write it in terms of the *three* dimensional momentum integral involving $d^3\boldsymbol{p}$ rather than d^4p. As usual, when we do the p^0 integration in the complex plane, we have poles at $p^0 = \pm(E_{\boldsymbol{p}} - i\epsilon)$. So, for $x^0 > 0$, we need to close the contour in the lower half plane to make the factor $\exp(-ip^0 x^0)$ converge. Going around the pole at $+[E_{\boldsymbol{p}} - i\epsilon]$ in the clockwise direction (see Fig. 5.1(a)), we get

$$iS(x) = (-i)i \int \frac{d^3\boldsymbol{p}}{(2\pi)^3} e^{-ipx} \frac{\gamma^a p_a + m}{2E_{\boldsymbol{p}}} \tag{5.143}$$

On the other hand, when $x^0 < 0$, we have to close the contour in the upper half plane but now we go around the pole at $-[E_{\boldsymbol{p}} - i\epsilon]$ in the anticlockwise direction. This leads to

$$iS(x) = i^2 \int \frac{d^3\boldsymbol{p}}{(2\pi)^3} e^{iE_{\boldsymbol{p}}x^0 + i\boldsymbol{p}\cdot\boldsymbol{x}} \frac{1}{-2E_{\boldsymbol{p}}}(-E_{\boldsymbol{p}}\gamma^0 - p^\alpha\gamma_\alpha + m) \tag{5.144}$$

[64]Of course, we need to add a positive imaginary part to m to get the correct $i\epsilon$ factor; this rule is universal.

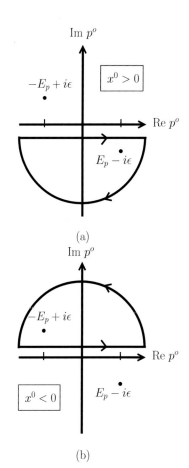

Figure 5.1: Contours to compute the p^0-integral for the propagator.

We now change the integration variable from $\boldsymbol{p}$ to $-\boldsymbol{p}$ to get the Lorentz invariant factors e^{ipx} and $\gamma^a p_a$. Then this contribution becomes:

$$iS(x) = -\int \frac{d^3\boldsymbol{p}}{(2\pi)^3} e^{ipx} \frac{1}{2E_{\boldsymbol{p}}} (\gamma^a p_a - m) \tag{5.145}$$

Putting these two together, we get:

$$iS(x) = \int \frac{d^3\boldsymbol{p}}{(2\pi)^3 (E_{\boldsymbol{p}}/m)} \left[\theta(x^0) \frac{\gamma^a p_a + m}{2m} e^{-ipx} - \theta(-x^0) \frac{\gamma^a p_a - m}{2m} e^{ipx} \right] \tag{5.146}$$

This suggests that, if we have to get the same propagator from the time-ordered product of the fields, we need to introduce a crucial *minus* sign in the time ordered product and define it as

$$T\psi(x)\bar\psi(0) \equiv \theta(x^0)\psi(x)\bar\psi(0) - \theta(-x^0)\bar\psi(0)\psi(x) \tag{5.147}$$

There is actually a simple reason why the time ordering operator for fermions needs to be defined with a minus sign. We know that fermionic annihilation operators, for example, anticommute for generic momenta and times. Therefore,

$$T\{a_p(t)a_q(t')\} = -T\{a_q(t')a_p(t)\} \tag{5.148}$$

So, if we try to define the time ordering operation as merely moving the operators around and putting them in time order, this equation will imply that time ordered product vanishes! To prevent this from happening, we should define the time ordering operator for fermions by keeping track of the minus signs which arise when we move the anticommuting operators through each other. So, the correct definition will require the extra minus sign as in:

$$T\{\psi(x)\chi(y)\} = \psi(x)\chi(y)\theta(x_0 - y_0) - \chi(y)\psi(x)\theta(y_0 - x_0) \tag{5.149}$$

Having motivated the minus sign from reverse engineering our result in Eq. (5.141), we could just take Eq. (5.147) to be our definition and compute the vacuum expectation value $\langle 0|T[\psi(x)\bar\psi(0)]|0\rangle$. An elementary calculation now gives, for $x^0 > 0$

$$\begin{aligned} iS(x) &= \langle 0|\psi(x)\bar\psi(0)|0\rangle = \int \frac{d^3\boldsymbol{p}}{(2\pi)^3 (E_{\boldsymbol{p}}/m)} \sum_s u(\boldsymbol{p}, s)\bar u(\boldsymbol{p}, s) e^{-ipx} \\ &= \int \frac{d^3\boldsymbol{p}}{(2\pi)^3 (E_{\boldsymbol{p}}/m)} \frac{\gamma^a p_a + m}{2m} e^{-ipx} \end{aligned} \tag{5.150}$$

For $x^0 < 0$, one needs to be slightly more careful, but it is again easy to obtain the result

$$\begin{aligned} iS(x) &= -\langle 0|\bar\psi(0)\psi(x)|0\rangle = -\int \frac{d^3\boldsymbol{p}}{(2\pi)^3 (E_{\boldsymbol{p}}/m)} \sum_s \bar v(\boldsymbol{p}, s)v(\boldsymbol{p}, s) e^{-ipx} \\ &= -\int \frac{d^3\boldsymbol{p}}{(2\pi)^3 (E_{\boldsymbol{p}}/m)} \left(\frac{\gamma^a p_a - m}{2m} \right) e^{-ipx} \end{aligned} \tag{5.151}$$

Comparing with the result in Eq. (5.146), we find that everything turns out fine once we have defined the time ordering operator with the minus sign.

Another closely related object we could compute is the anticommutator at *unequal* times, defined by

$$iS = \{\psi(x), \bar{\psi}(y)\} \tag{5.152}$$

The computation is straightforward and is very similar to the one we did above. The final result can be expressed in the form

$$iS(x-y) = (i\gamma^a \partial_a + m)[G(x-y) - G(y-x)]; \quad \partial_a \equiv \frac{\partial}{\partial x^a} \tag{5.153}$$

where

$$G(x-y) = \int \frac{d^3\boldsymbol{p}}{(2\pi)^3 2E_{\boldsymbol{p}}} e^{-ip(x-y)} \tag{5.154}$$

is the standard scalar field propagator. The other two anticommutators vanish

$$\{\psi(x), \psi(y)\} = \{\bar{\psi}(x), \bar{\psi}(y)\} = 0 \tag{5.155}$$

This result, at the face of it, should ring some alarm bells. In the case of the scalar field, we computed the corresponding quantity, viz. the *commutator* $[\phi(x), \phi(y)]$ which is exactly the term within the square bracket in Eq. (5.153). We also saw that $[\phi(x), \phi(y)] = 0$ for $(x-y)^2 < 0$, and made a song and dance of it, claiming that this fact — viz. the scalar fields at two spacelike separated events commute with each other — is an expression of causality. This doesn't hold for the Dirac field. Instead, what we get from the known structure of $G(x)$ is the result:

$$\{\psi(x), \bar{\psi}(y)\} = 0; \qquad (x-y)^2 < 0 \tag{5.156}$$

That is, the *anti-commutator* (rather than the commutator) vanishes outside the light cone. *So, what happened to causality in the presence of fermions?*

The only way to save the theory is to declare that $\psi(x)$ is not a direct observable in the laboratory.[65] But if that is the case, what *are* the observables for the fermionic fields? We would expect them to be built out of quadratic bilinear of the field, like

$$O(x) = \bar{\psi}_a(x) \, O_{ab}(x) \, \psi_b(x) \tag{5.157}$$

where $O_{ab}(x)$ is essentially a Dirac matrix consisting of c-numbers and differential operators.[66] We would like to have the *commutator*, (not the anticommutator), $[O(x), O(y)]$ of *these* observables to vanish for spacelike separations, otherwise we are sunk! Let us calculate this commutator which is given by

$$[O(x), O(y)] = O_{ab}(x) \, O_{cd}(y) \, [\bar{\psi}_a(x)\psi_b(x), \bar{\psi}_c(y)\psi_d(y)] \tag{5.158}$$

Using the operator identities

$$[A, BC] = [A, B] \, C + B \, [A, C]; \qquad [A, BC] = \{A, B\} \, C - B \, \{A, C\} \tag{5.159}$$

we can expand out the expression and relate it to the basic anticommutators of the fields at unequal times:

$$[\bar{\psi}_a(x)\psi_b(x), \bar{\psi}_c(y)\psi_d(y)] \tag{5.160}$$
$$= \bar{\psi}_a(x) \left(\{\psi_b(x), \bar{\psi}_c(y)\} \, \psi_d(y) - \bar{\psi}_c(y) \, \{\psi_b(x), \psi_d(y)\} \right)$$
$$+ \left(\{\bar{\psi}_a(x), \bar{\psi}_c(y)\} \, \psi_d(y) - \bar{\psi}_c(y) \, \{\bar{\psi}_a(x), \psi_d(y)\} \right) \psi_b(x)$$

[65] This might sound rather shady, but remember that spinors flip in sign if you rotate a coordinate by 2π. It is difficult to conceive of a measuring device in the lab which will become "the negative of itself" if we rotate the lab by 2π!

[66] A simple example of such an observable will be the Dirac current operator for which $O_{ab}^m = q(\gamma^m)_{ab}$.

On using the results in Eq. (5.153) and Eq. (5.155), we can reduce this to the form

$$
\begin{aligned}
\left[\bar{\psi}_a(x)\psi_b(x), \bar{\psi}_c(x)\psi_d(x)\right] &= i\mathcal{S}_{bc}(x-y)\bar{\psi}_a(x)\psi_d(y) \\
&\quad -i\mathcal{S}_{da}(y-x)\bar{\psi}_c(y)\psi_b(x) \quad (5.161)
\end{aligned}
$$

We now see that the vanishing of the *anti*-commutator $\mathcal{S}(x-y)$ at space-like separations (which we obtained earlier in Eq. (5.156)) now allows us to conclude that the *commutator* of the observables vanishes at spacelike separations:

$$
[O(x), O(y)] = 0 \qquad \text{for} \quad (x-y)^2 < 0 \qquad (5.162)
$$

Very gratifying, when you think about it. This is the conventional expression for causality.[67]

5.6.3   Propagator from a Fermionic Path Integral

In the study of the bosonic fields (like scalar or vector fields) we could also interpret the propagator using an external source. For example, if the Lagrangian for a free scalar field is taken to be $L = -(1/2)\phi D\phi$, then we could add a source term to the Lagrangian of the form $J\phi$ and perform the path integral over ϕ to define the vacuum persistence amplitude $Z[J] = \exp[-(1/2)JD^{-1}J]$. A double functional differentiation of $Z[J]$ allows us to define the propagator as D^{-1}.

In principle, we can do the same with the Dirac Lagrangian by adding suitable sources and performing the path integral. But if we treat ψ (and $\bar{\psi}$) as just c-number complex functions, we will get a result which is incorrect. This again has to do with the fact that the Dirac field anticommutes with itself, making $\psi^2 = 0$! While manipulating the Dirac field as a c-number function, one has to take into account this fundamental non-commutative nature of the fermionic fields. This can be done by introducing c-number quantities — called *Grassmannians* — which anticommute pairwise. We will now introduce the curious mathematics of Grassmannians and define the path integral for fermions using them.

In a way, Grassmannians are simpler than ordinary numbers. The fundamental property of these objects is that, for any two Grassmannians η and ξ, we have the relation $\eta\xi = -\xi\eta$ which immediately makes $\eta^2 = 0$ for any Grassmannian. As a result of this, the Taylor series expansion of any function $f(\eta)$ of a Grassmannian truncates with the linear term[68] giving $f(\eta) = a + b\eta$. As for integration over the Grassmannians, we will insist that the shifting of the dummy integration variable should not change the result. That is, we must have

$$
\int d\eta\, f(\eta + \xi) = \int d\eta\, f(\eta) \qquad (5.163)
$$

Using the fact that the functions on both sides are linear ones, we will get the condition that integral over η of $(b\xi)$ should vanish for all ξ. Since b is just an ordinary number, this leads to the first rule of Grassmannian integration

$$
\int d\eta = 0 \qquad (5.164)
$$

Further, given three Grassmannians, χ, η and ξ, we have $\chi(\eta\xi) = (\eta\xi)\chi$; i.e., the product $(\eta\xi)$ commutes with the Grassmannian χ. Based on this

---

[67] Note that causality arguments only prove that particles with integer spin should be quantized with commutation rules; it cannot be used to prove that particles with half-integer spins must be quantized using anticommutation rules. This is because observables for spinors are bilinear in spinors and essentially behave like integer spin particles.

[68] Recall the curious property of functions of Pauli matrices described by Eq. (5.9). You may now see a clear similarity between Grassmannians and structures involving Pauli or Dirac matrices.

motivation, we will declare that the product of any two anticommuting numbers should be an ordinary number and, in particular, the integral of η over η should be a pure number which — by convention — is taken to be unity. This leads to the second rule of Grassmannian integration

$$\int d\eta\, \eta = 1 \tag{5.165}$$

These two rules are sufficient to develop a consistent integral calculus involving the Grassmannians.

There are, however, some surprises in these integrals which you need to be aware of. For example, if $f(\eta)$ is a Grassmannian of the form $a + b\eta$, then the integral of f gives the result[69]

$$\int d\eta\, f(\eta) = \int d\eta\,(a + b\eta) = b \tag{5.166}$$

[69]So the integral of $f(\eta)$ over η is the same as the the the derivative of $f(\eta)$ with respect to η!

But if $f(\eta)$ is an ordinary number (so that b is a Grassmannian), the same integral gives you the result with a minus sign

$$\int d\eta\, f(\eta) = \int d\eta\,(a + b\eta) = -b \tag{5.167}$$

We can use these rules to compute Gaussian integrals over Grassmannians which is what we need in order to compute Fermionic path integrals. If we have a set of N Grassmannians denoted by $\eta = (\eta_1, \eta_2, \ldots \eta_N)$ and similarly for another set $\bar{\eta}$, then the Gaussian integral is given by

$$\int d\eta \int d\bar{\eta}\, e^{\bar{\eta}A\eta} = \det A \tag{5.168}$$

where $A = A_{ij}$ is an *anti-symmetric* $N \times N$ matrix. In the case of the Dirac Lagrangian, we can think of $\bar{\psi}$ and ψ as independent Grassmannian valued quantities and evaluate the generating functional in the form

$$\begin{aligned} Z &= \int D\psi \int D\bar{\psi}\, e^{i\int d^4x\, \bar{\psi}(i\gamma\partial - m + i\epsilon)\psi} \\ &= \det(i\gamma\partial - m + i\epsilon) = e^{\mathrm{tr}\,\log(i\gamma\partial - m + i\epsilon)} \end{aligned} \tag{5.169}$$

where we have omitted the overall multiplicative constant. This trace can be simplified using a trick (and the cyclic property of the trace) along the following lines:

$$\begin{aligned} \mathrm{tr}\,\log(i\gamma\partial - m) &= \mathrm{tr}\,\log\gamma^5(i\gamma\partial - m)\gamma^5 = \mathrm{tr}\,\log(-i\gamma\partial - m) \\ &= \frac{1}{2}\left[\mathrm{tr}\,\log(i\gamma\partial - m) + \mathrm{tr}\,\log(-i\gamma\partial - m)\right] \\ &= \frac{1}{2}\mathrm{tr}\,\log(\partial^2 + m^2) \end{aligned} \tag{5.170}$$

leading to the result

$$Z = \exp\left[\frac{1}{2}\mathrm{tr}\,\log(\partial^2 + m^2 - i\epsilon)\right] \tag{5.171}$$

In contrast, when we do the corresponding thing for a bosonic field, we end up getting the result

$$\begin{aligned} Z &= \int D\phi\, e^{i\int d^4x\,\frac{1}{2}[(\partial\phi)^2 - (m^2 - i\epsilon)\phi^2]} = \left(\frac{1}{\det[\partial^2 + m^2 - i\epsilon]}\right)^{1/2} \\ &= e^{-\frac{1}{2}\mathrm{tr}\,\log(\partial^2 + m^2 - i\epsilon)} \end{aligned} \tag{5.172}$$

So, essentially, there is a flip of sign in the generating function when we move from bosons to fermions.[70]

As for the propagator, we can now proceed very much along the way we did with the bosonic field. We first introduce two Grassmannian spinor sources η and $\bar{\eta}$ corresponding to ψ and $\bar{\psi}$. In the resulting generating functional, defined as,

$$Z(\eta, \bar{\eta}) = \int D\psi D\bar{\psi} \, \exp\left(i \int d^4x [\bar{\psi}(i\gamma\partial - m)\psi + \bar{\eta}\psi + \bar{\psi}\eta]\right) \quad (5.173)$$

we complete the square in the usual manner:

$$\bar{\psi}K\psi + \bar{\eta}\psi + \bar{\psi}\eta = (\bar{\psi} + \bar{\eta}K^{-1})K(\psi + K^{-1}\eta) - \bar{\eta}K^{-1}\eta \quad (5.174)$$

and perform the integration by shifting the integration measure. This leads to the (expected) result

$$Z(\eta, \bar{\eta}) = \exp\left(-i \int d^4x \, \bar{\eta}(i\gamma\partial - m)^{-1}\eta\right) \quad (5.175)$$

allowing us to identify the propagator as the inverse of the operator $(i\gamma\partial - m)$. This agrees with the result we obtained earlier.

Extending the path integral formalism to fermionic degrees of freedom brings within our grasp the technical prowess of the path integral to deal with issues in QED. We will describe two such applications.

5.6.4 Ward Identities

As a simple but important application of this, we will obtain a set of identities (usually called the *Ward-Takahashi identities*) which play a key role in QED. The importance of these identities lies in the fact that they relate to features of the exact, interacting, field theory.[71]

To derive these in a simple context, let us consider the effective action corresponding to the original action $S(\bar{\psi}, \psi, A_i)$ which depends on $(\bar{\psi}, \psi, A_i)$. As we described in Sect. 4.5, one can obtain the effective action as a Legendre transform by introducing a source term for the fields. In the example discussed in Sect. 4.5, we used just one scalar field ϕ and a source J with $\phi_c = \langle J|\phi|J\rangle$ denoting the expectation value in the state with source J. Now, since we have three fields $(\bar{\psi}, \psi, A_i)$, we need to introduce three sources $\bar{\eta}, \eta, J_i$ and define the effective action through a Legendre transform with respect to each of these sources. We will also have three background expectation values $\bar{\psi}_c, \psi_c, A_i^c$ which will act as the arguments for the effective action. In this case, the generating functional will be

$$Z[\bar{\psi}_c, \psi_c, A_i^c] \equiv \exp(iS_{\text{eff}}[\bar{\psi}_c, \psi_c, A_i^c]) \quad (5.176)$$

where $S_{\text{eff}}(\bar{\psi}_c, \psi_c, A_i^c)$ is the effective action obtained by the Legendre transform with respect to the sources $\bar{\eta}, \eta, J_i$. Let us suppose that we compute this effective action, say, order by order in a perturbation theory. To the lowest order, this would be given by the classical action which can be expressed in the form

$$S_{\text{eff}}^0[\bar{\psi}_c, \psi_c, A_i^c]$$
$$= \int d^4x d^4y \left[\frac{1}{2}A_c^i(x)D_{(0)ij}(x-y)A_c^j(y) + \bar{\psi}_c(x)S_{(0)}(x-y)\psi_c(y)\right]$$
$$- e \int d^4x \bar{\psi}_c(x)A_c^j(x)\gamma_j\psi_c(x) \quad (5.177)$$

where the subscript '0' indicates the free-field expressions for the propagators including the effect of any gauge fixing terms. When we compute the effective action, including higher order effects, we will expect it to have a form like

$$
S_{\text{eff}}[\bar{\psi}_c, \psi_c, A_c^m]
$$
$$
= \int d^4x \, d^4y \left[\frac{1}{2} A_c^m(x) D_{mn}(x-y) A_c^n(y) + \bar{\psi}_c(x) S(x-y) \psi_c(y) \right]
$$
$$
- e \int d^4x \, d^4y \, d^4z \, \bar{\psi}_c(x) A_c^m(y) \Lambda_m(x, y, z) \psi_c(z) + \cdots \quad (5.178)
$$

In this expression, D_{mn} and S would correspond to the photon and electron propagators incorporating the effect of interactions computed possibly upto a certain order in perturbation theory. Similarly, $\Lambda_m(x, y, z)$ incorporates the effect of interactions at the electron-photon vertex to the same order of approximation. (We will see an explicit computation of these quantities later in Sect. 5.7.)

We now use the fact that the properly defined interacting theory of QED must be gauge invariant. This, in turn, implies that the effective action itself must be invariant under the local gauge transformations, *except for any gauge fixing terms we might have added.* Let us assume for a moment that there are no gauge fixing terms in the action (which, of course, is not true and we will get back to it soon). Then the effective action must be invariant under the local gauge transformations given by

$$
\delta\psi = i\theta(z)\psi, \qquad \delta\bar{\psi} = -i\theta(z)\bar{\psi}, \qquad \delta A^m = \frac{1}{e}\partial^m\theta(z) \quad (5.179)
$$

This result can be expressed as

$$
0 = \delta S_{\text{eff}} = \int d^4x \left[\delta\bar{\psi}_c \frac{\delta S_{\text{eff}}}{\delta\bar{\psi}_c} + \frac{\delta S_{\text{eff}}}{\delta\psi_c}\delta\psi_c + \frac{\delta S_{\text{eff}}}{\delta A_c^m}\delta A_c^m \right]
$$
$$
= \int d^4z \left(-i\bar{\psi}_c \frac{\delta S_{\text{eff}}}{\delta\bar{\psi}_c} + i\psi_c \frac{\delta S_{\text{eff}}}{\delta\psi_c} - \frac{1}{e}\partial^m \frac{\delta S_{\text{eff}}}{\delta A_c^m} \right) \theta(z) \quad (5.180)
$$

where the last term has been rewritten by an integration by parts. Since θ is arbitrary, the expression within the brackets must vanish. The resulting relations are called the *Ward-Takahashi identities.*[72] We see that the term involving A^m will vanish only if the exact photon propagator satisfies the constraint

$$
\partial^m D_{mn}(x-y) = 0; \qquad k^m D_{mn}(k) = 0 \quad (5.181)
$$

where the second relation holds in the Fourier space.

Let us now get back to the issue of gauge fixing terms in the action. A gauge fixing term of the form

$$
-\frac{1}{2\lambda} \int d^4x (\partial_m A^m)(\partial_n A^n) = -\frac{1}{2\lambda} \int d^4x \, d^4y (\partial_m A^m)_x \delta(x-y)(\partial_n A^n)_y
$$
$$
= -\frac{1}{2\lambda} \int d^4x \, d^4y \, A^m(x)[\partial_m \partial_n \delta(x-y)] A^n
$$
$$
(5.182)
$$

will contribute a term proportional to $k_m k_n$ to $D_{mn}(k)$ in the Fourier space. When we repeat the calculation in Eq. (5.180) with a gauge fixing term

[72] In our case, we could have also obtained the same result more directly by performing a gauge transformation on Eq. (5.178).

added, δS_{eff} will not be zero but will be equal to the contribution from the gauge fixing term. This will change the result in Eq. (5.181) to the condition:

$$k^m D_{mn}(k) = -\frac{i\lambda k_n}{k^2} \tag{5.183}$$

This has an interesting implication. We know that, at the tree-level, the free-field photon propagator satisfies Eq. (5.183) (see Eq. (3.205)). So if we write the exact propagator, separating out the tree-level in the form

$$D_{mn}(k) = -\frac{i}{k^2}\left[\eta_{mn} - (1-\lambda)\frac{k_m k_n}{k^2}\right] + \mathcal{D}_{mn}(k) \tag{5.184}$$

then Eq. (5.183) implies that

$$k^m \mathcal{D}_{mn}(k) = 0 \tag{5.185}$$

In other words, the gauge parameter λ does not pick up any corrections due to interactions — which makes sense since we do not expect the interactions to take cognizance of an artificial gauge fixing term which we have added for our convenience. This result reinforces the fact that no physical phenomena in the theory can depend on the gauge parameter λ.

The constraint in Eq. (5.185), along with Lorentz invariance, tells you that the corrections to the photon propagator $\mathcal{D}_{mn}(k)$ must be expressible in the form $(k^2 g^{mn} - k^m k^n)\Pi(k^2)$ where $\Pi(k^2)$ is a function to be determined. We will see later that explicit computation does lead to this result.

As regards the terms involving $\bar{\psi}_c \psi_c$ in Eq. (5.180), we obtain the condition:

$$-iS(x-y)\delta(z-x) + iS(x-y)\delta(z-y) - \frac{\partial}{\partial z^m}\Lambda^m(x,z,y) = 0 \quad (5.186)$$

Fourier transforming this result, we obtain yet another identity in momentum space relating the exact electron propagator and the exact vertex function. Taking the limit of $p' \to p$, this result can be expressed in the momentum space in an equivalent form as

$$\Lambda_m(p,0,p) = \frac{\partial S_F}{\partial p^m} \tag{5.187}$$

We will have occasion to explicitly verify this relation later on.

5.6.5   Schwinger Effect for the Fermions

As a second application of the path integral over fermionic fields, we will compute the effective action for electromagnetism obtained by integrating out the electron field. This is in exact analogy with the corresponding result for the complex scalar field, obtained in Sect. 4.2. The idea again is to perform the integration over ψ and $\bar{\psi}$ in the full path integral:

$$Z = \int \mathcal{D}A_j \mathcal{D}\psi \mathcal{D}\bar{\psi} \, \exp(iS) \tag{5.188}$$

where

$$S = \int d^4x \left(-\frac{1}{4}F_{mn}^2 + \bar{\psi}(i\gamma D - m)\psi\right) \tag{5.189}$$

and we have not indicated the gauge fixing terms for simplicity. This will allow us to define the effective Lagrangian for the electromagnetic field as:

$$\int \mathcal{D}\psi \mathcal{D}\bar{\psi} \, \exp(iS) \equiv \exp iS_{\text{eff}}[A_j] = \exp i \int d^4x L_{\text{eff}}[A_j] \qquad (5.190)$$

This path integral can be calculated using a trick similar to the one which led to Eq. (5.171). We note that, from the result[73]

$$
\begin{aligned}
\text{Det}(i\gamma^a D_a - m) &= \exp\left[\text{Tr}\log(i\gamma^a D_a - m)\right] \\
&= \exp\left[\int d^4x \int \frac{d^4P}{(2\pi)^4} \text{Tr}\log(\gamma^a P_a - m)\right] \\
&= \exp\left[\int d^4x \int \frac{d^4P}{(2\pi)^4} \text{Tr}\log(-\gamma^a P_a - m)\right] \\
&= \text{Det}(-i\gamma^a D_a - m) \qquad (5.191)
\end{aligned}
$$

we have $\text{Tr}\langle x|\ln(i\gamma D - m)|x\rangle = \text{Tr}\langle x|\ln(-i\gamma D - m)|x\rangle$. Therefore, either expression is equal to half the sum. So:

$$\text{Tr}\langle x|\ln(i\gamma D - m)|x\rangle = \frac{1}{2}\text{Tr}\langle x|\ln(-(\gamma D)^2 - m^2)|x\rangle \qquad (5.192)$$

Therefore, the effective Lagrangian for the electromagnetic field is given by:

$$L_{\text{eff}} = -\frac{1}{4}F_{mn}^2 - \frac{i}{2}\text{Tr}\langle x|\ln(-(\gamma D)^2 - m^2)|x\rangle \qquad (5.193)$$

Expressing the logarithm using the integral representation, this becomes

$$L_{\text{eff}} = -\frac{1}{4}F_{mn}^2 + \frac{i}{2}\int_0^\infty \frac{ds}{s} e^{-ism^2}\text{Tr}\left[\langle x|e^{-is(\gamma D)^2}|x\rangle\right] \qquad (5.194)$$

Further, on using the result,

$$(\gamma D)^2 = D_m D^m + \frac{e}{2}F_{mn}\sigma^{mn} \qquad (5.195)$$

we can write

$$L_{\text{eff}} = -\frac{1}{4}F_{mn}^2 + \frac{i}{2}\int_0^\infty \frac{ds}{s} e^{-ism^2}\text{Tr}\left[\langle x|e^{i\left[\hat{p}-eA(\hat{x}))^2 - \frac{e}{2}F_{mn}\sigma^{mn}\right]s}|x\rangle\right] \qquad (5.196)$$

The rest of the calculation for computing the expectation value in Eq. (5.196) proceeds exactly as in the case of the complex scalar field, by reducing the problem to that of a harmonic oscillator. The final result can be expressed in a covariant form as:

$$L_{\text{eff}} = -\frac{1}{4}F_{mn}^2 - \frac{e^2}{8\pi^2}\int_0^\infty \frac{ds}{s} e^{-m^2 s}(a\cot eas)(b\coth ebs) \qquad (5.197)$$

where a and b are defined — as before, see Eq. (4.20) — by:

$$a^2 - b^2 = E^2 - B^2; \qquad ab = \boldsymbol{E}\cdot\boldsymbol{B} \qquad (5.198)$$

Here, and in what follows, ea, eb etc. stand for $|ea|$, $|eb|$ etc. which are positive definite.

As in the case of the scalar field, the expression for L_{eff} is divergent but these divergences are identical in structure to the ones obtained in the

[73] This is essentially a generalization of Eq. (5.170), proved here in a fancier way for fun!

Exercise 5.14: Obtain Eq. (5.197).

case of the scalar field. This can be seen by a perturbative expansion of the integrand in the coupling constant e, which will give

$$(a \cot eas)(b \coth ebs) = \frac{1}{e^2 s^2} - \frac{1}{3}(a^2 - b^2) - \frac{e^2 s^2}{45} \left[(a^2 - b^2) + 7(ab)^2 \right]$$
(5.199)

The first two terms lead to the divergences; the first one is an infinite constant independent of F_{ab} and can be ignored, while the second one — being proportion to $(a^2 - b^2) = (E^2 - B^2)$ — merely renormalizes the charge.[74] Subtracting out these two terms, we will get the finite part of the effective Lagrangian to be

$$L_{\text{eff}} = -\frac{1}{4} F_{mn}^2 - \frac{e^2}{32\pi^2} \int_0^\infty \frac{ds}{s} e^{-sm^2}$$
(5.200)

$$\times \left[4(a \cot eas)(b \coth ebs) - \frac{4}{e^2 s^2} - \frac{2}{3} F_{mn}^2 \right]$$

which is to be compared with Eq. (4.45). The last term in the square bracket combines with $-(1/4)F_{nm}^2$ to eventually lead to a running coupling constant exactly as in Eqns. (4.44), (4.47) and (4.55), but with Eq. (4.51) replaced by

$$Z = \frac{e^2}{12\pi^2} \int_0^\infty \frac{ds}{s} e^{-m^2 s}$$
(5.201)

This, in turn, will lead to a slightly different β-function — compared to the scalar field case — given by

$$\beta(e) = \mu \frac{\partial e}{\partial \mu} = \frac{e^3}{12\pi^2}$$
(5.202)

We will later obtain the same result (see Eq. (5.274)) from perturbation theory.

As in the case of the charged scalar field, the imaginary part of the effective Lagrangian will give you the pair production rate per unit volume per second through the relation

$$\left| e^{iA} \right|^2 = e^{iA} e^{-iA^*} = e^{i(A - A^*)} = e^{-2\text{Im}[A]} = e^{-2VT \text{Im} L_{\text{eff}}}$$
(5.203)

Calculating the imaginary part exactly as in the case of scalar field, we obtain the result

$$2\text{Im}(L_{\text{eff}}) = \frac{1}{4\pi} \sum_{n=1}^{\infty} \frac{1}{s_n^2} e^{-m^2 s_n} = \frac{\alpha E^2}{2\pi^2} \sum_{n=1}^{\infty} \frac{1}{n^2} \exp\left(\frac{-n\pi m^2}{eE} \right)$$
(5.204)

Except for numerical factors, this matches with the result in the case of scalar field and is again non-perturbative in the coupling constant e.

There is a somewhat more physical way of thinking about the effective Lagrangian, which is worth mentioning in this context. To do this, let us again begin with the standard QED Lagrangian

$$L = -\frac{1}{4} F_{mn}^2 + \bar{\psi}(i\gamma\partial - m)\psi - eA_m \bar{\psi}\gamma^m \psi$$
(5.205)

which describes the interaction between ψ and A_m. We are interested in studying the effect of fermions on the electromagnetic field in terms of an effective Lagrangian for the electromagnetic field. From general principles, we would expect such an effective Lagrangian to have the form

$$L_{\text{eff}} = -\frac{1}{4}F_{mn}^2 - eA_m J_A^m \tag{5.206}$$

where J_A^m is a suitable current. A natural choice for this current is the expectation value of the Dirac current in a quantum state $|A\rangle$ containing an electromagnetic field described by the vector potential A_m. That is, we think of the effective Lagrangian as given by Eq. (5.206) with

$$J_A^m \equiv \langle A|\bar{\psi}(x)\gamma^m\psi(x)|A\rangle \tag{5.207}$$

Computing this current is best done by using the following trick. We first note that, in the absence of any electromagnetic field, the vacuum expectation value in Eq. (5.207) can be written in the form

$$\begin{aligned} J_0^m(x) &= \langle 0|\bar{\psi}_b(x)\gamma_{ba}^m\psi_a(x)|0\rangle = -\text{Tr}\left[\langle 0|\psi_a(x)\bar{\psi}_b(x)\gamma_{ba}^m|0\rangle\right] \\ &\equiv -\text{Tr}\langle x|\hat{G}\gamma^m|x\rangle \end{aligned} \tag{5.208}$$

where we have introduced a Green function *operator* $\hat{G}$ such that the spinor Green function $G(x,y)$ can be expressed as the matrix element $\langle x|\hat{G}|y\rangle$ with

$$\hat{G} = \frac{i}{\gamma p - m + i\epsilon} \tag{5.209}$$

In the presence of an external electromagnetic field, Eq. (5.208) will get replaced by

$$J_A^m(x) = -\text{Tr}\langle x|\hat{G}_A\gamma^m|x\rangle \tag{5.210}$$

where $\hat{G}_A$ is the Green function operator in the presence of an external electromagnetic field. This is given by the usual replacement:

$$\hat{G}_A = \frac{i}{\gamma\hat{p} - e\gamma A(\hat{x}) - m + i\epsilon} \tag{5.211}$$

Multiplying the numerator and denominator by $\gamma(p - eA) + m$ and using the standard result

$$(\gamma p - e\gamma A(\hat{x}))^2 = (\hat{p} - eA(\hat{x}))^2 - \frac{e}{2}F_{mn}(\hat{x})\sigma^{mn} \tag{5.212}$$

we get:

$$\hat{G}_A = \frac{i(\gamma p - e\gamma A(\hat{x}) + m)}{(\hat{p} - eA(\hat{x}))^2 - \frac{e}{2}F_{mn}(\hat{x})\sigma^{mn} - m^2 + i\epsilon} \tag{5.213}$$

This means that the Dirac propagator, in the presence of an electromagnetic field, has the integral representation given by

$$\begin{aligned} G_A(x,y) &= \langle y|\frac{i}{\gamma p - e\gamma A - m + i\epsilon}|x\rangle \tag{5.214} \\ &= \int_0^\infty ds\, e^{-is(m^2 - i\epsilon)}\langle y|(\gamma p - e\gamma A(\hat{x}) + m)e^{-i\hat{H}s}|x\rangle \end{aligned}$$

where

$$\hat{H} = -(\hat{p}^m - eA^m(\hat{x}))^2 + \frac{e}{2}F_{mn}(\hat{x})\,\sigma^{mn} \tag{5.215}$$

As an aside, and just for comparison, let us consider the corresponding expressions for the complex scalar field. In this case, the relevant Lagrangian is

$$L = -\frac{1}{4}F_{mn}^2 - \phi^*(D^2 + m^2)\phi \qquad (5.216)$$

with $D_m = \partial_m + ieA_m$. We will now define a propagator in the presence of an external electromagnetic field as the expectation value of the time ordered field operators in the state $|A\rangle$ containing an electromagnetic field A_m:

$$G_A(x,y) = \langle A|T\{\phi(y)\phi^*(x)\}|A\rangle \qquad (5.217)$$

In the standard operator notation with $\partial_j \to -ip_j$, this expression can be thought of as the matrix element of the Green's function operator defined by:

$$\hat{G}_A = \frac{i}{(\hat{p} - eA(\hat{x}))^2 - m^2 + i\epsilon} \qquad (5.218)$$

Using the standard integral representation we can then write the Green's function in the form

$$\begin{aligned}
G_A(x,y) &= \langle y|\hat{G}_A|x\rangle = \langle y|\frac{i}{(\hat{p} - eA(\hat{x}))^2 - m^2 + i\epsilon}|x\rangle \\
&= \int_0^\infty ds\, e^{-is(m^2 - i\epsilon)}\langle y|e^{-i\hat{H}s}|x\rangle \qquad (5.219)
\end{aligned}$$

where the effective Hamiltonian is

$$\hat{H} = -(\hat{p} - eA(\hat{x}))^2 \qquad (5.220)$$

Comparing Eq. (5.215) with Eq. (5.220), we find that the only extra term is due to the spin of the particle, leading to $(e/2)F_{ab}\sigma^{ab}$.

We are now in a position to evaluate the current in Eq. (5.210). Using the integral representation for $\hat{G}_A$, we get

$$\begin{aligned}
J_A^m &= -\text{Tr}\left[\int_0^\infty ds\, e^{-is(m^2 - i\epsilon)}\langle x|\gamma^m(\gamma p - e\gamma A + m)e^{-i\hat{H}s}|x\rangle\right] \\
&= -\int_0^\infty ds\, e^{-is(m^2 - i\epsilon)} \\
&\qquad \times \langle x|\text{Tr}\left[\gamma^m(\gamma p - e\gamma A)e^{i\left((p-eA)^2 - \frac{e}{2}\sigma_{mn}F^{mn}\right)s}\right]|x\rangle \quad (5.221)
\end{aligned}$$

where we have used the fact that the trace of the product of an odd number of gamma matrices is zero. But it is obvious that this expression can be equivalently written in the form

$$J_A^m = -\frac{i}{2e}\frac{\partial}{\partial A_m}\int_0^\infty \frac{ds}{s}e^{-is(m^2 - i\epsilon)}\text{Tr}\left[\langle x|e^{-i\hat{H}s}|x\rangle\right] \qquad (5.222)$$

where $\hat{H}$ is given by Eq. (5.215). Integrating both sides with respect to A_m and substituting into Eq. (5.206), we find that the effective Lagrangian is given by

$$L_{\text{eff}} = -\frac{1}{4}F_{mn}^2 + \frac{i}{2}\int_0^\infty \frac{ds}{s}e^{-is(m^2 - i\epsilon)}\text{Tr}\left[\langle x|e^{-i\hat{H}s}|x\rangle\right] \qquad (5.223)$$

This result agrees with the expression we got earlier in Eq. (5.196) obtained by integrating out the spinor fields in the path integral. Further, it shows that the effective Lagrangian can indeed be interpreted as due to interaction with a spinor current in a state containing an external electromagnetic field, as indicated in Eq. (5.206). (The same idea also works for a complex scalar field. In this case, the current is more difficult to evaluate because of the $|\phi|^2 A_m A^m$ term in the Lagrangian, but it can be done.)

Incidentally, the corresponding expression for the complex scalar field is given by

$$L_{\text{eff}} = -\frac{1}{4} F_{mn}^2 - i \int_0^\infty \frac{ds}{s} e^{-is(m^2 - i\epsilon)} \langle x | e^{-i\hat{H}s} | x \rangle \qquad (5.224)$$

where $\hat{H} = -(\hat{p} - eA(\hat{x}))^2$. This differs from the expression in Eq. (5.223) as regards the sign of the second term, the factor half and the fermionic trace. Of these, the most crucial difference is the sign which is responsible for the difference in the sign of the zero point energy of the fermionic field vis-a-vis the bosonic field. This is most easily seen by evaluating the expressions in Eq. (5.223) and Eq. (5.224) in the absence of the electromagnetic field by setting $A_j = 0$. Such a calculation was done for the scalar field in Sect. 1.4.5 where we showed that it leads to the zero point energy of the harmonic oscillators making up the field. If you repeat the corresponding calculations for the fermionic fields, you will find that the fermionic trace, along with the factor $-(1/2)$, leads to an overall factor of $4(-1/2) = -2$ compared to the result for the complex scalar field. The factor 2 is consistent with the Dirac spinor having twice as many degrees of freedom as a complex scalar field; the extra minus sign is precisely the sign difference between fermionic and bosonic zero point energies, which we encountered earlier in Eq. (5.138).

5.6.6 Feynman Rules for QED

Having discussed some nonperturbative aspects of QED, we shall now take up the description of QED as a perturbation theory in the coupling constant e. The first step in that direction will be to derive the necessary Feynman diagrams in the momentum space for QED. This can be done exactly as in the case of $\lambda \phi^4$ theory by constructing the generating function and its Taylor expansion. The essential idea again is to expand the generating functional of an interacting theory as a power series in a suitable coupling constant and re-interpret each term by a diagram. We will not bother to derive these explicitly since — in this particular case — the rules are intuitively quite obvious; we merely state them here for future reference and for the sake of completeness:

(a) With each internal fermion line, we associate the propagator in the momentum space $iS(p)$. (See Fig. 5.2.)

(b) With each internal photon line, we associate a photon propagator $D_{mn}(p)$. Sometimes it is convenient to use a propagator with an infinitesimal mass for the photon or a gauge fixing term. Most of the time, we will use the Feynman gauge. (See Fig. 5.3.)

There is only one kind of vertex in QED which connects two fermion lines with a photon line. We conserve momentum at every vertex while drawing the diagram and labeling it.

(c) Each vertex where the two fermion and one photon lines meet, is associated with the factor $ie\gamma^m$. (See Fig. 5.4.)

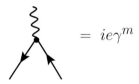

$$\frac{i(\gamma p + m)}{p^2 - m^2 + i\epsilon}$$

Figure 5.2: Electron propagator

$$\frac{-i\eta_{mn}}{p^2 + i\epsilon}$$

Figure 5.3: Photon propagator

$$= ie\gamma^m$$

Figure 5.4: Interaction vertex

$$= u^s(p)$$

$$= \bar{u}^s(p)$$

Figure 5.5: Incoming electron/positron

$$= \bar{v}^s(p)$$

$$= v^s(p)$$

Figure 5.6: Outgoing electron/positron

$$= \epsilon_m(p)$$

$$= \epsilon_m^*(p)$$

Figure 5.7: Incoming/outgoing photon

The external lines can represent electrons, positrons or photons. The rules for these are as follows.

(d) For an incoming fermion line, associate $u(p, s)$ and for an outgoing fermion line, associate $\bar{u}(p', s')$; for antifermions, use v and $\bar{v}$ instead of u and $\bar{u}$. (See Fig. 5.5 and Fig. 5.6.)

(e) With an incoming photon line, associate a polarization factor $\epsilon_m(p)$; with the outgoing photon line associate a polarization vector $\epsilon_m^*(p)$. (See Fig. 5.7.)

These essentially summarize the translational table between the diagrams and algebraic factors. Momenta associated with the internal lines should be integrated over, with the usual measure $d^4p/(2\pi)^4$. When several different diagrams contribute to a given process, we should introduce a (-1) factor if we interchange external, identical fermions to obtain a new Feynman diagram. This rule can be understood by looking at the structure of, say, a 4-point function

$$G(x_1, x_2, x_3, x_4) = \langle 0|T\{\psi(x_1)\bar{\psi}(x_3)\psi(x_2)\bar{\psi}(x_4)\}|0\rangle \qquad (5.225)$$

It is clear that $G(x_1, x_2, x_3, x_4) = -G(x_1, x_2, x_4, x_3)$ due to anticommutation rules. For a similar reason, we also need to introduce an overall factor of -1 for any closed fermion loop. We will use these rules extensively in what follows.

5.7   One Loop Structure of QED

As in the case of $\lambda\phi^4$ theory, the really non-trivial aspects of QED arise only when we go beyond the lowest order perturbation theory (viz., tree-level). This is because, when we consider a diagram involving integration over internal momenta (which could arise, e.g., in diagrams with one or more internal loops), we have to integrate over the momenta corresponding to these loops. The propagators corresponding to the internal loops will typically contribute a $(1/p^2)$ or $(1/p^4)$ factor to the integral at large momenta, and the integration measure will go as $p^3 dp$. This could lead to the contributions from the diagrams which require integration over the internal momenta, to diverge at large p. As in the $\lambda\phi^4$ theory, it is necessary to understand and reinterpret the perturbation theory when this occurs.[75]

[75]As we have stressed several times before, the process of renormalization, a priori, has very little to do with the existence or removal of divergences in perturbative field theory. The real reason for renormalization has to do with the fact that the parameters which occurred in the original Lagrangian have no direct physical meaning and the physically relevant parameters have to be defined operationally in terms of physical processes. These conceptual issues have been already highlighted in the case of $\lambda\phi^4$ theory and hence we will not go over them again.

In the case of QED, the bare Lagrangian contains the parameters representing the bare mass of the electron (m_e), the bare mass of the photon ($m_\gamma \equiv 0$) and the coupling constant which is the charge of the electron (e). Taking a cue from the $\lambda\phi^4$ theory, we would expect renormalization to change all these parameters to the physical values which are observed in the laboratory. We would expect that the physical mass of the photon, m_γ, should remain zero to all orders of perturbation theory. On the other hand, we do not have any strong constraints on the electron mass and the coupling constant except that they should correspond to the values observed in the lab.

As a simple example of how this works, consider the electronic charge. The charge of the electron can be operationally defined using the fact that the potential energy of electrostatic interaction between two electrons scales as e^2/r at sufficiently large distances; that is, at distances $r \gg (\hbar/mc)$ where m is the mass of the lightest charged particle in the theory. In

Fourier space, this is equivalent to the statement that the propagator scales as $V(k) = e^2/k^2$ for sufficiently small k, i.e., for $k \ll m$. (We have ignored the tensor indices η_{ab} of the propagator for simplicity.) This shows that the operational definition of electronic charge in our theory is determined by the limit:

$$e_R^2 = \lim_{k^2 \to 0} k^2 V(k) \qquad (5.226)$$

If the interactions involving the lightest charged fermion in the theory (which we will take to be the electron) modify $V(k)$ to the form $(e^2/k^2)[1 - e^2\Pi(k)^2]^{-1}$ where $\Pi(k^2)$ is a calculable function — and, as we shall see later, this is precisely what happens — then the physical charge measured in the laboratory (which we will call the renormalized charge) is given by

$$e_R^2 = \lim_{k^2 \to 0} k^2 V(k) = \frac{e^2}{1 - e^2\Pi(0)} \qquad (5.227)$$

In other words, if the interactions modify the nature of propagator (and hence Coulomb's law), they will also change the operationally defined value of the electronic charge in the lab. It will no longer be the parameter e^2 which you introduced into the Lagrangian but will be determined by the expression in Eq. (5.227); and it can be calculated[76] if we can calculate $\Pi(k)$.

More generally, you may decide to fix the coupling constant, not at $k = 0$ but at some other value, say $k = k_1$. That is, you define the coupling constant as $e_1 \equiv k_1^2 V(k_1^2)$. At tree-level, since $V \propto 1/k^2$, the $e_1^2 = e^2$ so defined is independent of the k_1 you have chosen. But when the propagator is modified to

$$V(k) = (e^2/k^2)[1 - e^2\Pi(k^2)]^{-1} \qquad (5.228)$$

this is no longer true, and the numerical value of the coupling constant $e_1 = e(k_1)$ is now a function of the energy scale which we used to define it, and is given by $e_1^2 = e^2[1 - e^2\Pi(k_1^2)]^{-1}$. Equivalently, e and e_1 are related by:

$$e^{-2} = e^{-2}(k_1) + \Pi(k_1^2) \qquad (5.229)$$

A crucial piece of information about the theory is contained in the form of this functional dependence.[77] In our case, if we evaluate the coupling constants e_1 and e_2 at two scales $k = k_1$ and $k = k_2$ and use the fact that, in Eq. (5.229), left hand side is the same at both $k = k_1$ and $k = k_2$, we get

$$\frac{1}{e_1^2} - \frac{1}{e_2^2} = \Pi(k_2^2) - \Pi(k_1^2) \qquad (5.230)$$

which relates the coupling constants defined at two different energy scales. We would expect such a relation, which connects directly observable quantities, to be well-defined and finite. (We will see later that this is indeed the case.)

As in the case of the $\lambda\phi^4$ theory, the renormalization program requires us to undertake the following steps. (a) Identify processes or Feynman diagrams which will lead to divergent contributions.[78] (b) Regularize this divergence in some sensible manner. (c) See whether all the divergent terms can be made to disappear by changing the original parameters λ_A^0 of the theory to the physically observable set λ_A^{ren}. (d) Re-express the theory and, in particular, the non-divergent parts of the amplitude in terms of

[76] Of course, since interactions modify the form of $V(k^2)$, they not only make the electronic charge observed in the laboratory, e_R^2, different from the parameter e^2 introduced into the theory, but also modify the actual form of the electromagnetic interaction.

[77] This is another way of thinking about the "running of the coupling constant".

[78] By and large, the procedure we have outlined deals with UV divergences of the theory arising from large values of the internal momentum during the integration. In QED (and, more generally, in theories involving massless bosons), one also encounters IR divergences which arise at arbitrarily low values of the internal momentum which is integrated over. While conceptually rather curious, the IR divergences arise due to different physical reasons compared to UV divergences and are considered relatively harmless. We will have occasion to mention IR divergences briefly later on. When we talk about divergences, they usually refer to the UV divergence unless otherwise specified.

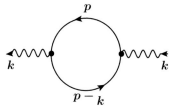

Figure 5.8: One loop correction to photon propagator

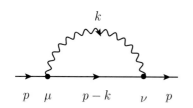

Figure 5.9: One loop correction to electron propagator

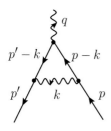

Figure 5.10: Correction to the electromagnetic vertex

[79]At this order, there are also other diagrams like the ones given in Fig. 5.11. These, however, can be thought of as correcting the propagators themselves and will not contribute to the magnetic moment of the electron.

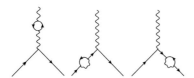

Figure 5.11: Sample of three diagrams at the same order.

[80]We need to carry out this program to all orders in perturbation theory in order to convince ourselves that the theory is well defined. This is indeed possible, but we shall concentrate on illustrating this procedure at the lowest order which essentially involves integration over a single internal momentum variable. Extending this approach to all orders in perturbation theory is technically more complicated but does not introduce any new conceptual features.

the physical parameters λ_A^{ren}. This will form the basis for comparing the theory with observations.

With the above motivation in mind, let us consider the diagrams in QED involving a single internal momentum integration, which lead to divergences. It is easy to verify that there are three such primitive diagrams.

(i) The first one is that in Fig. 5.8. This figure represents the lowest order correction to the photon propagator line (which will be a simple wavy line with a momentum labelled k originally.) You could think of this as a process in which a photon converts itself into a virtual electron-positron pair, which disappears, leading back to the photon. That is, this is a correction to the photon propagator due to an electron loop.

(ii) Figure 5.9 is the counterpart for the electron propagator (which, originally, would have been a straight line with momentum label p) due to the emission and subsequent absorption of a photon of momentum k by the electron. This provides the lowest order correction to the electron propagator.

(iii) Figure 5.10 is probably the most complicated of the three processes we need to consider. Here, a photon (with an internal momentum k) is emitted and reabsorbed while the electron is interacting with an external photon (momentum q). Unlike the other two diagrams, this involves *three* vertices rather than two. The net effect of such a process is to correct the coupling between the photon and electron in a — as we shall see — rather peculiar manner. You will find that the new coupling involves not only the charge but also the magnetic moment of the electron, thereby changing the value of the gyro-magnetic ratio.[79]

If you use the Feynman rules to write down the expressions for these three amplitudes, you will find that all of them are divergent. We need to regularize all these expressions and then show that the divergent parts can be incorporated into the parameters of the theory, modifying them to new values just as we did in the case of the $\lambda\phi^4$ theory. Once this is done, we will be able to predict new and non-trivial physical effects in terms of physical parameters of the theory. We will do this in the next few sections.[80]

5.7.1   Photon Propagator at One Loop

We will begin by computing the one loop correction to the photon propagator, described by the Feynman diagram in Fig. 5.8, and describing its physical consequences. We already know from the Ward-Takahashi identity [see Eq. (5.181)] that the amplitude corresponding to this diagram must have a form

$$i\Pi_{mn} = i[k^2\eta_{mn} - k_m k_n]e^2\Pi(k^2) \equiv iP_{mn}(k)e^2\Pi(k^2) \qquad (5.231)$$

We have scaled out a factor e^2 from $\Pi(k^2)$ since we know that Fig. 5.8 will have this factor arising from the two vertices. The rest of the structure of Π^{mn} is decided by Lorentz invariance and the condition $k^m\Pi_{mn} = 0$.

Incidentally, you can understand the condition $k^m\Pi_{mn} = 0$ and its relation to gauge invariance in a rather simple manner, which is worth mentioning. The interaction of an electron with an external potential is described, at the tree-level, by the diagram in Fig. 5.12(a); at the next order, this will get modified by the addition of the diagram in Fig. 5.12(b). Algebraically, this requires the addition of the two amplitudes given by

$$M_{\text{fi}}^{(1)} = e\bar{u}_f\gamma^m u_i A_m(q); \qquad M_{\text{fi}}^{(2)} = e\bar{u}_f\gamma^m u_i \frac{-i}{q^2 + i\epsilon} \, i\Pi_{mn}A^n(q) \quad (5.232)$$

(We are using the free-field photon propagator in Eq. (3.195), obtained by the gauge choice $\zeta = \lambda = 1$ in Eq. (3.203), etc. This will be our choice throughout.) If we now choose to describe the external electromagnetic field in a different gauge, so that $A_m(q) \to A_m(q) + iq_m\chi(q)$, then the gauge invariance of the second order term will require the condition

$$\Pi_{mn}(q)q^n = 0 \qquad (5.233)$$

Note that this holds for a virtual photon with $q^2 \neq 0$.

Getting back to Eq. (5.231), we note that the factor P_{ab} satisfies a useful identity, which can be directly verified:

$$P_{ab}(k)\left(\frac{\eta^{bl}}{k^2}\right)P_{lm}(k) = P_{am}(k) \qquad (5.234)$$

This allows us to sum up the corrections to the photon propagator coming from a geometric series of electron loops in a rather simple manner. (This is analogous to what we did to the scalar propagator in $\lambda\phi^4$ theory; see Eq. (4.145)) For example, the sum of the terms corresponding to the Feynman diagrams shown in Fig. 5.13. can be represented by

$$
\begin{aligned}
D_{rs}(k) &= \frac{-i\eta_{rs}}{k^2} + \frac{-i\eta_{ra}}{k^2} \, i\Pi_{ab}(k^2) \frac{-i\eta_{bs}}{k^2} \\
&\quad + \frac{-i\eta_{ra}}{k^2} i\Pi_{ab}(k^2) \frac{-i\eta_{bl}}{k^2} \Pi_{ln}(k^2) \frac{-i\eta_{ns}}{k^2} + \cdots \\
&= \frac{-i\eta_{rs}}{k^2} + \frac{i(k^2\eta_{rs} - k_r k_s)}{k^4}[e^2\Pi(k^2) + e^4\Pi^2(k^2) + \cdots] \\
&= \frac{-i\eta_{rs}}{k^2} + \frac{-i(k^2\eta_{rs} - k_r k_s)}{k^4} \frac{e^2\Pi(k^2)}{1 - e^2\Pi(k^2)} \\
&= \frac{-i\eta_{rs}}{k^2} \frac{1}{1 - e^2\Pi(k^2)} + \frac{ik_r k_s}{k^4} \frac{e^2\Pi(k^2)}{1 - e^2\Pi(k^2)} \qquad (5.235)
\end{aligned}
$$

In this expression, the second term with $k_r k_s$ can be ignored because it will never contribute to any physical process. This is because, in the computation of any physical amplitude, the propagator D_{rs} will be sandwiched between two conserved current terms, in the form $J^r(k)D_{rs}(k)J^s(k)$. Since gauge invariance demands that any current to which the photon couples needs to be conserved, we will necessarily have $k_a J^s(k) = 0$ in Fourier space. This means that terms with a free tensor index on the momentum, like k_r in the propagator, will not contribute to the physical processes; therefore, we shall hereafter ignore the second term in Eq. (5.235).

Before proceeding further, we will rescale the photon propagator by a factor e^2 and use the rescaled propagator $D_{rs} \to e^2 D_{rs}$. At this stage, you can think of it as just a redefinition. But this is far from a cosmetic exercise and has an important physical meaning, which is worth emphasizing. Conventionally, you would have written the Lagrangian describing QED in the form

$$L = \bar{\psi}[i\gamma^m(\partial_m - ieA_m) - m]\psi - \frac{1}{4}F_{mn}F^{mn} \qquad (5.236)$$

In this description, the photon field has nothing to do with the electric charge e. The charge appears as a property of the electron in the Dirac

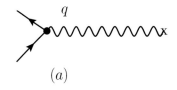

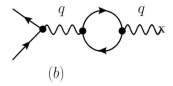

Figure 5.12: Scattering of an electron (a) by an external potential and (b) an one loop correction to the same.

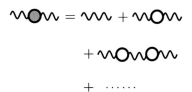

Figure 5.13: Diagrammatic representation of the geometric progression in Eq. (5.235).

sector when we couple ψ to A_j, which sounds reasonable. This theory is invariant under the gauge transformations given by $\psi(x) \to e^{ie\Lambda(x)}\psi(x)$ and $A_m(x) \to A_m(x) + \partial_m \Lambda(x)$. The gauge transformation of the Dirac field involves a phase term which depends on the charge e of the electron field, but the photon field is still oblivious of the electronic charge e.

We can, however, provide a completely equivalent description of this theory by rescaling $A_m \to (1/e)A_m$, thereby obtaining the Lagrangian:

$$L = \bar{\psi}[i\gamma^m(\partial_m - iA_m) - m]\psi - \frac{1}{4e^2}F_{mn}F^{mn} \qquad (5.237)$$

In this case, the charge has disappeared from the Dirac sector and appears as an overall constant $(1/e^2)$ in front of the electromagnetic action. This seems to suggest that the electric charge is more of a property of the photon field rather than the Dirac field! The gauge transformations are now given by $\psi \to e^{i\alpha}\psi$, $A_m \to A_m + \partial_m\alpha$ and are completely independent of the electronic charge. Obviously, this rescaling of the vector potential is equivalent to rescaling the photon propagator by $D_{rs} \to e^2 D_{rs}$ which is what we want to do. So we are going to work with the Lagrangian in Eq. (5.237) rather than Eq. (5.236).

Since the two descriptions are mathematically the same, you may wonder why we are making such a fuss about it and why we prefer the latter description. The reason has to do with the fact that the corrections which the photon propagator picks up due to the electron loops (as in Fig. 5.8) do depend on the electronic charge e which appears at the pair of vertices in Fig. 5.8. As we mentioned earlier (see Eq. (5.227)), one of the effects of the corrections to the photon propagator is to renormalize the value of e^2; that is, the e^2 you introduce into the Lagrangian and the e_R^2 which will be measured in the lab will be different due to renormalization. But how do we know that the renormalization changes both the charge of the electron and the proton in the same manner, especially since protons participate in strong interactions while electrons do not? To pose this question sharply, consider a theory involving protons, electrons and photons with a Dirac-like term added to the Lagrangian in Eq. (5.236), where the proton will be described with a different mass m_p and a charge having the same magnitude as the electron. (In addition, we can throw in the strong interaction terms for the proton.) We are keeping the charges of the electron and proton equal at the tree-level by fiat. But it is not a priori obvious that the corrections due to interactions will not lead to *different* renormalized charges for the electron and proton — which, of course, will be a disaster. We know that this does not happen, but this is not obvious when we write the Lagrangian in the form of Eq. (5.236). This fact is easier to understand if we transfer the burden of the electric charge to the photon field and write the Lagrangian as in Eq. (5.237). If we now add a proton field to this Lagrangian, it will not bring in any new charge parameter (but only a mass m_p). So clearly, what happens to e^2 is related to how the photon propagator gets modified due to all sorts of interactions, and the equality of the electron and proton charges are assured; after all, charge is a property of the electromagnetic field and not of the matter field[81] when we work with the Lagrangian in Eq. (5.237).

In this picture, which is conceptually better, the photon propagator describes the ease with which it propagates through a vacuum polarized by the existence of virtual, charged particle-antiparticle pairs. We will

[81]Of course, we get the same physical results even when we work with Eq. (5.236). We will come back to this aspect later in Sect. 5.8.

therefore work with the rescaled propagator written as

$$D_{rs} \to e^2 D_{rs}(k) \equiv -\frac{i\eta_{rs}}{k^2} \frac{e^2}{[1 - e^2\Pi(k^2)]} \qquad (5.238)$$

and have dropped the second term in Eq. (5.235). With future applications in mind, we will rewrite the factor which occurs in this expression in the following form:

$$\frac{e^2}{k^2[1 - e^2\Pi(k^2)]} = \frac{(e^2/k^2)}{[1 - e^2(\Pi(k^2) - \Pi(0)) - e^2\Pi(0)]}$$
$$= \frac{(e_R^2/k^2)}{[1 - e_R^2(\Pi(k^2) - \Pi(0))]} \qquad (5.239)$$

where we have defined the renormalized charge e_R^2 as in Eq. (5.227), by

$$e_R^2 \equiv \frac{e^2}{1 - e^2\Pi(0)} \qquad (5.240)$$

We have essentially written $\Pi(k^2)$ in Eq. (5.238) as $[\Pi(k^2) - \Pi(0)] + \Pi(0)$, pulled out the factor $1 - e^2\Pi(0)$ from both numerator and denominator and redefined the electric charge by Eq. (5.227). This exercise might appear a bit bizarre, but it has fairly sound physical motivation based on the operational definition of the charge, which led to Eq. (5.227).

When we explicitly compute $\Pi(k^2)$ — which we will do soon — we will see that the quantity $\Pi(0)$ is divergent while $[\Pi(k^2) - \Pi(0)]$ is finite and well defined. If we now define the renormalized charge as a finite quantity given by Eq. (5.240) (by absorbing a divergent term into the parameter e^2), the final propagator in Eq. (5.239) is finite and well defined when expressed in terms of e_R^2. This expression also tells you that in the limit of $k^2 \to 0$, (which corresponds to very large distances), the propagator in Eq. (5.239) reduces to the form e_R^2/k^2 which — in real space — will reproduce the Coulomb potential e_R^2/r. This is precisely the manner in which we would have defined the electronic charge at low energies (that is, at the $k^2 \to 0$ limit). So our renormalized charge in Eq. (5.240) is indeed the physically observed charge in the lab at arbitrarily low energies.[82]

Let us now compute the propagator by evaluating the Feynman diagram explicitly. We shall write down the contribution arising from the diagram in Fig. 5.8 as

$$i\Pi^{mn}(k) = -(-ie)^2 \int \frac{d^4p}{(2\pi)^4} \operatorname{tr} \gamma^m S(p)\gamma^n S(p-k) \qquad (5.241)$$

where $S(q) = i[\gamma q - m]^{-1}$ is the electron propagator. (We will write $\gamma^i q_i$, etc. as simply γq when no confusion is likely to arise.) Naive counting of powers shows that the integral will go as $(1/p^2)p^3 dp$ which is quadratically divergent. But we do know that the final result must be expressible in the form of Eq. (5.231), which suggests that the integral will actually go as $(1/p^4)p^3 dp$ and hence will diverge only logarithmically. But since the integral *is* divergent, we need to introduce a procedure for regularizing it before proceeding further.

One powerful technique we will use involves working in n dimensions with sufficiently small n to make the integrals converge and analytically continue the result back to $n = 4$. (We did this earlier in Sect. 4.7.2

[82]Note that everything we said above remains valid even if $\Pi(0)$ is a finite quantity. The physical charge is indeed measured as the coefficient of $(1/k^2)$ in the limit of $k \to 0$ and *is* given by Eq. (5.240). This is what an experimentalist will measure and not the e^2 you threw into the Lagrangian. From this point of view, renormalization of a parameter in the Lagrangian (like the charge) is a physical phenomena and has nothing to do, a priori, with any divergence — a fact we have stressed several times in our discussions. Because $\Pi(0)$ happens to be divergent in QED, this process *also* allows us to cure the divergence.

to tackle the divergences in the $\lambda\phi^4$ theory.) The divergences will now reappear but can be easily isolated in a controlled manner.

Since the technique of dimensional regularization requires our working with a generic dimension n, it is important to understand the dimensions of different quantities when we work in n dimensions. This is most easily done by examining different terms in the Lagrangian in Eq. (5.236) and ensuring that each of them remain dimensionless. (You will, of course, get the same results from Eq. (5.237).) To begin with, from the term involving $i\bar{\psi}\gamma^a\partial_a\psi$, it is clear that dim $[\psi] = (1/2)(n-1)$. This will ensure the correct dimensionality for $m\bar{\psi}\psi$ term as well, provided dim $[m] = 1$ for all n. From the $F^{ab}F_{ab}$ part, we discover that dim $[A_j] = (n/2) - 1$. (This will also ensure the correct dimension for all the gauge fixing terms when we add them.) Finally, in order to maintain dim $[eA^j\bar{\psi}\gamma_j\psi] = n$, we must have dim $[e] = 2 - (n/2)$. So, when we work in n-dimensions, we should find that we get a dimensionless amplitude only if e^2 has the dimension of μ^{4-n} where μ is an arbitrary mass scale.

There are two ways of incorporating this result when we do our analysis. The conventional procedure (followed in most textbooks) is to actually replace the coupling constant e^2 by $e^2(\mu)^{4-n}$, where μ is an arbitrary mass scale, thereby retaining the dimensionless status of e^2. (This is similar to what we did in the case of $\lambda\phi^4$ theory in Eq. (4.146).) The fact that e^2 acquires a dimension when $n \neq 4$ suggests that it is better to pull out a $e^2\mu^{n-4}$ factor out of $\Pi(k)$ in $n-$ dimensions (rather than just e^2) and rewrite Eq. (5.240) in the form

$$e_R^2 = \frac{e^2\mu^{n-4}}{1 - (e^2\mu^{n-4})\Pi(0)} \qquad (5.242)$$

This introduces an arbitrary mass scale μ, thereby keeping the renormalized coupling constant e_R^2 dimensionless. Rearranging the terms, we get

$$\left\{\frac{1}{e_R^2} + \Pi(0)\right\}\mu^{-2\epsilon} = \frac{1}{e^2}; \qquad 2\epsilon \equiv 4 - n \qquad (5.243)$$

Since the right hand side is independent of mass scale μ we have introduced, it follows that we must have the condition

$$\mu^{-2\epsilon}\left(\frac{1}{e_R^2} + \Pi(0)\right) = \quad \text{(independent of } \mu) \qquad (5.244)$$

This will prove to be useful later on — once we have evaluated $\Pi(0)$ — to determine the "running" of the coupling constant e_R^2 with respect to the mass scale μ.

In this approach, it is a little bit unclear how the final results are going to be independent of the parameter μ which we have introduced. Therefore we will *not* do this replacement right at the beginning but will carry on keeping e^2 to be a constant with dimension μ^{n-4}. We will see that, at a particular stage, we can introduce a mass scale into the theory in a manner which makes it obvious that *none* of the physical results will depend on μ.

So let us evaluate $\Pi^{mn}(k)$ in Eq. (5.241) keeping e^2 as it is. Proceeding to n dimensions and rationalizing the Dirac propagator will allow us to rewrite the amplitude as

$$i\Pi^{mn}(k) = -e^2 \int \frac{d^np}{(2\pi)^n} \frac{\text{tr}\{\gamma^m(m + \gamma p)\gamma^n(m + \gamma(p-k))\}}{(p^2 - m^2)((p-k)^2 - m^2)} \qquad (5.245)$$

where e^2 is defined in n - dimensions and is *not* dimensionless. The traces involved in the expression can be computed in a fairly straightforward manner[83] to give the numerator to be:

$$N \equiv \operatorname{tr}\{\cdots\} = \left\{p^m(p-k)^n + (p-k)^m p^n - [m^2 + p(p-k)]g^{mn}\right\}(\operatorname{tr}1) \tag{5.246}$$

In obtaining the result, we have left $(\operatorname{tr}1)$ unspecified, but as we shall see, this is irrelevant to our computation as long as $(\operatorname{tr}1) = 4$ in the four dimensional limit. The simplest way to compute the momentum integrals is to proceed as follows: (i) Analytically continue to the Euclidean sector (with $d^n p \to i d^n p_E, p^2 \to -p^2$ etc.). (ii) Write each of the factors in the denominator as exponential integrals using

$$\frac{1}{\mathcal{F}(p)} = \int_0^\infty dt\, e^{-t\mathcal{F}(p)} \tag{5.247}$$

(iii) Complete the square in the Gaussian and evaluate the momentum integrals as standard Gaussian integrals. (These are standard tricks; we used them earlier in Sect. 4.7.2.) The first two steps give

$$\Pi^{mn}(k) = -e^2 \int \frac{d_E^n p}{(2\pi)^n} \int_0^\infty ds\, dt\, e^{-s(p^2+m^2)}\, e^{-t(p-k)^2+m^2)}\, N \tag{5.248}$$

Completing the square requires shifting the momentum variable to $p' = p - [tk/(s+t)]$, which leads to

$$\Pi^{mn}(k) = -e^2(\operatorname{tr}1)\int_0^\infty ds\, dt\, e^{-(s+t)m^2} e^{-\frac{st}{s+t}k^2}\int \frac{d_E^n p'}{(2\pi)^n} \tag{5.249}$$

$$\left\{2p'^m p'^n - 2\frac{st}{(s+t)^2}k^m k^n - g^{mn}\left[m^2 + p'^2 - \frac{st}{(s+t)^2}k^2\right]\right\} e^{-(s+t)p'^2}$$

The Gaussian integrals are easy to do using the standard result:

$$\int \frac{d_E^n p}{(2\pi)^n} p^m p^n e^{-Ap^2} = g^{mn}\frac{1}{2A}\frac{1}{(4\pi A)^{n/2}} \tag{5.250}$$

and we obtain:

$$\Pi^{mn}(k) = -\frac{(\operatorname{tr}1)e^2}{(4\pi)^{n/2}}\int_0^\infty \frac{ds\, dt}{(s+t)^{n/2}}\, e^{-(s+t)m^2}\exp\left\{-\frac{st}{s+t}k^2\right\}$$

$$\left\{\frac{1}{s+t}g^{mn} - 2\frac{st}{(s+t)^2}k^m k^n - \left[m^2 + \frac{n}{2}\frac{1}{s+t} - \frac{st}{(s+t)^2}k^2\right]g^{mn}\right\} \tag{5.251}$$

At this stage, one resorts to yet another useful trick (see Eq. (4.165) of Sect. 4.7.2) to make further progress. We will introduce a factor of unity written in the form

$$1 = \int_0^\infty dq\, \delta(q - s - t) \tag{5.252}$$

and rescale the integration variables (s, t) to (α, β) where $s = \alpha q$, $t = \beta q$. Fairly straightforward manipulation now shows that the result indeed has the form advertised[84] in Eq. (5.231):

$$\Pi^{mn}(k) = (k^2 g^{mn} - k^m k^n)\mathcal{P}(k) \tag{5.253}$$

[83] The following relations which hold in a spacetime of even dimensionality n, are useful here and elsewhere.
(1) $\operatorname{tr}1 = 2^{n/2}$,
(2) $\operatorname{tr}\gamma^m\gamma^n = -2^{n/2}g^{mn}$,
(3) $\operatorname{tr}\gamma^m\gamma^n\gamma^r\gamma^s = 2^{n/2}(g^{mn}g^{rs} - g^{mr}g^{ns} + g^{ms}g^{nr})$
(4) $\operatorname{tr}\gamma^{m_1}\gamma^{m_2}\cdots\gamma^{m_k} = 0;\quad k = \text{odd}$

[84] The fact that $k_m\Pi^{mn}(k) = 0$ — which leads to this specific structure can be directly verified from Eq. (5.241). You first extend the integral to n-dimensions to make it convergent, and then take the dot product with k_m. If you now write $k\gamma = [\gamma(p+k) - m] - [\gamma p - m]$, you will find that the expression reduces to the difference between two integrals which are equal *when you shift the variable of integration* in one of them from p to $p + k$. It is this shifting which is invalid for the original divergent integral in $D = 4$; but we can do it after proceeding to n-dimensions.

where

$$\mathcal{P}(k) = -2\frac{(\mathrm{tr}\,1)}{(4\pi)^{n/2}}\Gamma\left(2-\frac{n}{2}\right)e^2\int_0^1 d\alpha\,\alpha(1-\alpha)[m^2+\alpha(1-\alpha)k^2]^{(n/2)-2} \tag{5.254}$$

Our aim now is to take the $n \to 4$ limit which will require us to take the limit of $\epsilon \to 0$ in a factor x^ϵ, with $\epsilon = n/2 - 2$ inside the integral. Writing $x^\epsilon = \exp(\epsilon \ln x)$ and taking the limit, we will obtain a term of the form $(1+\epsilon \ln x)$. For $\ln x$ to make sense, it is necessary that x is a dimensionless object. To take care of this fact, we will rewrite the Eq. (5.254) by multiplying and dividing the expression by a factor μ^{n-4} where μ is an unspecified finite energy scale.[85]

$$\mathcal{P}(k^2) = \ -\ 2\frac{(\mathrm{tr}\,1)}{(4\pi)^{n/2}}\Gamma\left(2-\frac{n}{2}\right)e^2\mu^{n-4} \times \tag{5.255}$$

$$\int_0^1 d\alpha\,\alpha(1-\alpha)\left[\frac{m^2-\alpha(1-\alpha)k^2}{\mu^2}\right]^{(n/2)-2} \equiv e^2\mu^{n-4}\Pi(k^2)$$

Since we know that e^2 in n- dimensions has the dimension μ^{4-n}, this keeps $e^2\mu^{n-4}$ dimensionless, as it should. We have also rotated back from the Euclidean to the Lorentzian space which changes the sign of k^2. The last equality defines $\Pi(k^2)$, after pulling out the factor $e^2\mu^{n-4}$.

Let us first evaluate $\Pi(0)$ using this expression, in the limit of $\epsilon \equiv [2-(n/2)] \to 0$. Setting $k=0$ and $tr1 = 4$, we have

$$\Pi(0) = -\frac{1}{2\pi^2}\left(\frac{\mu^2}{m^2}\right)^\epsilon\Gamma(\epsilon)\int_0^1 d\alpha\,\alpha(1-\alpha) \tag{5.256}$$

The integral has a value $(1/6)$. We now take the limit of $\epsilon \to 0$ using the expansion of the gamma function

$$\Gamma(\epsilon) \approx \frac{1}{\epsilon} - \gamma_E \tag{5.257}$$

(where γ_E is Euler's constant) and writing

$$\left(\frac{\mu}{m}\right)^{2\epsilon} = \exp\left[2\epsilon \ln\left(\frac{\mu}{m}\right)\right] \approx 1 + 2\epsilon \ln\left(\frac{\mu}{m}\right). \tag{5.258}$$

This leads to the result[86]

$$\Pi(0) = -\frac{1}{12\pi^2\epsilon}\left(1+\mathcal{O}(\epsilon)\right) \tag{5.259}$$

Once we know the structure of $\Pi(0)$ — which, of course, diverges as $(1/\epsilon)$ as $\epsilon \to 0$ — we can use it in Eq. (5.244). Substituting the form of $\Pi(0)$ into Eq. (5.244) and expanding the result to the relevant order, we demand that the expression

$$\left\{\frac{1}{e_R^2} - \frac{1}{12\pi^2\epsilon}\right\}\left(1 - 2\epsilon \ln\left(\frac{\mu}{m}\right)\right) = \frac{1}{e_R^2} + \frac{1}{6\pi^2}\ln\left(\frac{\mu}{m}\right) - \frac{1}{12\pi^2\epsilon} + \mathcal{O}(\epsilon) \tag{5.260}$$

remains independent of μ. Evaluating the right hand side at two values of μ, say $\mu = \mu_1$ and $\mu = \mu_2$ and subtracting one from the other, we get the result

$$\frac{1}{e_R^2(\mu_2)} = \frac{1}{e_R^2(\mu_1)} - \frac{1}{6\pi^2}\ln\left(\frac{\mu_2}{\mu_1}\right) \tag{5.261}$$

[85] This procedure (viz. multiplying by unity, $1 = (\mu^{n-4}/\mu^{n-4})$) clearly demonstrates that $\mathcal{P}(k^2)$ *is* independent of the mass scale μ as long as you do not do anything silly while regularizing the expression.

[86] To be precise, there is a bit of an ambiguity here. Because $\Pi(0)$ is a divergent quantity, you could have subtracted out any amount of finite terms and redefined it as a new $\Pi(0)$. Again, no physical result should depend on how you do this subtraction, but in higher order computations certain subtractions will turn out to be more advantageous than others. What we have done here is closely related to the *minimal subtraction* described in Sect. 4.8. Since our emphasis is on concepts rather than calculations, we will not elaborate on this again.

This allows us to relate the renormalized coupling constant at two different scales as:

$$e_R^2(\mu_2) = e_R^2(\mu_1) \left[1 - \frac{e_R^2(\mu_1)}{6\pi^2} \ln\left(\frac{\mu_2}{\mu_1}\right) \right]^{-1} \tag{5.262}$$

This is analogous to the result we obtained earlier nonperturbatively (see Eq. (4.55) of Sect. 4.2.2) and Eq. (5.202) for the charged scalar field and the Dirac field.[87] The sign of the coefficient of the logarithmic term is important. We see that, because of the negative sign, the effective charge becomes larger at short distances (i.e., at larger energies) and the coupling gets stronger.

Let us now proceed to evaluate the physical part of the propagator contributed by $\Pi_{\text{phy}}(k^2) \equiv \Pi(k^2) - \Pi(0)$. To do this, we first evaluate Eq. (5.255) in the limit of small ϵ by replacing x^ϵ by $(1 + \epsilon \ln x)$ to obtain:

$$\Pi(k^2) = -\frac{1}{2\pi^2} \Gamma(\epsilon) \int_0^1 d\alpha\, \alpha(1-\alpha) \left[1 + \epsilon \ln \frac{\mu^2}{m^2 - k^2\alpha(1-\alpha)} \right] \tag{5.263}$$

From this we can obtain, by direct subtraction, the result:

$$\Pi_{\text{phy}} \equiv \Pi(k^2) - \Pi(0) = -\frac{\epsilon}{2\pi^2} \Gamma(\epsilon) \int_0^1 d\alpha\, \alpha(1-\alpha) \ln \frac{m^2}{m^2 - k^2\alpha(1-\alpha)} \tag{5.264}$$

On using $\epsilon\Gamma(\epsilon) \to 1$ when $\epsilon \to 0$, we get:

$$\Pi_{\text{phy}}(k^2) = \frac{1}{2\pi^2} \int_0^1 d\alpha\, \alpha(1-\alpha) \ln \left[1 - \frac{k^2}{m^2}\alpha(1-\alpha) \right] \tag{5.265}$$

which is perfectly well-defined, independent of μ and, of course, finite.[88]

This quantity $\Pi_{\text{phy}}(k^2)$ does play an operational role in scattering amplitudes. To see this, consider the scattering of two very heavy fermions f_1 and f_2 which have the same charge as the electron but much larger masses $m_1 \gg m$ and $m_2 \gg m$. We want to study the lowest order scattering of these fermions at energies $m^2 \ll k^2 \ll (m_1^2, m_2^2)$. At these energies, the photon propagator will get modified due to virtual electron loops in exactly the manner described earlier. The lowest order scattering of these fermions, without incorporating electron loops, will be governed by the diagram in Fig. 5.14 which translates into the tree-level amplitude given by

$$\mathcal{M}_{\text{tree}} = (-ie)^2 \frac{-i\eta_{mn}}{q^2} [\bar{u}(p_1)\gamma^m u(k_1)][\bar{u}(p_2)\gamma^n u(k_2)]$$
$$\equiv (-ie)^2 \frac{-i\eta_{mn}}{q^2} j^m(p_1, k_1) j^n(p_2, k_2) \tag{5.266}$$

where q^i is the momentum transfer.[89] We, however, know that the photon propagator in the diagram in Fig. 5.14 gets modified due to the electron loops. (It will also get modified by, say, muon loops and even by loops of the heavy fermions f_1, f_2 we are studying. However, the most dominant effect will come from the lightest charged fermion, which we take to be the electron.) At the next order of approximation, the corrected amplitude will be given by

$$\mathcal{M}_{\text{loop}} = (-ie_R)^2 \frac{-i\eta_{mn}}{q^2(1 - e_R^2 \Pi_{\text{phy}}(q^2))} J^m(p_1, k_1) J^n(p_2, k_2) \tag{5.267}$$

[87]Since we are evaluating Π_{mn} perturbatively to first order in e^2, the question again arises as to whether it is legitimate to write an equation like Eq. (5.262) with a denominator containing $1 - e_R^2(\ldots)$ kind of factor. As we mentioned in the case of $\lambda\phi^4$ theory — see the discussion after Eq. (4.179) — the result is similar to the one in Eq. (5.238). The Eq. (5.238), in turn, was obtained by summing a subclass of diagrams in Fig. 5.13 as a geometric progression to all orders in the coupling constant. The expression in Eq. (5.262) thus captures the effect of summing over a class of diagrams on the running of the coupling constant.

[88]One can evaluate the integral explicitly, but it is not very transparent; see Eq. (5.277). We will discuss its explicit form later.

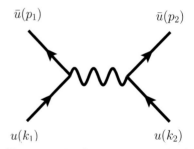

Figure 5.14: Diagram for tree-level scattering of two heavy fermions.

[89]This is a bit schematic in the sense that we have ignored spin labels etc. They are not important for the point we want to make.

where $J^m(p_1, k_1) = \bar{u}(p_1)\Lambda^m(p_1, k_1)u(k_1)$ etc. in which the vertex corrections have changed γ^m to Λ^m. (We will work this out explicitly in Sect. 5.7.3.) In the limit of $|q^2| \ll m_i^2$, however, we can still approximate the vertex functions by $\Lambda^m(p_1, k_1) \approx \gamma^m$, and also use the approximate form for the photon propagator. It is then obvious that the key effect of the 1-loop correction is to actually change the effective value of the electronic charge from e_R (fixed at large distances, i.e., at low energies) to the effective value $e_{\text{eff}}^2(q)$ (for a scattering involving the momentum transfer q^2), where:

$$\frac{1}{e_{\text{eff}}^2} = \frac{1}{e_R^2} - \Pi_{\text{phy}}(q^2) \tag{5.268}$$

and $\Pi_{\text{phy}}(q^2)$ is given by Eq. (5.265). It is easy to verify that

$$q^2 \frac{d}{dq^2}\left(\frac{1}{e_{\text{eff}}^2}\right) \approx \begin{cases} 0 & (-q^2 \ll m^2) \\[2mm] -\dfrac{1}{12\pi^2} & (-q^2 \gg m^2) \end{cases} \tag{5.269}$$

To a good approximation, e_{eff}^2 remains approximately constant for $|q^2| < m^2$ and it starts running at a constant rate for $|q^2| \gtrsim m^2$. This analysis shows that the concept of a running coupling constant is an *operationally* well defined result which is testable by scattering experiments.

There is another aspect of the running coupling constant which is worth mentioning at this stage. In Eq. (5.228), we determined how the Coulomb potential is modified by the electron loops by evaluating a sum over a set of diagrams in geometric progression. Suppose we do *not* perform such a sum and stick to the lowest order correction. Then Eq. (5.228) will read:

$$V(k) \approx \frac{e^2}{k^2}[1 + e^2\Pi(k^2) + \mathcal{O}(e^4)] \tag{5.270}$$

If we now define the coupling constant at some scale $k = k_1$ by $k_1^2 V(k_1) \equiv e^2(k_1)$, then we have $e_1^2 = e^2[1 + e^2\Pi(k_1^2)]$, which can be solved, to the same order of approximation we are working in, to give $e^2 = e_1^2[1 - e_1^2\Pi(k_1^2)]$. Substituting for e^2 in Eq. (5.270), we get:

$$k^2 V(k) \approx e^2(k_1)\left(1 - e^2(k_1)[\Pi(k_1^2) - \Pi(k^2)]\right) \tag{5.271}$$

So the corrected potential is

$$V(k) \approx \frac{e^2(k_1)}{k^2}\left[1 - \frac{e^2(k_1)}{2\pi^2}\int_0^1 d\alpha\,\alpha(1-\alpha)\ln\left[\frac{m^2 - k_1^2\alpha(1-\alpha)}{m^2 - k^2\alpha(1-\alpha)}\right]\right] \tag{5.272}$$

If we now consider the limit $k_1 \gg m$, this result reduces to:

$$V(k) \approx \frac{e^2(k_1)}{k^2}\left[1 - \frac{e^2(k_1)}{6\pi^2}\ln\frac{k_1}{k}\right] \tag{5.273}$$

We certainly do not want $V(k)$ to depend on the arbitrary scale k_1 we used to define our coupling constant; so we demand $k_1(dV/dk_1) = 0$. Working this out, to the order of accuracy[90] we are interested in, we get the differential equation:

$$k_1 \frac{de^2(k_1)}{dk_1} \equiv \beta(e^2) = \frac{e^4(k_1)}{6\pi^2} \tag{5.274}$$

[90] Terms like $e^3(k_1)[de(k_1)/dk_1] = \mathcal{O}(e^6)$ are consistently ignored to arrive at this result.

which again defines the beta-function of the theory. This is the same as Eq. (5.202) we obtained earlier from the Schwinger effect.

If we now integrate this differential equation *exactly*, we get the result that relates the coupling constant at two different energy scales by:

$$e^2(k_2) = e^2(k_1) \left[1 - \frac{e^2(k_1)}{6\pi^2} \ln\left(\frac{k_2}{k_1}\right) \right]^{-1} \qquad (5.275)$$

This is algebraically same as the result we found earlier in Eq. (5.262), but there is a subtle difference. In the above analysis, we started with a one loop result in Eq. (5.270) and did *not* sum over the geometric series of diagrams. We then took the large momentum approximation ($k_1 \gg m, k_2 \gg m$) which led to the potential in Eq. (5.273) in the 'leading log' approximation.[91] The demand that $V(k)$ should not depend on the arbitrary scale k_1 we have introduced, then led to the beta function of the theory in Eq. (5.274). We integrated this equation for the running coupling constant, pretending that it is exact; that is, we did not ignore terms of the order $\mathcal{O}(e^6)$, which were indeed ignored in the result in Eq. (5.270) which we started with. *This process has led to the same result as we would have obtained by summing over a specific set of diagrams as a geometric progression!* This is no big deal in 1-loop QED, but in more complicated cases or at higher orders, obtaining the beta function to a certain order of perturbation and integrating it *exactly*, allows us to bypass the (more) complicated summation of a subset of diagrams.

In this particular case of 1-loop QED, we can also use the form of $\Pi(k) - \Pi(k_1)$ for any two values of k and k_1 in Eq. (5.230) to obtain the coupling constants defined at these two scales as:

$$\frac{1}{e^2(k)} - \frac{1}{e^2(k_1)} = \Pi(k_1^2) - \Pi(k^2) \qquad (5.276)$$

$$= \frac{1}{2\pi^2} \int_0^1 d\alpha\, \alpha(1-\alpha) \ln\left[\frac{m^2 - k_1^2 \alpha(1-\alpha)}{m^2 - k^2 \alpha(1-\alpha)} \right]$$

When we do not invoke the approximation $k_1 \gg m, k_2 \gg m$, this result is more complicated.

Let us next consider the explicit form of $\Pi_{\mathrm{phy}}(q^2)$. The presence of the logarithm signals the fact that it is non-analytic in q^2. You can evaluate it by straightforward integration techniques to obtain the result[92]

$$\Pi_{\mathrm{phy}}(q^2) = \frac{e^2}{2\pi^2} \int_0^1 d\alpha\, \alpha(1-\alpha) \ln\left[1 - \alpha(1-\alpha)\frac{q^2}{m^2} \right]$$

$$= \frac{e^2}{4\pi^2} \left[-\frac{5}{9} - \frac{4}{3}\frac{m^2}{q^2} + \frac{1}{3}\left(1 + \frac{2m^2}{q^2} \right) f(q^2) \right] \qquad (5.277)$$

where $f(q^2)$ has different forms in different ranges and is given by

$$f(q^2) = \sqrt{1 - \frac{4m^2}{q^2}} \ln \frac{\sqrt{1 - \frac{4m^2}{q^2}} + 1}{\sqrt{1 - \frac{4m^2}{q^2}} - 1}, \qquad (q^2 < 0) \qquad (5.278)$$

$$= 2\sqrt{\frac{4m^2}{q^2} - 1}\, \tan^{-1}\frac{1}{\frac{4m^2}{q^2} - 1}, \qquad (0 < q^2 \le 4m^2)$$

$$= \sqrt{1 - \frac{4m^2}{q^2}} \ln \frac{1 + \sqrt{1 - \frac{4m^2}{q^2}}}{1 - \sqrt{1 - \frac{4m^2}{q^2}}} - i\pi\sqrt{1 - \frac{4m^2}{q^2}}, \qquad (4m^2 < q^2)$$

[91] Recall that something very similar happened with the running of λ in the $\lambda\phi^4$ theory which we commented upon in Sect.4.7.2 just after Eq. (4.179).

[92] The UV divergences which arise in a theory are usually proportional to polynomials in momenta. In a non-renormalizable theory, these polynomials — translating to higher and higher order derivatives in real space — lead to local short distance effects. In contrast, the *finite* part arising from the loops (even in a non-renormalizable theory) can have non-analytic momentum dependence which leads to residual, well defined, long distance interactions.

Exercise 5.15: Prove this result. (Hint: Integrate by parts to eliminate the logarithm and introduce the variable $v = 2\beta - 1$. The rest of it can be done by careful complex integration.)

We see that for $q^2 > 4m^2$, the function $\Pi_{\rm phy}(q^2)$ picks up an imaginary part. This is the threshold for the production of $e^+ - e^-$ pairs by the photon. If we interpret $(1 - \Pi_{\rm phy})$ as analogous to the dielectric function of a medium (in this case the vacuum), the negative imaginary part signals the absorption of the electromagnetic radiation. In fact, this analogy can be made more precise in the form of a dispersion relation

Exercise 5.16: Prove this result

$$\Pi_{\rm phy}(q^2) = \frac{1}{\pi} q^2 \int_0^\infty dq'^2 \, \frac{{\rm Im}\, \Pi_{\rm phy}(q'^2)}{q'^2(q'^2 - q^2 - i\epsilon)} \tag{5.279}$$

which tells you that the imaginary part of $\Pi_{\rm phy}$ determines the entire function.[93]

Yet another useful integral representation for $\Pi_{\rm phy}$ is given by the expression:

$$\Pi_{\rm phy}(k^2) = \frac{1}{4\pi} \int_{4m^2}^\infty ds \, r(s) \frac{k^2}{k^2 + s} \tag{5.280}$$

where

$$r(s) = \frac{1}{3\pi} \frac{1}{s} \sqrt{1 - \frac{4m^2}{s}} \left(1 + \frac{2m^2}{s}\right) \tag{5.281}$$

[93] As we described earlier in Sect. 3.7, the appearance for branch-cut (and imaginary part) in the propagators signals intermediate states with some specified mass threshold. In this particular case, the threshold $q^2 = 4m^2$ arises from an intermediate state $|e^+e^-\rangle$ containing an e^+e^- pair which will contribute a factor $|\langle 0|A_m|e^+e^-\rangle|^2$. Any such state with $\boldsymbol{p} = 0$ will be labelled by the momenta $p_1 = (\epsilon/2, \boldsymbol{k}/2)$, $p_2 = (\epsilon/2, -\boldsymbol{k}/2)$. The momentum transfer is now $q^2 = (p_1 + p_2)^2 = \epsilon^2 = |\boldsymbol{k}|^2 + 4m^2$. The factor $(1 - (4m^2/q^2))^{1/2}$ in Eq. (5.278) in fact arises from the phase space density $k^2 \, dk = kE\, dE \propto k\, d(q^2) = (q^2 - 4m^2)^{1/2} dq^2$. It is the creation of e^+e^- pairs which lead to the imaginary part in the propagator.

This form is useful to determine the modification of the Coulomb potential due to the vacuum polarization, described by $\Pi_{\rm phy}$. The electromagnetic field produced by a given source is always given by

$$\langle A_m(x)\rangle = \int d^4x' \, D_{mn}(x - x')\, J^n(x') \tag{5.282}$$

Since the propagator — which is the Green's function connecting the source to the field — is now modified by $\Pi_{\rm phy}$, it also affects the field produced by any source. Using Eq. (5.280), we can now write:

Exercise 5.17: Prove Eq. (5.280). [Hint: Write $\alpha(1 - \alpha)$ as $(d/d\alpha)[(\alpha^2/2) - (\alpha^3/3)]$. Substituting and integrating by parts, you should be able to get the relevant result with $r(s)$ expressed as an integral which can be easily evaluated.]

$$\langle A_m(x)\rangle = e^2 \int d^4x' \int \frac{d^4k}{(2\pi)^4} \left[\frac{1}{k^2} + \frac{e^2}{4\pi} \int_{4m^2}^\infty ds \frac{r(s)}{k^2 + s}\right] e^{ik(x-x')} J_m(x') \tag{5.283}$$

As an illustration, consider the Coulomb field produced by a point charge; in this case, the scalar potential evaluates to

$$\phi(\boldsymbol{r}) = \frac{q}{4\pi|\boldsymbol{r}|} \left[1 + \frac{e^2}{4\pi} \int_{4m^2}^\infty ds \, r(s) e^{-\sqrt{s}\,|\boldsymbol{r}|}\right] \tag{5.284}$$

Exercise 5.18: Prove Eq. (5.284), Eq. (5.285) and Eq. (5.286).

It is easy to obtain the limiting forms of this potential using the corresponding limitings forms of $r(s)$. We get

$$m|\boldsymbol{r}| \gg 1 : \qquad \phi(r) \simeq \frac{q}{4\pi|\boldsymbol{r}|} \left[1 + \frac{e^2}{4\pi^{3/2}} \frac{e^{-2m|\boldsymbol{r}|}}{(m|\boldsymbol{r}|)^{3/2}}\right] \tag{5.285}$$

and

[94] This change in the Coulomb potential does affect the atomic energy levels and contributes to the so called *Lamb shift* (which is the shift in the energy levels of $2S_{1/2}$ and $2P_{1/2}$ of the hydrogen atom, which have the same energy in Dirac's theory). There is, however, a much larger contribution to this energy shift which arises from the vertex terms. We shall not discuss this effect.

$$m|\boldsymbol{r}| \ll 1 : \qquad \phi(r) \simeq \frac{q}{4\pi|\boldsymbol{r}|} \left[1 + \frac{e^2}{6\pi^2} \ln\left(\frac{e^{-\gamma_E}}{2m|\boldsymbol{r}|}\right)\right] \tag{5.286}$$

It is obvious that the potential gets modified at distances of the order of the Compton wavelength of the electron which is precisely what one would have expected.[94]

5.7.2   Electron Propagator at One Loop

Just as the photon propagator is corrected by electron loops, the electron propagator is also corrected by emission and re-absorption of a virtual photon as indicated in the diagram of Fig. 5.9. The corresponding algebraic expression is given by

$$-i\Sigma(p) = (-ie\mu^{\epsilon})^2 \int \frac{d^n k}{(2\pi)^n} \gamma^m \frac{i}{\gamma(p-k)-m} \frac{-i\eta_{mn}}{k^2} \gamma^n \qquad (5.287)$$

where we have already gone over to $n-$ dimensions. We have also replaced e by $e\mu^{\epsilon}$ right at the beginning, thereby keeping e dimensionless. In exactly the same way as the photon propagator was corrected by the virtual electron loop, the process described above corrects the electron propagator. Just as in the case of photon propagator, we can also sum up an infinite number of photon loop diagram while evaluating the correction to the electron propagator. Taking care about ordering of factors, this will lead to the result

$$\begin{aligned} i\bar{S}(p) &\equiv S(p)[1 + (-i\Sigma(p))S(p) + (-i\Sigma(p)S(p))^2 + \cdots] \\ &= S(p)[1 + i\Sigma(p)S(p)]^{-1} = \frac{i}{\gamma p - m - \Sigma(p)} \qquad (5.288) \end{aligned}$$

So, the inverse propagator for the electron will get modified to the form

$$\bar{S}^{-1}(p) = \gamma p - m - \Sigma(p) \qquad (5.289)$$

Exercise 5.19: Fill in the details and obtain this result. (This is again similar to what we did for the scalar field to obtain Eq. (4.145).

To see what this could lead to, let us write $\Sigma(p)$ in a Taylor series expansion:

$$\Sigma(p) = A + B(\gamma p - m) + \Sigma^R(p)(\gamma p - m)^2 \qquad (5.290)$$

Substituting this into the propagator, and treating A, B, Σ^R as perturbative quantities of order e^2, we get

$$\begin{aligned} i\bar{S}(p) &= \frac{i}{\gamma p - m - A - B(\gamma p - m) - (\gamma p - m)^2\Sigma^R(p) + i\epsilon} \\ &\simeq \frac{i}{(\gamma p - m - A)(1 - B)\left(1 - (\gamma p - m)\Sigma^R(p)\right) + i\epsilon} \\ &\simeq \frac{(1 + B)i}{(\gamma p - m - A)\left(1 - (\gamma p - m)\Sigma^R(p)\right) + i\epsilon} \qquad (5.291) \end{aligned}$$

This expression shows that the propagator has been modified in two ways. First, the mass has shifted from m to $m + A$ which defines the renormalized mass. (We saw a similar feature in the case of $\lambda\phi^4$ theory in Sect. 4.7.1) Second, we have also acquired an overall multiplicative factor $Z_2 \equiv (1 + B)$ in the numerator.[95] As we shall see, both A and B will be divergent quantities. We can absorb the divergence in A by redefining the mass. To absorb the divergence in Z_2 (which is the residue of the pole of the propagator) we need to rescale the original field $\psi(x)$ and $\psi^{\dagger}(x)$ by $\psi' = Z_2^{-1/2}\psi$ etc. The propagators for the primed fields will now have unit residue at the pole, as they should. Once this is done, we have a finite quantity $\Sigma^R(p)$ which will describe non-trivial, finite effects of photon loops on the electron propagator. We will now see how this works out.

Expression in Eq. (5.287) can, again, be evaluated just as before.[96] The

[95]The notation $Z_1, Z_2, ...$ etc is of historical origin and contains no significance.

[96]The details are given in Mathematical Supplement 5.9.1.

final result is given by

$$\Sigma(p) = \frac{e^2}{16\pi^2\epsilon}(-\gamma p + 4m) + \frac{e^2}{16\pi^2}\left\{ \gamma p(1 + \gamma_E) - 2m(1 + 2\gamma_E) \right.$$

$$\left. + 2\int_0^1 d\alpha\,[\gamma p(1-\alpha) - 2m]\ln\left(\frac{\alpha m^2 - \alpha(1-\alpha)p^2}{4\pi\mu^2}\right) \right\} \quad (5.292)$$

where $n = 4 - 2\epsilon$.

The expression clearly has a divergence in the first term when $\epsilon \to 0$. This is the usual UV divergence which arises because the momentum of the virtual photon can be arbitrarily *high*. The cure for this UV divergence is based on the standard renormalization procedure in a manner similar to what we did for the photon propagator. We shall describe this procedure in a moment. However, before we do that, let us take note of the fact that there is another divergence lurking in Eq. (5.292) in the integral in the last term. One way to see this is to try and evaluate the coefficient B which appears in the Taylor series expansion in Eq. (5.290). By definition, B is given by $B \equiv \partial\Sigma/\partial(\gamma^j p_j)$ evaluated at $\gamma^j p_j = m$. Differentiating $\Sigma(p)$ in Eq. (5.292) with respect to $\gamma^j p_j$ and[97] evaluating the result at $\gamma^j p_j = m, p^2 = m^2$, we get:

[97]Recall that $\partial p^2/\partial(\gamma^j p_j) = 2\gamma^j p_j$

$$B = \left.\frac{\partial\Sigma}{\partial(\gamma^j p_j)}\right|_{\gamma p = m} = -\frac{e^2}{16\pi^2}\frac{1}{\epsilon} + \frac{e^2}{4\pi^2}\int_0^1 \frac{d\alpha}{\alpha} + \text{(finite terms)} \quad (5.293)$$

The first term diverges when $\epsilon \to 0$ and is the usual UV divergence. The second term diverges in the *lower* limit of the integration which is actually an infrared divergence. The reason for the *infrared divergence* is conceptually quite different from the UV divergence. It has to do with the fact that a process like the one in Fig. 5.9 is always accompanied by the emission of very low frequency photons. This can be taken care of mathematically by adding a small mass m_γ to the photons and taking the limit $m_\gamma \to 0$ right at the end of the calculation. One can easily show that changing k^2 to $(k^2 - m_\gamma^2)$ in Eq. (5.287) will replace αm^2 in Eq. (5.292) by $[\alpha m^2 + (1-\alpha)m_\gamma^2]$. Then the infrared divergence in Eq. (5.293) will disappear and, instead, we will get a term proportional to $\ln(m_\gamma/m)$. Since our interest is essentially in the UV divergence and renormalization, we will not discuss IR divergences in detail. (See Problem 18.)

Exercise 5.20: Work out explicitly the effect of m_γ.

From Eq. (5.292), we see that the UV divergent piece in the last term is given by

$$\Sigma_{\text{div}}(p) = -\frac{e^2}{(4\pi)^2}\frac{2}{n-4}(4m - \gamma p) \quad (5.294)$$

which allows us to write the divergent part of the inverse propagator as

$$\begin{aligned}\bar{S}_{\text{div}}^{-1}(p) &= \gamma p - m - \Sigma_{\text{div}}(p) \quad (5.295)\\[4pt] &= \left[1 - \frac{e^2}{(4\pi)^2}\frac{2}{n-4}\right]\left\{\gamma p - m\left[1 - \frac{e^2}{(4\pi)^2}\frac{6}{n-4}\right]\right\}\end{aligned}$$

To take care of these divergences, we need to do two things. First, we need to renormalize the mass of the electron to a physical value[98] by the relation

[98]This actually defines a *running* mass parameter $m_R(\mu)$ because e^2 depends on μ. From Eq. (5.244), we see that, to the lowest order, e^2 satisfies the relation $\mu(de^2/d\mu) = (n-4)e^2$. So we get $\mu(dm_R/d\mu) = -(3e^2/8\pi^2)m_R$. We will see later that the physical mass, defined by the location of the pole of the propagator, is independent of μ.

$$m = m_R\left[1 + \frac{e^2}{(4\pi)^2}\frac{6}{n-4}\right] \quad (5.296)$$

The correction to the mass δm arising from this process (while divergent) is proportional to m. In other words, $\delta m \to 0$ as $m \to 0$. This is actually related to the fact that the Dirac Lagrangian remains invariant under the transformation $\psi \to e^{-i\alpha\gamma_5}\psi$ in the $m = 0$ limit. This symmetry is retained to all orders of perturbation theories, which, in turn, implies that a non-zero mass should not be generated by the loop corrections. In other words, this symmetry — called the *chiral symmetry* — protects the electron mass from what is usually called the *hierarchy or fine-tuning problem*. You may recall that in the ϕ^4 theory, the one loop correction to the scalar propagator changes the mass in the form $m^2 = m_0^2 + (\text{constant})M^2$, where M is a high mass scale in the theory (see the margin note after Eq. (4.72)). We see that, in contrast to $\lambda\phi^4$ theory, the QED correction to the mass δm is determined by the mass scale m itself and is protected by chiral symmetry.

The second thing we have to do is to take care of the overall divergent factor in front of the curly bracket in Eq. (5.295). This is done by renormalizing ψ (and $\bar{\psi}$) themselves by rescaling them, so that we can write:

$$S(p) = Z_2 \, S_R(p); \quad Z_2 \equiv \left(1 + \frac{e^2}{(4\pi)^2} \frac{2}{n-4}\right) \tag{5.297}$$

where $S_R(p)$ is a finite renormalized propagator and Z_2 is a divergent constant. (This correction is also sometimes called the *wave function renormalization*.) Once this is done, we get a finite renormalized propagator given by

$$S_R^{-1}(p) = \gamma p - m_R - \Sigma_{\text{ren}}(p); \quad \Sigma_{\text{ren}}(p) = \Sigma(p) - \Sigma_{\text{div}}(p) \tag{5.298}$$

where

$$\begin{aligned} \Sigma_{\text{ren}}(p) &= -\frac{e^2}{(4\pi)^2} \int_0^1 d\alpha \left[4m - 2(1-\alpha)\gamma p\right] \ln\left[\frac{\alpha m^2 - \alpha(1-\alpha)p^2}{4\pi\mu^2}\right] \\ &\quad - \frac{e^2}{(4\pi)^2}\left[2(1+2\gamma_E)m - (1+\gamma_E)\gamma p\right] \end{aligned} \tag{5.299}$$

and we have taken the $n \to 4$ limit.

The *physical* mass of the electron, however, is determined by the location of the pole of the propagator at which $S_R^{-1}(p) = 0$ with $m_{\text{phy}} = \gamma^a p_a$. This leads to[99] the result:

$$\begin{aligned} m_{\text{phy}} &= m_R + \Sigma_{\text{ren}}(p)|_{\gamma^j p_j = m_R} \\ &= m_R\left\{1 - \frac{e^2}{(4\pi)^2}\left[3\ln\left(\frac{m_R^2}{4\pi\mu^2}\right) - 4 + 3\gamma\right]\right\} \end{aligned} \tag{5.300}$$

Re-expressing the propagator in terms of the physical mass, we get the final result to be

$$S^{-1}(p) = \gamma p - m_{\text{phy}} - \bar{\Sigma}(p) \tag{5.301}$$

where

$$\begin{aligned} \bar{\Sigma}(p) &= -\frac{e^2}{(4\pi)^2} \int_0^1 d\alpha \left[4m - 2(1-\alpha)\gamma p\right] \ln\left[\frac{\alpha m^2 - (1-\alpha)p^2}{\lambda m^2}\right] \\ &\quad - \frac{e^2}{(4\pi)^2}(-\gamma p + m)\left[\ln\left(\frac{m^2}{4\pi\mu^2}\right) + \gamma_E - 2\right] \end{aligned} \tag{5.302}$$

[99] You can now explicitly verify that m_{phy} satisfies the condition $dm_{\text{phy}}/d\mu = 0$ as it should.

Incidentally, the above calculations also illustrate several standard techniques used in renormalization theory, especially in higher order perturbation theory, QCD etc. For example, notice that we have actually introduced two different masses m_R and m_{phy}. The first one m_R was obtained by regularizing the divergences which appeared in the theory (see Eq. (5.296)). The divergent terms in this expression (as well as in the definition of Z_2 in Eq. (5.297)) were chosen without any finite parts. One could have always redefined the finite quantities which arise in Eq. (5.298) by adding some finite parts to the divergent quantities and redefining the terms.[100] Any one of these subtraction schemes allows us to define the renormalized mass m_R. As one can see from Eq. (5.296), the m_R does pick up radiative corrections and 'runs' with the renormalization scale μ. In contrast, the physical mass m_{phy} is always defined using the location of the pole of the electron propagator. In our scheme m_{phy} does not depend on the renormalization scale μ, as we have explicitly seen above.[101]

It should be stressed that the specific form of the divergence of $\Sigma(p)$ played a crucial role in our ability to carry out the renormalization. Because Σ contained two divergent pieces, one proportional to $\gamma^j p_j$ and the other proportional to m, we could write

$$
\begin{aligned}
\gamma p - m - \Sigma(p) &= (\gamma p - m) + \frac{e^2}{8\pi^2 \epsilon}(\gamma p - 4m) + \Sigma_{\text{fin}}(p) \\
&= Z_2^{-1}(\gamma p - (m + \delta m)) + \Sigma_{\text{fin}}(p) \qquad (5.303)
\end{aligned}
$$

This, in turn, allowed us to reabsorb the divergences by a mass renormalization and wave function renormalization. If, for example, we had picked up a divergent term like e.g., $p^2/m\epsilon$ in $\Sigma(p)$, we would be sunk; QED would not have been renormalizable.

Also note that the finite piece $\Sigma_{\text{fin}}(p)$ is *non-analytic* in p — unlike the divergent pieces which *are* analytic in p. This will turn out to be a general feature in all renormalization calculations. There is a branch-cut in the expression when the argument of the log becomes negative; i.e., when $p^2 > m^2/x$. In particular, at $x = 1$, there is a branch-cut at $p^2 = m^2$ and $\Sigma(p)$ will pick up an imaginary piece. This is similar to what we saw as regards the photon propagator (see Eq. (5.278)) and in Sect. 3.7.

5.7.3   Vertex Correction at One Loop

In this case, we need to compute the contribution from the diagram in Fig. 5.10 which translates into the algebraic expression

$$
-ie\mu^\epsilon \Lambda_m = (-ie\mu^\epsilon)^3 \int \frac{d^n k}{(2\pi)^n} \frac{-i\eta^{nr}}{k^2} \gamma_n \frac{i}{\gamma(p-k)-m} \gamma_m \frac{i}{\gamma(p-k)-m} \gamma_r \tag{5.304}
$$

The net effect of such a diagram will be to change the vertex factor from $-ie\gamma_m$ to $-ie\Lambda_m$ in $D = 4$. As mentioned earlier, it is convenient to pull out a dimensionful factor μ^ϵ and write (in $n = 4 - 2\epsilon$ dimensions) the relevant contribution as $-ie\mu^\epsilon \Lambda_m$; this will again keep e dimensionless.

Before we plunge into the calculation, (which is probably the most involved of the three we are performing, since this requires working with three vertices), we will pause for a moment to understand what such a modification of $\gamma_m \to \Lambda_m$ physically means.

Given the vectors which are available, the most general form for the modified vertex is given by[102]

---

[100]We encountered the same issues in the $\lambda\phi^4$ theory. As you will recall, when we subtract only the divergent terms, it is called the *minimal subtraction scheme* (MS). Occasionally, one includes a factor of $4\pi e^{-\gamma_E}$ into the subtracted term, which is the $\overline{\text{MS}}$ scheme.

[101]Numerically, the difference between m_{phy} and m_R will be small if one chooses the renormalization point $\mu \approx m_{\text{phy}}$; this is evident from Eq. (5.300). All these facts are not very pertinent to one loop QED, but the renormalization scheme and the subtraction procedure can make significant differences to the ease of calculation in QCD and in higher order perturbation theory.

[102]We do not have to worry about a term involving $\sigma_{mn}q^n$ because it can be expressed as a linear combination of the first two terms, in expressions of the form $\bar{u}(p')\Lambda_m u(p)$. (As usual, we will suppress the spin parameter in $u(p,s)$ and simply denote it by $u(p)$ etc.) We will only be concerned with the vertex function which occurs in this function.

$$\Lambda_m(p, q, p' = p + q) = \gamma_m A + (p' + p)_m B + (p' - p)_m C \qquad (5.305)$$

The coefficients A, B, C could contain expressions like γp or $\gamma p'$. But when sandwiched between the two spinors, in a term of the form $\bar{u}(p')\Lambda_m u(p)$, we can replace γp and $\gamma p'$ by m on-shell. Further, since $p^2 = p'^2 = m^2$, the only non-trivial quantity left for us to consider as an independent variable is $q^2 = 2m^2 - 2p \cdot p'$. Thus, A, B, C can be thought of as functions of q^2. There is one more simplification we can make. Gauge invariance requires that $q^m \bar{u}(p')\Lambda_m u(p) = 0$. This is trivially satisfied for the first term in Eq. (5.305) because of the Dirac equation and due to the identity $(p' - p) \cdot (p' + p) = p'^2 - p^2 = m^2 - m^2 = 0$ for the second term. But there is no reason for this to hold for the last term in Eq. (5.305) unless $C = 0$. Thus we conclude that, in the right hand side of Eq. (5.305) we can set $C = 0$ and need to retain only the first two terms.

It is conventional to replace the second term using the Gordon identity (see Eq. (5.125)) and write the resulting expression in the form

$$\bar{u}(p')\Lambda_m u(p) = \bar{u}(p') \left[\gamma_m F_1(q^2) + \frac{i\sigma_{mn}q^n}{2m} F_2(q^2) \right] u(p) \qquad (5.306)$$

where F_1 and F_2 are called the *form-factors*. When we do the computation, however, it is more convenient to re-write this result (again using the Gordon identity) in terms of $(p' + p)$ and $(p' - p)$. Doing this, we find that to leading order in the momentum transfer q, we can write this expression in the form

$$\bar{u}\Lambda^m u = \bar{u}(p') \left\{ \frac{(p' + p)^m}{2m} F_1(0) + \frac{i\sigma^{mn}q_n}{2m} [F_1(0) + F_2(0)] \right\} u(p) \quad (5.307)$$

By definition, the coefficient of the first term is the electric charge in the $q \to 0$ limit, and hence we must have $F_1(0) = 1$. This tells us that the primary effect of the vertex correction is to shift the magnetic moment from its Dirac value by a factor $1 + F_2(0)$. In other words, radiative corrections lead to the result that the gyro-magnetic ratio of the electron is actually $g = 2[1 + F_2(0)]$. This is a clear prediction which can be compared with observations once we have computed Λ_m and extracted $F_2(0)$ from it. We shall now proceed to do this.

We compute the expression in Eq. (5.304) by the usual tricks to reduce it to $\Lambda_m = \Lambda_m^{(1)} + \Lambda_m^{(2)}$ where the first term has a UV divergent piece with the structure[103]

$$\Lambda_m^{(1)} = \frac{e^2}{16\pi^2} \frac{1}{\epsilon}\gamma_m + \text{finite}; \quad 2\epsilon = 4 - n \qquad (5.308)$$

and the second term is finite and is given by

$$\Lambda_m^{(2)} = \frac{e^2}{16\pi^2} \int_0^1 d\alpha \int_0^{1-\alpha} d\beta \qquad (5.309)$$

$$\times \frac{\gamma_n(\gamma^j p'_j(1 - \beta) - \alpha\gamma^j p_j + m)\gamma_m(\gamma^j p_j(1 - \alpha) - \beta\gamma^j p'_j + m)\gamma^n}{[-m^2(\alpha + \beta) + \alpha(1 - \alpha)p^2 + \beta(1 - \beta)p'^2 - \alpha\beta 2p' \cdot p]}$$

Once again, we need to take care of the divergent part $\Lambda_m^{(1)}$ using a renormalization prescription. But let us first compute the finite part and extract from it the value of $F_2(0)$ which will give us the anomalous magnetic

[103] The details are given in Mathematical Supplement 5.9.2. This term also has an IR divergent piece which can be regulated by adding a small mass to the photon. We will be primarily concerned with the UV divergence.

moment of the electron. A somewhat lengthy calculation (given in Mathematical Supplement 5.9.2) shows that the finite part, when sandwiched between two Dirac spinors, can be expressed in the form

$$\bar{u}(p')\Lambda_m^{(2)}u(p) = \bar{u}(p')\frac{-e^2}{16\pi^2m^2}\int_0^1 d\alpha \int_0^{1-\alpha} d\beta \frac{1}{(\alpha+\beta)^2}$$
$$\times\Big(-2\gamma_m m^2[2(\beta+\alpha)+(\alpha+\beta)^2-2]$$
$$-2i\sigma_{mn}q^n m[(\alpha+\beta)-(\alpha+\beta)^2]\Big)u(p) \qquad (5.310)$$

Right now, we are only interested in the part which is the coefficient of $i\sigma_{mn}/2m$; this is given in terms of the two easily evaluated integrals

$$\int_0^1 d\alpha \int_0^{1-\alpha} d\beta \frac{1}{(\alpha+\beta)} = 1; \qquad \int_0^1 d\alpha \int_0^{1-\alpha} d\beta = \frac{1}{2} \qquad (5.311)$$

The term which we are after now reduces to

$$\frac{i\sigma_{mn}q^n}{2m}\left[\frac{e^2}{8\pi^2}\right] = \frac{i\sigma_{mn}q^n}{2m}\left[\frac{\alpha}{2\pi}\right] \qquad (5.312)$$

so that $F_2(0) = \alpha/2$. This leads to the famous result by Schwinger, viz., that the gyro-magnetic ratio of the electron is given by

$$g = 2\left(1+\frac{\alpha}{2\pi}\right) \qquad (5.313)$$

5.8 QED Renormalization at One Loop

We have now computed the corrections to the photon propagator, the electron propagator and the vertex term at the one loop level. Each of the results involved a divergent piece and finite terms. To clarify the renormalization program, we will now summarize how the divergent pieces were handled.

Let us begin by re-writing our results, isolating the divergences in the form[104] (see Eq. (5.294), Eq. (5.259), Eq. (5.308):

$$\Sigma(p) = \frac{e^2}{16\pi^2\epsilon}(-\gamma^j p_j + 4m) + \text{finite}$$

$$\Pi_{mn}(k) = \frac{e^2}{12\pi^2\epsilon}(k_m k_n - k^2 g_{mn}) + \text{finite}$$

$$\Lambda_m^{(1)}(p,q,p') = \frac{e^2}{16\pi^2\epsilon}\gamma_m + \text{finite} \qquad (5.314)$$

Let us begin by eliminating the divergence in the electron propagator which actually has two separate pieces, one proportional to $\gamma^j p_j$ and another proportional to m. To take care of these, we will now add a counter-term to the Lagrangian of the form

$$\Delta L = iB\bar{\psi}\gamma^j \partial_j \psi - A\bar{\psi}\psi \qquad (5.315)$$

where A and B are undetermined at present. These counter-terms will lead to two more Feynman rules requiring the factors $iB(\gamma^j p_j)$ and $-iA$ in the

[104]Note that the divergent $(1/\epsilon)$ terms satisfy the identity $\Lambda_m = -\partial\Sigma(p)/\partial p^m$. This is implied by the Ward identity in Eq. (5.187) which should hold to all orders in the perturbation theory. Here, we have explicitly verified it to the one loop order.

respective diagrams. So, all the amplitudes in the modified theory (with the counter-term in the Eq. (5.315) added) will now have the contribution

$$-i\Sigma(p) - iA + iB\gamma^j p_j = -i\frac{e^2}{16\pi^2\epsilon}(-\gamma^j p_j + 4m) + \text{finite} - iA + iB\gamma^j p_j$$
(5.316)

We now need to choose A and B such that this amplitude is finite. This is easily done with the choices:

$$A = -\frac{me^2}{4\pi^2\epsilon}, \qquad B = -\frac{e^2}{16\pi^2\epsilon}$$
(5.317)

The trick now is to match the fermionic part in the two Lagrangians: The modified physical Lagrangian $L + \Delta L$ and the original Lagrangian L_0 with bare parameters which are indicated with a subscript zero. That is, we demand

$$i(1 + B)\bar{\psi}\gamma^j \partial_j \psi - (m + A)\bar{\psi}\psi = i\bar{\psi}_0\gamma^j \partial_j \psi_0 - m_0\bar{\psi}_0\psi_0$$
(5.318)

As indicated in the earlier discussion, this can be accomplished by a mass renormalization and a wave function renormalization of the form:

$$\psi_0 = \sqrt{Z_2}\,\psi, \qquad m_0 = m + \delta m$$
(5.319)

where

$$
\begin{aligned}
Z_2 &= 1 + B = 1 - \frac{e^2}{16\pi^2\epsilon}, \\
m_0 &= Z_2^{-1}(m + A) = m\left(1 + \frac{e^2}{16\pi^2\epsilon}\right)\left(1 - \frac{e^2}{4\pi^2\epsilon}\right) \\
&= m\left(1 - \frac{3e^2}{16\pi^2\epsilon}\right) = m + \delta m
\end{aligned}
$$
(5.320)

We had already commented on the fact that the correction δm to the mass is proportional to m (unlike in the $\lambda\phi^4$ theory).

Let us next consider the divergent term in the photon propagator which we hope to tame by adding a counter-term to the electromagnetic Lagrangian. We are working with the Lagrangian of the form[105]

$$L = -\frac{1}{4}F_{mn}F^{mn} - \frac{1}{2}(\partial_m A^m)^2 = \frac{1}{2}A^m(\eta_{mn}\Box)A^n$$
(5.321)

which includes the gauge fixing term. Hence, a priori, one could have added a counter-term of the kind

$$\Delta L = -\frac{C}{4}F_{mn}F^{mn} - \frac{E}{2}(\partial_m A^m)^2$$
(5.322)

This is, however, a little bit disturbing since you would not like a pure gauge fixing term to pick up corrections due to renormalization. Let us, however, proceed keeping these terms general at this stage and try to determine C and E. These counter-terms in the Lagrangian will lead to two more Feynman rules requiring the inclusion of the factors $-iC(k^2\eta_{mn} - k_m k_n)$ and $-iEk^2\eta_{mn}$ in the respective diagrams. But we see that the divergent term in the propagator has the structure:

$$i\Pi_{mn}(k) = -i\frac{e^2}{12\pi^2\epsilon}(k^2\eta_{mn} - k_m k_n)$$
(5.323)

[105] We can use either the Lagrangian in Eq. (5.237) or the one in Eq. (5.236); we will use the latter since it will clarify an important point. Exercise: Redo this analysis using the Lagrangian in Eq. (5.237).

which tells you that we can take care of everything with the choice

$$C = -\frac{e^2}{12\pi^2\epsilon}; \qquad E = 0 \tag{5.324}$$

In other words, the gauge fixing term does not invite corrections at the one loop level. With this addition, the electromagnetic part of the Lagrangian takes the form[106]

$$L + \Delta L = -\left(\frac{1+C}{4}\right) F_{mn}F^{mn} - \frac{1}{2}(\partial_m A^m)^2 \equiv -\frac{Z_3}{4}F_{mn}F^{mn} - \frac{1}{2}(\partial_m A^m)^2 \tag{5.325}$$

where

$$Z_3 = 1 + C = 1 - \frac{e^2}{12\pi^2\epsilon} \tag{5.326}$$

As usual, we identify $L + \Delta L$ with the original Lagrangian plus any gauge terms which you choose to add. This is easily taken care of by another wave function renormalization of the form

$$A_0^m = \sqrt{Z_3}\,A^m \tag{5.327}$$

Having handled the pure Dirac sector and the pure electromagnetic sector, we finally take up the vertex correction which modifies the interaction term. Playing the same game once again, we now add a counter-term of the form

$$\Delta L = -De\mu^\epsilon \bar{\psi}\gamma^j A_j \psi \tag{5.328}$$

which will lead to a Feynman rule of the form $-iDe\mu^\epsilon\gamma_m$. (As described earlier, we have kept e to be dimensionless and introduced a mass scale μ.) We now need to choose D such that $-ie\mu^\epsilon(\Lambda_m^{(1)} + D\gamma_m)$ is finite, where $\Lambda_m^{(1)} = (e^2/16\pi^2\epsilon)\gamma_m$ plus finite terms. This is easily taken care of by the choice $D = -(e^2/16\pi^2\epsilon)$. The full Lagrangian will then become

$$L + \Delta L = -(1+D)e\mu^\epsilon A^m\bar{\psi}\gamma_m\psi \equiv -Z_1 e\mu^\epsilon A^m\bar{\psi}\gamma_m\psi \tag{5.329}$$

with

$$Z_1 = 1 + D = 1 - \frac{e^2}{16\pi^2\epsilon} \tag{5.330}$$

We thus find that we need 3 multiplicative renormalization factors for the wave functions given by:

$$Z_1 = Z_2 = 1 - \frac{e^2}{16\pi^2\epsilon}, \qquad Z_3 = 1 - \frac{e^2}{12\pi^2\epsilon} \tag{5.331}$$

Equating the modified Lagrangian to the bare one as regards the interaction term, we obtain the condition:

$$-Z_1 e\mu^\epsilon A^m\bar{\psi}\gamma_m\psi = -e_0 A_0^m \bar{\psi}_0 \gamma_m \psi_0 = -e_0 \sqrt{Z_3}\left(\sqrt{Z_2}\right)^2 A^m\bar{\psi}\gamma_m\psi \tag{5.332}$$

where we have related the bare fields to the renormalized fields.

Let us now put these results together. The renormalization prescription used in conjunction with dimensional regularization relates the bare and renormalized quantities through the relations

$$A_m = \frac{1}{\sqrt{Z_3}}A_m^{(0)}, \quad \psi = \frac{1}{\sqrt{Z_2}}\psi^{(0)}, \quad m_R = \frac{1}{Z_m}m^{(0)}, \quad e_R = \frac{1}{Z_e}\mu^{\frac{d-4}{2}}e^{(0)} \tag{5.333}$$

[106]This matches with the conclusion we obtained earlier in Eq. (5.183) and Eq. (5.184), from the Ward identity, viz. that the gauge fixing parameter does not pick up radiative corrections. This, however, does not mean that the structure of the propagator at higher orders will have the same form as the original one in whatever gauge you are working with. But, the subsequent calculations at higher order can always be done in the same gauge.

leading to a QED Lagrangian of the form

$$L = -\frac{1}{4}Z_3 F_{mn}^2 + iZ_2\bar{\psi}\gamma\partial\psi - m_R Z_2 Z_m\bar{\psi}\psi - \mu^{\frac{4-d}{2}}e_R Z_e Z_2\sqrt{Z_3}\bar{\psi}\gamma A\psi$$
(5.334)

In this rescaling, we have kept e_R and the various Zs as dimensionless. If we write every Z in the form $Z_i = 1 + \delta_i$, then we know from our previous studies that the divergent parts of δ_i are given by

$$\delta_2 = \frac{e_R^2}{16\pi^2}\left[-\frac{1}{\epsilon}\right], \ \delta_3 = \frac{e_R^2}{16\pi^2}\left[-\frac{4}{3\epsilon}\right], \ \delta_m = \frac{e_R^2}{16\pi^2}\left[-\frac{3}{\epsilon}\right], \ \delta_e = \frac{e_R^2}{16\pi^2}\left[\frac{2}{3\epsilon}\right]$$
(5.335)

to $\mathcal{O}(e_R^2)$. This, in turn, allows us to obtain the running of the coupling constants directly from the Lagrangian. For example, since the renormalized Lagrangian depends on μ but not the bare Lagrangian, we must have

$$0 = \mu\frac{d}{d\mu}e_0 = \mu\frac{d}{d\mu}\left[\mu^\epsilon e_R Z_e\right] = \mu^\epsilon e_R Z_e\left[\epsilon + \frac{\mu}{e_R}\frac{d}{d\mu}e_R + \frac{\mu}{Z_e}\frac{dZ_e}{d\mu}\right] \quad (5.336)$$

We know that, to leading order in e_R, we have $Z_e = 1$ giving

$$\mu\frac{d}{d\mu}e_R = -\epsilon\, e_R \tag{5.337}$$

while, to the next order:

$$\mu\frac{d}{d\mu}Z_e = \mu\frac{d}{d\mu}\left(1 + \frac{e_R^2}{16\pi^2}\frac{2}{3\epsilon}\right) = \frac{1}{\epsilon}\frac{e_R}{12\pi^2}\left(\mu\frac{d}{d\mu}e_R\right) = -\frac{e_R^2}{12\pi^2} \quad (5.338)$$

We therefore get

$$\beta(e_R) \equiv \mu\frac{d}{d\mu}e_R = -\epsilon\, e_R + \frac{e_R^3}{12\pi^2} \to \frac{e_R^3}{12\pi^2} \tag{5.339}$$

where the last expression is obtained in the limit of $\epsilon \to 0$. This agrees with our previous result for the β function in e.g., Eq. (5.202) or in Eq. (5.274). (Note that the β's differ by a trivial factor 2 depending on whether we define β as $\mu(de_R/d\mu)$ or as $\mu(de_R^2/d\mu)$. But we have now calculated it using only the counter-terms without summation of logs etc.[107] In a similar manner, we can also work out the "running" of the electron mass. Since the bare mass must be independent of μ, we have the result

$$0 = \mu\frac{d}{d\mu}m_0 = \mu\frac{d}{d\mu}\left(Z_m m_R\right) = Z_m m_R\left[\frac{\mu}{m_R}\frac{dm_R}{d\mu} + \frac{\mu}{Z_m}\frac{dZ_m}{d\mu}\right] \quad (5.340)$$

It is convenient to define a quantity $\gamma_m \equiv \left[(\mu/m_R)(dm_R/d\mu)\right]$ (which is called the *anomalous dimension*). Using the fact that Z_m depends on μ only through e_R, we have

$$\gamma_m = -\frac{\mu}{Z_m}\frac{dZ_m}{d\mu} = -\frac{1}{Z_m}\frac{dZ_m}{de_R}\mu\frac{de_R}{d\mu} \tag{5.341}$$

Using the known expressions for Z_m and $\beta(e_R)$ correct to one loop order, we find that

$$\gamma_m = -\frac{1}{1+\delta_m}\left(\frac{2}{e_R}\delta_m\right)\left(-\frac{\epsilon}{2}e_R\right) = \delta_m\epsilon = -\frac{3e_R^2}{8\pi^2} \tag{5.342}$$

[107] Notice that the β function depends on the combination $Z_e = Z_1/(Z_2\sqrt{Z_3})$. Here, Z_1 arises from the electron-photon vertex, Z_3 from the vacuum polarization and Z_2 from the electron self-energy. In QED (but not in other theories like QCD), $Z_1 = Z_2$ and hence the β function can be calculated directly from Z_3 itself. The fact that $Z_1 = Z_2$ is not a coincidence but is implied, again, by the Ward identity in QED. Consequently, the charge renormalization only depends on Z_3 which is the photon field renormalization factor. (This would have been obvious if we had used the Lagrangian in Eq. (5.237); we used the form in Eq. (5.236) to explicitly show that we still get the correct result.) The procedure of computing the running coupling constant through the μ dependence of the bare Lagrangian is more useful in theories when $Z_1 \neq Z_2$.

This matches with the result obtained earlier from Eq. (5.296).

The fact that $Z_1 = Z_2$ is also closely related to the charge being a property of the photon, along the lines we described earlier. To see this more explicitly, consider a theory with a quark having the charge $Q_q = 2/3$ and an electron with charge $Q_e = -1$. A Lagrangian which includes both the fields will be given by

$$
L = -\frac{1}{4}Z_3 F_{mn}^2 + iZ_{2e}\bar{\psi}_e\gamma\partial\psi_e - e_R Z_{1e}\bar{\psi}_e\gamma A\psi_e + iZ_{2q}\bar{\psi}_q\gamma\partial\psi_q
$$
$$
+ \frac{2}{3}e_R Z_{1q}\bar{\psi}_q\gamma A\psi_q \tag{5.343}
$$

Because $Z_{1e} = Z_{2e}$ and $Z_{1q} = Z_{2q}$, this Lagrangian can be rewritten in the form:

$$
L = -\frac{1}{4}Z_3 F_{mn}^2 + Z_{2e}\bar{\psi}_e(i\gamma\partial - e_R\gamma A)\psi_e + Z_{2q}\bar{\psi}_q\left(i\gamma\partial + \frac{2}{3}e_R\gamma A\right)\psi_q
$$
$$
\tag{5.344}
$$

In other words, the relative coefficient between $i\gamma\partial$ and $e_R(\gamma A)$ does not pick up any radiative correction. So the ratio of the charges of the electron and the quark does not change due to radiative corrections, which of course is vital.[108] In our way of describing the Lagrangian, we would have rescaled the vector potential by $A_m \to A_m/e_R$. Then, the above Lagrangian becomes:

$$
L = -\frac{1}{4e_R^2}Z_3 F_{mn}^2 + Z_{2e}\bar{\psi}_e\left(i\gamma\partial - \frac{Z_{1e}}{Z_{2e}}\gamma A\right)\psi_e
$$
$$
+ Z_{2q}\bar{\psi}_q\left(i\gamma\partial + \frac{Z_{1q}}{Z_{2q}}\frac{2}{3}\gamma A\right)\psi_q \tag{5.345}
$$

At the tree-level, all the Z-factors are unity and this Lagrangian is invariant under the gauge transformations of the form

$$
\psi_q \to e^{(2/3)i\alpha}\psi_q, \qquad \psi_e \to e^{-i\alpha}\psi_e, \qquad A_m \to A_m + \partial_m\alpha \tag{5.346}
$$

These gauge transformations only involve the numbers -1 and $2/3$ but not e_R. Further, such a transformation has nothing to do with perturbation theory. As long as we do everything correctly using a regulator which preserves gauge invariance, the loop corrections, counter-terms etc. will all respect this symmetry and we must have $Z_1 = Z_2$ to all orders of perturbation theory. This shows the overall consistency of the formalism.

The above discussion involved dealing with the bare parameters of a Lagrangian, a set of counter-terms and the final Lagrangian in terms of physical, renormalized quantities. Both the bare terms and the counter-terms are divergent but the final renormalized expressions are finite. There is an alternative way of viewing this procedure, which is algebraically equivalent and conceptually better. In this approach, we *start* with a QED Lagrangian written in the form

$$
L = -\frac{1}{4}Z_3 F_{mn}^2 + iZ_2\bar{\psi}\gamma\partial\psi - Z_2 Z_m m_R\bar{\psi}\psi - e_R Z_1\bar{\psi}\gamma A\psi \tag{5.347}
$$

where we have used the abbreviation $Z_1 \equiv Z_e Z_2\sqrt{Z_3}$. We next expand the parameters around the tree-level values as

$$
Z_1 \equiv 1 + \delta_1, \quad Z_2 \equiv 1 + \delta_2, \quad Z_3 \equiv 1 + \delta_3, \quad Z_m \equiv 1 + \delta_m, \quad Z_e \equiv 1 + \delta_e
$$
$$
\tag{5.348}
$$

[108]This is an explicit demonstration of the features we described just after Eq. (5.237). We used the example of quark + electron here to emphasize that the charges need not have the same magnitude at the tree-level.

where $\delta_e = \delta_1 - \delta_2 - (1/2)\delta_3 + \mathcal{O}(e_R^4)$. This allows us to separate the Lagrangian into one involving the physical fields and the other involving counter-terms, in the form[109]

$$L = -\frac{1}{4}F_{mn}^2 + i\bar{\psi}\gamma\partial\psi - m_R\bar{\psi}\psi - e_R\bar{\psi}\gamma A\psi - \frac{1}{4}\delta_3 F_{mn}^2 \quad (5.349)$$
$$+ i\delta_2\bar{\psi}\gamma\partial\psi - (\delta_m + \delta_2)m_R\bar{\psi}\psi - e_R\delta_1\bar{\psi}\gamma A\psi$$

The counter-terms will lead to new Feynman rules shown in Fig 5.15. We can now work out a perturbation theory based on this renormalized Lagrangian. This has the virtue that, even though the counter-terms are arbitrarily large numbers — scaling as $(1/\epsilon)$ in dimensional regularization — they are defined by their Taylor expansions in powers of e_R starting at $\mathcal{O}(e_R^2)$. So, the regularized perturbation expansion (with finite but arbitrarily small ϵ) can be formally justified when e_R is small. This should be contrasted with the previous discussion in which we used the perturbation expansion in the bare coupling e_0 which itself diverges as ϵ and hence leads to a somewhat dubious procedure.

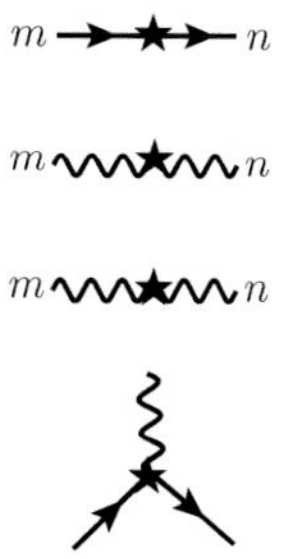

Finally, we comment on the relation between perturbation theory and the computation of effective action. You may be wondering what is the relation between the order-by-order perturbation theory based on renormalized coupling constants, say, and the effective Lagrangians we computed in Sect. 5.6.5 (or in Sect. 4.2). The answer to this question is somewhat complicated but we will briefly mention some relevant aspects.

To begin with, the key idea behind the calculation of the effective Lagrangian is to capture the effect of higher order Feynman diagrams (say, the one involving several loops) by a tree-level Lagrangian. In other words, a tree-level computation using the effective Lagrangian incorporates the effects of summing over a class of Feynman diagrams, depending on the nature of approximations used in the computation of the effective Lagrangian. The way the effective Lagrangian achieves this is by modifying the structure of the vertices in a specific manner. For example, the computation of the Euler-Heisenberg effective Lagrangian did not use the photon propagator at all. Nevertheless, the effects of complicated Feynman diagrams involving photon propagators can be reproduced from the effective Lagrangian by a suitable modification of the vertex.

In fact, the situation is better than this at least in a formal sense. In computing the Euler-Heisenberg Lagrangian, we integrated out the electron field but assumed the existence of a background electromagnetic field A_j. One could have in fact computed an effective Lagrangian $\Gamma[A, \bar{\psi}, \psi]$ by assuming background values for both the vector potential and the electron field and integrating out the quantum fluctuations of all the fields. Such an effective Lagrangian, when used at the tree-level will, in principle, give the two-point function containing all the quantum corrections. The price we pay for getting exact results for a tree-level effective action is that it will be highly non-local.

[109]This is similar to what we did in Sect. 4.8 for the $\lambda\phi^4$ theory.

$m \longrightarrow\!\!\bigstar\!\!\longrightarrow n = i[\gamma p\delta_2 - (\delta_m + \delta_2)m_R]$

$m \sim\!\!\bigstar\!\!\sim n = i\delta_3(p^2 g^{mn} - p^m p^n)$

$m \sim\!\!\bigstar\!\!\sim n = -i\delta_3 p^2 g^{mn}$

$= -ie_R\delta_1\gamma^m$

Figure 5.15: Diagrams generated by the counter-terms and their algebraic equivalents.

5.9 Mathematical Supplement

5.9.1 Calculation of the One Loop Electron Propagator

The integrals involved in Eq. (5.287) can be evaluated exactly as before by writing the propagators in exponential form, completing the square and performing the momentum integral. However, we will do it in a slightly different manner to introduce a technique you will find in many other textbooks.

This will involve combining the two denominators in a particular way which is completely equivalent to writing the propagators in exponential form.[110] We again begin with the expression:

$$
\begin{aligned}
-i\Sigma(p) &= (-ie\mu^\epsilon)^2 \int \frac{d^n k}{(2\pi)^n} \gamma^m \frac{i}{\gamma^j p_j - \gamma^j k_j - m} \frac{-i\eta_{mn}}{k^2} \gamma^n \\
&= -e^2\mu^{2\epsilon} \int \frac{d^n k}{(2\pi)^n} \frac{\gamma_m(\gamma^j p_j - \gamma^j k_j + m)\gamma^m}{[(p-k)^2 - m^2]k^2}
\end{aligned}
\tag{5.350}
$$

We now combine the two denominators using the identity:

$$
\frac{1}{AB} = \int_0^1 d\alpha \frac{1}{[A + (B-A)\alpha]^2}
\tag{5.351}
$$

We then get:

$$
-i\Sigma(p) = -e^2\mu^{2\epsilon} \int_0^1 d\alpha \int \frac{d^n k}{(2\pi)^n} \frac{\gamma_m(\gamma^j p_j - \gamma^j k_j + m)\gamma^m}{[\alpha(p-k)^2 - \alpha m^2 + (1-\alpha)k^2]^2}
\tag{5.352}
$$

We next introduce the variable $k' = k - \alpha p$ which eliminates any $k' \cdot p$ terms from the denominator. This allows us to ignore terms which are odd under $k' \to -k'$ in the numerator. This leads to:

$$
\begin{aligned}
-i\Sigma(p) &= -e^2\mu^{2\epsilon} \int_0^1 d\alpha \int \frac{d^n k'}{(2\pi)^n} \frac{\gamma_m[(1-\alpha)\gamma^j p_j - \gamma^j k'_j + m]\gamma^m}{[k'^2 - \alpha m^2 + \alpha(1-\alpha)p^2]^2} \\
&= -e^2\mu^{2\epsilon} \int_0^1 d\alpha \gamma_m[(1-\alpha)\gamma^j p_j + m]\gamma^m \\
&\quad \times \int \frac{d^n k'}{(2\pi)^n} \frac{1}{[k'^2 - \alpha m^2 + \alpha(1-\alpha)p^2]^2}
\end{aligned}
\tag{5.353}
$$

We now rotate to the Euclidean sector and perform the k'_0 integration, thereby obtaining

$$
\begin{aligned}
-i\Sigma(p) &= -ie^2\mu^{2\epsilon} \frac{\Gamma(2 - n/2)}{(4\pi)^{n/2}} \int_0^1 d\alpha \gamma_m[(1-\alpha)\gamma^j p_j + m] \\
&\qquad \times \gamma^m[\alpha m^2 - \alpha(1-\alpha)p^2]^{n/2-2} \\
&= -ie^2\mu^{2\epsilon} \frac{\Gamma(2 - n/2)}{(4\pi)^{n/2}} \int_0^1 d\alpha[(1-\alpha)\gamma^j p_j(2-n) + mn] \\
&\qquad \times [\alpha m^2 - \alpha(1-\alpha)p^2]^{n/2-2} \\
&= -ie^2 \frac{\Gamma(\epsilon)}{16\pi^2} \int_0^1 d\alpha[(1-\alpha)\gamma^j p_j(-2+2\epsilon) + m(4-2\epsilon)] \\
&\qquad \times \left[\frac{\alpha m^2 - \alpha(1-\alpha)p^2}{4\pi\mu^2} \right]^{-\epsilon}
\end{aligned}
$$

$$
\tag{5.355}
$$

Taking the $\epsilon \to 0$ limit and using $\Gamma(\epsilon) \simeq \epsilon^{-1} - \gamma_E$, we find that:

$$-i\Sigma(p) = -ie^2 \frac{(1/\epsilon) - \gamma_E}{16\pi^2} \int_0^1 d\alpha [(1-\alpha)\gamma^j p_j(-2 + 2\epsilon) + m(4 - 2\epsilon)]$$

$$\times \left[1 - \epsilon \ln \left(\frac{\alpha m^2 - \alpha(1-\alpha)p^2}{4\pi\mu^2} \right) \right]$$

$$= -i\frac{e^2}{16\pi^2\epsilon}(-\gamma^j p_j + 4m) - i\frac{e^2}{16\pi^2} \left\{ \gamma^j p_j(1 + \gamma_E) - 2m(1 + 2\gamma_E) \right.$$

$$\left. + 2\int_0^1 d\alpha \left[\gamma^j p_j(1-\alpha) - 2m \right] \ln \left(\frac{\alpha m^2 - \alpha(1-\alpha)p^2}{4\pi\mu^2} \right) \right\}$$

$$(5.356)$$

This is the result used in the text.

5.9.2 Calculation of the One Loop Vertex Function

The calculation proceeds exactly as before except that it is more complicated. We first reduce the expression in Eq. (5.304) to the form

$$
\begin{aligned}
-ie\mu^\epsilon \Lambda_m &= (ie\mu^\epsilon)^3 \int \frac{d^n k}{(2\pi)^n} \frac{-i\eta^{nr}}{k^2} \gamma_n \frac{1}{\gamma(p-k) - m} \gamma_m \frac{1}{\gamma(p-k) - m} \gamma_r \\
&= -(e\mu^\epsilon)^3 \int \frac{d^n k}{(2\pi)^n} \frac{\gamma_n(\gamma(p-k) + m)\gamma_m(\gamma(p-k) + m)\gamma^n}{k^2[(p'-k)^2 - m^2][(p-k)^2 - m^2]} \\
&= -2(e\mu^\epsilon)^3 \int_0^1 d\alpha \int_0^{1-\alpha} d\beta \int \frac{d^n k}{(2\pi)^n} \\
&\quad \times \frac{\gamma_n(\gamma(p-k) + m)\gamma_m(\gamma(p-k) + m)\gamma^n}{[k^2 - m^2(\alpha + \beta) - 2k \cdot (\alpha p + \beta p') + \alpha p^2 + \beta p'^2]^3}
\end{aligned}
$$

$$(5.357)$$

Next, we combine the denominators using a slightly generalized version of the formula used earlier:

$$\frac{1}{abc} = 2\int_0^1 d\alpha \int_0^{1-\alpha} d\beta \frac{1}{[a(1-\alpha-\beta) + \alpha b + \beta c]^3} \tag{5.358}$$

Shifting the variable to $k \to k - \alpha p - \beta p'$ reduces it to the expression

$$\Lambda_m = -2i(e\mu^\epsilon)^2 \int_0^1 d\alpha \int_0^{1-\alpha} d\beta \int \frac{d^n k}{(2\pi)^n}$$

$$\times \frac{\gamma_n(\gamma p(1-\beta) - \alpha\gamma p - \gamma k + m)\gamma_m(\gamma p(1-\alpha) - \beta\gamma p - \gamma k + m)\gamma^n}{[k^2 - m^2(\alpha+\beta) + \alpha(1-\alpha)p^2 + \beta(1-\beta)p'^2 - \alpha\beta 2p' \cdot p]^3}$$

$$\equiv \Lambda_m^{(1)} + \Lambda_m^{(2)} \tag{5.359}$$

The $\Lambda_m^{(1)}$ contains two factors of k^a in the numerator and is divergent; the $\Lambda_m^{(2)}$ has no k^a in the numerator and is finite. (Note that terms with one k^a vanish due to $k \to -k$ symmetry.) Let us first evaluate $\Lambda_m^{(1)}$.

Writing $\gamma_n \gamma^j k_j \gamma_m \gamma^j k_j \gamma^n = k_r k_s \gamma_n \gamma^r \gamma_m \gamma^s \gamma^n$ and using a standard integral

Exercise 5.21: Prove this.

$$\int \frac{d^n l}{(2\pi)^n} \frac{l_m l_n}{[l^2 + M^2 + 2l \cdot q]^A} = i(-1)^{n/2} \frac{1}{(4\pi)^{n/2}\Gamma(A)} \tag{5.360}$$

$$\times \left[\frac{1}{2} g_{mn} \frac{\Gamma(A - 1 - n/2)}{(M^2 - q^2)^{A-1-n/2}} + \frac{\Gamma(A - n/2)q_m q_n}{(M^2 - q^2)^{A-n/2}} \right]$$

we can evaluate the integral with two factors of k to obtain

$$\int \frac{d^n k}{(2\pi)^n} \frac{k_r k_s}{[k^2 - m^2(\alpha + \beta) + \alpha(1 - \alpha)p^2 + \beta(1 - \beta)p'^2 - \alpha\beta 2p' \cdot p]^3}$$

$$= i(-1)^{n/2} \frac{1}{(4\pi)^{n/2}\Gamma(3)} \frac{1}{2} g_{rs}$$

$$\times \frac{\Gamma(2 - n/2)}{[-m^2(\alpha + \beta) + \alpha(1 - \alpha)p^2 + \beta(1 - \beta)p'^2 - \alpha\beta 2p' \cdot p]^{2-n/2}}$$

$$= i \frac{1}{4(4\pi)^{n/2}} g_{rs}$$

$$\times \frac{\Gamma(\epsilon)}{[m^2(\alpha + \beta) - \alpha(1 - \alpha)p^2 - \beta(1 - \beta)p'^2 + \alpha\beta 2p' \cdot p]^\epsilon} \tag{5.361}$$

Using now the result

$$g_{rs}\gamma_n \gamma^r \gamma_m \gamma^s \gamma^n = \gamma_n \gamma^r \gamma_m \gamma_r \gamma^n = \gamma_n (2 - n)\gamma_m \gamma^n = (2 - n)^2 \gamma_m \tag{5.362}$$

we can easily isolate the singular part and obtain

$$\Lambda_m^{(1)} = -2ie^2 \int_0^1 d\alpha \int_0^{1-\alpha} d\beta \left[i \frac{1}{4(4\pi)^2 \epsilon} \right] 4\gamma_m + \text{finite}$$

$$= \frac{e^2}{16\pi^2} \frac{1}{\epsilon} \gamma_m + \text{finite} \tag{5.363}$$

The evaluation of the non-singular part makes use of the integral:

$$\int \frac{d^n l}{(2\pi)^n} \frac{1}{[l^2 + M^2 + 2l \cdot q]^A} = i(-1)^{n/2} \frac{\Gamma(A - n/2)}{(4\pi)^{n/2}\Gamma(A)(M^2 - q^2)^{A-n/2}} \tag{5.364}$$

This will lead to the result in Eq. (5.309) of the text.

To proceed further and obtain $F_2(0)$, we first note that sandwiching the above expression between the Dirac spinors gives

$$\bar{u}(p')\Lambda_m^2 u(p) = \bar{u}(p') \left\{ \frac{e^2}{16\pi^2} \int_0^1 d\alpha \int_0^{1-\alpha} d\beta \right. \tag{5.365}$$

$$\left. \times \frac{\gamma_n(\gamma^j p'_j(1 - \beta) - \alpha\gamma^j p_j + m)\gamma_m(\gamma^j p_j(1 - \alpha) - \beta\gamma^j p'_j + m)\gamma^n}{[-m^2(\alpha + \beta) + \alpha(1 - \alpha)p^2 + \beta(1 - \beta)p'^2 - \alpha\beta 2p' \cdot p]} \right\} u(p)$$

To simplify the numerator, we use the standard identities:

$$\gamma_n \gamma^j a_j \gamma^j b_j \gamma^n = 4a \cdot b, \qquad \gamma_n \gamma^j a_j \gamma^j b_j \gamma^j c_j \gamma^n = -2\gamma^j c_j \gamma^j b_j \gamma^j a_j,$$

$$\gamma_n \gamma^j a_j \gamma^n = -2\gamma^j a_j \tag{5.366}$$

which gives

$$\gamma_n(\gamma^j p'_j(1 - \beta) - \alpha\gamma^j p_j + m)\gamma_m(\gamma^j p_j(1 - \alpha) - \beta\gamma^j p'_j + m)\gamma^n \tag{5.367}$$

$$= -2(\gamma^j p_j(1 - \alpha) - \beta\gamma^j p'_j)\gamma_m(\gamma^j p'_j(1 - \beta) - \alpha\gamma^j p_j)$$

$$+ 4m(p_m(1 - \alpha) - \beta p'_m) + 4m(p'_m(1 - \beta) - \alpha p_m) - 2\gamma_m m^2$$

$$= -2(\gamma^j p_j(1 - \alpha) - \beta m)\gamma_m(\gamma^j p'_j(1 - \beta) - \alpha m)$$

$$+ 4m(p_m(1 - \alpha) - \beta p'_m) + 4m(p'_m(1 - \beta) - \alpha p_m) - 2\gamma_m m^2$$

To arrive at the last expression, we have used the on-shell condition arising from the Dirac equation. We next use the easily provable results:

$$\begin{aligned}
\gamma^j p_j \gamma_m \gamma^j p'_j &= 2(p_m + p'_m)m - 3m^2 \gamma_m, \\
\gamma^j p_j \gamma_m &= 2p_m - \gamma_m \gamma^j p_j \to 2p_m - \gamma_m m, \\
\gamma_m \gamma^j p'_j &= 2p'_m - \gamma^j p'_j \gamma_m \to 2p'_m - \gamma_m m.
\end{aligned} \tag{5.368}$$

These allow us to reduce the numerator to the form:

$$N = -2\left[(1-\alpha)(1-\beta)(2(p+p')_m m - 3m^2 \gamma_m) \right.$$

$$\left. -\beta(1-\beta)m(2p'_m - m\gamma_m) - \alpha(1-\alpha)m(2p_m - m\gamma_m) + \alpha\beta m^2 \right]$$

$$+4m\left[p'_m(1-2\beta) + p_m(1-2\alpha) \right] - 2\gamma_m m^2$$

$$= -2\gamma_m m^2 \left[-3(1-\alpha)(1-\beta) + \beta(1-\beta) + \alpha(1-\alpha) + \alpha\beta + 1 \right]$$

$$+4p_m m[\beta - \alpha\beta - \alpha^2] + 4p'_m m[\alpha - \beta\alpha - \beta^2] \tag{5.369}$$

while the denominator becomes $D = -m^2(\alpha+\beta)^2$. The structure of the denominator shows that we can re-write the numerator, as far as the integral goes, in the equivalent form:

$$N = -2\gamma_m m^2[4(\beta+\alpha) - (\alpha+\beta)^2 - 2] + 2(p_m + p'_m)m[(\alpha+\beta) - (\alpha+\beta)^2] \tag{5.370}$$

Plugging in the numerator and denominator into Eq. (5.365), we get:

$$\bar{u}(p')\Lambda_m^{(2)} u(p) = \bar{u}(p')\frac{-e^2}{16\pi^2 m^2} \int_0^1 d\alpha \int_0^{1-\alpha} d\beta \frac{1}{(\alpha+\beta)^2}$$

$$\times \left(-2\gamma_m m^2[4(\beta+\alpha) - (\alpha+\beta)^2 - 2] \right.$$

$$\left. +2(p_m + p'_m)m[(\alpha+\beta) - (\alpha+\beta)^2] \right) u(p)$$

$$= \bar{u}(p')\frac{-e^2}{16\pi^2 m^2} \int_0^1 d\alpha \int_0^{1-\alpha} d\beta \frac{1}{(\alpha+\beta)^2}$$

$$\times \left(-2\gamma_m m^2[4(\beta+\alpha) - (\alpha+\beta)^2 - 2] \right.$$

$$\left. +2[2m\gamma_m - i\sigma_{mn}q^n]m[(\alpha+\beta) - (\alpha+\beta)^2] \right) u(p)$$

$$= \bar{u}(p')\frac{-e^2}{16\pi^2 m^2} \int_0^1 d\alpha \int_0^{1-\alpha} d\beta \frac{1}{(\alpha+\beta)^2}$$

$$\times \left(-2\gamma_m m^2[2(\beta+\alpha) + (\alpha+\beta)^2 - 2] \right.$$

$$\left. -2i\sigma_{mn}q^n m[(\alpha+\beta) - (\alpha+\beta)^2] \right) u(p) \tag{5.371}$$

This is the expression used in the text.

A Potpourri of Problems

Problem 1.   Evaluation of $\langle 0|\phi(x)\phi(y)|0\rangle$ for Spacelike Separation

Find $\langle 0|\phi(x)\phi(y)|0\rangle$ in the coordinate space for $(x-y)^2 = -r^2 < 0$. Pay special attention to the convergence properties of the integral.

Solution: We need to evaluate the integral

$$G(x-y) = \int \frac{d^3\boldsymbol{p}}{(2\pi)^3} \frac{1}{2\omega_{\boldsymbol{p}}} e^{-ip\cdot(x-y)} = \frac{-i}{2(2\pi)^2 r} \int_{-\infty}^{\infty} dp \frac{p e^{ipr}}{\sqrt{p^2+m^2}} \qquad \text{(P.1)}$$

where we have performed the angular part of the integration. This integral is not convergent as it stands. One can distort the integration contour in to the complex plane, but the contribution from the semicircle at infinity in the complex plane will not vanish. One possibility is to explicitly add a convergence factor like $\exp(-\epsilon p)$ and take the limit $\epsilon \to 0$ at the end of the calculation. A simpler trick is to re-write the integral in the form

$$\int_{-\infty}^{\infty} dp \frac{p e^{ipr}}{\sqrt{p^2+m^2}} = \int_{-\infty}^{\infty} dp \frac{(p - \sqrt{p^2+m^2}) e^{ipr}}{\sqrt{p^2+m^2}} + 2\pi\delta(r) \qquad \text{(P.2)}$$

where the Dirac delta function on the right hand side is cancelled by the contribution from the second term in the integrand. The integrand now falls as $(1/p^2)$ for large p and the contour integration trick will work for *this* integral. This tells you that the usual regulators miss the $2\pi\delta(r)$ contribution; but for strictly spacelike separations we can take $r > 0$ (since $r = 0$ will be on the light cone). So everything is fine and we can use the standard integral representation for an avatar of the Bessel function

$$\int_{u}^{\infty} \frac{x e^{-\mu x} dx}{\sqrt{x^2 - u^2}} = u K_1(u\mu) \qquad [u > 0, \text{ Re } \mu > 0] \qquad \text{(P.3)}$$

to get the final answer.

Problem 2.   Number Density and the Non-relativistic Limit

Consider a free, massive, scalar field $\phi(x)$ which is expanded in the terms of creation and annihilation operators in the standard manner as

$$\phi(x) = \frac{1}{(2\pi)^3} \int \frac{d^3\boldsymbol{k}}{\sqrt{2\omega_{\boldsymbol{k}}}} \left(a(\boldsymbol{k}) e^{-ik\cdot x} + a^\dagger(\boldsymbol{k}) e^{ik\cdot x} \right) \equiv \phi^{(+)}(x) + \phi^{(-)}(x) \qquad \text{(P.4)}$$

Since the total number of particles is an integral over $a_{\mathbf{k}}^{\dagger} a_{\mathbf{k}}$, it makes sense to define a number density operator $\mathcal{N}(x^i)$ in the real space through the relations

$$N \equiv \int \frac{d^3 \mathbf{k}}{(2\pi)^3} a^{\dagger}(\mathbf{k}) a(\mathbf{k}) \equiv \int d^3 x \, \mathcal{N}(x^i) \tag{P.5}$$

(a) Find an expression for $\mathcal{N}$ in terms of $\phi^{\pm}$.

(b) In the text, we lamented the fact that localized particle states cannot be introduced in a meaningful fashion in QFT. To see this in action once again, let us study the expectation value of $\mathcal{N}$ in one-particle states. Define a one-particle state $|\psi\rangle$ through the relations

$$|\psi\rangle \equiv \int \frac{d^3 \mathbf{k}}{(2\pi)^3} \psi(\mathbf{k}) a^{\dagger}(\mathbf{k}) |0\rangle \equiv \int \frac{d^3 \mathbf{k}}{(2\pi)^3} \psi(\mathbf{k}) |1_{\mathbf{k}}\rangle, \tag{P.6}$$

with

$$\langle \psi | \psi \rangle = \int \frac{d^3 \mathbf{k}}{(2\pi)^3} |\psi(\mathbf{k})|^2 = 1 \tag{P.7}$$

Compute the expectation value $\langle \psi | \mathcal{N}(0, \mathbf{x}) | \psi \rangle$ in terms of the "wavefunction" $\psi(\mathbf{x})$. What do you find?

(c) Show that things work out fine in the nonrelativistic limit.

(d) What happens if you choose $\psi(\mathbf{k}) \propto \sqrt{\omega_{\mathbf{k}}}$ or if you choose $\psi(\mathbf{k}) \propto [1/\sqrt{\omega_{\mathbf{k}}}]$? Calculate the expectation value of the Hamiltonian for a general $\psi(\mathbf{k})$ and explain why the above choices do not really lead to localization of a particle. What happens in the non-relativistic case?

Solution: (a) It is easy to verify that a possible choice for $\mathcal{N}$ is given by the expression

$$\mathcal{N}(x^i) = i\phi^{(-)}(x^i) \overset{\leftrightarrow}{\frac{\partial}{\partial t}} \phi^{(+)}(x^i) \tag{P.8}$$

(b) The evaluation of the expectation value is straightforward. We get

$$\langle \psi | \mathcal{N}(0, \mathbf{x}) | \psi \rangle = i \int \frac{d^3 \mathbf{k}}{(2\pi)^3} \frac{d^3 \mathbf{k}'}{(2\pi)^3} \psi^*(\mathbf{k}') \psi(\mathbf{k}) \tag{P.9}$$

$$\times \, \langle \mathbf{k}' | \phi^{(-)}(\mathbf{x}) \dot{\phi}^{(+)}(\mathbf{x}) - \dot{\phi}^{(-)}(\mathbf{x}) \phi^{(+)}(\mathbf{x}) | \mathbf{k}' \rangle = \mathrm{Re} \left(\chi_1^*(\mathbf{x}) \chi_2(\mathbf{x}) \right)$$

involving *two* distinct "wavefunctions" given by

$$\chi_1(\mathbf{x}) \equiv \int \frac{d^3 \mathbf{k}}{(2\pi)^3} \frac{1}{\sqrt{\omega_{\mathbf{k}}}} \psi(\mathbf{k}) e^{i\mathbf{k} \cdot \mathbf{x}}; \quad \chi_2(\mathbf{x}) \equiv \int \frac{d^3 \mathbf{k}}{(2\pi)^3} \sqrt{\omega_{\mathbf{k}}} \, \psi(\mathbf{k}) e^{i\mathbf{k} \cdot \mathbf{x}} \tag{P.10}$$

The fact that you cannot work with a *single*, unique wavefunction is again related to the fact that particle states are not localizable.

(c) When $c \to \infty$, the two wavefunctions reduce, except for a normalization, to

$$\chi_1(\mathbf{x}) \to \frac{1}{\sqrt{m}\, c} \psi(\mathbf{x}), \qquad \chi_2(\mathbf{x}) \to \sqrt{m}\, c \psi(\mathbf{x}) \tag{P.11}$$

with

$$\psi(\mathbf{x}) = \frac{1}{(2\pi)^3} \int d^3 \mathbf{k} \, \psi(\mathbf{k}) e^{i\mathbf{k} \cdot \mathbf{x}} \tag{P.12}$$

In this case, the expectation value of the number density becomes the standard probability density $|\psi|^2$ of NRQM. This is the best one can do.

(d) If we choose $\psi(\boldsymbol{k}) \propto \sqrt{\omega_{\boldsymbol{k}}}$, then $\chi_1(\boldsymbol{x})$ is proportional to the Dirac delta function $\delta(\boldsymbol{x})$ suggesting that one of the two wavefunctions describes a particle localized at the origin. But in this case, χ_2 is proportional to $G_+(\boldsymbol{x})$ and is spread over a Compton wavelength of the particle. [The situation is similar for the other choice $\psi(\boldsymbol{k}) \propto [1/\sqrt{\omega_{\boldsymbol{k}}}]$.] More importantly, the expectation value of the energy density operator is non-zero all over the space. The expectation value of the normal ordered Hamiltonian

$$\mathcal{H}(0, \boldsymbol{x}) = \frac{1}{2} : \left(\dot{\phi}(0, \boldsymbol{x})^2 + |\boldsymbol{\nabla}\phi(0, \boldsymbol{x})|^2 + m^2 \phi(0, \boldsymbol{x})^2 \right) : \qquad \text{(P.13)}$$

is given by

$$\langle \psi | \mathcal{H}(0, \boldsymbol{x}) | \psi \rangle = \frac{1}{2} \left(|\chi_2(\boldsymbol{x})|^2 + c^2 \boldsymbol{\nabla} \chi_1^*(\boldsymbol{x}) \cdot \boldsymbol{\nabla} \chi_1(\boldsymbol{x}) + m^2 c^4 |\chi_1(\boldsymbol{x})|^2 \right)$$
$$\text{(P.14)}$$

Again, things work out in the nonrelativistic limit when we have

$$\langle \psi | \mathcal{H}(0, \boldsymbol{x}) | \psi \rangle \to mc^2 |\psi(\boldsymbol{x})|^2 + \frac{1}{2m} |\boldsymbol{\nabla}\psi(\boldsymbol{x})|^2 \qquad \text{(P.15)}$$

If you integrate this over all space, the first term gives mc^2 while the second term gives the expectation value of the nonrelativistic kinetic energy operator.

Problem 3. From $[\phi, \pi]$ to $[a, a^\dagger]$

Using the standard mode expansion of the field $\phi(x)$ in terms of the creation and annihilation operators, prove that the equal time commutation rule, $[\phi(\boldsymbol{x}), \pi(\boldsymbol{y})] = i\delta^{(3)}(\boldsymbol{x} - \boldsymbol{y})$, implies the commutation rules $[a_{\boldsymbol{p}}, a_{\boldsymbol{p}'}^\dagger] = (2\pi)^3 \delta^{(3)}(\boldsymbol{p} - \boldsymbol{p}')$.

Solution: Using the relations

$$\phi(\boldsymbol{x}) = \int \frac{d^3 \boldsymbol{p}}{(2\pi)^3} \frac{1}{\sqrt{2\omega_{\boldsymbol{p}}}} \left(a_{\boldsymbol{p}} + a_{-\boldsymbol{p}}^\dagger \right) e^{i\boldsymbol{p} \cdot \boldsymbol{x}}$$

$$\pi(\boldsymbol{y}) = \int \frac{d^3 \boldsymbol{p}'}{(2\pi)^3} (-i) \sqrt{\frac{\omega_{\boldsymbol{p}'}}{2}} \left(a_{\boldsymbol{p}'} - a_{-\boldsymbol{p}'}^\dagger \right) e^{i\boldsymbol{p}' \cdot \boldsymbol{y}} \qquad \text{(P.16)}$$

which are valid at $t = 0$, one can determine the creation and annihilation operators in terms of $\phi(\boldsymbol{x})$ and $\pi(\boldsymbol{y})$. A straightforward computation will then lead to the necessary result. An alternative route is the following: Use Eq. (P.16) to compute the commutator $[\phi(\boldsymbol{x}), \pi(\boldsymbol{y})]$ in terms of the different sets of the commutators of creation and annihilation operators. If you now write the Dirac delta function in Fourier space, you should be able to argue that the only consistent solutions for the commutators of the creation and annihilation operators are the standard ones.

Problem 4. Counting the Modes between the Casimir Plates

Prove the counting used in the text for the degrees of freedom associated with different wavenumbers of the electromagnetic wave modes between the Casimir plates.

Solution: Choose a coordinate system with one boundary at $z = 0$ and consider the region $0 \le x \le L_x$, $0 \le y \le L_y$, $0 \le z \le L_z$. From Maxwell's equations, it is straightforward to show that $B_\perp = 0$, $\boldsymbol{E}_\parallel = 0$ where $\perp$ and $\parallel$ indicate the components perpendicular and parallel to the surface respectively. These translate to the conditions $\boldsymbol{A}_\parallel = 0$, $\partial A_z / \partial z = 0$. If we take the plane wave modes to have the standard form $\boldsymbol{A}(t, \boldsymbol{x}) = \boldsymbol{\epsilon} \, e^{i(\boldsymbol{k} \cdot \boldsymbol{x} - \omega t)}$ with $\boldsymbol{\epsilon} \cdot \boldsymbol{k} = 0$, where the wave vectors satisfy the standard boundary conditions

$$k_x = \frac{\pi n_x}{L_x}, \quad k_y = \frac{\pi n_y}{L_y}, \quad k_z = \frac{\pi n_z}{L_z}, \quad (n_\alpha = 0, 1, 2, \ldots) \tag{P.17}$$

then we can write the expansion for A_x in the form

$$A_x(x, y, z) = \sum_{n_x=0}^{\infty} \sum_{n_y=0}^{\infty} \sum_{n_z=0}^{\infty} \epsilon_x(\boldsymbol{k}) \cos(k_x x) \sin(k_y y) \sin(k_z z) \tag{P.18}$$

The expansions for A_y and A_z follow the same pattern. If all the three n_α are non-zero, then the gauge condition puts one constraint; therefore only two of the coefficients in $\epsilon_\alpha(\boldsymbol{k})$ can be chosen independently. If only two of the n_α's are non-zero, then the situation is a little tricky. Consider, for example, the case with $n_z = 0$ with $n_x \ne 0$, $n_y \ne 0$. Since A_x and A_y contain the factor $\sin(k_z z)$, they both vanish leaving only one independent solution which will be proportional to $\epsilon_z(\boldsymbol{k})$. So there are *two* harmonic oscillator modes when all of n_α are non-zero but *only one* oscillator mode for every triplet in which only two components are non-zero. This is the result which was used in the text.

Problem 5.   Effective Potential for $m = 0$

Consider the effective potential V_{eff} of the $\lambda \phi^4$ theory in the limit of $m_0^2 \to 0$. We know that the theory exhibits SSB when $m_0^2 < 0$ while, it has a single minimum (at $\phi = 0$) when $m_0^2 > 0$. The question arises as to what happens to the theory due to higher order corrections when $m_0^2 = 0$. In this case, it is convenient to define the physical mass and coupling constant by the conditions

$$m_{\mathrm{phys}}^2 = \left(\frac{d^2 V}{d\phi^2} \right)_{\phi=0}; \qquad \lambda = \left(\frac{d^4 V}{d\phi^2} \right)_{\phi=\Lambda}. \tag{P.19}$$

where Λ is an arbitrary energy scale. Work out the effective potential using this prescription in terms of Λ.

Solution: In this case, the effective potential (worked out in the text) becomes

$$V_{\mathrm{eff}} = \frac{\lambda}{4!} \phi^4 + \frac{\lambda^2}{(16\pi)^2} \phi^4 \ln \frac{\lambda \phi^2}{4\mu^2} \tag{P.20}$$

From Eq. (P.19), we find that

$$\lambda = \lambda + \left(\frac{5}{8\pi} \lambda \right)^2 + \frac{24}{(16\pi)^2} \lambda^2 \ln \frac{\lambda \Lambda^2}{4\mu^2}. \tag{P.21}$$

allowing us to express $(4\mu^2 / \lambda)$ in terms of Λ:

$$\ln \frac{\lambda \Lambda^2}{4\mu^2} = -\frac{25}{6} \tag{P.22}$$

Substituting back into Eq. (P.20), we get:

$$V_{\text{eff}} = \frac{\lambda}{4!}\phi^4 + \frac{\lambda^2}{(16\pi)^2}\phi^4\left[\ln\frac{\phi^2}{\Lambda^2} - \frac{25}{6}\right] \tag{P.23}$$

The constant $(25/6)$, of course, can be reabsorbed into $\ln\Lambda^2$; it is conventional to leave it as it is.

Once again, we have obtained a finite V_{eff} without any cut-off dependence, but it has an apparent dependence on Λ. We have already seen that this is only apparent because the coupling constant λ is now defined at the energy scale Λ. If we change this scale Λ to Λ', then we can retain the form of V_{eff} by changing λ to λ', such that

$$\lambda' = \lambda + \frac{3\lambda^2}{16\pi^2}\ln\frac{\Lambda'}{\Lambda} \tag{P.24}$$

Under $\Lambda \to \Lambda', \lambda \to \lambda'$ transformations, V_{eff} is invariant: $V_{\text{eff}}(\lambda',\Lambda') = V(\lambda,\Lambda) + \mathcal{O}(\lambda^3)$. This is exactly the "running" of the coupling constant we have seen earlier.

Problem 6.   Anomalous Dimension of the Coupling Constant

Consider the massless $\lambda\phi^4$ theory in $d = 4$ with the Lagrangian $L = (1/2)(\partial\phi)^2 + (\lambda/4!)\phi^4$. The resulting action is invariant under the scale transformation $x^m \to s^{-1}x^m$, $\phi \to s\phi$ with λ remaining invariant. Figure out how the effective coupling constant of the *renormalized* theory changes under the same scaling transformation.

Solution: This is fairly straight forward. Using the analysis given in the text, you should be able to show that, to the lowest order,

$$\lambda_{\text{eff}} \to \lambda_{\text{eff}}\exp\left(\frac{3\lambda_{\text{eff}}}{16\pi^2}\ln s\right) = \lambda_{\text{eff}}s^{(3\lambda/16\pi^2)} \equiv s^{\gamma}\lambda_{\text{eff}} \tag{P.25}$$

The constant γ is called the anomalous scaling dimension of the coupling constant.[1]

Problem 7.   Running of the Cosmological Constant

One key result we arrived at in the study of renormalization (both in the case of the $\lambda\phi^4$ theory and QED) is that the renormalized constants appearing in the Lagrangian can depend on an arbitrary energy scale μ and "run with it". This idea will be applicable even to a constant term (say, $-\Lambda_0$) added to the standard scalar field Lagrangian.[2] We found in Sect. 1.4.5 that the coincidence limit of the Green's function, $G(x;x)$, can be interpreted in terms of the zero point energy of the infinite number of oscillators. This zero point energy will combine with the bare parameter Λ_0 to lead to a renormalized cosmological constant Λ. Regularize the relevant expressions by the standard methods and show that the renormalized cosmological constant Λ runs with the scale μ according to the equation

$$\mu\frac{d\Lambda}{d\mu} = (4-n)\Lambda - \frac{1}{2}\frac{m^4}{(4\pi)^2} \tag{P.26}$$

[1] This is one of the many examples in QFT where a classical symmetry is lost in the quantum theory. One way of understanding this result is to note that our regularization involves using the arbitrary dimension d. The original symmetry existed only for $d = 4$ and hence one can accept dimensional regularization leading to a final result in which a dimension dependent symmetry is not respected.

[2] Because of historical reasons, Λ is called the cosmological constant; for our purpose it is just another bare parameter in the original, unrenormalized Lagrangian like m_0, λ_0 etc..

where m is the mass of the scalar field. Also show that the regularized zero point energy is given by the expression

$$E = \frac{1}{4}\frac{m^4}{(4\pi)^2}\left[\ln\left(\frac{m^4}{4\pi\mu^2}\right) + \gamma_E - \frac{3}{2}\right] - \Lambda(\mu) \qquad (P.27)$$

where the running of $\Lambda(\mu)$ ensures that E is independent of μ.

Solutions: Most of the work for this problem is already done in the text. Just for fun, we will derive it using a more formal approach along the following lines. From the Euclidean vacuum-to-vacuum amplitude of the theory, expressed in the form

$$Z \equiv e^{-W} = \int [\mathcal{D}\phi]\exp\left\{-\int(d^n x_E)\left[\frac{1}{2}(\partial_m\phi)^2 + \frac{1}{2}m_0^2\phi^2 - \Lambda_0\right]\right\} \quad (P.28)$$

we immediately obtain the relation

$$\frac{\partial}{\partial m_0^2}\ln Z = -\frac{1}{2}\int(d^n x_E)G_E(x,x) \qquad (P.29)$$

where

$$G_E(x,x) = \int\frac{(d^n k_E)}{(2\pi)^n}\frac{1}{k^2 + m_0^2} \qquad (P.30)$$

is the coincidence limit of the Green's function. We are working in Euclidean n-dimensional space in anticipation of dimensional regularization being used to handle Eq. (P.30). Using the standard Schwinger trick, we can evaluate Eq. (P.30) to obtain

$$G_E(x,x) = \frac{(m_0^2)^{[(n/2)-1]}}{(4\pi)^{n/2}}\,\Gamma\left(1 - \frac{n}{2}\right) \qquad (P.31)$$

which allows us to integrate Eq. (P.29) and obtain:

$$Z = \exp\left\{-\int(d^n x_E)E\right\}; \quad E = \frac{m_0^n}{(4\pi)^{n/2}}\frac{1}{n}\Gamma\left(1 - \frac{n}{2}\right) - \Lambda_0 \quad (P.32)$$

(compare with Eq. (2.84)). Taking the limit $n \to 4$ and isolating the divergences, we find that the renormalized cosmological constant Λ and the bare constant Λ_0 are related by

$$\Lambda_0 = \mu^{n-4}\left\{\frac{1}{2}\frac{m_0^4}{(4\pi)^2}\frac{1}{n-4} + \Lambda\right\} \qquad (P.33)$$

The standard condition $\mu(d\Lambda_0/d\mu) = 0$ now gives us the running of the cosmological constant in Eq. (P.26). In the free theory, physical mass is the same as the bare mass and this identification has been made in this context.[3] Substituting back Eq. (P.33) into Eq. (P.32), we get Eq. (P.27).

[3]For a more challenging task, you can try computing the running of Λ and E in the *interacting* $\lambda\phi^4$ theory. You will find that, at $\mathcal{O}(\lambda)$, the Eq. (P.26) does not change!

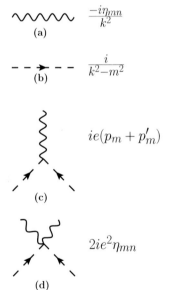

$$\sim\!\!\sim\!\!\sim\!\!\sim \qquad \frac{-i\eta_{mn}}{k^2}$$
(a)

$$\text{-- -}\!\!\blacktriangleright\!\text{-- --} \qquad \frac{i}{k^2-m^2}$$
(b)

$$ie(p_m + p'_m)$$
(c)

$$2ie^2\eta_{mn}$$
(d)

$$-i\lambda$$
(e)

Figure P.1: Feynman rules for the scalar QED.

Problem 8. Feynman Rules for Scalar Electrodynamics

In Sect. 4.2, we studied the scalar electrodynamics non-perturbatively in order to compute the Schwinger effect. This theory is described by the action

$$A = \int d^4x\left[-\frac{1}{4}F_{mn}F^{mn} + (D_m\phi)^*(D^m\phi) - m^2\phi^*\phi - \frac{\lambda}{4}(\phi^*\phi)^2\right] \quad (P.34)$$

where $D_m(x) = \partial_m - ieA_m(x)$ is the standard gauge covariant derivative. Work out and write down the Feynman rules for this theory.

Solution: The Feynman diagram with equivalent algebraic expressions are given in Fig. P.1. Figures P.1(a) and (b) are straightforward and represent the photon propagator and the scalar propagator. The last figure (Fig. P.1(e)) represents the self-interaction vertex of the scalar field and has nothing to do with the electromagnetic coupling. Figures P.1(c) and (d) represent the photon-scalar vertices of the theory. Since there are both linear and quadratic couplings to the vector potential, we get two kinds of vertices — one with 3 lines and the other with 4 lines. The first one is similar to the one in standard QED while the second one is new to the scalar field interaction.

Problem 9. Two-loop Contribution to the Propagator in the $\lambda\phi^4$ Theory

In Sect. 4.8 we described the renormalized perturbation theory for the $\lambda\phi^4$ theory. We found that, at the one loop order, we needed counter-terms to take care of the renormalization of the mass and the coupling constant. But we did not require any wave function renormalization. The purpose of this problem is to compute the corrections to the propagator to $\mathcal{O}(\lambda_R^2)$ and show that everything works out fine with a renormalization of the wavefunction. To do this, it is convenient to relate the bare and renormalized quantities through the equations:

$$\phi_B = Z_\phi^{1/2}\phi_R, \quad m_B^2 = \frac{1}{Z_\phi}(m_R^2 + \delta m^2), \quad \lambda_B = \frac{1}{Z_\phi^2}(\lambda_R + \delta\lambda) \quad \text{(P.35)}$$

where Z_ϕ is the wavefunction renormalization. The Lagrangian can now be separated into the renormalized part and the counter-terms in the form

$$L = \frac{1}{2}(\partial_\mu\phi_R)^2 - \frac{1}{2}m_R^2\phi_R^2 - \frac{\lambda_R}{4!}\phi_R^4 + \frac{1}{2}(Z_\phi - 1)(\partial_\mu\phi_R)^2 - \frac{1}{2}\delta m^2\phi_R^2 - \frac{\delta\lambda}{4!}\phi_R^4$$
$$\text{(P.36)}$$

We will expand the parameters as a power series in the coupling constant by

$$Z_\phi - 1 = z_1\lambda_R + z_2\lambda_R^2 + \cdots, \quad \delta m^2 = m_R^2(a_1\lambda_R + a_2\lambda_R^2 + \cdots),$$
$$\delta\lambda = b_2\lambda_R^2 + b_3\lambda_R^3 + \cdots \quad \text{(P.37)}$$

We already know from our study correct to $\mathcal{O}(\lambda_R)$, in the main text, that:

$$z_1 = 0; \quad a_1 = \frac{\mu^{-2\epsilon}}{2(4\pi)^2\epsilon}; \quad b_2 = \frac{3\mu^{-2\epsilon}}{2(4\pi)^2\epsilon} \quad \text{(P.38)}$$

Draw the relevant diagrams with two-loops for the corrections to the propagator and compute the resulting expression. Use the dimensional regularization and show that the new divergences can be removed with the choices

$$z_2 = \frac{\mu^{-4\epsilon}}{24(4\pi)^4\epsilon}, \quad a_2 = \frac{\mu^{-4\epsilon}}{4(4\pi)^4}\left(\frac{2}{\epsilon^2} - \frac{1}{\epsilon}\right) \quad \text{(P.39)}$$

Solution: This requires fairly involved computation, but you should be able to fill in the details to the following steps. The relevant Feynman diagrams

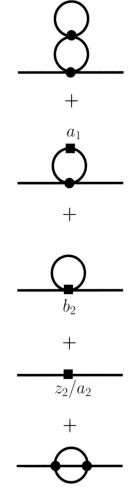

Figure P.2: Two-loop diagrams that contribute to the propagator.

are shown in Fig. P.2. The contribution from the first one is given by the integral

$$
\begin{aligned}
I_1 &= \frac{1}{4}(-i\lambda_R)^2 \int \frac{id^n p}{(2\pi)} \frac{-i}{p^2 + m_R^2} \int \frac{id^n k}{(2\pi)} \left(\frac{-i}{k^2 + m_R^2}\right)^2 \\
&= \frac{i\lambda_R^2 m_R^2 \mu^{-4\epsilon}}{4(4\pi)^4} \left(\frac{4\pi\mu^2}{m_R^2}\right)^{2\epsilon} \Gamma(\epsilon)\Gamma(\epsilon - 1)
\end{aligned}
\tag{P.40}
$$

where the integrals are in Euclidean space. Standard dimensional regularization techniques will now give the divergent contribution to be

$$
I_1 = -\frac{i\lambda_R^2 m_R^2 \mu^{-4\epsilon}}{4(4\pi)^4} \left[\frac{1}{\epsilon^2} + \frac{1}{\epsilon}\left(1 - 2\gamma_E + 2\log\frac{4\pi\mu^2}{m_R^2}\right)\right]
\tag{P.41}
$$

where we have used the MS scheme. The contribution from the second diagram is given by:

$$
\begin{aligned}
I_2 &= \frac{1}{2}(-i\lambda_R)(-ia_1\lambda_R m_R^2) \int \frac{id^n p}{(2\pi)} \left(\frac{-i}{k^2 + m_R^2}\right)^2 \\
&= \frac{ia_1\lambda_R^2 m_R^2 \mu^{-2\epsilon}}{2(4\pi)^2} \left(\frac{4\pi\mu^2}{m_R^2}\right)^{\epsilon} \Gamma(\epsilon)
\end{aligned}
\tag{P.42}
$$

where we know from the result in the text that

$$
a_1 = \frac{\mu^{-2\epsilon}}{2(4\pi)^2 \epsilon}
\tag{P.43}
$$

Again, the divergent contribution can be isolated by the MS scheme, leading to

$$
I_2 = \frac{i\lambda_R^2 m_R^2 \mu^{-4\epsilon}}{4(4\pi)^4} \left[\frac{1}{\epsilon^2} + \frac{1}{\epsilon}\left(-\gamma_E + \log\frac{4\pi\mu^2}{m_R^2}\right)\right]
\tag{P.44}
$$

(Interestingly, you will find that the worst divergences proportional to ϵ^{-2} cancel when we add I_1 and I_2.) The third contribution is given by

$$
I_3 = \frac{1}{2}(-ib_2\lambda_R^2) \int \frac{id^n p}{(2\pi)} \frac{-i}{p^2 + m_R^2} = -\frac{ib_2\lambda_R^2 m_R^2 \mu^{-2\epsilon}}{2(4\pi)^2} \left(\frac{4\pi\mu^2}{m_R^2}\right)^{\epsilon} \Gamma(\epsilon - 1)
\tag{P.45}
$$

where we know from the result in the text that

$$
b_2 = \frac{3\mu^{-2\epsilon}}{2(4\pi)^2 \epsilon}
\tag{P.46}
$$

This leads to:

$$
I_3 = \frac{3i\lambda_R^2 m_R^2 \mu^{-4\epsilon}}{4(4\pi)^4} \left[\frac{1}{\epsilon^2} + \frac{1}{\epsilon}\left(1 - \gamma_E + \log\frac{4\pi\mu^2}{m_R^2}\right)\right]
\tag{P.47}
$$

Diagram 4 is completely straightforward to compute and gives the contribution $(iz_2\lambda_R^2 p^2 - ia_2\lambda_r^2 m_R^2)$.

Let us next get on to diagram 5 which is possibly the most complicated one we need to calculate. Here we need to evaluate the integral

$$
I_5 = \frac{i\lambda_R^2}{6} \int \frac{d^n k}{(2\pi)} \frac{d^n q}{(2\pi)} \frac{1}{k^2 + m_R^2} \frac{1}{q^2 + m_R^2} \frac{1}{(p + k + q)^2 + m_R^2}
\tag{P.48}
$$

We will obtain the divergent parts by a series of tricks rather than attempt a brute force computation. There are two divergent contributions at $\mathcal{O}(p^0)$ and $\mathcal{O}(p^2)$ if we expand I_5 as a power series in p. We will evaluate the coefficients of the divergences separately.

For the first one, we can set $p = 0$ and write the integral, say, as $I(m_R^2)$, scaling out the factor $i\lambda_R^2/6$. It is a little easier to compute the derivative $I'(m_R^2)$ rather than $I(m_R^2)$ itself. With a little bit of guess work, you can separate out $I'(m_R^2)$ into two terms of which the second one is finite and the first one is given by

$$I'(m_R^2) = -3 \int \frac{d^n k}{(2\pi)^d} \frac{d^n q}{(2\pi)^d} \frac{1}{(k^2 + m_R^2)(q^2 + m_R^2)^2(k+q)^2} \qquad (P.49)$$

This is a standard integral and its evaluation leads to the result

$$I'(m_R^2) = -\frac{3(m_R^2)^{-2\epsilon}}{2(4\pi)^{4-2\epsilon}} \frac{[\Gamma(\epsilon)]^2}{1-\epsilon} \qquad (P.50)$$

Integrating $I'(m_R^2)$, we find the divergent part to be

$$I(m_R^2) = -\frac{3m_R^2\mu^{-4\epsilon}}{2(4\pi)^4} \left[\frac{1}{\epsilon^2} + \frac{1}{\epsilon}\left(3 - 2\gamma_E + 2\log\frac{4\pi\mu^2}{m_R^2} \right) \right] \qquad (P.51)$$

This takes care of the divergence of $\mathcal{O}(p^0)$.

To determine the second divergence of $\mathcal{O}(p^2)$, it is again good to combine the denominators of the integral in I_5 by the Schwinger-Feynman trick. One can then pick out the divergent pieces which are proportional to p^2. This requires a fair amount of algebraic ingenuity but the final answer is given by

$$D = -\frac{p^2\mu^{-4\epsilon}}{2(4\pi)^4\epsilon} \int_0^\infty dx\,dy\,dz\,\delta(x+y+z-1)\frac{xyz}{(xy+yz+zx)^3} \qquad (P.52)$$

This triple integral, in spite of the appearances, has the simple value of $(1/2)$. This is most easily done by writing the denominator using the Schwinger trick, obtaining

$$D = \frac{1}{2}\int_0^\infty dx\,dy\,dz\,\delta(x+y+z-1)\,xyz \int_0^\infty dt\,t^2 e^{-t(xy+yz+zx)} \qquad (P.53)$$

If you now rescale x, y, z by $\sqrt{t}$ and change variables to $u = yz$, $v = zx$, $w = xy$ (with the Jacobian $du\,dv\,dw = 2xyz\,dx\,dy\,dz$), then you will find that the expression reduces to

$$D = \frac{1}{2}\int_0^\infty du\,dv\,dw\,e^{-(u+v+w)} = \frac{1}{2} \qquad (P.54)$$

Therefore, the final answer for the divergent part is given by

$$I_5 = \frac{i\lambda_R^2 m_R^2\mu^{-4\epsilon}}{4(4\pi)^4} \left[-\frac{1}{\epsilon^2} + \frac{1}{\epsilon}\left(-3 + 2\gamma_E - 2\log\frac{4\pi\mu^2}{m_R^2} - \frac{p^2}{6m_R^2} \right) \right] \qquad (P.55)$$

Adding up all the divergent contributions, we get our final result to be

$$I = \frac{i\lambda_R^2 m_R^2\mu^{-4\epsilon}}{4(4\pi)^4} \left(\frac{2}{\epsilon^2} - \frac{1}{\epsilon} - \frac{p^2}{6m_R^2\epsilon} \right) + iz_2\lambda_R^2 p^2 - ia_2\lambda_R^2 m_R^2 \qquad (P.56)$$

You would have noticed that there were several cancellations, as to be expected. The remaining divergences can be eliminated if we choose

$$z_2 = \frac{\mu^{-4\epsilon}}{24(4\pi)^4\epsilon}, \qquad a_2 = \frac{\mu^{-4\epsilon}}{4(4\pi)^4}\left(\frac{2}{\epsilon^2}-\frac{1}{\epsilon}\right) \tag{P.57}$$

Clearly, we now need a non-zero wavefunction renormalization.

Problem 10.   Strong Field Limit of the Effective Lagrangian in QED

In the text, we computed the QED effective Lagrangian resulting from integrating out the charged scalar and fermionic particles. The real part of the effective Lagrangian was computed for weak fields in a Taylor series expansion. Do the same for strong fields and show that the resulting corrections have the asymptotic forms given by

$$L_\phi^{\text{eff}} \equiv -\frac{e^2}{192\pi^2}(\boldsymbol{E}^2 - \boldsymbol{B}^2)\log\left[-4e^2\frac{(\hbar c)^3}{(mc^2)^4}(\boldsymbol{E}^2 - \boldsymbol{B}^2)\right] + \cdots \tag{P.58}$$

for the scalar field and

$$L_\psi^{\text{eff}} \equiv -\frac{e^2}{48\pi^2}(\boldsymbol{E}^2 - \boldsymbol{B}^2)\log\left[-4e^2\frac{(\hbar c)^3}{(mc^2)^4}(\boldsymbol{E}^2 - \boldsymbol{B}^2)\right] + \cdots \tag{P.59}$$

for the fermionic field.

Solution: This is completely straightforward and can be obtained by evaluating the parametric integral for L_{eff} using the saddle point approximation.

Problem 11.   Structure of the Little Group

[4] See Sect. 5.4.7.

Let $|\boldsymbol{p}; a, b\rangle$ be an eigenstate of the operators A and B with eigenvalues a and b related to the Little Group in the massless case.[4] Show that one can now construct another eigenstate $|\boldsymbol{p}; a, b, \theta\rangle$ of A and B with θ being a *continous* parameter, leading[5] to a continous set of eigenvalues if $a \neq 0$ or $b \neq 0$.

[5] As we said in the main text, nobody has found a physical situation in which θ is relevant. So we consider only states with $a = b = 0$.

Solution: The required state can be defined through the relation $|\boldsymbol{p}; a, b, \theta\rangle \equiv e^{-i\theta J^3}|\boldsymbol{p}; a, b\rangle$. Using the usual trick of writing

$$A\, e^{-i\theta J^3}|\boldsymbol{p}; a, b\rangle = e^{-i\theta J^3}\left(e^{i\theta J^3} A\, e^{-i\theta J^3}\right)|\boldsymbol{p}; a, b\rangle \tag{P.60}$$

and the commutation rule $e^{i\theta J^3} A\, e^{-i\theta J^3} = A\cos\theta - B\sin\theta$, it is easy to show that $|\boldsymbol{p}; a, b, \theta\rangle$ is indeed an eigenket of A and B with eigenvalues $(a\cos\theta - b\sin\theta)$ and $(a\sin\theta + b\cos\theta)$ respectively.

Problem 12.   Path Integral for the Dirac Propagator

In the text, we showed that the propagator $G(x, y)$ for a spinless relativistic particle can be obtained by a parametric integration over the proper time s of a 5-dimensional kernel $K(x, y; s)$. The latter can be thought of as a quantum mechanical propagator of the form $\langle x|\exp(-isH)|y\rangle$ with $H = p^2 + m^2$. This allows one to obtain a path integral representation for $G(x,y)$ which is manifestly Lorentz invariant. Show that one can carry out similar steps for a spin-half Dirac particle as well. What kind of paths should one sum over in the spacetime in this case?

Solution: There is a general trick to obtain any propagator of the form

$$\mathcal{G}(x,y) \equiv \int \frac{d^4p}{(2\pi)^4} \frac{e^{ip\cdot(x-y)}}{H(p)} \qquad \text{(P.61)}$$

in terms of a parametric integration. To do this, we use the Schwinger trick to write the denominator $H(p)$ in an integral representation and use the fact that $\langle p|y\rangle = \exp(ipy)$. This gives

$$
\begin{aligned}
\mathcal{G}(x,y) &= i\int_0^\infty ds \int \frac{d^4p}{(2\pi)^4} e^{-isH(p)-ip\cdot(x-y)} = i\int_0^\infty ds\langle x|e^{-isH(\hat{p})}|y\rangle \\
&\equiv i\int_0^\infty ds\, K(x,y;s) \qquad \text{(P.62)}
\end{aligned}
$$

In the final expression, we think of $H(\hat{p})$ as a quantum mechanical operator with the usual representation $\hat{p}_a = i\partial_a$. (It is assumed that H has a regularizer $-i\epsilon$ to make the integral converge in the upper limit.) The kernel $K(x,y;s)$ has the obvious path integral representation given by

$$K(x,y;s) = \int \mathcal{D}p \int \mathcal{D}x \,\exp i\int_0^s d\tau\,[p_a\dot{x}^a - H(p)] \qquad \text{(P.63)}$$

In the case of a spin half particle, we can take $H(p) = \gamma^a p_a - m + i\epsilon$, resulting in the path integral representation for the Dirac propagator

$$S(x,y) = \int_0^\infty ds \int \mathcal{D}^4x \int \mathcal{D}^4p\, e^{iA} \qquad \text{(P.64)}$$

where

$$A = \int_0^s d\tau\,[p_a\dot{x}^a - (\gamma^a p_a - m)] \qquad \text{(P.65)}$$

Notice that, unlike the bosonic case, the Hamiltonian is now (formally) *linear* in the momentum. (The result is formal because both the kernel and the Dirac propagator are 4×4 matrices.) If we proceed in the standard manner, the functional integration over the momentum will lead to a Dirac delta functional of the form $\delta[\dot{x}^a - \gamma^a]$! If we interpret this result in terms of the eigenvalues, this would require every bit of the path to have a velocity equal to ± 1. So the paths we need to sum over are on the light cone everywhere.[6]

Problem 13. Dirac Propagator as a Green Function

Show that the Dirac propagator satisfies the differential equation for the Green's function, viz.

$$(i\gamma^m\partial_m - m)S(x-y)_{ab} = i\delta^{(4)}(x-y)\delta_{ab} \qquad \text{(P.66)}$$

by explicitly working it out in the coordinate representation.

[6]One can actually compute such a path integral and obtain the Dirac propagator. Stated without any preamble, this construction will look *unnecessarily* mysterious. But, as we have emphasized, once you know the propagator — say, from the field theory — it is trivial to do a reverse engineering and obtain this path integral expression for the Dirac propagator. (The details of this calculation are available, for example, in J.V. Narlikar, 'Path Amplitudes for Dirac Particles', (1972), *Jour.Indian Math. Soc*, **36**, p.9.) Unfortunately, trying to get the propagators for higher spin particles using path integrals — even for a massive spin one particle — does not lead to any better insight than what is provided by the field theory.

Solution: This is straightforward to do in Fourier space but may be fun to work it out in real space to see how exactly it comes about. We need to start with the expression in

$$S(x-y)_{ab} = \theta(x^0-y^0)\langle 0|\psi_a(x)\bar{\psi}_b(y)|0\rangle - \theta(y^0-x^0)\langle 0|\bar{\psi}_b(y)\psi_a(x)|0\rangle \quad \text{(P.67)}$$

and act on it with the operator $(i\gamma^m\partial_m - m)$. The computations are simplified by noticing the following fact: The partial derivative ∂_m will act on the correlator as well as on the θ function. Of these, the terms arising from ∂_m acting on the correlator will lead to terms like $(i\gamma^m\partial_m - m)\langle 0|\psi_a(x)\bar{\psi}_b(y)|0\rangle$ which identically vanish due to the Dirac equation. So, we need to only worry about terms arising from the θ function, leading to Dirac delta functions. These terms give, through straightforward computation, the result:

$$\begin{aligned}
(i\gamma^m\partial_m - m)S(x-y)_{ab} &= i\gamma^0\delta(x^0-y^0)\langle 0|\{\psi_a(x),\bar{\psi}_b(y)\}|0\rangle, \\
&= i\gamma^0\delta(x^0-y^0)\langle 0|\{\psi_a(x),\psi_b^\dagger(y)\gamma^0\}|0\rangle, \\
&= i(\gamma^0)^2\delta(x^0-y^0)\{\psi_a(x),\psi_b^\dagger(y)\}\langle 0|0\rangle, \\
&= i\delta(x^0-y^0)\delta^{(3)}(\boldsymbol{x}-\boldsymbol{y})\delta_{ab} \quad \text{(P.68)}
\end{aligned}$$

Problem 14. An Alternative Approach to the Ward Identities

Consider a field theory with an action $A(\phi)$ which is invariant under the transformation $\phi \to \phi' = \phi + \delta_\epsilon\phi$ where ϵ is a infinitesimal constant parameter. Let j^m be the associated conserved current in the classical theory.

(a) Show that in quantum theory, $\partial_m\langle j^m\rangle = 0$ where $\langle\cdots\rangle$ denotes the path integral average.

(b) Let $O(\phi)$ denote a member of a class of local operators which change as $O(\phi) \to O(\phi + \delta_\epsilon\phi) = O(\phi) + \delta_\epsilon O$ where $\delta_\epsilon O := (\delta_\epsilon\phi)(\partial O/\partial\phi)$, when the field changes by $\phi \to \phi' = \phi + \delta_\epsilon\phi$. Prove that

$$\partial_\mu\langle j^\mu(x)\prod_{i=1}^n O_i(x_i)\rangle = -\sum_{i=1}^n \delta^D(x-x_i)\langle\delta O_i(x_i)\prod_{j\neq i} O_j(x_j)\rangle \quad \text{(P.69)}$$

That is, the divergence of a correlation function involving the current j^m and a product of n local operators vanishes everywhere except at the locations of the operator insertions.

(c) Consider the transformations $\psi \to \psi' = e^{i\alpha}\psi$, $\bar{\psi} \to \bar{\psi}' = e^{-i\alpha}\bar{\psi}$, $A_m \to A'_m = A_m$ in standard QED where we have *not*[7] changed A_j. Promote this to a local transformation (again without changing A_j) and obtain the resulting Ward identity related to the operator $\langle j^m(x)\psi(x_1)\bar{\psi}(x_2)\rangle$.

[7]So this is not a gauge transformation.

Solutions: (a) Since the theory is invariant when $\epsilon = $ constant, the variation of the path integral to the lowest order must be proportional to $\partial_m\epsilon$ when ϵ is promoted as a spacetime dependent function. Therefore, to the lowest order, we must have[8]

[8]The current may now include a possible contribution from the change in the path integral measure as well.

$$Z = \int \mathcal{D}\phi' e^{-A[\phi']} = \int \mathcal{D}\phi\, e^{-A[\phi]}\left[1 - \int_M j^m\partial_m\epsilon\, d^D x\right] \quad \text{(P.70)}$$

Since the zeroth order term on both sides of Eq. (P.70) are equal, and $\epsilon(x)$ is arbitrary, an integration by parts leads to the result $\partial_m\langle j^m\rangle = 0$.

(b) Start with the result

$$\int \mathcal{D}\phi\, e^{-A[\phi]} \prod_{i=1}^{n} O_i(\phi(x_i)) = \int \mathcal{D}\phi'\, e^{-A[\phi']} \prod_{i=1}^{n} O_i(\phi'(x_i)) = \int \mathcal{D}\phi\, e^{-A[\phi]}$$

$$\times \left[1 - \int_M j^m \partial_m \epsilon\, d^D x\right] \left[\prod_{i=1}^{n} O_i(x_i) + \sum_{i=1}^{n} \delta_\epsilon O_i(x_i) \prod_{j \neq i} O_j(x_j)\right]$$

$$(P.71)$$

The first equality is trivial and the second follows by writing $\phi' = \phi + \delta_\epsilon \phi$ and expanding $\mathcal{D}\phi' \exp(-A[\phi'])$ and the operators to first order in $\epsilon(x)$. This leads to the result

$$\int_M d^D x\, \epsilon(x) \partial_m \left(\langle j^m(x) \prod_{i=1}^{n} O_i(x_i)\rangle\right) = -\sum_{i=1}^{n} \langle \delta_\epsilon O_i(x_i) \prod_{j \neq i} O_j(x_j)\rangle$$

$$(P.72)$$

We now use a simple trick of writing

$$\delta_\epsilon O_i(x_i) = \int_M d^D x\, \delta^D(x - x_i)\, \epsilon(x)\, \delta O_i(x_i) \qquad (P.73)$$

so that all the terms in Eq. (P.72) become proportional to $\epsilon(x)$. This leads to the result quoted in the question.

(c) You should first verify that the transformation stated in the text with constant α is indeed a quantum symmetry. The QED action is definitely invariant but you need to verify that the Jacobian of the path integral measure is indeed unity, which turns out to be the case.

Consider now the resulting current $j^m = \bar{\psi}\gamma^m\psi$ and the correlation function $\langle\psi(x_1)\bar{\psi}(x_2)\rangle$. Since we now have $\delta\psi \propto \psi$, the Ward identity obtained in part (b) above becomes

$$\partial_m \langle j^m(x)\psi(x_1)\bar{\psi}(x_2)\rangle = -\delta^D(x-x_1)\langle\psi(x_1)\bar{\psi}(x_2)\rangle + \delta^D(x-x_2)\langle\psi(x_1)\bar{\psi}(x_2)\rangle$$

$$(P.74)$$

If you now introduce the Fourier transforms (in $D = 4$) by:

$$M^m(p, k_1, k_2) := \int d^4 x\, d^4 x_1\, d^4 x_2\, e^{ip \cdot x}\, e^{ik_1 \cdot x_1}\, e^{-ik_2 \cdot x_2} \langle j^m(x)\psi(x_1)\bar{\psi}(x_2)\rangle$$

$$M_0(k_1, k_2) := \int d^4 x_1\, d^4 x_2\, e^{ik_1 \cdot x_1}\, e^{-ik_2 \cdot x_2} \langle \psi(x_1)\bar{\psi}(x_2)\rangle \qquad (P.75)$$

then we get the momentum space Ward identity

$$ip_m M^m(p, k_1, k_2) = M_0(k_1 + p, k_2) - M_0(k_1, k_2 - p) \qquad (P.76)$$

This is precisely the one obtained in the text.

Problem 15. Chiral Anomaly

In the text, we computed the expectation value of the electromagnetic current in a quantum state, $|A\rangle$, hosting an external vector potential A_j. In a similar fashion, one can define a pseudoscalar current as the expectation value $J_5 \equiv \langle A|\bar\psi(x)\gamma^5\psi(x)|A\rangle$. As usual, define a and b by $\boldsymbol{E}\cdot\boldsymbol{B}=ab$, $E^2-B^2=a^2-b^2$.

(a) Compute J_5 and show that it is given by

$$J_5 = i\frac{e^2}{4\pi^2 m}\,ab \tag{P.77}$$

(b) When $m=0$, the Dirac Lagrangian is invariant under the the chiral transformation $\psi\to\exp(i\alpha\gamma^5)\,\psi$, leading to a conserved current J^{m5}. This current is no longer conserved when $m\neq 0$. Show that

$$\langle A|\partial_m J^{m5}|A\rangle = -\frac{e^2}{2\pi^2}\,ab \tag{P.78}$$

Solutions: (a) The procedure is almost identical to what we did for the scalar and fermionic currents in the text. We first note that J_5 can be expressed in the form

$$
\begin{aligned}
J_5 &= \langle A|\bar\psi(x)\gamma^5\psi(x)|A\rangle = -\text{Tr}\left[\langle x|G_A\gamma^5|x\rangle\right]\\
&= -\text{Tr}\left[\int_0^\infty ds\, e^{-ism^2}\langle x|(\gamma p - e\gamma A + m)\gamma_5 e^{i(\gamma p - e\gamma A)^2 s}|x\rangle\right]\\
&= -m\int_0^\infty ds\, e^{-ism^2}\text{Tr}\left[\langle x|\gamma_5 e^{-i\hat H s}|x\rangle\right]
\end{aligned}
\tag{P.79}
$$

The evaluation of the matrix element proceeds exactly as before, using:

$$\langle x|e^{-i\hat H s}|x\rangle = \langle x;0|x;s\rangle = \frac{i}{16\pi^2}e^{i(1/2)es\sigma_{mn}F^{mn}}\frac{(es)^2 ab}{\text{Im}\,\cosh(es(b+ia))} \tag{P.80}$$

and

$$(\sigma_{mn}F^{mn})^2 = 4(b^2 - a^2) + 8i\gamma_5 ab \tag{P.81}$$

leading to

$$\text{Tr}\left[\gamma_5 e^{i(1/2)es\sigma F}\right] = -4i\,\text{Im}\,\cosh(es(b+ia)) \tag{P.82}$$

These results should be enough for you to obtain the result quoted in the question.

(b) Recall that the standard QED Lagrangian is invariant under the chiral transformation $\psi\to\exp(i\alpha\gamma^5)\,\psi$ in the limit of $m\to 0$, with a conserved current $J^{k5}=\bar\psi\gamma^k\gamma^5\psi$. When $m\neq 0$, this axial current satisfies the condition

$$\partial_k J^{k5} = 2im\bar\psi\gamma^5\psi \tag{P.83}$$

Using our result for $\langle A|\bar\psi(x)\gamma^5\psi(x)|A\rangle$, we find the result quoted in the question.

Problem 16. Compton Scattering: A Case Study

We never completely worked out any probability amplitude using the tree-level Feynman diagrams in the main text. The purpose of this exercise is to encourage you to do this for Compton scattering, with a bit of guidance in the form of a solution.

One possible experimental set up for Compton scattering is as follows. A photon with energy ω, travelling along the z-axis, hits an electron at rest. After the scattering, the photon carries an energy ω' and travels at an angle θ to the z-axis. You can assume that the photon is initially unpolarized (so that you should average over the initial polarizations) and the electron is in a mixed spin state (so that you average over the initial spins). In the final state, we usually measure the momenta of the electron and photon but not the final polarization or spin. (So you can sum over the final polarization and spins as well.)

(a) Draw the relevant tree-level Feynman diagrams and translate them into an algebraic expression for the amplitude $\mathcal{M}$.

(b) Show that

$$\frac{1}{2}\sum_{s,s'}\frac{1}{2}\sum_{\text{pol}}|\mathcal{M}|^2 \qquad\qquad\qquad\qquad\text{(P.84)}$$

$$= 2e^4\left[\frac{p_{12}}{p_{14}} + \frac{p_{14}}{p_{12}} + 2m^2\left(\frac{1}{p_{12}} - \frac{1}{p_{14}}\right) + m^4\left(\frac{1}{p_{12}} - \frac{1}{p_{14}}\right)^2\right]$$

where, in the left hand side, you have averaged over the polarizations and the spins of both the initial and the final states, and p_{AB} is a shorthand for $p_A - p_B$.

Solutions: (a) This part is easy. There are two contributing diagrams shown in Fig. P.3. Writing down the relevant algebraic expressions, we find that the net tree-level amplitude is given by

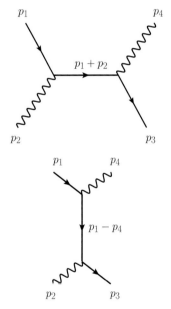

$$\begin{aligned}i\mathcal{M} &= -ie^2\epsilon_m(p_2)\epsilon_n^*(p_4)\,\bar{u}_s(p_3)\left[\frac{\gamma^n(\gamma p_1 + \gamma p_2 + m)\gamma^m}{(p_1+p_2)^2 - m^2}\right.\\ &\left.\quad + \frac{\gamma^m(\gamma p_1 - \gamma p_4 + m)\gamma^n}{(p_1-p_4)^2 - m^2}\right]u_{s'}(p_1)\\ &\equiv -ie^2\epsilon_m(p_2)\epsilon_n^*(p_4)\mathcal{M}^{mn} \qquad\qquad\text{(P.85)}\end{aligned}$$

Figure P.3: Feynman diagram for the Compton scattering

(b) This part involves lots and lots of work but is a fairly standard calculation in QED. You should be able to fill in the details of the steps outlined below. To begin with, we need to compute $|\mathcal{M}|^2$. So calculate $(\mathcal{M}^{mn})^* = (\mathcal{M}^{mn})^\dagger$. Evaluating this expression from the previous result, we get

$$\begin{aligned}(\mathcal{M}^{mn})^\dagger &= \bar{u}_{s'}^\dagger(p_1)\left[\frac{\gamma^n(\gamma p_1 - \gamma p_4 + m)\gamma^m}{(p_1-p_4)^2 - m^2}\right.\\ &\left.\quad + \frac{\gamma^m(\gamma p_1 + \gamma p_2 + m)\gamma^n}{(p_1+p_2)^2 - m^2}\right]u_s(p_3)\qquad\text{(P.86)}\end{aligned}$$

We can now write down an expression for $|\mathcal{M}|^2$. It is easiest to average over the initial and final polarizations at this stage. This will give

$$\begin{aligned}\frac{1}{2}\sum_{\text{pol}}|\mathcal{M}|^2 &= \frac{1}{2}e^4\sum_{\text{pol}}\epsilon_m(p_2)\epsilon_a^*(p_2)\sum_{\text{pol}}\epsilon_n^*(p_4)\epsilon_b(p_4)\mathcal{M}^{mn}\left(\mathcal{M}^{ab}\right)^\dagger\\ &= \frac{1}{2}e^4(-g_{ma})(-g_{nb})\mathcal{M}^{mn}\left(\mathcal{M}^{ab}\right)^\dagger = \frac{1}{2}e^4\mathcal{M}^{mn}\mathcal{M}^\dagger_{mn}\quad\text{(P.87)}\end{aligned}$$

The next step is to average over the spins. The summing over s' is fairly easy, and we get:

$$\frac{1}{2}\sum_{s\,s'}\frac{1}{2}\sum_{\text{pol}}|\mathcal{M}|^2 = \frac{1}{4}e^4\sum_s \bar{u}_s(p_3)Mu_s(p_3) \qquad \text{(P.88)}$$

with

$$M \equiv \left[\frac{\gamma^n(\gamma p_1 + \gamma p_2 + m)\gamma^m}{(p_1+p_2)^2 - m^2} + \frac{\gamma^m(\gamma p_1 - \gamma p_4 + m)\gamma^n}{(p_1-p_4)^2 - m^2}\right](\gamma p_1 + m)$$
$$\times \left[\frac{\gamma_n(\gamma p_1 - \gamma p_4 + m)\gamma_m}{(p_1-p_4)^2 - m^2} + \frac{\gamma_m(\gamma p_1 + \gamma p_2 + m)\gamma_n}{(p_1+p_2)^2 - m^2}\right] \qquad \text{(P.89)}$$

Plugging everything in, the problem reduces to the calculation of a bunch of traces of the gamma matrices in the expression

$$\frac{1}{2}\sum_{s\,s'}\frac{1}{2}\sum_{\text{pol}}|\mathcal{M}|^2$$

$$= \frac{1}{4}e^4\text{Tr}\left[\left[\frac{\gamma^n(\gamma p_1 + \gamma p_2 + m)\gamma^m}{(p_1+p_2)^2 - m^2} + \frac{\gamma^m(\gamma p_1 - \gamma p_4 + m)\gamma^n}{(p_1-p_4)^2 - m^2}\right](\gamma p_1 + m)\right.$$

$$\left.\times \left[\frac{\gamma_n(\gamma p_1 - \gamma p_4 + m)\gamma_m}{(p_1-p_4)^2 - m^2} + \frac{\gamma_m(\gamma p_1 + \gamma p_2 + m)\gamma_n}{(p_1+p_2)^2 - m^2}\right](\gamma p_3 + m)\right]$$

$$\text{(P.90)}$$

This is fairly tedious and the best way to do this is to separate it out term by term. For example, the term which has a factor m^4 reduces to

$$\begin{aligned} Q &= \frac{1}{4}e^4 m^4 \text{Tr}\left[\left(\frac{\gamma^n\gamma^m}{(p_1+p_2)^2 - m^2} + \frac{\gamma^m\gamma^n}{(p_1-p_4)^2 - m^2}\right)\right. \\ &\qquad\qquad \left.\times \left(\frac{\gamma_n\gamma_m}{(p_1-p_4)^2 - m^2} + \frac{\gamma_m\gamma_n}{(p_1+p_2)^2 - m^2}\right)\right] \\ &= 4e^4 m^4\left[\frac{1}{p_{12}\,p_{14}} + \frac{1}{p_{12}^2} + \frac{1}{p_{14}^2}\right] \qquad \text{(P.91)} \end{aligned}$$

If you similarly calculate all the rest and add them up, you should get the result quoted in the question.

Problem 17.   Photon Mass from Radiative Corrections in $d=2$

We saw that the radiative corrections preserve the condition $m=0$ for the photon in standard QED in $d=4$. This depended crucially on the fact that the $\Pi(p^2)$ — computed by summing the electron loop contribution to the photon propagator — has no pole at $p=0$. Curiously enough, the situation is different in $d=2$ where $\Pi(p^2)$ acquires a pole, and as a result, the photon acquires a mass! Prove this result — originally obtained by Schwinger — by computing $\Pi_{mn}(p^2)$ in $d=2$.

Solution: We now need to compute in $d=2$ the standard integrals for $\Pi_{mn}(p)$ and regularize them. We will choose dimensional regularization and set $D=2-\epsilon$. Then the relevant integral is given by

$$-i\Pi_{mn}(p) = (ie)^2(-i^2)\int \frac{d^D q}{(2\pi)^D}\,\frac{\text{Tr}\left[(\gamma q - \gamma p)\gamma_n\gamma q\gamma_m\right]}{q^2(q-p)^2} \qquad \text{(P.92)}$$

We will use, for the traces, the relations[9]

$$\text{Tr}(\gamma_m \gamma_n) = 2g_{mn}; \quad \text{Tr}(\gamma_m \gamma_n \gamma_r \gamma_s) = 2\left(g_{mn}g_{rs} - g_{mr}g_{ns} + g_{ms}g_{rn}\right) \tag{P.93}$$

The integral can be evaluated exactly as we did in the main text for $d = 4$, and will now lead to the result

$$
\begin{aligned}
-i\Pi_{mn} &= -\frac{2ie^2 \pi^{D/2}}{(2\pi)^D} \int_0^1 dx \left[2\left(\frac{x^2 p_m p_n}{(-p^2 x + p^2 x^2)^{1+\epsilon/2}} \Gamma\left(1 + \frac{\epsilon}{2}\right) \right. \right. \\
&\quad \left. - \frac{1}{2} \frac{g_{mn}}{(-p^2 x + p^2 x^2)^{\epsilon/2}} \Gamma\left(\frac{\epsilon}{2}\right) \right) \\
&\quad - g_{mn} \left(\frac{x^2 p^2}{(-p^2 x + p^2 x^2)^{1+\epsilon/2}} \Gamma\left(1 + \frac{\epsilon}{2}\right) \right. \\
&\quad \left. - \frac{2-\epsilon}{2} \frac{1}{(-p^2 x + p^2 x^2)^{\epsilon/2}} \Gamma\left(\frac{\epsilon}{2}\right) \right) \\
&\quad - 2\frac{x p_m p_n}{(-p^2 x + p^2 x^2)^{1+\epsilon/2}} \Gamma\left(1 + \frac{\epsilon}{2}\right) \\
&\quad \left. + g_{mn} \frac{p^2 x}{(-p^2 x + p^2 x^2)^{1+\epsilon/2}} \Gamma\left(1 + \frac{\epsilon}{2}\right) \right] \tag{P.94}
\end{aligned}
$$

Taking the limit $D \to 2$, $\epsilon \to 0$, we find that

$$-i\Pi_{mn}(p) = i(p_m p_n - p^2 g_{mn})\Pi(p^2) = -\frac{ie^2}{\pi p^2}(p_m p_n - p^2 g_{mn}) \tag{P.95}$$

This is a finite quantity in $d = 2$. The sum of the electron loops will now lead to a photon propagator of the form

$$iD_{mn}(p) = -\frac{i\left(g_{mn} - (p_m p_n/p^2)\right)}{p^2\left(1 - \Pi(p^2)\right)} \tag{P.96}$$

Because $\Pi_{mn}(p^2) \propto (1/p^2)$, the resulting propagator has the structure

$$iD_{mn}(p) = -\frac{i\left(g_{mn} - (p_m p_n/p^2)\right)}{p^2 - (e^2/\pi)} \tag{P.97}$$

showing that the photon acquires a mass $m_\gamma = e/\sqrt{\pi}$.

Problem 18. Electron Self-Energy with a Massive Photon

Compute $\Sigma(p, m)$ for the one loop electron self-energy graph when the photon has a mass m_γ. This can be done using dimensional regularization but for some additional practice, try it out using a technique called the Pauli-Villars regularization. The idea behind this regularization is to replace the $(1/k^2)$ of the photon propagator by

$$\frac{1}{k^2} \to \frac{1}{k^2} - \frac{1}{k^2 - \Lambda^2} \tag{P.98}$$

where Λ is a large regulator mass scale. Obviously, $\Lambda \to \infty$ gives back the correct photon propagator, but the modified photon propagator in

[9]The coefficient 2 on the right hand side could have been replaced by any analytic function $F(D)$ with $F(2) = 2$. We simplify the notation by setting it to 2 right from the start.

Eq. (P.98) has better UV convergence which will make the integrals finite. Show that you can now express Σ in the form $\Sigma(p, m) = A(p^2)(\gamma p) + mB(p^2)$, where:

$$A(p^2) = +\frac{2e^2}{(4\pi)^2} \int_0^1 d\alpha \, (1 - \alpha) \ln\left(\frac{(1 - \alpha)\Lambda^2}{-\alpha(1 - \alpha)p^2 + \alpha m^2 + (1 - \alpha)m_\gamma^2} \right)$$

$$B(p^2) = -\frac{4e^2}{(4\pi)^2} \int_0^1 d\alpha \, \ln\left(\frac{(1 - \alpha)\Lambda^2}{-\alpha(1 - \alpha)p^2 + \alpha m^2 + (1 - \alpha)m_\gamma^2} \right) \qquad \text{(P.99)}$$

Solution: This is completely straightforward, though a bit tedious. You can perform the calculation exactly as in the case of $m_\gamma = 0$. The photon propagator can be taken as

$$D_{mn} = \frac{-i\eta_{mn}}{k^2 - m_\gamma^2} \qquad \text{(P.100)}$$

in the generalization of a Feynman gauge.

Annotated Reading List

There are several excellent text books about QFT, with more being written every year, if not every month! It is therefore futile to give a detailed bibliography on such a well developed subject. I shall content myself with sharing my experience as regards some of the books in this subject, which the students might find useful.

1. I learnt quantum field theory, decades back, from the following two books:

   - Landau L.D. and Lifshitz E.M., *Relativistic quantum theory, Parts I and II*, (Pergamon Press, 1974).

     The first three chapters of Part I are masterpieces of — concise but adequate — description of photons, bosons and fermions. There is a later avatar of this Volume 4 of the Course of Theoretical Physics, in which the discussion of some of the "outdated" topics are omitted, and hence is much less fun to read!

   - Roman P., *Introduction to Quantum Field Theory*, (John Wiley, 1969).

     The discussion of the formal aspects of QFT (LSZ, Wightman formalism, ...) is presented in a human readable form. Unfortunately, the author does not discuss QED.

2. The following two books are very different in style, contents and intent, but I enjoyed both of them.

   - Zee A., *Quantum Field Theory in a Nutshell*, (Princeton University Press, 2003).

     Possibly the best contemporary introduction to QFT, with correct balance between concepts and calculational details and delightful to read.

   - Schwinger J., *Particles, Sources and Fields, Vol. I, II and III*, (Perseus Books, 1970).

     An extraordinarily beautiful approach which deserves to be better known among students — and also among professors but (alas!) they are usually too opinionated to appreciate this work — for clarity, originality and efficiency of calculations.

   You will find the influence of the above two books throughout the current text!

3. A conventional treatment of QFT, covering topics more extensive (and different) compared to the present one, from a modern perspective can be found in the following two books:

- Peskin M. E. and Schroeder D.V., *An introduction to Quantum Field Theory*, (Addison-Wesley Publishing Company, 1995).
- Alvarez-Gaume L. and Vazquez-Mozo M. A., *An Invitation to Quantum Field Theory*, (Springer, 2012).

Another modern treatment, clear and to the point, is available in

- Maggiore Michele, *A Modern Introduction to Quantum Field Theory*, (Oxford University Press, 2005).

There are, doubtless, several other books of similar genre, but these are among my personal favourites.

4. Very down-to-earth, unpretentious discussion, with rich calculational details can be found in the books by Greiner. Three of them which are of particular relevance to QFT are:

- Greiner W., *Relativistic Quantum Mechanics Wave Equations*, (Springer, 2000).
- Greiner W. and Reinhardt J., *Quantum Electrodynamics*, (Springer, 1994).
- Greiner W. and Reinhardt J., *Field Quantization*, (Springer, 1996).

These are very student-friendly and give details of the calculations which you may not find in many other books.

Index

Printed in the United States
By Bookmasters